BIODIVERSITY

E. O. Wilson, *Editor*
Frances M. Peter, *Associate Editor*

NATIONAL ACADEMY PRESS
Washington, D.C. 1988

NATIONAL ACADEMY PRESS · 2101 Constitution Ave, NW · Washington, DC 20418

The National Academy of Sciences was chartered by Congress in 1863 as a private, nonprofit organization dedicated to the furtherance of science and engineering for the public welfare. In 1916 the National Research Council was organized, enabling the Academy to draw upon the entire American scientific and technical community in the pursuit of its mandate to provide independent advice to the nation on critical scientific and technical questions.

The Smithsonian Institution was created by act of Congress in 1846 in accordance with the will of the Englishman, James Smithson, who in 1826 bequeathed his property to the United States of America, "to found at Washington, under the name of the Smithsonian Institution, an establishment for the increase and diffusion of knowledge among men." The Smithsonian has since evolved into an institution devoted to public education, research and national service in the arts, sciences and history. This independent federal establishment is the world's largest museum complex and is responsible for public and scholarly activities, exhibitions and research projects nationwide and overseas.

The National Forum on BioDiversity was developed by the Board on Basic Biology of the National Research Council's Commission on Life Sciences and by the Smithsonian Institution's Directorate of International Activities.

The views expressed in this book are solely those of the individual authors and are not necessarily the views of the National Academy of Sciences or of the Smithsonian Institution.

Library of Congress Cataloging-in-Publication Data

National Forum on Biodiversity (1986: Washington, D.C.)
  Biodiversity/Edward O. Wilson, editor, Frances M. Peter,
associate editor.
    p. cm.
  "Papers from the National Forum on BioDiversity held September
21–25, 1986, in Washington, D.C., under the cosponsorship of the
National Academy of Sciences and the Smithsonian Institution."
  Includes index.
  ISBN 0-309-03783-2.   ISBN 0-309-03739-5   (pbk.)
  1. Biological diversity conservation—Congresses.   2. Biological
diversity—Congresses.   I. Wilson, Edward Osborne, 1929–
II. Peter, Frances M.   III. National Academy of Sciences (U.S.)
IV. Smithsonian Institution.   V. Title.
QH75.A1N32 1986
333.7'2—dc19

First Printing, March 1988
Second Printing, May 1988
Third Printing, October 1988
Fourth Printing, May 1989
Fifth Printing, September 1989
Sixth Printing, May 1990

Seventh Printing, November 1990
Eighth Printing, August 1991
Ninth Printing, March 1992
Tenth Printing, January 1993
Eleventh Printing, January 1994
Twelfth Printing, April 1995

Printed in the United States of America

The National Forum on BioDiversity, on which this book is based, was developed by the Board on Basic Biology of the National Research Council's Commission on Life Sciences and by the Smithsonian Institution's Directorate of International Activities.

EDITOR'S FOREWORD

The diversity of life forms, so numerous that we have yet to identify most of them, is the greatest wonder of this planet. The biosphere is an intricate tapestry of interwoven life forms. Even the seemingly desolate arctic tundra is sustained by a complex interaction of many species of plants and animals, including the rich arrays of symbiotic lichens. The book before you offers an overall view of this biological diversity and carries the urgent warning that we are rapidly altering and destroying the environments that have fostered the diversity of life forms for more than a billion years.

The source of the book is the National Forum on BioDiversity, held in Washington, D.C., on September 21–24, 1986, under the auspices of the National Academy of Sciences and Smithsonian Institution. The forum was notable for its large size and immediately perceived impact on the public. It featured more than 60 leading biologists, economists, agricultural experts, philosophers, representatives of assistance and lending agencies, and other professionals. The lectures and panels were regularly attended by hundreds of people, many of whom participated in the discussions, and various aspects of the forum were reported widely in the press. On the final evening, a panel of six of the participants conducted a teleconference downlinked to an estimated audience of 5,000 to 10,000 at over 100 sites, most of them hosted by Sigma Xi chapters at universities and colleges in the United States and Canada.

The forum coincided with a noticeable rise in interest, among scientists and portions of the public, in matters related to biodiversity and the problems of international conservation. I believe that this increased attention, which was evident by 1980 and had steadily picked up momentum by the time of the forum, can be ascribed to two more or less independent developments. The first was the accumulation of enough data on deforestation, species extinction, and tropical biology to bring global problems into sharper focus and warrant broader public exposure. It is no coincidence that 1986 was also the year that the Society for Conservation Biology was founded. The second development was the growing awareness of the close linkage between the conservation of biodiversity and economic development. In the United States and other industrial countries, the two

v

are often seen in opposition, with environmentalists and developers struggling for compromise in a zero-sum game. But in the developing nations, the opposite is true. Destruction of the natural environment is usually accompanied by short-term profits and then rapid local economic decline. In addition, the immense richness of tropical biodiversity is a largely untapped reservoir of new foods, pharmaceuticals, fibers, petroleum substitutes, and other products.

Because of this set of historical circumstances, this book, which contains papers from the forum, should prove widely useful. It provides an updating of many of the principal issues in conservation biology and resource management. It also documents a new alliance between scientific, governmental, and commercial forces—one that can be expected to reshape the international conservation movement for decades to come.

The National Forum on BioDiversity and thence this volume were made possible by the cooperative efforts of many people. The forum was conceived by Walter G. Rosen, Senior Program Officer in the Board on Basic Biology—a unit of the Commission on Life Sciences, National Research Council/National Academy of Sciences (NRC/NAS). Dr. Rosen represented the NRC/NAS throughout the planning stages of the project. Furthermore, he introduced the term *biodiversity*, which aptly represents, as well as any term can, the vast array of topics and perspectives covered during the Washington forum. Edward W. Bastian, Smithsonian Institution, mobilized and orchestrated the diverse resources of the Smithsonian in the effort. Drs. Rosen and Bastian were codirectors of the forum. Michael H. Robinson (Director of the National Zoological Park) served as chairman of the Program Committee, organized one of the forum panels, and served as general master of ceremonies. The remainder of the Program Committee consisted of William Jordan III, Thomas E. Lovejoy III, Harold A. Mooney, Stanwyn Shetler, and Michael E. Soulé.

The various panels of the forum were organized and chaired by F. William Burley, William Conway, Paul R. Ehrlich, Michael Hanemann, William Jordan III, Thomas E. Lovejoy III, Harold A. Mooney, James D. Nations, Peter H. Raven, Michael H. Robinson, Ira Rubinoff, and Michael E. Soulé. David Johnson at the New York Botanical Garden was very helpful in verifying some of the botanical terms used in this book. Helen Taylor and Kathy Marshall of the NRC staff and Anne Peret of the Smithsonian Institution assisted with the wide variety of arrangements necessary to the successful conduct of the forum. Linda Miller Poore, also of the NRC staff, entered this entire document on a word processer and was responsible for formatting and checking the many references. Richard E. Morris of the National Academy Press guided this book through production.

The National Forum on BioDiversity was supported by the National Research Council Fund and the Smithsonian Institution, with supplemental support from the Town Creek Foundation, the Armand G. Erpf Fund, and the World Wildlife Fund. The National Research Council Fund is a pool of private, discretionary, nonfederal funds that is used to support a program of Academy-initiated studies of national issues in which science and technology figure significantly. The NRC Fund consists of contributions from a consortium of private foundations including the Carnegie Corporation of New York, the Charles E. Culpeper Foundation, the William and Flora Hewlett Foundation, the John D. and Catherine T. MacArthur Foundation, the Andrew W. Mellon Foundation, the Rockefeller Foundation, and the Alfred P. Sloan Foundation; the Academy Industry Program, which seeks annual contributions from companies that are concerned with the health of U.S. science and technology and with public policy issues with technological content; and the National Academy of Sciences and the National Academy of Engineering endowments. The publication of this volume was supported by the National Research Council Dissemination Fund, with supplemental support from the World Wildlife Fund. We are deeply grateful to all these organizations for making this project possible.

Finally, and far from least, Frances M. Peter marshalled the diverse contributions in the present volume and was essential to every step of the manuscript editing process. The cover for *Biodiversity* was derived from a forum poster designed by artist Robert Goldstrom.

E. O. WILSON

CONTENTS

PART 9
ALTERNATIVES TO DESTRUCTION

PART 10
POLICIES TO PROTECT DIVERSITY

PART 11
PRESENT PROBLEMS AND FUTURE PROSPECTS

BIODIVERSITY

THE CURRENT STATE OF BIOLOGICAL DIVERSITY

E. O. WILSON

Frank B. Baird, Jr. Professor of Science, Harvard University,
Museum of Comparative Zoology, Cambridge, Massachusetts

Biological diversity must be treated more seriously as a global resource, to be indexed, used, and above all, preserved. Three circumstances conspire to give this matter an unprecedented urgency. First, exploding human populations are degrading the environment at an accelerating rate, especially in tropical countries. Second, science is discovering new uses for biological diversity in ways that can relieve both human suffering and environmental destruction. Third, much of the diversity is being irreversibly lost through extinction caused by the destruction of natural habitats, again especially in the tropics. Overall, we are locked into a race. We must hurry to acquire the knowledge on which a wise policy of conservation and development can be based for centuries to come.

To summarize the problem in this chapter, I review some current information on the magnitude of global diversity and the rate at which we are losing it. I concentrate on the tropical moist forests, because of all the major habitats, they are richest in species and because they are in greatest danger.

THE AMOUNT OF BIOLOGICAL DIVERSITY

Many recently published sources, especially the multiauthor volume *Synopsis and Classification of Living Organisms*, indicate that about 1.4 million living species of all kinds of organisms have been described (Parker, 1982; see also the numerical breakdown according to major taxonomic category of the world insect fauna prepared by Arnett, 1985). Approximately 750,000 are insects, 41,000 are vertebrates, and 250,000 are plants (that is, vascular plants and bryophytes). The remainder consists of a complex array of invertebrates, fungi, algae, and microorganisms (see Table 1-1). Most systematists agree that this picture is still very incomplete except

TABLE 1-1 Numbers of Described Species of Living Organisms[a]

Kingdom and Major Subdivision	Common Name	No. of Described Species	Totals
Virus			
	Viruses	1,000 (order of magnitude only)	1,000
Monera			
Bacteria	Bacteria	3,000	
Myxoplasma	Bacteria	60	
Cyanophycota	Blue-green algae	1,700	4,760
Fungi			
Zygomycota	Zygomycete fungi	665	
Ascomycota (including 18,000 lichen fungi)	Cup fungi	28,650	
Basidiomycota	Basidiomycete fungi	16,000	
Oomycota	Water molds	580	
Chytridiomycota	Chytrids	575	
Acrasiomycota	Cellular slime molds	13	
Myxomycota	Plasmodial slime molds	500	46,983
Algae			
Chlorophyta	Green algae	7,000	
Phaeophyta	Brown algae	1,500	
Rhodophyta	Red algae	4,000	
Chrysophyta	Chrysophyte algae	12,500	
Pyrrophyta	Dinoflagellates	1,100	
Euglenophyta	Euglenoids	800	26,900
Plantae			
Bryophyta	Mosses, liverworts, hornworts	16,600	
Psilophyta	Psilopsids	9	
Lycopodiophyta	Lycophytes	1,275	
Equisetophyta	Horsetails	15	
Filicophyta	Ferns	10,000	
Gymnosperma	Gymnosperms	529	
Dicotolydonae	Dicots	170,000	
Monocotolydonae	Monocots	50,000	248,428
Protozoa			
	Protozoans: Sarcomastigophorans, ciliates, and smaller groups	30,800	30,800
Animalia			
Porifera	Sponges	5,000	
Cnidaria, Ctenophora	Jellyfish, corals, comb jellies	9,000	
Platyhelminthes	Flatworms	12,200	
Nematoda	Nematodes (roundworms)	12,000	
Annelida	Annelids (earthworms and relatives)	12,000	

TABLE 1-1 Continued

Kingdom and Major Subdivision	Common Name	No. of Described Species	Totals
Mollusca	Mollusks	50,000	
Echinodermata	Echinoderms (starfish and relatives)	6,100	
Arthropoda	Arthropods		
Insecta	Insects	751,000	
Other arthropods		123,161	
Minor invertebrate phyla		9,300	989,761
Chordata			
Tunicata	Tunicates	1,250	
Cephalochordata	Acorn worms	23	
Vertebrata	Vertebrates		
Agnatha	Lampreys and other jawless fishes	63	
Chrondrichthyes	Sharks and other cartilaginous fishes	843	
Osteichthyes	Bony fishes	18,150	
Amphibia	Amphibians	4,184	
Reptilia	Reptiles	6,300	
Aves	Birds	9,040	
Mammalia	Mammals	4,000	43,853
TOTAL, all organisms			1,392,485

[a]Compiled from multiple sources.

in a few well-studied groups such as the vertebrates and flowering plants. If insects, the most species-rich of all major groups, are included, I believe that the absolute number is likely to exceed 5 million. Recent intensive collections made by Terry L. Erwin and his associates in the canopy of the Peruvian Amazon rain forest have moved the plausible upper limit much higher. Previously unknown insects proved to be so numerous in these samples that when estimates of local diversity were extrapolated to include all rain forests in the world, a figure of 30 million species was obtained (Erwin, 1983). In an even earlier stage is research on the epiphytic plants, lichens, fungi, roundworms, mites, protozoans, bacteria, and other mostly small organisms that abound in the treetops. Other major habitats that remain poorly explored include the coral reefs, the floor of the deep sea, and the soil of tropical forests and savannas. Thus, remarkably, we do not know the true number of species on Earth, even to the nearest order of magnitude (Wilson, 1985a). My own guess, based on the described fauna and flora and many discussions with entomologists and other specialists, is that the absolute number falls somewhere between 5 and 30 million.

A brief word is needed on the meaning of species as a category of classification. In modern biology, species are regarded conceptually as a population or series of populations within which free gene flow occurs under natural conditions. This means that all the normal, physiologically competent individuals at a given time are capable of breeding with all the other individuals of the opposite sex belonging to the same species or at least that they are capable of being linked genetically to them through chains of other breeding individuals. By definition they do not breed freely with members of other species.

This biological concept of species is the best ever devised, but it remains less than ideal. It works very well for most animals and some kinds of plants, but for some plant and a few animal populations in which intermediate amounts of hybridization occur, or ordinary sexual reproduction has been replaced by self-fertilization or parthenogenesis, it must be replaced with arbitrary divisions.

New species are usually created in one or the other of two ways. A large minority of plant species came into existence in essentially one step, through the process of polyploidy. This is a simple multiplication in the number of gene-bearing chromosomes—sometimes within a preexisting species and sometimes in hybrids between two species. Polyploids are typically not able to form fertile hybrids with the parent species. A second major process is geographic speciation and takes much longer. It starts when a single population (or series of populations) is divided by some barrier extrinsic to the organisms, such as a river, a mountain range, or an arm of the sea. The isolated populations then diverge from each other in evolution because of the inevitable differences of the environments in which they find themselves. Since all populations evolve when given enough time, divergence between all extrinsically isolated populations must eventually occur. By this process alone the populations can acquire enough differences to reduce interbreeding between them should the extrinsic barrier between them be removed and the populations again come into contact. If sufficient differences have accumulated, the populations can coexist as newly formed species. If those differences have not yet occurred, the populations will resume the exchange of genes when the contact is renewed.

Species diversity has been maintained at an approximately even level or at most a slowly increasing rate, although punctuated by brief periods of accelerated extinction every few tens of millions of years. The more similar the species under consideration, the more consistent the balance. Thus within clusters of islands, the numbers of species of birds (or reptiles, or ants, or other equivalent groups) found on each island in turn increases approximately as the fourth root of the area of the island. In other words, the number of species can be predicted as a constant X (island area)$^{0.25}$, where the exponent can deviate according to circumstances, but in most cases it falls between 0.15 and 0.35. According to this theory of island biogeography, in a typical case (where the exponent is at or near 0.25) the rule of thumb is that a 10-fold increase in area results in a doubling of a number of species (MacArthur and Wilson, 1967).

In a recent study of the ants of Hispaniola, I found fossils of 37 genera (clusters of species related to each other but distinct from other such clusters) in amber from the Miocene age—about 20 million years old. Exactly 37 genera exist on the island today. However, 15 of the original 37 have become extinct, while 15 others not present in the Miocene deposits have invaded to replace them, thus sustaining the original diversity (Wilson, 1985b).

On a grander scale, families—clusters of genera—have also maintained a balance within the faunas of entire continents. For example, a reciprocal and apparently symmetrical exchange of land mammals between North and South America began

3 million years ago, after the rise of the Panamanian land bridge. The number of families in South America first rose from 32 to 39 and then subsided to the 35 that exist there today. A comparable adjustment occurred in North America. At the generic level, North American elements dominated those from South America: 24 genera invaded to the south whereas only 12 invaded to the north. Hence, although equilibrium was roughly preserved, it resulted in a major shift in the composition of the previously isolated South American fauna (Marshall et al., 1982).

Each species is the repository of an immense amount of genetic information. The number of genes range from about 1,000 in bacteria and 10,000 in some fungi to 400,000 or more in many flowering plants and a few animals (Hinegardner, 1976). A typical mammal such as the house mouse (*Mus musculus*) has about 100,000 genes. This full complement is found in each of its myriad cells, organized from four strings of DNA, each of which comprises about a billion nucleotide pairs (George D. Snell, Jackson Laboratory, Maine, personal communication, 1987). (Human beings have genetic information closer in quantity to the mouse than to the more abundantly endowed salamanders and flowering plants; the difference, of course, lies in what is encoded.) If stretched out fully, the DNA would be roughly 1-meter long. But this molecule is invisible to the naked eye because it is only 20 angstroms in diameter. If we magnified it until its width equalled that of wrapping string, the fully extended molecule would be 960 kilometers long. As we traveled along its length, we would encounter some 20 nucleotide pairs or "letters" of genetic code per inch, or about 50 per centimeter. The full information contained therein, if translated into ordinary-size letters of printed text, would just about fill all 15 editions of the *Encyclopaedia Britannica* published since 1768 (Wilson, 1985a).

The number of species and the amount of genetic information in a representative organism constitute only part of the biological diversity on Earth. Each species is made up of many organisms. For example, the 10,000 or so ant species have been estimated to comprise 10^{15} living individuals at each moment of time (Wilson, 1971). Except for cases of parthenogenesis and identical twinning, virtually no two members of the same species are genetically identical, due to the high levels of genetic polymorphism across many of the gene loci (Selander, 1976). At still another level, wide-ranging species consist of multiple breeding populations that display complex patterns of geographic variation in genetic polymorphism. Thus, even if an endangered species is saved from extinction, it will probably have lost much of its internal diversity. When the populations are allowed to expand again, they will be more nearly genetically uniform than the ancestral populations. The bison herds of today are biologically not quite the same—not so interesting—as the bison herds of the early nineteenth century.

THE NATURAL LONGEVITY OF SPECIES

Within particular higher groups of organisms, such as ammonites or fishes, species have a remarkably consistent longevity. As a result, the probability that a given species will become extinct in a given interval of time after it splits off from other species can be approximated as a constant, so that the frequency of species surviving through time falls off as an exponential decay function; in other words, the percentage (but not the absolute number) of species going extinct in each period of time stays the same (Van Valen, 1973).[1] These regularities, such as they are, have been interrupted during the past 250 million years by major episodes of extinction that have been recently estimated to occur regularly at intervals of 26 million years (Raup and Sepkoski, 1984).

Because of the relative richness of fossils in shallow marine deposits, the longevity of fish and invertebrate species living there can often be determined with a modest degree of confidence. During Paleozoic and Mesozoic times, the average persistence of most fell between 1 and 10 million years: that is, 6 million for echinoderms, 1.9 million for graptolites, 1.2 to 2 million for ammonites, and so on (Raup, 1981, 1984).

These estimates are extremely interesting and useful but, as paleontologists have generally been careful to point out, they also suffer from some important limitations. First, terrestrial organisms are far less well known, few estimates have been attempted, and thus different survivorship patterns might have occurred (although Cenozoic flowering plants, at least, appear to fall within the 1- to 10-million-year range). More importantly, a great many organisms on islands and other restricted habitats, such as lakes, streams, and mountain crests, are so rare or local that they could appear and vanish within a short time without leaving any fossils. An equally great difficulty is the existence of sibling species—populations that are reproductively isolated but so similar to closely related species as to be difficult or impossible to distinguish through conventional anatomical traits. Such entities could rarely be diagnosed in fossil form. Together, all these considerations suggest that estimates of the longevity of natural species should be extended only with great caution to groups for which there is a poor fossil record.

RAIN FORESTS AS CENTERS OF DIVERSITY

In recent years, evolutionary biologists and conservationists have focused increasing attention on tropical rain forests, for two principal reasons. First, although these habitats cover only 7% of the Earth's land surface, they contain more than half the species in the entire world biota. Second, the forests are being destroyed so rapidly that they will mostly disappear within the next century, taking with them hundreds of thousands of species into extinction. Other species-rich biomes are in danger, most notably the tropical coral reefs, geologically ancient lakes, and coastal wetlands. Each deserves special attention on its own, but for the moment the rain forests serve as the ideal paradigm of the larger global crisis.

Tropical rain forests, or more precisely closed tropical forests, are defined as habitats with a relatively tight canopy of mostly broad-leaved evergreen trees

---

[1]Van Valen's original formulation, whose difficulties and implications are revealed by more recent research, has been discussed by Raup (1975) and by Lewin (1985). These studies deal with the clade, or set of populations descending through time after having split off as a distinct species from other such populations. They do not refer to the chronospecies, which is just a set of generations of the same species that is subjectively different from sets of generations.

sustained by 100 centimeters or more of annual rainfall. Typically two or more other layers of trees and shrubs occur beneath the upper canopy. Because relatively little sunlight reaches the forest floor, the undergrowth is sparse and human beings can walk through it with relative ease.

The species diversity of rain forests borders on the legendary. Every tropical biologist has a favorite example to offer. From a single leguminous tree in the Tambopata Reserve of Peru, I recently recovered 43 species of ants belonging to 26 genera, about equal to the entire ant fauna of the British Isles (Wilson, 1987). Peter Ashton found 700 species of trees in 10 selected 1-hectare plots in Borneo, the same as in all of North America (Ashton, Arnold Arboretum, personal communication, 1987). It is not unusual for a square kilometer of forest in Central or South America to contain several hundred species of birds and many thousands of species of butterflies, beetles, and other insects.

Despite their extraordinary richness, tropical rain forests are among the most fragile of all habitats. They grow on so-called wet deserts—an unpromising soil base washed by heavy rains. Two-thirds of the area of the forest surface consists of tropical red and yellow earths, which are typically acidic and poor in nutrients. High concentrations of iron and aluminum form insoluble compounds with phosphorus, thereby decreasing the availability of phosphorus to plants. Calcium and potassium are leached from the soil soon after their compounds are dissolved from the rain. As little as 0.1% of the nutrients filter deeper than 5 centimeters beneath the soil surface (NRC, 1982). An excellent popular account of rain forest ecology is given by Forsyth and Miyata (1984).

During the 150 million years since its origin, the principally dicotyledonous flora has nevertheless evolved to grow thick and tall. At any given time, most of the nonatmospheric carbon and vital nutrients are locked up in the tissue of the vegetation. As a consequence, the litter and humus on the ground are thin compared to the thick mats of northern temperate forests. Here and there, patches of bare earth show through. At every turn one can see evidence of rapid decomposition by dense populations of termites and fungi. When the forest is cut and burned, the ash and decomposing vegetation release a flush of nutrients adequate to support new herbaceous and shrubby growth for 2 or 3 years. Then these materials decline to levels lower than those needed to support a healthy growth of agricultural crops without artificial supplements.

The regeneration of rain forests is also limited by the fragility of the seeds of the constituent woody species. The seeds of most species begin to germinate within a few days or weeks, severely limiting their ability to disperse across the stripped land into sites favorable for growth. As a result, most sprout and die in the hot, sterile soil of the clearings (Gomez-Pompa et al., 1972). The monitoring of logged sites indicates that regeneration of a mature forest might take centuries. The forest at Angkor (to cite an anecdotal example) dates back to the abandonment of the Khmer capital in 1431, yet is still structurally different from a climax forest today, 556 years later. The process of rain forest regeneration is in fact so generally slow that few extrapolations have been possible; in some zones of greatest combined damage and sterility, restoration might never occur naturally (Caufield, 1985; Gomez-Pompa et al., 1972).

Approximately 40% of the land that can support tropical closed forest now lacks it, primarily because of human action. By the late 1970s, according to estimates from the Food and Agricultural Organization and United Nations Environmental Programme, 7.6 million hectares or nearly 1% of the total cover is being permanently cleared or converted into the shifting-cultivation cycle. The absolute amount is 76,000 square kilometers (27,000 square miles) a year, greater than the area of West Virginia or the entire country of Costa Rica. In effect, most of this land is being permanently cleared, that is, reduced to a state in which natural reforestation will be very difficult if not impossible to achieve (Mellilo et al., 1985). This estimated loss of forest cover is close to that advanced by the tropical biologist Norman Myers in the mid-1970s, an assessment that was often challenged by scientists and conservationists as exaggerated and alarmist. The vindication of this early view should serve as a reminder always to take such doomsday scenarios seriously, even when they are based on incomplete information.

A straight-line extrapolation from the first of these figures, with identically absolute annual increments of forest-cover removal, leads to 2135 A.D. as the year in which all the remaining rain forest will be either clear-cut or seriously disturbed, mostly the former. By coincidence, this is close to the date (2150) that the World Bank has estimated the human population will plateau at 11 billion people (The World Bank, 1984). In fact, the continuing rise in human population indicates that a straight line estimate is much too conservative. Population pressures in the Third World will certainly continue to accelerate deforestation during the coming decades unless heroic measures are taken in conservation and resource management.

There is another reason to believe that the figures for forest cover removal present too sanguine a picture of the threat to biological diversity. In many local areas with high levels of endemicity, deforestation has proceeded very much faster than the overall average. Madagascar, possessor of one of the most distinctive floras and faunas in the world, has already lost 93% of its forest cover. The Atlantic coastal forest of Brazil, which so enchanted the young Darwin upon his arrival in 1832 ("wonder, astonishment & sublime devotion, fill & elevate the mind"), is 99% gone. In still poorer condition—in fact, essentially lost—are the forests of many of the smaller islands of Polynesia and the Caribbean.

HOW MUCH DIVERSITY IS BEING LOST?

No precise estimate can be made of the numbers of species being extinguished in the rain forests or in other major habitats, for the simple reason that we do not know the numbers of species originally present. However, there can be no doubt that extinction is proceeding far faster than it did prior to 1800. The basis for this statement is not the direct observation of extinction. To witness the death of the last member of a parrot or orchid species is a near impossibility. With the exception of the showiest birds, mammals, or flowering plants, biologists are reluctant to say with finality when a species has finally come to an end. There is always the chance (and hope) that a few more individuals will turn up in some remote forest remnant or other. But the vast majority of species are not monitored at all. Like the dead of Gray's "Elegy Written in a Country Churchyard," they pass from the Earth without notice.

Instead, extinction rates are usually estimated indirectly from principles of biogeography. As I mentioned above, the number of species of a particular group of organisms in island systems increases approximately as the fourth root of the land area. This has been found to hold true not just on real islands but also on habitat islands, such as lakes in a "sea" of land, alpine meadows or mountaintops surrounded by evergreen forests, and even in clumps of trees in the midst of a grassland (MacArthur and Wilson, 1967).

Using the area-species relationship, Simberloff (1984) has projected ultimate losses due to the destruction of rain forests in the New World tropical mainland. If present levels of forest removal continue, the stage will be set within a century for the inevitable loss of 12% of the 704 bird species in the Amazon basin and 15% of the 92,000 plant species in South and Central America.

As severe as these regional losses may be, they are far from the worst, because the Amazon and Orinoco basins contain the largest continuous rain forest tracts in the world. Less extensive habitats are far more threatened. An extreme example is the western forest of Ecuador. This habitat was largely undisturbed until after 1960, when a newly constructed road network led to the swift incursion of settlers and clear-cutting of most of the area. Now only patches remain, such as the 0.8-square-kilometer tract at the Rio Palenque Biological Station. This tiny reserve contains 1,033 plant species, perhaps one-quarter of which are known only to occur in coastal Ecuador. Many are known at the present time only from a single living individual (Gentry, 1982).

In general, the tropical world is clearly headed toward an extreme reduction and fragmentation of tropical forests, which will be accompanied by a massive extinction of species. At the present time, less than 5% of the forests are protected within parks and reserves, and even these are vulnerable to political and economic pressures. For example, 4% of the forests are protected in Africa, 2% in Latin America, and 6% in Asia (Brown, 1985). Thus in a simple system as envisioned by the basic models of island biogeography, the number of species of all kinds of organisms can be expected to be reduced by at least one-half—in other words, by hundreds of thousands or even (if the insects are as diverse as the canopy studies suggest) by millions of species. In fact, the island-biogeographic projections appear to be conservative for two reasons. First, tropical species are far more localized than those in the temperate zones. Consequently, a reduction of 90% of a tropical forest does not just reduce all the species living therein to 10% of their original population sizes, rendering them more vulnerable to future extinction. That happens in a few cases, but in many others, entire species are eliminated because they happened to be restricted to the portion of the forest that was cut over. Second, even when a portion of the species survives, it will probably have suffered significant reduction in genetic variation among its members due to the loss of genes that existed only in the outer portions.

The current reduction of diversity seems destined to approach that of the great natural catastrophes at the end of the Paleozoic and Mesozoic eras—in other words, the most extreme in the past 65 million years. In at least one important respect, the modern episode exceeds anything in the geological past. In the earlier mass extinctions, which some scientists believe were caused by large meteorite strikes, most of the plants survived even though animal diversity was severely reduced. Now, for the first time, plant diversity is declining sharply (Knoll, 1984).

HOW FAST IS DIVERSITY DECLINING?

The area-species curves of island systems, that is, the quantitative relationship between the area of islands and the number of species that can persist on the islands, provide minimal estimates of the reduction of species diversity that will eventually occur in the rain forests. But how long is "eventually"? This is a difficult question that biogeographers have attacked with considerable ingenuity. When a forest is reduced from, say, 100 square kilometers to 10 square kilometers by clearing, some immediate extinction is likely. However, the new equilibrium will not be reached all at once. Some species will hang on for a while in dangerously reduced populations. Elementary mathematical models of the process predict that the number of species in the 10-square-kilometer plot will decline at a steadily decelerating rate, i.e., they will decay exponentially to the lower level.

Studies by Jared Diamond and John Terborgh have led to the estimation of the decay constants for the bird faunas on naturally occurring islands (Diamond, 1972, 1984; Terborgh, 1974). These investigators took advantage of the fact that rising sea levels 10,000 years ago cut off small land masses that had previously been connected to South America, New Guinea, and the main islands of Indonesia. For example, Tobago, Margarita, Coiba, and Trinidad were originally part of the South American mainland and shared the rich bird fauna of that continent. Thus they are called land-bridge islands. In a similar manner, Yapen, Aru, and Misol were connected to New Guinea. In the study of the South American land-bridge islands, Terborgh found that the smaller the island, the higher the estimated decay constant and hence extinction rate. Terborgh then turned to Barro Colorado Island, which was isolated for the first time by the rise of Gatun Lake during the construction of the Panama Canal. Applying the natural land-bridge extinction curve to an island of this size (17 square kilometers) and fitting the derived decay constant to the actual period of isolation (50 years), Terborgh predicted an extinction of 17 bird species. The actual number known to have vanished as a probable result of insularization is 13, or 12% of the 108 breeding species originally present. The extinction rates of bird species on Barro Colorado Island were based on careful studies by E. O. Willis and J. R. Karr and have been recently reviewed by Diamond (1984).

Several other studies of recently created islands of both tropical and temperate-zone woodland have produced similar results, which can be crudely summarized as follows: when the islands range from 1 to 25 square kilometers—the size of many smaller parks and reserves—the rate of extinction of bird species during the first 100 years is 10 to 50%. Also as predicted, the extinction rate is highest in the smaller patches, and it rises steeply when the area drops below 1 square kilometer. To take one example provided by Willis (1979), three patches of subtropical forest isolated (by agricultural clearing) in Brazil for about a hundred years varied from 0.2 to 14 square kilometers, and, in reverse order, their resident bird species suffered 14 to 62% extinction rates.

What do these first measurements tell us about the rate at which diversity is being reduced? No precise estimate can be made for three reasons. First, the number of species of organisms is not known, even to the nearest order of magnitude. Second, because even in a simple island-biogeographic system, diversity reduction depends on the size of the island fragments and their distance from each other—factors that vary enormously from one country to the next. Third, the ranges of even the known species have not been worked out in most cases, so that we cannot say which ones will be eliminated when the tropical forests are partially cleared.

However, scenarios of reduction can be constructed to give at least first approximations if certain courses of action are followed. Let us suppose, for example, that half the species in tropical forests are very localized in distribution, so that the rate at which species are being eliminated immediately is approximately this fraction multiplied by the rate-percentage of the forests being destroyed. Let us conservatively estimate that 5 million species of organisms are confined to the tropical rain forests, a figure well justified by the recent upward adjustment of insect diversity alone. The annual rate of reduction would then be $0.5 \times 5 \times 10^6 \times 0.007$ species, or 17,500 species per year. Given 10 million species in the fauna and flora of all the habitats of the world, the loss is roughly one out of every thousand species per year. How does this compare with extinction rates prior to human intervention? The estimates of extinction rates in Paleozoic and Mesozoic marine faunas cited earlier (Raup, 1981, 1984; Raup and Sepkoski, 1984; Van Valen, 1973) ranged according to taxonomic group (e.g., echinoderms versus cephalopods) from one out of every million to one out of every 10 million per year. Let us assume that on the order of 10 million species existed then, in view of the evidence that diversity has not fluctuated through most of the Phanerozoic time by a factor of more than three (Raup and Sepkoski, 1984). It follows that both the per-species rate and absolute loss in number of species due to the current destruction of rain forests (setting aside for the moment extinction due to the disturbance of other habitats) would be about 1,000 to 10,000 times that before human intervention.

I have constructed other simple models incorporating the quick loss of local species and the slower loss of widespread species due to the insularization effect, and these all lead to comparable or higher extinction rates. It seems difficult if not impossible to combine what is known empirically of the extinction process with the ongoing deforestation process without arriving at extremely high rates of species loss in the near future. Curiously, however, the study of extinction remains one of the most neglected in ecology. There is a pressing need for a more sophisticated body of theories and carefully planned field studies based on it than now exist.

WHAT CAN BE DONE?

The biological diversity most threatened is also the least explored, and there is no prospect at the moment that the scientific task will be completed before a large fraction of the species vanish. Probably no more than 1,500 professional systematists in the world are competent to deal with the millions of species found in the humid tropic forests. Their number may be dropping, due to decreased professional opportunities, reduced funding for research, and the assignment of a higher priority to other disciplines. Data concerning the number of taxonomists, as well as detailed arguments for the need to improve research in tropical countries, are given by NRC (1980). The decline has been accompanied by a more than 50% decrease in the number of publications in tropical ecology from 1979 to 1983 (Cole, 1984).

The problem of tropical conservation is thus exacerbated by the lack of knowledge and the paucity of ongoing research. In order to make precise assessments and recommendations, it is necessary to know which species are present (recall that the great majority have not even received a scientific name) as well as their geographical ranges, biological properties, and possible vulnerability to environmental change.

It would be a great advantage, in my opinion, to seek such knowledge for the entire biota of the world. Each species is unique and intrinsically valuable. We cannot expect to answer the important questions of ecology and other branches of evolutionary biology, much less preserve diversity with any efficiency, by studying only a subset of the extant species.

I will go further: the magnitude and control of biological diversity is not just a central problem of evolutionary biology; it is one of the key problems of science as a whole. At present, there is no way of knowing whether there are 5, 10, or 30 million species on Earth. There is no theory that can predict what this number might turn out to be. With reference to conservation and practical applications, it also matters why a certain subset of species exists in each region of the Earth, and what is happening to each one year by year. Unless an effort is made to understand all of diversity, we will fall far short of understanding life in these important respects, and due to the accelerating extinction of species, much of our opportunity will slip away forever.

Lest this exploration be viewed as an expensive Manhattan Project unattainable in today's political climate, let me cite estimates I recently made of the maximum investment required for a full taxonomic accounting of all species: 25,000 professional lifetimes (4,000 systematists are at work full or part time in North America today); their final catalog would fill 60 meters of library shelving for each million species (Wilson, 1985a). Computer-aided techniques could be expected to cut the effort and cost substantially. In fact, systematics has one of the lowest cost-to-benefit ratios of all scientific disciplines.

It is equally true that knowledge of biological diversity will mean little to the vast bulk of humanity unless the motivation exists to use it. Fortunately, both scientists and environmental policy makers have established a solid linkage between economic development and conservation. The problems of human beings in the tropics are primarily biological in origin: overpopulation, habitat destruction, soil deterioration, malnutrition, disease, and even, for hundreds of millions, the uncertainty of food and shelter from one day to the next. These problems can be solved in part by making biological diversity a source of economic wealth. Wild species are in fact both one of the Earth's most important resources and the least utilized. We have come to depend completely on less than 1% of living species for our existence, the remainder waiting untested and fallow. In the course of history, according to estimates made by Myers (1984), people have utilized about 7,000 kinds of plants for food; predominant among these are wheat, rye, maize, and about a dozen other highly domesticated species. Yet there are at least 75,000 edible plants in existence, and many of these are superior to the crop plants in widest use. Others are potential sources of new pharmaceuticals, fibers, and petroleum substitutes. In addition, among the insects are large numbers of species that are potentially superior as crop pollinators, control agents for weeds, and parasites and predators of insect pests. Bacteria, yeasts, and other microorganisms are likely to continue yielding new medicines, food, and procedures of soil restoration. Biologists have begun to fill volumes with concrete proposals for the further exploration and better use of diversity, with increasing emphasis on the still unexplored portions of the tropical biota. Some of the most recent and useful works on this subject include those by Myers (1984), NRC (1975), Office of Technology Assessment (1984), Oldfield (1984), and the U.S. Department of State (1982). In addition, an excellent series of specialized publications on practical uses of wild species have been produced during the past 10 years by authors and panels commissioned by the Board on Science and Technology for International Development (BOSTID) of the National Research Council.

In response to the crisis of tropical deforestation and its special threat to biological diversity, proposals are regularly being advanced at the levels of policy and research. For example, Nicholas Guppy (1984), noting the resemblance of the lumbering of rain forests to petroleum extraction as the mining of a nonrenewable resource for short-term profit, has recommended the creation of a cartel, the Organization of Timber-Exporting Countries (OTEC). By controlling production and prices of lumber, the organization could slow production while encouraging member states to "protect the forest environment in general and gene stocks and special habitats in particular, create plantations to supply industrial and fuel wood, benefit indigenous tribal forest peoples, settle encroachers, and much else." In another approach, Thomas Lovejoy (1984) has recommended that debtor nations with forest resources and other valuable habitats be given discounts or credits for undertaking conservation programs. Even a small amount of forgiveness would elevate the sustained value of the natural habitats while providing hard currency for alternatives to their exploitation.

Another opportunity for innovation lies in altering somewhat the mode of direct economic assistance to developing countries. A large part of the damage to tropical forests, especially in the New World, has resulted from the poor planning of road systems and dams. For example, the recent settlement of the state of Rondonia and construction of the Tucurui Dam, both in Brazil, are now widely perceived by ecologists and economists alike as ill-conceived (Caufield, 1985). Much of the responsibility of minimizing environmental damage falls upon the international agencies that have the power to approve or disapprove particular projects.

The U.S. Congress addressed this problem with amendments to the Foreign Assistance Act in 1980, 1983, and 1986, which call for the development of a strategy for conserving biological diversity. They also mandate that programs funded through the U.S. Agency for International Development (USAID) include an assessment of environmental impact. In implementing this new policy, USAID has recognized that "the destruction of humid tropical forests is one of the most important environmental issues for the remainder of this century and, perhaps, well into the next," in part because they are "essential to the survival of vast numbers of species of plants and animals" (U.S. Department of State, 1985). In another sphere, The World Bank and other multinational lending agencies have come under increasing pressure to take a more active role in assessing the environmental impact of the large-scale projects they underwrite (Anonymous, 1984).

In addition to recommendations for international policy initiatives, there has recently been a spate of publications on the linkage of conservation and economic use of tropical forests. Notable among them are *Research Priorities in Tropical Biology* (NRC, 1980), based on a study of the National Research Council; *Technologies to Sustain Tropical Forest Resources* (OTA, 1984), prepared by the Office of Technology Assessment for the U.S. Congress; and the *U.S. Strategy on the Conservation of Biological Diversity* (USAID, 1985), a report to Congress by an interagency task force. Most comprehensive of all—and in my opinion the most encouraging in its implications—is the three-part series *Tropical Forests: A Call for Action*, released by the World Resources Institute, The World Bank, and the United Nations Development Programme (1985). The report makes an assessment of the problem worldwide and reviews case histories in which conservation or restoration have contributed to economic development. It examines the needs of every tropical country with important forest reserves. The estimated cost to make an impact on tropical deforestation over the next 5 years would be U.S. $8 billion—a large sum but surely the most cost-effective investment available to the world at the present time.

In the end, I suspect it will all come down to a decision of ethics—how we value the natural worlds in which we evolved and now, increasingly, how we regard our status as individuals. We are fundamentally mammals and free spirits who reached this high a level of rationality by the perpetual creation of new options. Natural philosophy and science have brought into clear relief what might be the essential paradox of human existence. The drive toward perpetual expansion—or personal freedom—is basic to the human spirit. But to sustain it we need the most delicate, knowing stewardship of the living world that can be devised. Expansion and stewardship may appear at first to be conflicting goals, but the opposite is true. The depth of the conservation ethic will be measured by the extent to which each of the two approaches to nature is used to reshape and reinforce the other. The paradox can be resolved by changing its premises into forms more suited to ultimate survival, including protection of the human spirit. I recently wrote in synecdochic form about one place in South America to give these feelings more exact expression:

> To the south stretches Surinam eternal, Surinam serene, a living treasure awaiting assay. I hope that it will be kept intact, that at least enough of its million-year history will be saved for the reading. By today's ethic its value may seem limited, well beneath the pressing concerns of daily life. But I suggest that as biological knowledge grows the ethic will shift fundamentally so that everywhere, for reasons that have to do with the very fiber of the brain, the fauna and flora of a country will be thought part of the national

heritage as important as its art, its language, and that astonishing blend of achievement and farce that has always defined our species (Wilson, 1984).

REFERENCES

Anonymous. 1984. Critics fault World Bank for ecological neglect. Conserv. Found. News. Nov.-Dec.:1–7.

Arnett, R. H. 1985. General considerations. Pp. 3–9 in American Insects: A Handbook of the Insects of America North of Mexico. Van Nostrand Reinhold, New York.

Brown, R. L., ed. 1985. State of the World 1985: A Worldwatch Institute Report on Progress Toward a Sustainable Society. W. W. Norton, New York. 301 pp.

Caufield, C. 1985. In the Rainforest. A. A. Knopf, New York. 283 pp.

Cole, N. H. A. 1984. Tropical ecology research. Nature 309:204.

Diamond, J. M. 1972. Biogeographic kinetics: Estimation of relaxation times for avifaunas of Southwest Pacific islands. Proc. Natl. Acad. Sci. USA 69:3199–3203.

Diamond, J. M. 1984. "Normal" extinctions of isolated populations. Pp. 191–246 in M. H. Nitecki, ed. Extinctions. University of Chicago Press, Chicago.

Erwin, T. L. 1983. Beetles and other insects of tropical forest canopies at Manaus, Brazil, sampled by insecticidal fogging. Pp. 59–75 in S. L. Sutton, T. C. Whitmore, and A. C. Chadwick, eds. Tropical Rain Forest: Ecology and Management. Blackwell, Edinburgh.

Forsyth, A., and K. Miyata. 1984. Tropical Nature: Life & Death in the Rain Forests of Central & South America. Scribner's, New York. 272 pp.

Frankel, O. H., and M. E. Soulé. 1981. Conservation and Evolution. Cambridge University Press, Cambridge, Mass. 327 pp.

Gentry, A. H. 1982. Patterns of Neotropical plant-species diversity. Evol. Biol. 15:1–85.

Gomez-Pompa, A., C. Vazquez-Yanes, and S. Guevara. 1972. The tropical rain forest: A nonrenewable resource. Science 177:762–765.

Guppy, N. 1984. Tropical deforestation: A global view. Foreign Affairs 62:928–965.

Hinegardner, R. 1976. Evolution of genome size. Pp. 179–199 in F. J. Ayala, ed. Molecular Evolution. Sinauer Associates, Sunderland, Mass.

Knoll, A. H. 1984. Patterns of extinction in the fossil record of vascular plants. Pp. 21–68 in M. H. Nitecki, ed. Extinctions. University of Chicago Press, Chicago.

Lewin, R. 1985. Red Queen runs into trouble? Science 227:399–400.

Lovejoy, T. E. 1984. Aid debtor nations' ecology. New York Times, October 4.

MacArthur, R. H., and E. O. Wilson. 1967. The Theory of Island Biogeography. Princeton University Press, Princeton, N.J. 203 pp.

Marshall, L. G., S. D. Webb, J. J. Sepkoski, Jr., and D. M. Raup. 1982. Mammalian evolution and the great American interchange. Science 215:1351–1357.

Melillo, J. M., C. A. Palm, R. A. Houghton, G. M. Woodwell, and N. Myers. 1985. A comparison of two recent estimates of disturbance in tropical forests. Environ. Conserv. 12:37–40.

Myers, N. 1983. A Wealth of Wild Species: Storehouse for Human Welfare. Westview Press, Boulder, Colo. 300 pp.

Myers, N. 1984. The Primary Source: Tropical Forests and Our Future. W. W. Norton, New York. 399 pp.

NRC (National Research Council). 1975. Underexploited Tropical Plants with Promising Economic Value. Board on Science and Technology for International Development Report 16. National Academy of Sciences, Washington, D.C. 187 pp.

NRC (National Research Council). 1979. Tropical Legumes: Resources for the Future. Board on Science and Technology for International Development Report 25. National Academy of Sciences, Washington, D.C. 331 pp.

NRC (National Research Council). 1980. Research Priorities in Tropical Biology. National Academy of Sciences, Washington, D.C. 116 pp.

NRC (National Research Council). 1982. Ecological Aspects of Development in the Humid Tropics. National Academy Press, Washington, D.C. 297 pp.

Oldfield, M. L. 1984. The Value of Conserving Genetic Resources. National Park Service, U.S. Department of the Interior, Washington, D.C. 360 pp.

OTA (Office of Technology Assessment). 1984. Technologies to Sustain Tropical Forest Resources.

Congress of the United States, Office of Technology Assessment, Washington, D.C. 344 pp.

Parker, S. P., ed. 1982. Synopsis and Classification of Living Organisms. McGraw-Hill, New York. 2 vols.

Raup, D. M. 1975. Taxonomic survivorship curves and Van Valen's Law. Paleobiology 1:82–86.

Raup, D. M. 1981. Extinction: Bad genes or bad luck? Acta Geol. Hisp. 16:25–33.

Raup, D. M. 1984. Evolutionary radiations and extinction. Pp. 5–14 in H. D. Holland and A. F. Trandell, eds. Patterns of Change in Evolution. Dahlem Konferenzen, Abakon Verlagsgesellschaft, Berlin.

Raup, D. M., and J. J. Sepkoski, Jr. 1984. Periodicity of extinctions in the geologic past. Proc. Natl. Acad. Sci. USA 81:801–805.

Selander, R. K. 1976. Genic variation in natural populations. Pp. 21–45 in F. J. Ayala, ed. Molecular Evolution. Sinauer Associates, Sunderland, Mass.

Simberloff, D. S. 1984. Mass extinction and the destruction of moist tropical forests. Zh. Obshch. Biol. 45:767–778.

Terborgh, J. 1974. Preservation of natural diversity: The problem of extinction-prone species. BioScience 24:715–722.

USAID (U.S. Agency for International Development). 1985. U.S. Strategy on the Conservation of Biological Diversity. An Interagency Task Force Report to Congress. U.S. Agency for International Development, Washington, D.C. 52 pp.

U.S. Department of State. 1982. Proceedings of the U.S. Strategy Conference on Biological Diversity. November 16–18, 1981, Washington, D.C. Publication No. 9262. U.S. Department of State, Washington, D.C. 126 pp.

U.S. Department of State. 1985. Humid Tropical Forests: AID Policy and Guidance. U.S. Department of State Memorandum. Government Printing Office, Washington, D.C. 3 pp.

Van Valen, L. 1973. A new evolutionary law. Evol. Ther. 1:1–30.

Willis, E. O. 1979. The composition of avian communities in remanescent woodlots in southern Brazil. Papeis Avulsos Zool. 33:1–25.

Wilson, E. O. 1971. The Insect Societies. Belknap Press of Harvard University Press, Cambridge, Mass. 548 pp.

Wilson, E. O. 1984. Biophilia. Harvard University Press, Cambridge, Mass. 176 pp.

Wilson, E. O. 1985a. The biological diversity crisis: A challenge to science. Issues Sci. Technol. 2:20–29.

Wilson, E. O. 1985b. Invasion and extinction in the West Indian ant fauna: Evidence from the Dominican amber. Science 229:265–267.

Wilson, E. O. 1987. The arboreal ant fauna of Peruvian Amazon forests: A first assessment. Biotropica 2:245–251.

World Bank. 1984. World Development Report 1984. Oxford University Press, New York. 286 pp.

World Resources Institute, The World Bank, and United Nations Development Programme. 1985. Tropical Forests: A Call for Action. World Resources Institute, Washington, D.C. 3 vols.

CHALLENGES TO THE PRESERVATION
OF BIODIVERSITY

Trans-Amazon Highway being cut through the
rain forest near Altamira, Brazil—one
example of the deforestation that takes place
along with traditional frontier expansion.
Photo courtesy of Nigel J. H. Smith.

THE LOSS OF DIVERSITY
Causes and Consequences

PAUL R. EHRLICH

Professor of Biological Sciences, Stanford University, Stanford, California

Discussions of the current extinction crisis all too often focus on the fates of prominent endangered species, and in many cases on deliberate overexploitation by human beings as the cause of the endangerment. Thus black rhinos are disappearing from Africa, because their horns are in demand for the manufacture of ceremonial daggers for Middle Eastern puberty rites; elephants are threatened by the great economic value of ivory; spotted cats are at risk because their hides are in demand by furriers; and whales are rare because, among other things, they can be converted into pet food.

Concern about such direct endangerment is valid and has been politically important, because public sympathy seems more easily aroused over the plight of furry, cuddly, or spectacular animals. The time has come, however, to focus public attention on a number of more obscure and (to most people) unpleasant truths, such as the following:

• The primary cause of the decay of organic diversity is not direct human exploitation or malevolence, but the habitat destruction that inevitably results from the expansion of human populations and human activities.

• Many of the less cuddly, less spectacular organisms that *Homo sapiens* is wiping out are more important to the human future than are most of the publicized endangered species. People need plants and insects more than they need leopards and whales (which is not to denigrate the value of the latter two).

• Other organisms have provided humanity with the very basis of civilization in the form of crops, domestic animals, a wide variety of industrial products, and many important medicines. Nonetheless, the most important anthropocentric reason for preserving diversity is the role that microorganisms, plants, and animals

21

play in providing free ecosystem services, without which society in its present form could not persist (Ehrlich and Ehrlich, 1981; Holdren and Ehrlich, 1974).

• The loss of genetically distinct populations within species is, at the moment, at least as important a problem as the loss of entire species. Once a species is reduced to a remnant, its ability to benefit humanity ordinarily declines greatly, and its total extinction in the relatively near future becomes much more likely. By the time an organism is recognized as endangered, it is often too late to save it.

• Extrapolation of current trends in the reduction of diversity implies a denouement for civilization within the next 100 years comparable to a nuclear winter.

• Arresting the loss of diversity will be extremely difficult. The traditional "just set aside a preserve" approach is almost certain to be inadequate because of factors such as runaway human population growth, acid rains, and climate change induced by human beings. A quasi-religious transformation leading to the appreciation of diversity for its own sake, apart from the obvious direct benefits to humanity, may be required to save other organisms and ourselves.

Let us examine some of these propositions more closely. While a mere handful of species is now being subjected to purposeful overexploitation, thousands are formally recognized in one way or another as threatened or endangered. The vast majority of these are on the road to extinction, because humanity is destroying habitats: paving them over, plowing them under, logging, overgrazing, flooding, draining, or transporting exotic organisms into them while subjecting them to an assault by a great variety of toxins and changing their climate.

As anyone who has raised tropical fishes knows, all organisms require appropriate habitats if they are to survive. Just as people cannot exist in an atmosphere with too little oxygen, so neon tetras (*Paracheirodon innesi*) cannot survive in water that is 40F (4.4C) or breed in highly alkaline water. Trout, on the other hand, cannot breed in water that is too warm or too acid. And the bacteria that produce the tetanus toxin cannot reproduce in the *presence* of oxygen. In order to persist, Bay checkerspot butterflies (*Euphydryas editha bayensis*) must have areas of serpentine grassland (to support the growth of plants that serve as food for their caterpillars and supply nectar to the adults). Whip-poor-wills, red-eyed vireos, Blackburnian warblers, scarlet tanagers, and dozens of other North American birds must have mature tropical forest in which to overwinter (see Terborgh, 1980, for example). Black-footed ferrets (*Mustela nigripes*) require prairie that still supports the prairie dogs on which the ferrets dine.

This utter dependence of organisms on appropriate environments (Ehrlich, 1986) is what makes ecologists so certain that today's trends of habitat destruction and modification—especially in the high-diversity tropical forest (where at least one-half of all species are believed to dwell)—are an infallible recipe for biological impoverishment. Those politicians and social scientists who have questioned the extent of current extinctions are simply displaying their deep ignorance of ecology;

habitat modification and destruction and the extinction of populations and species go hand in hand.

The extent to which humanity has already wreaked havoc on Earth's environments is shown indirectly by a recent study of human appropriation of the products of photosynthesis (Vitousek et al., 1986). The food resource of the animals in all major ecosystems is the energy that green plants bind into organic molecules in the process of photosynthesis, minus the energy those plants use for their own life processes—growth, maintenance, and reproduction. In the jargon of ecologists, that quantity is known as the net primary production (NPP). Globally, this amounts to a production of about 225 billion metric tons of organic matter annually, nearly 60% of it on land.

Humanity is now using directly (e.g., by eating, feeding to livestock, using lumber and firewood) more than 3% of global NPP, and about 4% of that on land. This is a minimum estimate of human impact on terrestrial systems. Since *Homo sapiens* is one of (conservatively) 5 million species, this may seem an excessive share of the food resource. But considering that human beings are perhaps a million times the weight of the average animal (since the overwhelming majority of animals are small insects and mites) and need on the order of a million times the energy per individual, this share might not be too unreasonable.

Yet human beings can be thought of as co-opting NPP not only by direct use but also by indirect use. Thus if we chalk up to the human account not only the NPP directly consumed, but such other categories as the amount of biomass consumed in fires used to clear land, the parts of crop plants not consumed, the NPP of pastureland (converted from natural habitat) not consumed by livestock, and so on, the human share of terrestrial NPP climbs to a staggering 30%. And if we add to that the NPP foregone when people convert more productive natural systems to less productive ones (such as forest to farm or pasture, grassland to desert, marsh to parking lot), the total *potential* NPP on land is reduced by 13%, and the human share of the unreduced potential NPP reaches almost 40%. There is no way that the co-option by one species of almost two-fifths of Earth's annual terrestrial food production could be considered reasonable, in the sense of maintaining the stability of life on this planet.

These estimates alone both explain the basic causes and consequences of habitat destruction and alteration, and give reason for great concern about future trends. Most demographers project that *Homo sapiens* will double its population within the next century or so. This implies a belief that our species can safely commandeer upwards of 80% of terrestrial NPP, a preposterous notion to ecologists who already see the deadly impacts of today's level of human activities. Optimists who suppose that the human population can double its size again need to contemplate where the basic food resource will be obtained.

A standard fool's answer to that question is that indefinite expansion of the human population will be supported by the immeasurable riches of the sea. Unhappily for that notion, the riches of the sea have been quite carefully measured and found wanting. People now use about 2% of the NPP of the sea, and the prospects even for doubling that yield are dim. The basic reason is that efficient harvesting of the sea requires the exploitation of concentrated pools of resources—schools of fishes and larger invertebrates. People cannot efficiently harvest much of the NPP that resides in tiny phytoplankton (the green plants of the sea) or in the zooplankton (animals too small to swim against the currents). Humanity appears to be already utilizing about as much of oceanic NPP as it can on a sustainable basis.

This discrepancy in the ability of *Homo sapiens* to exploit terrestrial and oceanic NPP is reflected in the general lack of an extinction crisis in the seas. Except for such organisms as some whales and fishes that are threatened by direct exploitation, animals that spend their entire lives in the open sea are relatively secure. Aside from some limited environments, such as certain coral reefs, the effects of habitat destruction are relatively small away from shorelines and estuaries. This situation could, of course, change rapidly if marine pollution increases—a distinct possibility.

The extirpation of populations and species of organisms exerts its primary impact on society through the impairment of ecosystem services. All plants, animals, and microorganisms exchange gases with their environments and are thus directly or indirectly involved in maintaining the mix of gases in the atmosphere. Changes in that mix (such as increases in carbon dioxide, nitrogen oxides, and methane) can lead to rapid climate change and, in turn, agricultural disaster. As physicist John Holdren put it, a carbon dioxide-induced climatic change could lead to the deaths by famine of as many as a billion people before 2020. Destroying forests deprives humanity not only of timber but also of dependable freshwater supplies and furthermore increases the danger of floods. Destruction of insects can lead to the failure of crops that depend upon insect pollination. Extermination of the enemies of insect pests (a usual result of ad lib pesticide spraying) can terminate the pest control services of an ecosystem and often leads to severe pest outbreaks. The extinction of subterranean organisms can destroy the fertility of the soil. Natural ecosystems maintain a vast genetic library that has already provided people with countless benefits and has the potential for providing many, many more.

These examples can be multiplied manyfold—the basic point is that organisms, most of which are obscure to nonbiologists, play roles in ecological systems that are essential to civilization. When a population playing a certain role is wiped out, ecosystem services suffer, even if many other populations of the same organism are still extant. If the *population* of Engelmann spruce trees (*Picea engelmanni*) in the watershed above your Colorado home is chopped down, you could be killed in a resulting flood, even though the *species* of spruce is not endangered. Equally, if that were the last population and it were reduced to just a dozen trees (so that, technically, the species still existed), you would not be spared the flood, and chance events would likely finish off the Engelmann spruce eventually anyway.

In most cases, numerous genetically diverse populations are necessary to ensure the persistence of a species in the face of inevitable environmental changes that occur naturally. The existence of many populations spreads the risk so that un-favorable conditions in one or a few habitats do not threaten the entire species. And the presence of abundant genetic variation within a species (virtually assured if its populations are living in different geographic areas) increases its potential for successfully evolving in response to long-term environmental changes. Today, this genetic diversity *within* species is declining precipitously over much of Earth's land surface—an unheralded loss of one of humanity's most vital resources. That resource is largely irreplaceable. Along with fossil fuels, rich soils, ancient groundwater, and mineral deposits, genetic diversity is part of the inheritance of capital that *Homo sapiens* is rapidly squandering.

What then will happen if the current decimation of organic diversity continues? Crop yields will be more difficult to maintain in the face of climatic change, soil erosion, loss of dependable water supplies, decline of pollinators, and ever more serious assaults by pests. Conversion of productive land to wasteland will accelerate; deserts will continue their seemingly inexorable expansion. Air pollution will increase, and local climates will become harsher. Humanity will have to forego many of the direct economic benefits it might have withdrawn from Earth's once well-stocked genetic library. It might, for example, miss out on a cure for cancer; but that will make little difference. As ecosystem services falter, mortality from respiratory and epidemic disease, natural disasters, and especially famine will lower life expectancies to the point where cancer (largely a disease of the elderly) will be unimportant. Humanity will bring upon itself consequences depressingly similar to those expected from a nuclear winter (Ehrlich, 1984). Barring a nuclear conflict, it appears that civilization will disappear some time before the end of the next century—not with a bang but a whimper.

Preventing such a denouement will prove extremely difficult at the very least; it may well prove to be impossible. Earth's habitats are being nickeled and dimed to death, and human beings have great difficulty perceiving and reacting to changes that occur on a scale of decades. Our nervous systems evolved to respond to short-term crises—the potential loss of a mate to a rival, the sudden appearance of a bear in the mouth of the cave. For most of human evolutionary history there was no reason for natural selection to tune us to recognize easily more gradual trends, since there was little or nothing one could do about them. The human lineage evolved in response to changes in the ecosystems in which our ancestors lived, but individuals could not react adaptively to those changes, which usually took place slowly. The depletion of organic diversity and the potential destruction of civilization may, ironically, be an inevitable result of our evolutionary heritage.

If humanity is to avoid becoming once again a species consisting of scattered groups practicing subsistence agriculture, dramatic steps will be necessary. They can only be briefly outlined here. Simply setting aside preserves in the remaining relatively undisturbed ecosystems will no longer suffice. In most parts of the planet such areas are too scarce, and rapid climatic changes may make those preserves impossible to maintain (Peters and Darling, 1985). Areas already greatly modified by human activities must be made more hospitable for other organisms; for example, the spewing of toxins into the environment (leading to intractable problems like acid deposition) must be abated.

Above all, the growth of the human population must be halted, since it is obvious that if the scale of human activities continues to increase for even a few more decades, the extinction of much of Earth's biota cannot be avoided. Indeed, since *Homo sapiens* is now living largely on its inherited capital and in the future will have to depend increasingly on its income (NPP), one can argue persuasively that the size of the human population and the scale of human activities should be gradually *reduced* below present levels. Reducing that scale will be an especially difficult task, since it means that the environmental impacts of the rich must be enormously curtailed to permit the poor a chance for reasonable development.

Although improvements in the technologies used to support human life and affluence can of course help to ameliorate the extinction crisis, and to a limited extent technologies can substitute for lost ecosystem services, it would be a dangerous miscalculation to look to technology for the answer (see, for example, Ehrlich and Mooney, 1983). In my opinion, only an intensive effort to make those improvements and substitutions, combined with a revolution in attitudes toward other people, population growth, the purpose of human life, and the intrinsic values of organic diversity, is likely to prevent the worst catastrophe ever to befall the human lineage. Curiously, scientific analysis points toward the need for a quasi-religious transformation of contemporary cultures. Whether such a transformation can be achieved in time is problematic, to say the least.

We must begin this formidable effort by increasing public awareness of the urgent need for action. People everywhere should understand the importance of the loss of diversity not only in tropical forests, coastal zones, and other climatically defined regions of the world but also in demographically delineated regions such as areas of urbanization. The geological record can tell us much about catastrophic mass extinctions of the past. That, and more intensive studies of the living biota, can provide hints about what we might expect in the future. At the present time, data on the rates and direction of biodiversity loss remain sparse and often uncertain. As a result, estimates of the rate of loss, including the number and variety of species that are disappearing, vary greatly—in some cases, as pointed out by E. O. Wilson in Chapter 1, by as much as an order of magnitude. Moreover, scientists have also differed in their predictions of the eventual impact that will result from the diminishing biodiversity. Some aspects of these challenges are explored in the following five chapters comprising this section and are reflected throughout this volume.

REFERENCES

Ehrlich, A. H. 1984. Nuclear winter. A forecast of the climatic and biological effects of nuclear-war. Bull. At. Sci. 40(4):S1–S15.

Ehrlich, P. R. 1986. The Machinery of Nature. Simon and Schuster, New York. 320 pp.

Ehrlich, P. R., and A. H. Ehrlich. 1981. Extinction: The Causes and Consequences of the Disappearance of Species. Random House, New York. 305 pp.

Ehrlich, P. R., and H. A. Mooney. 1983. Extinction, substitution, and ecosystem services. BioScience 33(4):248–254.

Holdren, J. P., and P. R. Ehrlich. 1974. Human population and the global environment. Am. Sci. 62:282–292.

Peters, R. L., and J. D. S. Darling. 1985. The greenhouse effect and nature reserves. BioScience 35(11):707–717.

Terborgh, J. W. 1980. The conservation status of neotropical migrants: Present and future. Pp. 21–30 in A. Keast and E. S. Morton, eds. Migrant Birds in the Neotropics: Ecology, Behavior, Distribution, and Conservation. A symposium held at the Conservation and Research Center, National Zoological Park, Smithsonian Institution. Smithsonian Institution Press, Washington, D.C.

Vitousek, P. M., P. R. Ehrlich, A. H. Ehrlich, and P. M. Matson. 1986. Human appropriation of the products of photosynthesis. BioScience 36(6):368–373.

TROPICAL FORESTS AND THEIR SPECIES
Going, Going . . . ?

NORMAN MYERS

Consultant in Environment and Development, Oxford, United Kingdom

There is strong evidence that we are into the opening stages of an extinction spasm. That is, we are witnessing a mass extinction episode, in the sense of a sudden and pronounced decline worldwide in the abundance and diversity of ecologically disparate groups of organisms.

Of course extinction has been a fact of life since the emergence of species almost 4 billion years ago. Of all species that have ever existed, possibly half a billion or more, there now remain only a few million. But the natural background rate of extinction during the past 600 million years, the period of major life, has been on the order of only one species every year or so (Raup and Sepkoski, 1984). Today the rate is surely hundreds of times higher, possibly thousands of times higher (Ehrlich and Ehrlich, 1981; Myers, 1986; Raven, 1987; Soulé, 1986; Western and Pearl, in press; Wilson, 1987). Moreover, whereas past extinctions have occurred by virtue of natural processes, today the virtually exclusive cause is *Homo sapiens*, who eliminates entire habitats and complete communities of species in super-short order. It is all happening in the twinkling of an evolutionary eye.

To help us get a handle on the situation, let us take a lengthy look at tropical forests. These forests cover only 7% of Earth's land surface, yet they are estimated to contain at least 50% of all species (conceivably a much higher proportion [see Erwin, Chapter 13 of this volume]). Equally important, they are being depleted faster than any other ecological zone.

TROPICAL FORESTS

There is general agreement that remaining primary forests cover rather less than 9 million square kilometers, out of the 15 million or so that may once have existed according to bioclimatic data. There is also general agreement that between 76,000 and 92,000 square kilometers are eliminated outright each year, and that at least a further 100,000 square kilometers are grossly disrupted each year (FAO and UNEP, 1982; Hadley and Lanley, 1983; Melillo et al., 1985; Molofsky et al., 1986; Myers, 1980, 1984). These figures for deforestation rates derive from a data base of the late 1970s; the rates have increased somewhat since then. This means, roughly speaking, that 1% of the biome is being deforested each year and that more than another 1% is being significantly degraded.

The main source of information lies with remote-sensing surveys, which constitute a thoroughly objective and systematic mode of inquiry. By 1980 there were remote-sensing data for approximately 65% of the biome, a figure that has risen today to 82%. In all countries where remote-sensing information has been available in only the past few years—notably Indonesia, Burma, India, Nigeria, Cameroon, Guatemala, Honduras, and Peru—we find there is greater deforestation than had been supposed by government agencies in question.

Tropical deforestation is by no means an even process. Some areas are being affected harder than others; some will survive longer than others. By the end of the century or shortly thereafter, there could be little left of the biome in primary status with a full complement of species, except for two large remnant blocs, one in the Zaire basin and the other in the western half of Brazilian Amazonia, plus two much smaller blocs, in Papua New Guinea and in the Guyana Shield of northern South America. These relict sectors of the biome may well endure for several decades further, but they are little likely to last beyond the middle of next century, if only because of sheer expansion in the numbers of small-scale cultivators.

Rapid population growth among communities of small-scale cultivators occurs mainly through immigration rather than natural increase, i.e., through the phenomenon of the shifted cultivator. As a measure of what ultrarapid growth rates can already impose on tropical forests, consider the situation in Rondonia, a state in the southern sector of Brazilian Amazonia. Between 1975 and 1986, the population grew from 111,000 to well over 1 million, i.e., a 10-times increase in little more than 10 years. In 1975, almost 1,250 square kilometers of forest were cleared. By 1982, this amount had grown to more than 10,000 square kilometers, and by late 1985, to around 17,000 square kilometers (Fearnside, 1986).

It is this broad-scale clearing and degradation of forest habitats that is far and away the main cause of species extinctions. Regrettably, we have no way to know the actual current rate of extinction, nor can we even come close with accurate estimates. But we can make substantive assessments by looking at species numbers before deforestation and then applying the analytic techniques of island biogeography. To help us gain an insight into the scope and scale of present extinctions, let us briefly consider three particular areas: the forested tracts of western Ecuador, Atlantic-coast Brazil, and Madagascar. Each of these areas features, or rather featured, exceptional concentrations of species with high levels of endemism. Western Ecuador is reputed to have once contained between 8,000 and 10,000 plant species with an endemism rate somewhere between 40 and 60% (Gentry, 1986). If we suppose, as we reasonably can by drawing on detailed inventories in sample plots, that there are at least 10 to 30 animal species for every one plant species, the species complement in western Ecuador must have amounted to 200,000 or more in all. Since 1960, at least 95% of the forest cover has been destroyed to make way for banana plantations, oil exploiters, and human settlements of various sorts. According to the theory of island biogeography, which is supported by abundant and diversified evidence, we can realistically expect that when a habitat has lost 90% of its extent, it will eventually lose half its species. Precisely how many species have actually been eliminated, or are on the point of extinction, in western Ecuador is impossible to say. But ultimate accuracy is surely irrelevant, insofar as the number must total tens of thousands at least, conceivably 50,000— all eliminated or at least doomed in the space of just 25 years.

Very similar baseline figures for species totals and endemism levels, and a similar story of forest depletion (albeit for different reasons and over a longer time period), apply to the Atlantic-coastal forest of Brazil, where the original 1 million square kilometers of forest cover have been reduced to less than 50,000 square kilometers (Mori et al., 1981). Parallel data apply also to Madagascar, where only 5% of the island's primary vegetation remains undisturbed—and where the endemism levels are rather higher (Rauh, 1979).

So in these three tropical forest areas alone, with their roughly 600,000 species, the recent past must have witnessed a sizeable fallout of species. Some may not have disappeared as yet, due to the time lag in equilibration, i.e., delayed fallout effects stemming from habitat depletion. But whereas the ultimate total of extinctions in these areas in the wake of deforestation to date will presumably amount to some 150,000 species, we may realistically assume that already half, some 75,000 species, have been eliminated or doomed.

Deforestation in Brazil's Atlantic-coastal forest and Madagascar has been going on for several centuries, but the main damage has occurred during this century, especially since 1950, i.e., since the spread of broad-scale industrialization and plantation agriculture in Brazil and since the onset of rapid population growth in Madagascar. This all means that as many as 50,000 species have been eliminated or doomed in these areas alone during the last 35 years. This works out to a crude average of almost 1,500 species per year—a figure consistent with the independent assessment of Wilson (1987), who postulates an extinction rate in all tropical forests of perhaps 10,000 species per year. Of course many reservations attend these calculations. More species than postulated may remain until a new equilibrium is established and causes their disappearance. Conversely, more species will presumably have disappeared during the later stages of the 35-year period than during the opening stage. Whatever the details of the outcome, we can judiciously use the figures and conclusions to form a working appraisal of the extent that an extinction spasm is already under way.

EXTINCTION RATES: FUTURE

The outlook for the future seems all the more adverse, though its detailed dimensions are even less clear than those of the present. Let us look again at tropical forests. We have seen what is happening to three critical areas. We can identify a good number of other sectors of the biome that feature exceptional concentrations of species with exceptional levels of endemism and that face exceptional threat of depletion, whether quantitative or qualitative. They include the Choco forest of Colombia; the Napo center of diversity in Peruvian Amazonia, plus seven other centers (out of 20-plus centers of diversity in Amazonia) that lie around the fringes of the basin and hence are unusually threatened by settlement programs and various other forms of development; the Tai Forest of Ivory Coast; the montane forests of East Africa; the relict wet forest of Sri Lanka; the monsoon forests of the Himalayan foothills; northwestern Borneo; certain lowland areas of the Philippines; and several islands of the South Pacific (New Caledonia, for instance, is 16,100 square kilometers, almost the size of New Jersey, and contains 3,000 plant species, 80% of them endemic).

These various sectors of the tropical forest biome amount to roughly 1 million square kilometers (2.5 times the size of California), or slightly more than one-tenth of the remaining undisturbed forests. As far as we can best judge from their documented numbers of plant species, and by making substantiated assumptions about the numbers of associated animal species, we can estimate that these areas surely harbor 1 million species (could be many more)—and in many of the areas, there is marked endemism. If present land-use patterns and exploitation trends persist (and they show every sign of accelerating), there will be little left of these forest tracts, except in the form of degraded remnants, by the end of this century or shortly thereafter. Thus forest depletion in these areas alone could well eliminate large numbers of species, surely hundreds of thousands, within the next 25 years at most.

Looking at the situation another way, we can estimate, on the basis of what we know about plant numbers and distribution together with what we can surmise about their associated animal communities, that almost 20% of all species occur in forests of Latin America outside of Amazonia and that another 20% are present in forests of Asia and Africa outside the Zaire basin (Raven, 1987). That is, these forests contain some 1 million species altogether, even if we estimate that the planetary total is only 5 million. All the primary forests in which these species occur may well disappear by the end of this century or early in the next. If only half the species in these forests disappear, this will amount to several hundred thousand species.

What is the prognosis for the longer-term future? Could we eventually lose at least one-quarter, possibly one-third, or conceivably an even larger share of all extant species? Let us take a quick look at Amazonia (Simberloff, 1986). If deforestation continues at present rates until the year 2000, but then comes to a complete halt, we could anticipate an ultimate loss of about 15% of the plant species and a similar percentage of animal species. If Amazonia's forest cover were to be ultimately reduced to those areas now set aside as parks and reserves, we could anticipate that 66% of the plant species will eventually disappear together with almost 69% of bird species and similar proportions of all other major categories of species.

Of course we may learn how to manipulate habitats to enhance survival prospects. We may learn how to propagate threatened species in captivity. We may be able to apply other emergent conservation techniques, all of which could help to relieve the adverse repercussions of broad-scale deforestation. But in the main, the damage will have been done. For reasons of island biogeography and equilibration, some extinctions in Amazonia will not occur until well into the twenty-second century, or even further into the future. So a major extinction spasm in Amazonia is entirely possible, indeed plausible if not probable.

TROPICAL FOREST AND CLIMATIC CHANGE

Protected areas are not likely to provide a sufficient answer for reasons that reflect climatic factors. In Amazonia, for instance, it is becoming apparent that if as much as half the forest were to be safeguarded in some way or another (e.g., through multiple-use conservation units as well as protected areas), but the other half of the forest were to be developed out of existence, there could soon be at work a hydrological feedback mechanism that would allow a good part of Amazonia's moisture to be lost to the ecosystem (Salati and Vose, 1984). The remaining forest would likely be subjected to a steady desiccatory process, until the moist forest became more like a dry forest, even a woodland—with all that would mean for the species communities that are adapted to moist forest habitats. Even with a set of forest safeguards of exemplary type and scope, Amazonia's biotas would be more threatened than ever.

Still more widespread climatic changes with yet more marked impact are likely to occur within the foreseeable future. By the first quarter of the next century, we may well be experiencing the climatic dislocations of a planetary warming, stemming from a buildup of carbon dioxide and other so-called greenhouse gases in the global atmosphere (Bolin and Doos, 1986; DoE, 1985). The consequences for protected areas will be pervasive and profound. The present network of protected areas, grossly inadequate as it is, has been established in accord with present-day needs. Yet its ultimate viability will be severely threatened in the wake of a greenhouse effect as vegetation zones start to migrate away from the equator with all manner of disruptive repercussions for natural environments (Peters and Darling, 1985; Peters, Chapter 51 of this volume).

These, then, are some dimensions of the extinction spasm that we can reasonably assume will overtake the planet's biotas within the next few decades (unless of course we do a massively better job of conservation). In effect we are conducting an irreversible experiment on a global scale with Earth's stock of species.

REPERCUSSIONS FOR THE FUTURE OF EVOLUTION

The foreseeable fallout of species, together with their subunits, is far from the entire story. A longer-term and ultimately more serious repercussion could lie in a disruption of the course of evolution, insofar as speciation processes will have to work with a greatly reduced pool of species and their genetic materials. We are probably being optimistic when we call it a disruption; a more likely outcome is that certain evolutionary processes will be suspended or even terminated. In the graphic phrasing of Soulé and Wilcox (1980), "Death is one thing; an end to birth is something else."

From what little we can discern from the geologic record, a normal recovery time may require millions of years. After the dinosaur crash, for instance, between 50,000 and 100,000 years elapsed before there started to emerge a set of diversified and specialized biotas, and another 5 to 10 million years went by before there were bats in the skies and whales in the seas (Jablonski, 1986). Following the crash during the late Permian Period, when marine invertebrates lost about half their families, as many as 20 million years elapsed before the survivors could establish even half as many families as they had lost (Raup, 1986).

The evolutionary outcome this time around could prove even more drastic. The critical factor lies with the likely loss of key environments. Not only do we appear ready to lose most if not virtually all tropical forests, but there is also progressive depletion of coral reefs, wetlands, estuaries, and other biotopes with exceptional biodiversity. These environments have served in the past as preeminent powerhouses of evolution, in that they have supported the emergence of more species than have other environments. Virtually every major group of vertebrates and many other large categories of animals have originated in spacious zones with warm, equable climates, notably tropical forests. In addition, the rate of evolutionary diversification—whether through proliferation of species or through the emergence of major new adaptations—has been greatest in the tropics, again most notably in tropical forests.

Of course tropical forests have been severely depleted in the past. During drier phases of the recent Ice Ages (Pleistocene Epoch), they have been repeatedly reduced to only a small fraction, occasionally as little as one-tenth, of their former expanse. Moreover, tropical biotas seem to have been unduly prone to extinction. But the remnant forest refugia usually contained sufficient stocks of surviving species to recolonize suitable territories when moister conditions returned (Prance, 1982). Within the foreseeable future, by contrast, it seems all too possible that most tropical forests will be reduced to much less than one-tenth of their former expanse, and their pockets of holdout species will be much less stocked with potential colonizers.

Furthermore, the species depletion will surely apply across most if not all major categories of species. This is almost axiomatic, if extensive environments are eliminated wholesale. The result will contrast sharply with the end of the Cretaceous Period, when not only placental mammals survived (leading to the adaptive radiation of mammals, eventually including humans), but also birds, amphibians, and crocodiles, among other nondinosaurian reptiles. In addition, the present extinction spasm looks likely to eliminate a sizeable share of terrestrial plant species, at least one-fifth within the next half century and a good many more within the following half century. By contrast, during most mass-extinction episodes of the prehistoric past, terrestrial plants have survived with relatively few losses (Knoll, 1984). They have thus supplied a resource base on which evolutionary processes could start to generate replacement animal species forthwith. If this biotic substrate is markedly depleted within the foreseeable future, the restorative capacities of evolution will be all the more reduced.

In sum, the evolutionary impoverishment of the impending extinction spasm, plus the numbers of species involved and the telescoped time scale of the phenomenon, may result in the greatest single setback to life's abundance and diversity since the first flickerings of life almost 4 billion years ago.

REFERENCES

Bolin, B., and B. R. Doos, eds. 1986. The Greenhouse Effect: Climatic Change and Ecosystems. Wiley, New York. 541 pp.

DoE (U.S. Department of Energy). 1985. Direct Effects of Increasing Carbon Dioxide on Vegetation. U.S. Department of Energy, Washington, D.C.

Ehrlich, P. R., and A. H. Ehrlich. 1981. Extinction: The Causes and Consequences of the Disappearance of Species. Random House, New York. 305 pp.

FAO and UNEP (Food and Agriculture Organization and United Nations Environment Programme). 1982. Tropical Forest Resources. Food and Agriculture Organization of the United Nations, Rome, Italy, and United Nations Environment Programme, Nairobi, Kenya. 106 pp.

Fearnside, P. M. 1986. Human Carrying Capacity of the Brazilian Rain Forest. Columbia University Press, New York. 293 pp.

Gentry, A. H. 1986. Endemism in tropical versus temperate plant communities. Pp. 153–181 in M. E. Soul, ed. Conservation Biology: The Science of Scarcity and Diversity. Sinauer Associates, Sunderland, Mass. 584 pp.

Hadley, M., and J. P. Lanley. 1983. Tropical forest ecosystems: Identifying differences, seeing similarities. Nat. Resour. 19:2–19.

Jablonski, D. 1986. Causes and consequences of mass extinction: A comparative approach. Pp. 183–230 in D. K. Elliott, ed. Dynamics of Extinction. Wiley Interscience, New York.

Knoll, A. H. 1984. Patterns of extinction in the fossil record of vascular plants. Pp. 21–68 in M. H. Nitecki, ed. Extinctions. University of Chicago Press, Chicago.

Melillo, J. M., C. A. Palm, R. A. Houghton, G. M. Woodwell, and N. Myers. 1985. A comparison of recent estimates of disturbance in tropical forests. Environ. Conserv. 12(1):37–40.

Molofsky, J., C. A. S. Hall, and N. Myers. 1986. A Comparison of Tropical Forest Surveys. U.S. Department of Energy, Washington, D.C.

Mori, S. A., B. M. Boom, and G. T. Prance. 1981. Distribution patterns and conservation of eastern Brazilian coastal forest tree species. Brittonia 33(2):233–245.

Myers, N. 1980. Conversion of Tropical Moist Forests. A report prepared for the Committee on Research Priorities in Tropical Biology of the National Research Council. National Academy of Sciences, Washington, D.C. 205 pp.

Myers, N. 1984. The Primary Source: Tropical Forests and Our Future. W. W. Norton, New York. 399 pp.

Myers, N. 1986. Tackling Mass Extinction of Species: A Great Creative Challenge. Albright Lecture, University of California, Berkeley. 40 pp.

Peters, R. L., and J. D. S. Darling. 1985. The greenhouse effect and nature reserves. BioScience 35(11):707–717.

Prance, G. T., ed. 1982. Biological Diversification in the Tropics. Proceedings of the Fifth International Symposium of the Association for Tropical Biology, held at Macuto Beach, Caracas, Venezuela, February 8–13, 1979. Columbia University Press, New York. 714 pp.

Rauh, W. 1979. Problems of biological conservation in Madagascar. Pp. 405–421 in D. Bramwell, ed. Plants and Islands. Academic Press, London, U.K.

Raup, D. M. 1986. Biological extinction in earth history. Science 231:1528–1533.

Raup, D. M., and J. J. Sepkoski. 1984. Periodicity of extinction in the geologic past. Proc. Natl. Acad. Sci. USA 81:801–805.

Raven, P. H. 1987. We're Killing Our World. Keynote Paper Presented to Annual Conference of the American Association for the Advancement of Science, Chicago, February 1987. Missouri Botanical Garden, St. Louis.

Salati, E., and P. B. Vose. 1984. Amazon basin: A system in equilibrium. Science 225:129–138.

Simberloff, D. 1986. Are we on the verge of a mass extinction in tropical rain forests? Pp. 165–180 in D. K. Elliott, ed. Dynamics of Extinction. Wiley, New York.

Soulé, M. E. 1986. Conservation Biology, The Science of Scarcity and Diversity. Sinauer Associates, Sunderland, Mass.

Soulé, M. E., and B. A. Wilcox, eds. 1980. Conservation Biology: An Evolutionary-Ecological Perspective. Sinauer Associates, Sunderland, Mass. 395 pp.

Western, D., and M. Pearl, eds. In press. Conservation 2100. Proceedings of International Conference on Threatened Wildlife and Species, Manhattan, October 1986, organized by the New York Zoological Society. Oxford University Press, New York.

Wilson, E. O. 1987. Biological diversity as a scientific and ethical issue. Pp. 29–48 in Papers Read at a Joint Meeting of the Royal Society and the American Philosophical Society. Volume 1. Meeting held April 24, 1986, in Philadelphia. American Philosophical Society, Philadelphia.

4

ECOLOGICAL DIVERSITY IN
COASTAL ZONES AND OCEANS

G. CARLETON RAY

Department of Environmental Sciences, University of Virginia, Charlottesville

Near the center of Charlottesville, Virginia, stands a heroic statue to Meriwether Lewis and William Clark, the two men Thomas Jefferson sent across our continent nearly two centuries ago. At its base, they are described as "bold and farseeing pathfinders who carried the flag of the young republic to the western ocean and revealed an unknown empire to the uses of mankind." There soon followed an exploitative horde and a loss of landscape diversity as great as for any place on Earth during the history of mankind. How anachronistic the words on that statue sound today. Yet the seeking of empires for "the uses of mankind" is the principal factor that has led to the present marine revolution (Ray, 1970). What loss of coastal and marine biodiversity may soon result, no one can presently say. But it is my view that the coastal zone is being altered just as fast as tropical forests.

The intent of this chapter is not to describe details of the biodiversity of coasts and oceans; rather, it is to examine the challenges we face in addressing this subject. The first of these is to define diversity. Slobodkin (1986, p. 263) has pointed out, "On occasion, metaphors have replaced the empirical world as foci of discussions, while precise meanings and derivations have been forgotten in the process." I have the impression that the word diversity is in some danger of this—that it sometimes is used to reinforce preexisting bias. In the introduction to *Diversity* (Patrick, 1983, p.1), this concept is defined as a "variety or multiformity, a condition of being different in character and quality," but the papers in that volume demonstrate that there is no single way to evaluate diversity. It surely is not merely species variety, as some of the public may be led to believe. Nor is it bound to dry land.

Those of us who practice ocean science must wonder about the oceanless world that often confronts us. Witness the cataloging of the diversity of "Realms, Biomes, and Biogeographical Provinces of the World" in the recent assessment by the World Resources Institute (1986). This "world" leaves oceanic space simply blank! This is the same biogeography that is repeated in many textbooks, conservation circles, and international aid agencies. Unfortunately, this world view is also that of the majority of society. Therefore, we first face the challenge of differentiating what sort of world-planet the Earth is against the backdrop of our bias.

COASTS AND OCEANS—A WORLD VIEW

J. E. Lovelock firmly grasped the world view when he said, "Less than a third of the Earth's surface is land. This may be why the biosphere has been able to contend with the radical transformations wrought by agriculture and animal husbandry, and will probably continue to strike a balance as our numbers grow and farming becomes ever more intensive. We should not, however, assume that the sea, and especially the arable regions of the continental shelves, can be farmed with the same impunity. Indeed, no one knows what risks are run when we disturb this key area of the biosphere. That is why I believe that our best and most rewarding course is to sail with Gaia[1] in view, to remind us throughout the voyage and in all our explorations that the sea is a vital part of her" (Lovelock, 1979, p. 106). I interpret this to say that biodiversity is the result of global as well as regional and local processes and that to conserve the biodiversity of one biogeographic realm might require the conservation of processes of others as well, both wet and dry.

Let us carry this a bit further. Our evolution as giant, terrestrial mammals causes us to draw hard lines on maps between land and sea. In fact, land maps do not usually include the sea; for that, we turn to charts, which do not include land. Despite the cartographers, from an ecological perspective there can be no sharp distinction. The coastal zone unifies the two, but it is not merely a narrow transition between dry and wet; on paleoecological, geological, and biological grounds, it is distinct in its own right (Figure 4-1). The coastal zone includes at least the extents of continental plains and continental shelves (Ketchum, 1972), that is, more than 8% of Earth, or about an Africa and a half. In volume, the wet portion alone comprises approximately 3 million cubic kilometers, just about the same volume occupied by all terrestrial life! It includes coastal forests and marshes as well as watersheds, in some cases quite far inland, and is as productive as any place on Earth—one reason for the fact that more than 50% of all humans live within it and take more than 90% of their marine-living resources from it. How species-rich it is, I cannot say, nor am I inclined to believe that species accounting should warp our view of it one way or the other. Nevertheless, the major objective is to define coastal zone ecosystems and their ecological characteristics.

---

[1]The concept of Mother Earth, as named by the ancient Greeks. See Lovelock, Chapter 56 in this volume.

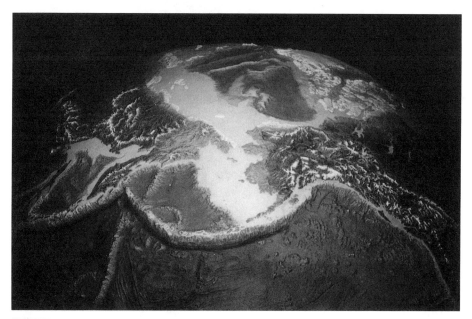

FIGURE 4-1 Coastal zones are clearly a separate but unifying region between land and sea. Photo by G. Carleton Ray.

This leads inevitably to a tripartite view of Earth in which biogeographic patterns fall within upland, open ocean, and coastal zone realms, all about equally distinct. This requires readjustments of our world view. Terrestrial realms, biomes, and provinces should not be carried to the water's edge (e.g., Udvardy, 1975). Furthermore, our perceptions of biogeographical patterns will have to change if we are to see our planet as it really is.

LIFE ACCORDING TO THE BOOK OF TAXONOMY

Wilson (1985) wondered why there are so many species and pointed out that most are in tropical forests. There can be little doubt that tropical forests hold a major proportion of species (see also Myers, Chapter 3 in this volume). It is generally supposed that our present knowledge provides a rough approximation of the relative numbers of species in the world's ecosystems and that about 80% of all species are terrestrial. This proportion may be seriously in error. According to the recent research of J. F. Grassle of Woods Hole Oceanographic Institution and his associates, "quantitative samples [in the deep sea] represent a fauna that rivals the tropical forests in diversity of species" (Grassle, personal communication, 1987). Only the future will tell how many species there are and which environments are most diverse.

Nevertheless, the measure of species presents but one dimension of diversity. At the other end of the taxonomic scale stand phyla. With help from Barnes (1963), Grzimek (1974), and Margulis and Schwartz (1982), we may count more than 70 phyla of all life from bacteria to vertebrates. Those that are exclusively marine number about 20; 18 are exclusively terrestrial. Twenty-three other phyla contain marine species, whereas only 10 more contain terrestrial species. In short, diversity of the oceans is about double the land's if it is phyla that we consider. What can we make of this? Looking further, we see that protists and invertebrates predominate in oceans and higher plants predominate on land. That is to say, these environments are vastly different in community composition, making biologically dubious any attempt to compare diversity among them simply by counting taxa. This same difficulty exists on the level of species. The tropics contain more species than do polar regions, but there are hardly any walruses in the Amazon, nor parrots in Antarctica. The species and phylum content of environments is an essential fact of ecology, but simply knowing which environments have more or fewer may be misleading and must be subject to further interpretation.

This leads to an examination of life form, that is, distinguishing species by means of verbs (describing what they do) instead of nouns (indicating what they are); this approach gathers life into functional, ecological groupings not necessarily related to their taxonomy. It is instructive to compare the aquatic and terrestrial realms from this viewpoint. I cannot think of a terrestrial life form that does not have an aquatic equivalent, but counterparts of several marine life forms are so rare on land that cartoonists have to invent them; see, for example, the sit-and-wait, deception-bait gulper-predator in Figure 4-2. The goosefish (Figure 4-3) is one example of this life form that is common in the sea. A life-style that is totally absent from land is filter feeding—an activity practiced by numerous aquatic life forms, from sponges to whales. There may be some distant terrestrial equivalents of filter feeding. I have been reminded by Dr. Eugene Morton of the Smithsonian Institution that swallows and swifts are analogs of filter feeders because they scoop high-flying "planktonic" insects from the air. But these isolated examples do not

FIGURE 4-2   A predatory life form. Cartoon by Gary Larson. This Far Side cartoon is reprinted by permission of Chronicle Features, San Francisco.

FIGURE 4-3   The goosefish (*Lophius americanus*)—a sit-and-wait, deception-bait gulper-predator. Photo by M. A. deCamp.

alter the fact of the predominance of some life forms in oceans that are rare or nonexistent on land.

We must therefore conclude that accounting of species alone can be highly misleading as a yardstick of diversity. It may also mislead us genetically. The genetic diversity of both land and sea species can be striking, as, for example, the variation among the hamlet fishes, *Hypoplectrus* (Figure 4-4). But does a family of thousands of species contain more or less genetic uniqueness than a phylum comprising one to a dozen? Some marine phyla contain very few species, but their evolutionary history is long and their species are unique; the horseshoe crab, *Limulus*, is an example. In sum, a major challenge in examining diversity lies in our perceptions and interpretations of it, taxonomically and functionally.

ECOLOGICAL DIVERSITY

A great diversity of life forms implies that there is an equally great diversity of food webs and trophic relationships, i.e., food supply and demand, and requirements for nutrients. For example, filter feeders, especially zooplankton, create extra levels in aquatic food chains that do not exist on land. In the oceans, there is also much greater diversity in body sizes than on land—from picoplankton to whales—and much larger ranges of ecological time-space relationships. Consequently, aquatic food webs tend to be more complex than terrestrial ones and there are more trophic levels in food chains. Unraveling this complexity is made all the more challenging because we are almost infinitely less knowledgeable about the nature of marine systems than we are about terrestrial systems. A good many oceanographers still adhere to the concept that marine organisms are pushed around like billiard balls by the physics and chemistry of their environments. There is too little recognition that large predatory marine animals can have marked effects on the structure of their communities, and hence on nutrient cycling, and that physical and biotic processes are, no doubt, strongly linked in a cybernetic network. For terrestrial systems, this biotic influence has become obvious. For all systems, an important question is how to distinguish between biotic and physical control mechanisms. This is but one critical area where marine science lags.

Returning to the subject of biogeography, the realms, biomes, and provinces of the coastal zones and open oceans exhibit a remarkable array of environments.

FIGURE 4-4   Hamlet fish (*Hypoplectrus unicolor*) occur in a variety of colors. ©John Douglass 1987. From Robins et al., 1986.

Figure 4-5 depicts our recent attempt to classify them. This is a world made even more complex by its strong three-dimensionality, which is not shown in the figure. In concordance with this classification are distinct biotic assemblages. In tropical reefs, we find many species in a wide taxonomic array, similar to the variety of tropical forests (Figure 4-6). Temperate marine communities have fewer species but generally higher productivity, well illustrated by commercial fishes, which constitute a very large biomass (Figure 4-7). There is also high productivity in polar areas where sea ice is annual, but marine birds and mammals—the largest concentrations of them on Earth—predominate there (Figure 4-8). Which ecosystems are more diverse seems almost irrelevant in this context. Rather, let us say that each has its own "characteristic" diversity. The description of characteristic diversity—including indicator and keystone species—must be our immediate focus, and the preservation of that diversity our ultimate challenge.

What does characteristic diversity imply? The study of island biogeography tells us that the geographic size of ecosystems is a factor in species richness. Of course, this does not mean that one species will be part of any community in perpetuity. Some will come and some will go, but functionally, the ecosystem processes might remain fundamentally the same. That is, the demise of Southern Ocean whales does not seem to have altered that system much, their roles being more or less assumed now by penguins and seals—or perhaps by the krill fishery. Also, we are all aware of the formidable amount of paper that has been consumed by publications discussing whether diversity somehow confers stability to ecosystems. I trust this has become a nonquestion for scientists, but perhaps it lingers on in some circles. I suspect that diversity per se has little to do with the stability of most marine systems, i.e., the nondiverse systems are just as stable as those that are diverse. More to the point is whether characteristic diversity confers some predictability to ecosystems. Behind this important question lies our definition of a system. Ecosystems are far from chance physical-biotic associations or mere heuristic creations of ecologists; they are functional units in every sense of that term. But defining them presents great challenges. Figure 4-9 shows a simplified concept of the components of coastal zone ecosystems. Following are some major factors that control coastal processes and that must be considered in defining the boundaries of these ecosystems:

- watershed and receiving basin morphology
- terrestrial and marine climates
- winds, waves, currents, and tides
- fluvial discharge, bedload, suspended load, and dissolved load
- terrestrial and marine biota
- human use of land or sea

Even from this simple characterization, we see that ecosystem definition requires intensive field research coupled with complex analysis. Without such an effort, one cannot reach conclusions about diversity.

We must not forget that productivity is what interests most of humanity. Is diversity a factor here? We must distinguish productivity needed to sustain ecosystems from productivity that benefits human beings (Figure 4-10). Coral reefs

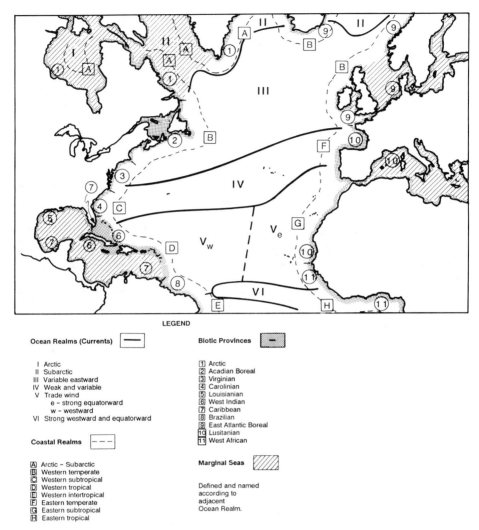

LEGEND

Ocean Realms (Currents) ───

I Arctic
II Subarctic
III Variable eastward
IV Weak and variable
V Trade wind
    e – strong equatorward
    w – westward
VI Strong westward and equatorward

Coastal Realms ─ ─ ─

Ⓐ Arctic – Subarctic
Ⓑ Western temperate
Ⓒ Western subtropical
Ⓓ Western tropical
Ⓔ Western intertropical
Ⓕ Eastern temperate
Ⓖ Eastern subtropical
Ⓗ Eastern tropical

Biotic Provinces

1 Arctic
2 Acadian Boreal
3 Virginian
4 Carolinian
5 Louisianian
6 West Indian
7 Caribbean
8 Brazilian
9 East Atlantic Boreal
10 Lusitanian
11 West African

Marginal Seas

Defined and named
according to
adjacent
Ocean Realm.

FIGURE 4-5 Classification of North Atlantic coastal and marine environments (after Hayden et al., 1984, with modifications for arctic and subarctic realms after Dunbar, 1985). This is a symbolic representation, not drawn to scale, especially for coastal realms. Ocean realms are for surface waters only. Coastal realms are highly variable, especially for temperate areas, which contain attributes of both subarctic and subtropical coastal waters.

FIGURE 4-6   Coral reefs contain ecological diversity as extensive as that found in tropical forests. Photo by G. Carleton Ray.

are productive and diverse, but they are not nearly as useful for food production as are temperate seas, where schooling fishes predominate and can be easily caught over extensive banks and shelves. There is a negative correlation between diversity and productivity in these cases. By analogy, farming on land is most productive for humans when systems are simplified. One of the greatest challenges for marine science is the prediction of consequences that would result from the loss of diversity in the increasing number of coastal systems that are being farmed through aquaculture. Will this lead to the loss of the characteristic diversity of coastal systems and thus to the loss of system predictability? This is the danger of not heeding Lovelock's warning quoted above.

Perhaps the greatest challenge of all lies in determining which characteristic species contribute most to their ecosystem, to productivity, to predictability. Are some species more essential than others from a functional, ecological point of view? In the present state of our ignorance, an attempt to answer this might lead to some nasty choices. Surely some species are more important to their ecosystems than are others, as indicators of ecological processes or as keystones that influence community structure. But which are these? We know pitifully few of them for coastal and ocean systems. So when some decision-maker asks which species might be sacrificed, we cannot say. The immense diversity of life seems simply redundant to many who are in the position of having to decide about environmental matters— and we might have to admit that some species may indeed be redundant. But when asked to identify such redundancies, we may react like the young Mozart when told by Emperor Josef II that his sonata contained too many notes. He replied that it contained "exactly the necessary number."

CONCLUSIONS IN PROSPECT

The ultimate challenge lies in detecting the loss of biodiversity in coastal and marine systems. The last fallen mahogany would lie perceptibly on the landscape, and the last black rhino would be obvious in its loneliness, but a marine species may disappear beneath the waves unobserved and the sea would seem to roll on the same as always. Extinction rates in the coastal zone and oceans are not known. Very few species seem to have gone. Some relicts, such as Steller's sea cow, are gone, as are some especially vulnerable species, such as the Labrador duck and the great auk. I wonder how much effort would be spent on ensuring their survival today. Would we dare pull the plug as some would do for the California condor so that our attention and limited resources could be turned toward other equally pressing matters? Or would we use these species, like the panda is being used, to raise funds for conservation efforts?

Though the bulk of humanity lives in coastal zones, the wet portion of our planet still seems distantly remote—out of sight and out of mind to most people.

FIGURE 4-7 Temperate Atlantic Ocean school of amberjack (*Seriola dumerili*). Photo by M. A. deCamp.

Not so long ago in our history, the ocean was regarded primarily as a surface for commerce. Now there is more awareness that we are only beginning to know and understand the oceans. The astonishing rates at which new marine life and processes are being discovered testify to this. The phylum Loricifera was described only in 1983 as a result of the discovery of a single species, *Nanaloricus mysticus*, a small organism that lives in the sediment (Kristensen, 1983). The 5-meter-long mega-mouth shark *Megachasma pelagios* is known from but two specimens caught in only the last decade. An entirely new habitat—ocean vents, such as the sulfide chimneys called "black smokers"—contains species that were unknown until the last half decade or so. The productivity of some marine systems may have been underes-timated by half due to our ignorance of the role played by bacterioplankton and to the lack of appropriate methods of measurement. Also, it has recently been revealed that wave energy creates the most productive ecosystems yet discovered, twice that of the most productive tropical forests (Leigh et al., 1987). How must we respond to all this? Clearly, we must intensify our research and communicate our findings rapidly to the public.

The goal of future efforts to address biodiversity must not be merely the com-pilation of lists of species. Though one must be sympathetic to intensive efforts to find out how much species diversity exists, there is no substitute for learning how systems work, the implications of their characteristic diversity, and the role in-dividual species play. That is, I see our task not as species inventory, but more as ecological discovery. The description of species is not sufficient. Rather, we need to identify the species that are important contributors to ecosystem processes, that

FIGURE 4-8   Walruses (*Odobenus rosmarus*) in the Bering Sea. Birds and mammals such as these predominate in polar regions. Photo by G. Carleton Ray.

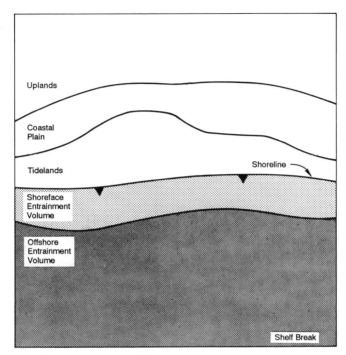

Uplands

Coastal
Plain

Tidelands

Shoreline

Shoreface
Entrainment
Volume

Offshore
Entrainment
Volume

Shelf Break

FIGURE 4-9   The ecology of the coastal zone may be influenced by the distant and nearby environments identified in the figure. Courtesy B. P. Hayden.

help structure their communities, enhance productivity, and help recycle essential nutrients. Marine scientists have been looking into such questions for some time and have decided that for some systems zooplankton is the key group. For the Southern Ocean, for example, is it the krill or the whales and seals and penguins that matter most? Or is it all of them? This is a most pragmatic question in the expensive world of marine science.

I do not wish to challenge those who would save some particular portion of this planet as a high priority simply because of its diversity. As members of the biological community, we have a common goal: the preservation of as much of this whole planet's diversity as possible. Nevertheless, I feel compelled to emphasize that diversity often lies in the eye of the beholder. There is little question that some ecosystems have more species than others. But it does not follow that any one number of species or biomass conveys on any ecosystem more value than on any other, nor can the value of species be ranked on strictly taxonomic grounds; that is, are whales or plankton most worth saving? We are slowly growing out of our bias toward species that are most like us—the warm-blooded animals that cause our anthropocentric senses to soar and our hands to reach for our checkbooks. Furthermore, the point is often made that since the potential medical or economic value of a species cannot often be predicted, we must save them all. This is clearly impossible, and it may also be illogical.

FIGURE 4-10 Sustenance for mankind. Cartoon by Gary Larson. This Far Side cartoon is reprinted by permission of Chronicle Features, San Francisco.

Pogo once observed that he had "seen the enemy, and it is us." Most creatures of the sea are cold-blooded and strange to us—even forbidding. Great white sharks (*Carcharodon carcharias*) can be crudely slaughtered with scarcely a peep from the conservation community. New hordes of people occupy the coastal zone yearly. Some hope to farm it; most just unwittingly pollute it, while increasingly drawing upon marine systems for resources and leisure. Despite dramatic advances in awareness, coastal zones and oceans continue to receive compassionate neglect. We seem to love the sea, the romance of it, the symbols (such as whales) that swim in it, and the coral reefs that we swim over. But the oceans remain foreign to most, and the concept of coastal zones as the broad systems they are continues to go largely unrecognized. Thus, the principal challenge, when addressing coastal and marine diversity, lies in recognizing its global role. If Lovelock is correct in his perception of Gaia, the coastal zone may be the single most important portion of our planet. The loss of its biodiversity may have repercussions far beyond our worst fears. Addressing this need will take an intensive research effort backed by intensive political persuasion.

We might start by giving the coastal zone and oceans equal time. The Forum on BioDiversity, whose participants have contributed to this volume, demonstrates a need in this respect. The brochure announcing the Forum's program (and used as the jacket illustration of this book) depicts 13 insects, 6 mammals, 6 birds, 3 amphibians, a fish, and a reptile, and but three marine critters, all starfishes. Among the contributors to this publication are about 25 terrestrial scientists, overwhelmingly tropical, scatterings of economists and philosophers, about two-and-a-half classified as coastal or marine biologists, and perhaps one or two whose focus is the polar regions. A film presented during the Forum, "The Frozen Ocean," is not merely a misnomer—there is, after all, lots of water beneath the far from continuous and mostly seasonal ice—but the program's description of the film refers to "the unexpected ecological riches of the Arctic." This demonstrates that biases die hard, since such riches are "unexpected" only to those who have never been there! Meanwhile, there continues to be a benign intolerance in some conservation and development circles for supporting the basic research and concept development necessary for preservation of biodiversity. We are reminded to address problems of the "real world." But whose real world? For want of so-called esoteric knowledge, we watch helplessly as exploited species become rare, the rare endangered, and perhaps the endangered extinct, without knowing why or what to do. Science and conservation clearly need to be joined in a much more comprehensive alliance.

Time to act to preserve the characteristic diversities of coastal and marine systems grows short. Our decision-makers and the public at large seem intent, for example, blissfully to use what is called the assimilative capacity of coastal and ocean waters as a receptacle for our wanton creation of toxic waste and garbage. We are told that this is an economic necessity, but we have few defenses that are ecological. Both the research and conservation communities must intensify their efforts to understand the relationships underlying the ecological processes that result in each ecosystem's characteristic biodiversity. In conservation and management quarters, this requires new perceptions of research and new definitions of real worlds. This need is especially important for coastal zones and oceans, where we are so far behind. Perhaps this requires no less than a government office of biodiversity that would allocate two-thirds of its time, space, and effort to coastal and marine systems, reflecting their global proportions. This should be backed up by specific mandates for research within the National Science Foundation and elsewhere. The status quo can only result in the unwitting recording of extinction.

REFERENCES

Barnes, R. D. 1963. Invertebrate Zoology. W. B. Saunders Co., Philadelphia. 632 pp.

Dunbar, M. J. 1985. The Arctic marine ecosystem. Pp. 1–35 in F. R. Engelhardt, ed. Petroleum Effects in the Arctic Environment. Elsevier Applied Science Publishers, London and New York.

Grzimek, B., ed. 1974. Grzimek's Animal Life Encyclopedia. 13 volumes. Van Nostrand Reinhold, New York.

Hayden, B. P., G. C. Ray, and R. Dolan. 1984. Classification of coastal and marine environments. Environ. Conserv. 11(3):199–207.

Ketchum, B. H., ed. 1972. The Water's Edge. Critical Problems of the Coastal Zone. Coastal Zone workshop 1972, Woods Hole, Mass. The MIT Press, Cambridge, Mass. 393 pp.

Kristensen, R. M. 1983. Loricifera, a new phylum with Aschelminthes characters from the meiobenthos. Z. Zool. Syst. 21(3):163–180.

Leigh, E. G., Jr., R. T. Paine, J. F. Quinn, and T. H. Suchanek. 1987. Wave energy and intertidal productivity. Proc. Natl. Acad. Sci. USA 84:1314–1318.

Lovelock, J. E. 1979. Gaia: A New Look at Life on Earth. Oxford University Press, New York. 157 pp.

Margulis, L., and K. V. Schwartz. 1982. Five Kingdoms: An Illustrated Guide to the Phyla of Life on Earth. W. H. Freeman, San Francisco. 338 pp.

Patrick, R., ed. 1983. Diversity. Benchmark Papers in Ecology/13. Hutchinson Ross, Stroudsbourg, Pa. 413 pp.

Ray, G. C. 1970. Ecology, law, and the "Marine Revolution." Biol. Conserv. 3(1):7–17.

Robins, C. R., G. C. Ray, and J. Douglass. 1986. A Field Guide to Atlantic Coast Fishes of North America. Houghton Mifflin, Boston. 354 pp.

Slobodkin, L. B. 1986. The role of minimalism in art and science. Am. Nat. 127(3):257–265.

Udvardy, M. D. F. 1975. A Classification of the Biogeographical Provinces of the World. IUCN Occasional Paper No. 18. International Union for the Conservation of Nature and Natural Resources, Morges, Switzerland. 49 pp.

Wilson, E. O. 1985. Time to revive systematics. Science 230(4731):1227.

World Resources Institute. 1986. World Resources 1986. An Assessment of the Resource Base That Supports the Global Economy. Basic Books, New York. 353 pp.

CHAPTER

5

DIVERSITY CRISES IN THE GEOLOGICAL PAST

DAVID M. RAUP

Sewell L. Avery Distinguished Service Professor, Department of Geophysical
Sciences, University of Chicago, Chicago, Illinois

The geological record of the past several hundred million years contains a wealth of information about species extinction. With these data we can place our knowledge of present-day extinctions in the larger time context of the global evolution of life.

Are the present and projected extinctions in the moist tropics unusual in the history of life? What have been the evolutionary consequences of past extinction events, especially the mass extinctions? How resilient is the global biota when confronted with the elimination of large numbers of species within a short time?

In our attempt to tackle these and related questions, there are serious problems of scale. In most cases, the geologist is forced to work on time scales measured in millions of years. And it is rarely possible in the fossil record to do a large-scale, synoptic analysis at the population or species levels. Limited fossilization usually coarsens the analysis into higher taxonomic levels: genera, families, and even orders. Interpolation of the results back down to the level of species is possible but often difficult.

Within the overall framework of geological time, the paleontologist can operate in two rather distinct time frames. The first is so-called deep time, which includes the history of life since the emergence of complex metazoans (multicellular organisms with differentiated tissues) near the beginning of the Cambrian period, about 600 million years ago. The interval since this initial metazoan proliferation, generally called Phanerozoic time (comprising the Paleozoic, Mesozoic, and Cenozoic eras), contains most of the extinction data and yields estimates of background rates plus glimpses of the mass extinctions that so nearly ended life on Earth.

The second time frame is shallow time: the record of the past few hundred thousand years during which plants and animals were essentially modern. Data from shallow time can be tied directly to present-day biogeography and diversity. Such data include a record of the effects of climatic change in the tropics during the Pleistocene epoch (approximately the last 2 million years), which are especially critical to the modeling of present and future changes.

THE PHANEROZOIC RECORD OF EXTINCTION

Complex life as we know it became firmly established on Earth toward the end of the Precambrian era and in the early Cambrian period. The exponential increase in diversity of multicellular organisms came after almost 3 billion years of surprisingly sluggish evolution of smaller, simpler organisms. The trigger for the diversification of higher organisms is not known for sure, but no matter what the cause, the fossil record record shows an epidemic of diversification. Most of the major phyla originated during this phase, and numbers of species increased dramatically. Ironically, the major groups of most interest to us now, including land vertebrates, insects, and higher plants, did not develop until somewhat later. But these late-comers did not profoundly affect global biology, except from our own anthropo-centric viewpoint.

Following the initial diversification, species extinction was and continued to be almost as common as species origination. Average durations of species were generally less than 10 million years, and the biological composition of Earth, at least at the species level, changed completely many times. Phanerozoic time included a number of profound perturbations: the mass extinctions. The most serious of these, near the end of the Permian period (250 million years ago), eliminated an estimated 52% of the families of the marine animals then living and had significant though lesser effects on plants and terrestrial organisms. Published attempts to interpolate the 52% rate of family extinction to the level of species kill have yielded estimates ranging from 77 to 96% extinction for the marine animal species then living. If these estimates are even reasonably accurate, global biology (for higher organisms at least) had an extremely close brush with total destruction.

Another four or five Phanerozoic events are also usually classed as mass extinctions, including the Cretaceous-Tertiary event 65 million years ago. Each of these large extinctions probably eliminated at least half the animal species then living.

In the times between the big mass extinctions, there have been many smaller events, which have been used by geologists to subdivide the Phanerozoic time into periods, epochs, and smaller time units. It is not yet clear whether the smaller events are most properly lumped into a general phenomenon called background extinction, which is qualitatively different from mass extinction, or whether the smaller extinctions differ only in size from the mass extinctions. Although the biggest mass extinctions do show a qualitatively different picture of selective survival than the intervening extinctions, there is increasing evidence that even the smaller extinctions are short-lived, point events (see Raup, 1986, for review). The terms episodic and stepwise extinction have been applied to this interpretation, that is, relatively long periods of biological stability, perhaps measured in hundreds of thousands of years, punctuated by short bursts of species kill. This is rapidly becoming an important area for research, because it speaks to the problem of whether

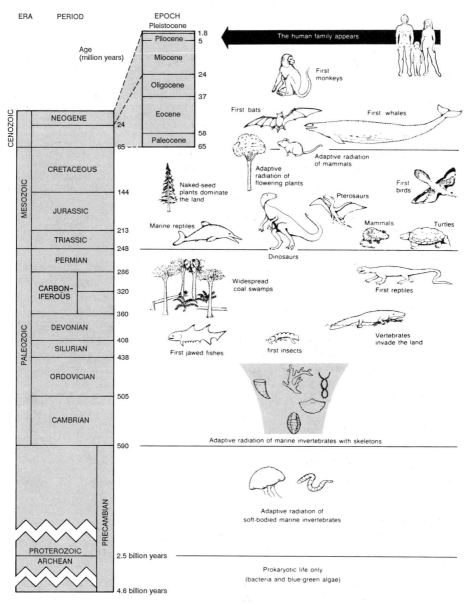

Reprinted from Stanley, 1986, with permission.

plant and animal species are fundamentally fragile and subject to elimination throughout their existence or whether they are effectively immune to extinction except during short periods of extreme stress.

EXTINCTION RATES IN DEEP TIME

It is a simple matter to compute average rates of extinction for large portions of the Phanerozoic fossil record, but there are some serious problems of interpretation. For the entire Phanerozoic time, the average species extinction rate has been estimated to be 9% per million years (Raup, 1978). This translates into 0.000009% per year, or about one species lost every 5 years in a biosphere containing 2 million living species. This number is probably low by at least a factor of 10 because the paleontologist is generally not able to see local endemic species. But even if we increase the average extinction rate by an order of magnitude, to two species every year, the rate is trivial in comparison to the extinction presumably being caused by habitat destruction and other human activities at present.

The main problem with the average rate calculations is that they lump together times of high and low extinction. If, as is urged by the proponents of mass extinction by comet or asteroid impact, the Cretaceous-Tertiary extinctions took place over a time as short as a single year, then calculations of long-term rates become meaningless: during short intervals of extreme physical environmental stress, extinction rates were nearly infinite, whereas between these events, extinction rates may have been virtually zero. In this connection, it is interesting to note that there is no statistical correlation between the durations of the standard time units in the Phanerozoic eon and the numbers of extinctions known to occur during those units. Although this does not prove that the incidence of extinction is independent of elapsed time, it is compatible with the view that extinctions are point events rather than the result of a time-continuous process.

Until more solid research is done on the detailed timing of extinctions in the fossil record, we will not know for sure whether the extinctions now projected for the contemporary moist tropics are typical of the history of life.

EVOLUTIONARY CONSEQUENCES OF
PAST EXTINCTIONS

Little is known about the condition of the biosphere immediately following mass extinctions other than the tautological inference that biodiversity must have been less than immediately before the events. For a few of the larger extinctions, however, the recovery time is sufficiently protracted that the postextinction milieu can be studied.

For at least 5 million years following the mass extinction of the Permian period, marine assemblages were clearly depauperate. Biological groups that were dominant in Permian seas are either absent altogether or are represented by just one or a few species. Often these few surviving species are surprisingly abundant. Several large class- and phylum-level groups are completely absent, even though they are known to have survived the extinctions because of their appearance later in the Mesozoic record—a phenomenon Jablonski (1986) has called the Lazarus effect. That is, they "rose again" after apparent extinction.

The Lazarus taxa provide a special challenge for students of the fossil record, because there are two equally plausible explanations for major gaps in the fossil record: species diversity may have been so low that the organisms were not preserved as fossils, or sedimentary environments conducive to fossilization may have been absent. The choice between these two explanations is difficult to make, and no unequivocal case has yet been made for either of them. However, the presence of a few abundant species immediately following the Permian period argues in favor of the lowered diversity theory.

Other consequences of mass extinction are somewhat clearer. Many of the extinction events were followed by major shifts in dominance of biological groups and by the evolutionary radiation of new innovations. A classic example is the diversification of the mammals following the extinction of the dinosaurs. Mammals had been present in moderate numbers throughout most of the time of dinosaur dominance, but it was not until the removal of the dinosaurs during the mass extinction at the end of the Cretaceous period that mammals became truly diversified. It is presumed, though difficult to prove absolutely, that the diversification of mammals, and ultimately the evolution of *Homo sapiens*, was possible because of the newly available ecological space in terrestrial habitats.

Other examples of replacement resulting from extinction involve tropical reef communities. The builders of reef frameworks, now dominated by stony corals of the Scleractinia order, switched roles repeatedly during Phanerozoic time. Reefs have been built at various times by molluscs, bryozoans, calcareous algae, or coral groups only distantly related to modern corals. It is clear that the extinction-replacement phenomenon has been largely responsible for these changeovers. This is important in broader evolutionary terms, because it suggests that the evolution of communities, and the changing dominance of certain kinds of plants and animals, is not a simple progression based on species-species competition. Rather, the changes may occur simply as a result of the filling of voids left by the demise of previously dominant groups. And the extinctions of the previously dominant groups, if caused by rare conditions of extreme stress, may have little or nothing to do with adaptive level or general efficiency. Thus, there is no reason to believe that the present dominance of scleractinian corals in most tropical reefs implies anything about the fitness of these animals to that environment relative to previous occupants.

One can go further and suggest that without the perturbing effect of the extinction-replacement events, evolution as we know it would have been very different. It is easy to imagine that diversification and innovation in evolution would have come to a stop early in Phanerozoic time, the occupants of most ecological niches or adaptive zones maintaining a stable, steady state. From this viewpoint, extinction, and especially mass extinction, can be seen as a vital ingredient in the evolution of complex life as we know it. This must remain somewhat speculative, of course, because the evolution of life cannot be replayed under different conditions.

EXTINCTION IN SHALLOW TIME:
THE PLEISTOCENE EXPERIENCE

Within the past million years, the Earth's climate and biosphere have been strongly influenced by changes associated with the Pleistocene glaciations. In the context of the current concerns about extinction in the moist tropics, special attention should be paid to the effects on tropical diversity of the recent climatic fluctuations.

Many geologists and biogeographers have argued that the tropical rain forests of South America and Africa were largely replaced by dry savannas during the glacial advances. It has been postulated that the rain forests were reduced to a few small refugia, and the locations and extents of these remnant patches have been mapped in both South America and Africa (see Beven et al., 1984; Mayr and O'Hara, 1986; Simberloff, 1986).

If the refugium maps are accurate, they have profound implications for the effects of changes in tropical habitats. From theory, one would expect that total number of species would be reduced due to the greatly decreased habitable area for rain forest species and because of the elimination of the habitat of many geographically restricted species. The current estimates of present reductions in diversity caused by habitat destruction in the tropics are comparable to reductions estimated in the refugium model for the glacial intervals.

Also, if the refugium maps are accepted as reliable, it is difficult to explain the recovery of tropical diversity to present levels in the extremely short time since the last glacial advance. If present insect diversities are as great as recent estimates suggest, how did all the local endemics develop by speciation in such a short time?

There are major problems in applying the refugium model to the Pleistocene history of the tropics. The geological evidence for the climatic change comes mainly from scattered and generally inadequate data on fossil pollen. The biogeographical evidence is inferred from present-day distributions: the argument is that the refugia of the past are reflected now in concordant ranges of living species. That is, the near-coincident geographical ranges of species delineate the refuge patches from which diversification and geographical spreading occurred since the return of warm, moist conditions. There has been much argument in the recent biogeographical literature both for and against the refuge reconstructions. For both South America and Africa, strong cases have been made for opposing conclusions.

Another major problem with the refugium model is the extreme difficulty of documenting Pleistocene extinctions in the affected areas. The fossil record in present rain forest areas is notoriously poor because of the paucity of good rock exposures from which collections can be made. Furthermore, the organisms of most interest in this context—land animals, plants, and insects—have very low fossilization potentials and thus there are poor geological records, even under good circumstances. It is therefore difficult to determine from existing data whether or not the Pleistocene glaciations were accompanied by mass extinctions in the tropics. On a global scale, the Pleistocene epoch was not a time of mass extinction, but it is certainly possible that there were extensive species kills in rain forest areas.

RESEARCH FOR THE FUTURE

The fossil record has great untapped potential for contributing to our under-standing of contemporary extinction. This is true for shallow as well as deep time. In deep time, considering Phanerozoic time as a whole, the most pressing and relevant priorities are closer investigation of the timing of the great mass extinctions (Did the major events take place in a matter of days, years, or millions of years?) and more analysis of the biological selectivity of extinction (Who were the sur-vivors, who were the victims, and why?).

In shallow time, concentrating on the last few hundred thousand years, we need more direct, empirical data on the physical environmental history of the Pleistocene epoch and the biological consequences, with emphasis on species extinction, of the environmental changes. If we can substantially increase our knowledge of the Pleistocene record, we will be in a much better position to evaluate the conse-quences of the activities of humans in tropical regions.

Without consideration of the time perspective available from the geological record, a full evaluation of the contemporary extinction problem may prove as difficult as would be the case if a land-use planner were to attempt projections without benefit of historical experience or if an epidemiologist were to treat an infectious disease without medical records.

REFERENCES

Beven, S., E. F. Connor, and K. Beven. 1984. Avian biogeography in the Amazon basin and the biological model of diversificatic.a. J. Biogeogr. 11(5):383–399.

Jablonski, D. 1986. Causes and consequences of mass extinctions: A comparative approach. Pp. 183–229 in D. K. Elliott, ed. Dynamics of Extinction. Wiley, New York.

Mayr, E., and R. J. O'Hara. 1986. The biogeographic evidence supporting the Pleistocene refuge hypothesis. Evolution 40(1):55–67.

Raup, D. M. 1978. Cohort analysis of generic survivorship. Paleobiology 4(1):1–15.

Raup, D. M. 1986. Biological extinction in Earth history. Science 231:1528–1533.

Simberloff, D. S. 1986. Are we on the verge of a mass extinction in the tropical rain forests? Pp. 165–180 in D. K. Elliott, ed. Dynamics of Extinction. Wiley, New York.

Stanley, S. M. 1986. Earth and Life Through Time. W. H. Freeman, New York. 690 pp.

ESTIMATING REDUCTIONS IN THE DIVERSITY OF TROPICAL FOREST SPECIES

ARIEL E. LUGO

Project Leader, Institute of Tropical Forestry, U.S. Department of Agriculture, Forest Service, Southern Forest Experiment Station, Rio Piedras, Puerto Rico

T his chapter focuses on the empirical basis of estimates for species extinctions in tropical environments. The variation in estimates commonly cited (Table 6-1) points to inconsistencies that require discussion. I also call attention to examples in the tropics that suggest ecosystem resiliency in the conservation of species diversity. My intention is not to diminish in any way the sense of urgency that resource managers and government agencies should have about the progressive increment of loss and onerous consequences of a reduction in the number of species. Instead, I hope to stimulate a more critical and balanced scientific analysis of the issue.

The need for a balanced and rigorous analysis of the loss-of-species issue stems from the unquantifiable importance of species diversity to life support on a global scale. Scientists must be as precise as possible when communicating such important phenomena to the public and its governmental representatives. A loss of scientific credibility can seriously hamper continuing efforts to develop lasting popular support for the conservation of ecological diversity. Also, the time, money, and talent needed to address the ecological problems of the tropics are very limited, and their allocation is affected by public perception of the situation. Errors of perception lead to waste of resources and loss of opportunity to achieve solutions.

THE ACCEPTED VIEW

The numbers cited for species decline and used to gain public support for the conservation of species diversity are impressive. According to Myers (1979), the

TABLE 6-1   Estimates of Potential Species Extinction in the Tropics

Estimate	Basis of Estimate	Source
1 species/day to 1 species/hour between 1970s and 2000	Unknown	Myers, 1979
33–50% of all species between the 1970s and 2000	A concave relationship between percent of forest area loss and percent of species loss (see Table 6-2)	Lovejoy, 1980
A million species or more by end of this century	If present land-use trends continue	National Research Council, 1980
As high as 20% of all species	Unknown	Lovejoy, 1981
50% of species by the year 2000 or by the beginning of next century	Different assumptions and an exponential function (see Table 6-2)	Ehrlich and Ehrlich, 1981
Several hundred thousand species in just a few decades	Unknown	Myers, 1982
25–30% of all species, or from 500,000 to several million by end of this century	Unknown	Myers, 1983
500,000–600,000 species by the end of this century	Unknown	Oldfield, 1984
0.75 million species by the end of this century	All tropical forests will disappear and half their species will become extinct	Raven, Missouri Botanical Gardens, personal communication to WRI and IIED, 1986
33% or more of all species in the 21st century	Present rates of forest loss will continue	Simberloff, 1983
20–25% of existing species by the next quarter of century	Present trends will continue	Norton, 1986
15% of all plant species and 2% of all plant families by the end of this century	Forest regression will proceed as predicted until 2000 and then stop completely	Simberloff, 1986

world was losing one species per day in the 1970s, and by the mid-1980s, the loss will increase to about one species per hour. By the end of this century, our planet could lose anywhere from 20 to 50% of its species (Table 6-1). Humans are the basic cause of these losses, because in the process of securing a living from the land, people modify it. The human population is growing at a faster rate in tropical latitudes than anywhere else, and this results in more habitat destruction in the tropics. In fact, the greatest losses of species are reported to occur in the tropics, which contain half of the world's remaining forests. Some writers suggest that present tropical forests will be destroyed by the beginning of the next century and that because these forests are the world's richest in terms of species numbers, their destruction becomes the primary source of a global loss of species.

How are these scenarios derived? What are the bases of these calculations? How firm are they? To develop such scenarios, three types of data are needed: the relative distribution of species in each type of tropical forest, the rate of change in the area of each type of tropical forest, and the relationship between change in forest area and change in species numbers.

Most published projections of species extinctions resulting from deforestation in the tropics do not include the basis for their estimates in ways that can be examined independently (Table 6-1). Exceptions are the estimates of Lovejoy (1980) in the Global 2000 Report, that of Ehrlich and Ehrlich (1981) in their classic book on extinctions, and the recent paper by Simberloff (1986).

NUMBER OF SPECIES IN THE TROPICS

Estimating the total species richness of the tropical biome is probably beyond the means of scientific endeavor at this time. Total species inventory of a single tropical ecosystem does not even exist. Insufficient information handicaps any effort to estimate the number of species extinctions. Myers (1979) discussed the problems of estimating species numbers and concluded that of the 3 to 10 million species that exist globally, approximately 70% occur in the tropics. The World Resources Institute and the International Institute for Environment and Development (WRI and IIED, 1986) reported between 3.7 and 8.7 million species in the tropics (the actual number depending on whether the world has 5 or 10 million species), of which 0.6 million are known to science. Taxonomists estimate that only 1.5 to 1.7 million species are presently known to science (Raven, 1977; WRI and IIED, 1986). Clearly, scientific understanding of total numbers of species is still fragmentary. For this reason, it is best to use relative distributions of species in different forest types when making global estimates of species extinctions.

RATE OF CHANGE IN TROPICAL FOREST AREAS

The rate of change in tropical forests of all kinds has been discussed in depth only by Lanly (1982), who made an effort to document the rate of increase in the area of secondary forests (by reforestation, afforestation, and natural regeneration; see Figure 6-1) as well as the rate of forest loss. Other attempts usually emphasize conversion or modification of mature forests with little or no analysis of recovery (Myers, 1980). Lanly's data show that of the 11.3 million hectares of mature forest land deforested annually, 5.1 million hectares are converted to secondary forest fallow. He estimated that the total area of this forest type is 409 million hectares and that almost 1 million hectares of secondary forest is created annually on unforested land through natural regeneration or human intervention. Such large forest areas cannot be dismissed as irrelevant to the conservation of species diversity because they support an extensive biota (discussed below) and because under certain conditions, they are capable of supporting more complex biota than the mature system they replace (Ewel, 1983).

Lanly's data also show that deforestation rates are higher in closed than in open forests (Figure 6-1). Within closed forests, a large fraction of the conversion involves logged forests—forests that have previously been modified by human activ-

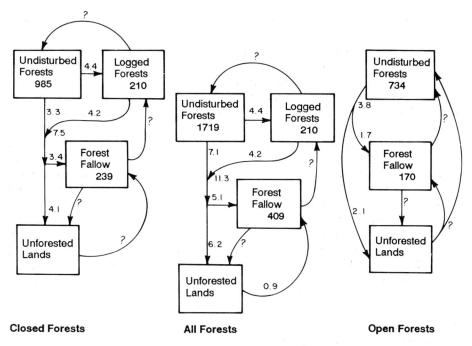

FIGURE 6-1   Pathways of land conversion in tropical forest lands. Data were derived from the Food and Agriculture Organization (1981) and from Lanly (1982), and are presented in millions of hectares. The numbers inside the boxes represent total area in 1980; those on the lines ending in arrowheads represent the annual rate of conversion. Closed forests have complete canopy cover; open forests do not and therefore support a grass understory.

ity. Because the dynamics of these changes in land use as well as the species richness of the forests also change according to country, region, and economic conditions, it behooves scientists to be extremely careful when projecting local experiences to global scales.

DIVERSITY OF FOREST TYPES IN THE TROPICS

The Holdridge Life Zone Classification System identifies some 120 ecological life zones in the world, 68 of which are tropical or subtropical (Holdridge, 1967). Thirty-two of the tropical and subtropical life zones are capable of supporting forests. About 19 million square kilometers of mature forests exist in the tropics and are distributed as follows: 42% in the dry forest life zones, 25% in the wet and rain forest life zones, and 33% in the moist forest life zones (Brown and Lugo, 1982). Statistically significant relationships suggest that life zone conditions relate to characteristic numbers of tree species (Holdridge et al., 1971), biomass and rate of primary productivity (Brown and Lugo, 1982), and capacity to resist and recover from disturbance (Ewel, 1977). These relationships are based on climatic data.

Some parameters increase while others decrease with water availability and temperature.

Quantitative studies of the richness of tree species and its association with environmental factors show that the total number of tree species increases linearly with rainfall (Gentry, 1982) and correlates negatively with the ratio of potential evapotranspiration to rainfall (Holdridge et al., 1971; Lugo and Brown, 1981). Gentry found a 3.5-fold difference (40-140 species per 0.1-hectare plot) in the number of tree species with a diameter at breast height (dbh) greater than 2.5 centimeters along a rainfall gradient of 1,000 to 3,000 millimeters. For every 1,000 millimeters of rainfall, the community gained about 50 tree species. Gentry indicated that species richness doubles from dry to moist forests and triples from dry to wet forests. Quantitative studies such as these are extremely important for obtaining accurate estimates of potential species extinctions resulting from forest loss. In an earlier publication, Gentry (1979) discussed such a phytogeographical approach to demonstrate that the number of species lost when forests are destroyed depends on the type of life zone environment being destroyed. Recognizing that tropical forests are diverse in terms of their ecological and species richness is critical for global estimates of species extinctions. Generalizations applied to all the tropics that are based on fragmented, qualitative studies are at best of limited utility.

An additional complicating factor is that life zones are subjected to different deforestation and regeneration rates (Tosi, 1980; Figure 6-1). In tropical America, for example, most human populations are clustered in dry and moist forest life zones that consequently suffer the greatest impacts of human activity (Tosi, 1980; Tosi and Voertman, 1964). The very wet life zones support the highest number of plant species and are subjected to the lowest rate of deforestation (particularly those in inaccessible locations; see, for example, Lugo et al., 1981). The fact that the intensities and consequent impacts of human activity vary among life zones has important implications for the reliability of species extinction estimates.

In summary, those who calculate species extinction rates must not assume that all tropical forests are subjected to equal rates of deforestation, respond uniformly to reductions in area, contain the same density of species, or turn into sterile pavement once converted. They must recognize and account for the diversity of forest types when making such calculations if estimates are to be considered reliable. Moreover, recovering secondary forests are potential foster ecosystems for endangered species, and their role in species conservation must also be considered. (This is discussed further later in this chapter.)

RELATIONSHIP BETWEEN DEFORESTATION RATE AND LOSS OF SPECIES

The nature of the relationship between deforestation rate and loss of species is not known. However, any calculation of the reduction of diversity must include this relationship. Myers (1983) suggested that islands on which 90% of the forest are "grossly disrupted" and the remaining 10% of their forests are protected stand to lose 50% of their species. Lovejoy (1980) discussed five possible functions that could be assumed in determining the relationship between forest area loss and loss of species and used a gradually increasing rate function to arrive at the extinction estimates in the Global 2000 Report (Table 6-2). Ehrlich and Ehrlich (1981, p. 280) assumed that "the diversity of species will be lost more rapidly than the forest itself" and used an exponential function to estimate depletion of species. They assigned a constant rate of increase to the rate of depletion based on human population growth (1.5% per year), human impacts in overdeveloped countries (1% per year), and forest loss (1% per year). The total rate of increase (3.5% per year) plus an assumed current rate of species extinction (1% per year in one calculation and 2% in another) were substituted in the exponential function to obtain the estimate of species depletion (Table 6-2).

The rates of deforestation used in both estimates discussed above are 3.8 to 5.5 times higher than the rates obtained by Lanly (1982). If Lanly's values are sub-stituted in Lovejoy's analysis (Table 6-3), the estimate of species extinctions by the year 2000 would be almost 9% of the total biota instead of 33 to 50%. The high estimate of Ehrlich and Ehrlich would be halved simply by changing the assumed fraction of the biota presently undergoing depletion. The function used by the Ehrlichs is very sensitive to changes in assumptions because of its exponential nature and the absence of any negative feedback to stabilize its response. Therefore, any change in the value of any of the factors contributing to the rate of increase or the rate of extinctions would change the prediction significantly (Table 6-3). For example, if the estimated rate of avian extinctions in Puerto Rico (discussed below) is substituted for the total rate of extinctions, the expected species depletion by the year 2010 would be reduced to 4%. The estimates by the Ehrlichs also suffer from not taking into account the heterogeneity of destruction and regeneration among different forest types. Are these definitive estimates? Clearly not!

Correcting for differences in species richness of forests, forest recovery rates, and differential human impact by forest type will certainly lower any of the estimates that now lack consideration of mitigating factors. Furthermore, the functions used to relate forest loss to species loss are still to be established experimentally. When and if this comes about, the results may be either more or less conservative than those assumed by either Lovejoy or Ehrlich and Ehrlich.

Lower extinction rates for plants (Table 6-1) were estimated by Simberloff (1986) by using a species-area relationship, conservative assumptions about the fraction of forest area loss, a Z factor (an exponent of the forest area lost) of 0.25, and various scenarios of forest conservation. Simberloff could not derive a mass ex-tinction of plant species by the year 2000 comparable to those of the geological past, even though his analysis does not correct for forest recovery after conversion. However, his estimates of extinction are lower than those discussed above, even though the function he used usually accounts for only 44.8% of the variation in species when area changes.

SEEKING A BETTER ESTIMATE

I believe that to estimate the reduction in the number of species in the tropics it is necessary to consider the effect of forest types on species abundance, the spatially selective (life zone) intensity of human activity, the role of secondary

TABLE 6-2 Extinction of Species in Tropical Forests as Implied by Lovejoy and by Ehrlich and Ehrlich[a]

Source of Data and Region	Species Present (thousands)	Projected Deforestation (%)		Loss of Species (%)		Species Extinction (thousands)	
		Low Case	High Case	Low Case	High Case	Low Case	High Case
Lovejoy							
Latin America	300–1,000	50	67	33	50	100–333	150–500
Africa	150–500	20	67	13	50	20–65	75–250
South and Southeast Asia	300–1,000	60	67	43	50	129–430	150–500
All tropics	750–2,500	47	67	33	50	249–828	375–1,250
Ehrlich and Ehrlich (1981)							
Total (Annual rate of increase, 3.5%; current rate of extinctions, 1%)			50[b]	100[c]			
Total (Annual rate of increase, 2%; current rate of extinctions, 3.5%)			50[d]	100[e]			

[a] $D = \dot{Q}_o(e^{rt-1})$, where D = depletion of diversity, $\dot{Q}_o$ = rate of depletion as a fraction of remaining diversity, r = rate of increase of $\dot{Q}_o$, e = constant, and t = time interval in years.

[b] By early part of next century

[c] By 2025.

[d] By 2000.

[e] By 2010.

TABLE 6-3 Extinction of Species in Tropical Forests When Lanly's Data Are Substituted in the Calculations Used by Lovejoy and by Ehrlich and Ehrlich

Region and Author of Original Calculations	Rates When the Data of Lanly (1982) Are Substituted in Original Calculations			
	Species Present[a] (thousands)	Projected Deforestation, 1980–2000 (%)	Loss of Species (%)	Extinctions (thousands)
Lovejoy (1980)				
Latin America	300–1,000	17.1	10	30–100
Africa	150–500	8.9	4	6–20
Asia	300–1,000	15.1	10	30–100
All tropics	750–2,500	12.3[a]	8.8[a]	66–220
Ehrlich and Ehrlich (1981, Table 1)				
Total (Annual rate of increase, 3.5%; current rate of extinction, 0.62%)			25[b]	

[a]Weighted average.
[b]By 2010.

forests as species refugia, and the role of natural disturbances in maintaining regional species richness. At a regional level, one also has to consider the importance of exotic species in the maintenance of species richness, particularly in ecosystems subjected to the impact of human activity. This approach seeks balance by considering factors that maintain species richness as well as those that decrease it. Considerable research is required to provide sound estimates based on this approach, because critical data concerning ecosystem function are not available in enough breadth to support enlightened management or policy making.

CALLING ATTENTION TO THE POSITIVE TERMS IN THE SPECIES EXTINCTION ISSUE

Most calculations of species extinction rates emphasize the negative aspects of the problem, and this can have beneficial effects in terms of public awareness of environmental issues. I call attention to the positive terms of this issue, using examples from the Caribbean. These examples must be used with caution, because natural conditions in the Caribbean (particularly the frequency of hurricanes) select for resilient ecosystem, and it could be argued that this selective force invalidates the examples given. However, human impacts have been so intense in the Caribbean that the region remains as a test case for theories that emphasize island fragility. And besides, the essence of my argument is that in the development of any prediction involving biotic phenomena (whether it is species extinction, global carbon cycle, or acid rain effects), it is necessary to include the plethora of checks and balances that typify ecosystem function. In the Caribbean example, ecosystems must cope with hurricanes and intensive human-induced disturbances, whereas elsewhere, periodic fire, earthquakes, frost, or landslides may play the natural role of ecosystem stressor.

Forest ecosystems of Caribbean islands have proven to be more resilient than one would assume on the basis of the relationships used in Table 6-2 or the idea of the fragility of island biota (Carlquist, 1974; Soulé, 1983). The Caribbean islands are densely populated (100 to 500 people per square kilometer, or about 10 times more densely populated than surrounding continental tropical lands (Lugo et al., 1981), and their lands have been intensively used and degraded for centuries. All the ills that Carlquist (1974) and Soulé (1983) described for islands (e.g., the introduction of exotic species, intensive predation, and habitat destruction) are present in this region. There are many examples of catastrophic waste of natural resources in the Caribbean islands [see, for example, *Ambio* 10(6) 1981, which was dedicated to environmental problems of the Caribbean], but there are also examples to give us some hope; these are the ones I am emphasizing.

In Puerto Rico, human activity reduced the area of primary forest by 99%, but because of extensive use of coffee shade trees in the coffee region and secondary forests, forest cover was never less than 10 to 15%. This massive forest conversion did not lead to a correspondingly massive species extinction, certainly nowhere near the 50% alluded to by Myers (1983). In an analysis of the bird fauna, Brash (1984) concluded that seven bird species (four of them endemic) became extinct after 500 years of human pressure (this is equivalent to an 11.6% loss of the bird fauna) and that exotic species enlarged the species pool. In the 1980s more birds are present on the island (97 species) than were present in pre-Columbian times (60 species). The resiliency of the bird fauna was attributed to its generalist survival strategy (a characteristic of island fauna) and to the location of secondary forests and coffee plantations on mountaintops along the east-west axis of the island, which acted as refugia.

Secondary forests in Puerto Rico have served as refugia for primary forest tree species as well (Wadsworth and Birdsey, 1982; R. O. Woodbury, University of Puerto Rico, personal communication, 1986). After 20 to 30 years of growth, the understory of these ecosystems is supporting species characteristic of mature forests. A random survey of 4,500 trees in secondary forests of two life zones (moist and wet forests) resulted in a tally of 189 tree species (Birdsey and Weaver, 1982). This survey excluded four of the six forested life zones in the island and the species-rich mature publicly owned forests. Yet it is important that 25% of the tree species identified on the island were recorded in this survey of secondary forests. (Puerto Rico has 750 tree species, 203 of which are naturalized; Little et al., 1974.) Dominant species in these secondary forests owe their dominance to human activity, and many of the native species that are typical of mature forests are rare in the forest canopy (142 tree species accounted for 16% of the total basal area of secondary forests) but are now beginning to appear as pole-size individual trees in these forest sites. Secondary forests in high-impact regions obviously require time to fulfill their role as foster ecosystem for endangered species, but in due time, a wide variety of tree species appear to return to forest lands.

An extreme example of the importance of species conservation and of human-dominated habitats acting as foster ecosystems for endangered species is that of the Chinese maiden hair tree (*Ginkgo biloba*). No one has ever seen a wild individual of this species. This primitive tree was preserved in courtyards of temples in China and is considered to be the first species saved by humans (Stebbins, 1979).

In the United States where extensive human-caused deforestation and subsequent forest recovery have occurred, remnant secondary forest islands account for a large portion of landscape species diversity (Burgess and Sharpe, 1981). As a group, these secondary forest islands constitute a landscape with greater species richness than found in a landscape dominated only by climax forests. Clearly, secondary forests require more scientific attention before their role and value in landscapes affected by human activity can be properly assessed.

Catastrophic natural events may also be deleterious to the maintenance of species diversity, particularly to those species already at the edge of extinction. However, these catastrophic events are natural phenomena with predictable rates of recurrence to which the biota as a whole is adapted. Evidence is mounting to show that tropical forest ecosystems have endured catastrophic events for millennia, e.g., periodic fires in the moist forests of the Amazon (Sanford et al., 1985) and in Borneo (Leighton, 1984). In the Caribbean, hurricanes appear to be important in the maintenance of species diversity. Long-term studies in areas of the Luquillo Experimental Forest Biosphere Reserve have shown that there are progressive reductions in tree species between hurricane events (Crow, 1980; Weaver, 1986). The effects of periodic hurricanes maintain a diverse mix of successional and climax species on a given site. Without hurricanes, successional species would be more restricted. Sanford et al. (1985) suggest that fire performs the same function in Amazonian moist forests; Sepkoski and Raup (1986) expanded this idea to the effects of global perturbations on the history of life on the planet.

Studies of regeneration strategies for mature forests have indicated that disturbance is usually associated with the early phases of seedling germination and establishment in most forest types, including tropical forests (Pickett and White, 1985). This has led Pickett and White to propose the concept of "patch dynamics" as a focus of scientific inquiry aimed at understanding ecosystem dynamics. The relevance of this to the maintenance of species diversity is that environmental change and disturbance may be required to maintain a species-rich tropical landscape.

Because humans have facilitated immigration and created new environments, exotic (nonnative) species have successfully become established in the Caribbean islands. This has resulted in a general increase in total species inventories of birds and trees. Some of these exotic species are pests and thus are called biological pollutants (CEQ, 1980). However, many exotic species have become so well integrated into the natural landscape that most islanders consider them native.

Although conservationists and biologists have an aversion to exotic species such as predatory mammals and pests (with good reason!), this may not be totally justified if the full inventory of exotic fauna and flora and certain ecological arguments are taken into consideration. For example, the growth of exotic plant species is usually an indication of disturbed environments, and under these conditions, exotic species compete successfully (Vermeij, 1986). They accumulate and process carbon and nutrients more efficiently than do the native organisms they replace. In so doing, many exotic species improve soil and site quality and either pave the way for the succession of native species or form stable communities themselves. There is no biological criterion on which to judge a priori the smaller or greater value of one species against that of another, and if exotic species are occupying environments that are unavailable to native species, it would probably be too costly or impossible to pursue their local extinction.

The paradox of exotic species invasions of islands with high levels of endemism is discussed by Vitousek in Chapter 20. He correctly points out that if the invasion of exotic species is at the expense of the extinction of local endemics, the total species richness of the biosphere decreases and the Earth's biota is homogenized since most of the invading exotics are cosmopolitan.

NEED FOR BETTER LAND AND RESOURCE MANAGEMENT

In summary, strong evidence can be assembled to document the resiliency of the functional attributes of some types of tropical ecosystems (including their ability to maintain species richness) when they are subjected to intensive human use. Initial human intervention results in the loss of a few, highly vulnerable species. Massive forest destruction is probably required to remove more widely distributed species. Because massive species extinctions may be possible if human destruction of forests continues unabated, the evidence for ecosystem resiliency is not to be construed as an excuse for continued abuse of tropical environments. Rather, ecosystem resiliency is an additional tool available to managers if they choose to manage tropical resources prudently.

We cannot tell the needy of the tropical world that they must cease and desist in their struggle for survival to prevent a catastrophe whose dimensions, consequences, or mitigating conditions we cannot define with any certainty. It may turn out that the public call for conserving natural diversity is also an expression of frustration over the poor use of the natural resources of the tropics and our apparent inability to do something about it. Scientists have the responsibility of focusing the debate. Its fundamental essence, I believe, is the need for better land and resource management.

Experience in the Luquillo Experimental Forest Biosphere Reserve in Puerto Rico has demonstrated that species richness can be partially restored to lands previously used heavily for agriculture, that growing timber need not eliminate all natural species richness on site, and that tropical lands respond to sensible care through management. I know of no technical reason why sensible land management in tropical areas cannot lead to the success that is usually associated with temperate zones. The obstacles to progress are social and rooted in poor training and education programs, lack of facilities and infrastructure, weak institutions, misguided foreign aid programs, lack of commitment to forestry research and to enforcement of regulations, and the absence of a land conservation ethic. A strategy for forest and species conservation in tropical regions should focus on the restoration of forest production on former forest lands where food production is not sustainable. This, and sensible use of secondary forests and tree plantations, will reduce pressure on forest lands with mature forests or with unique ecological characteristics and set us on a course to meet the needs of the needy while protecting species diversity.

ACKNOWLEDGMENTS

In this article, I benefited from the comments of S. Brown, P. Kangas, E. Medina, O. Solbrig, R. Waide, C. Asbury, J. Lodge, W. Lawrence, and colleagues at the Institute. I thank all of them. This work was done in cooperation with the University of Puerto Rico.

REFERENCES

Ambio 10(6) 1981. An entire issue devoted to environmental problems of the Caribbean.

Birdsey, R. A., and P. L. Weaver. 1982. The Forest Resources of Puerto Rico. USDA Forest Service Southern Forest Experiment Station. Resource Bulletin SO-85. New Orleans, U.S. Department of Agriculture. 59 pp.

Brash, A. R. 1984. Avifaunal Reflections of Historical Landscape Ecology in Puerto Rico. Tropical Resources Institute. Yale University, New Haven, Conn. 24 pp.

Brown, S., and A. E. Lugo. 1982. The storage and production of organic matter in tropical forests and their role in the global carbon cycle. Biotropica 14:161–187.

Burgess, R. L., and D. M. Sharpe, eds. 1981. Forest Island Dynamics in Man-Dominated Landscapes. Ecological Studies 41. Springer-Verlag, New York. 310 pp.

Carlquist, S. J. 1974. Island Biology. Columbia University Press, New York. 660 pp.

CEQ (Council on Environmental Quality). 1980. Environmental Quality-1980. The Eleventh Annual Report of the CEQ. U.S. Government Printing Office, Washington, D.C. 497 pp.

Crow, T. 1980. A rain forest chronicle: A 30-yr record of change in structure and composition at El Verde, Puerto Rico. Biotropica 12:42–55.

Ehrlich, P., and A. Ehrlich. 1981. Extinction. The Causes of the Disappearance of Species. Random House, New York. 305 pp.

Ewel, J. J. 1977. Differences between wet and dry successional tropical ecosystems. Geo-Eco-Trop 1:103–177.

Ewel, J. J. 1983. Succession. Pp. 217–223 in F. B. Golley, ed. Tropical Rain Forest Ecosystems, Structure and Function. Elsevier, Amsterdam.

Food and Agriculture Organization. 1981. Los Recursos Forestales de la America Tropical. Informe Tecnico 1; Forest resources of tropical Asia, Technical Report 2; Forest Resources of Tropical Africa, parts 1 and 2, Technical Report 3. UN32/6.1301–78–04, Food and Agriculture Organization, Rome. 4 volumes.

Gentry, A. H. 1979. Extinction and conservation of plant species in tropical America: A phytogeographical perspective. Pp. 110–126 in I. Hedberg, ed. Systematic Botany, Plant Utilization, and Biosphere Conservation. Proceedings of a symposium held in Uppsala in commemoration of the 500th anniversary of the university. Almquist and Wiksell International, Stockholm.

Gentry, A. H. 1982. Patterns of neotropical plant-species diversity. Evol. Biol. 15:1–85.

Holdridge, L. R. 1967. Life Zone Ecology. Tropical Science Center, San Jose, Costa Rica. 206 pp.

Holdridge, L. R., W. C. Grenke, W. H. Hatheway, T. Liang, and J. A. Tosi. 1971. Forest Environments in Tropical Life Zones, a Pilot Study. Pergamon, New York. 747 pp.

Lanly, J. P. 1982. Tropical Forest Resources. FAO Forestry Paper 30. Food and Agriculture Organization, Rome. 106 pp.

Leighton, M. 1984. Effects of drought and fire on primary rain forest in eastern Borneo. P. 48 in B. C. Klein-Helmuth and J. L. Hufnagel, compilers. Abstracts of Papers. AAAS Meeting, New York. 24–29 May, 1984. American Association for the Advancement of Science, Washington, D.C.

Little, E. L., R. O. Woodbury, and F. H. Wadsworth. 1974. Trees of Puerto Rico and the Virgin Islands. USDA Forest Service, Agricultural Handbook 449, Vol. 2. Washington, D.C. 1,024 pp.

Lovejoy, T. E. 1980. A projection of species extinctions. Pp. 328–331, Vol. 2 in G. O. Barney (study director). The Global 2000 Report to the President. Entering the Twenty-First Century. Council on Environmental Quality, U.S. Government Printing Office, Washington, D.C.

Lovejoy, T. E. 1981. Prepared statement. Pp. 175–180 in Tropical Deforestation, An Overview, the Role of International Organizations, the Role of Multinational Corporations. Hearings before the Subcommittee on International Organizations of the Committee on Foreign Affairs. House of Representatives, 96th Congress, second session, May 7, June 19, and September 19, 1980. U.S. Government Printing Office, Washington, D.C.

Lugo, A. E., and S. Brown. 1981. Tropical lands: Popular misconceptions. Mazingira 5(2):10–19.

Lugo, A. E., R. Schmidt, and S. Brown. 1981. Tropical forests in the Caribbean. Ambio 10:318–324.

Myers, N. 1979. The Sinking Ark. A New Look at the Problem of Disappearing Species. Pergamon, New York. 307 pp.

Myers, N. 1980. Conversion of Tropical Moist Forests. National Academy of Sciences, Washington, D.C. 205 pp.

Myers, N. 1982. Forest refuges and conservation in Africa with some appraisal of survival prospects for tropical moist forests throughout the biome. Pp. 658–672 in G. T. Prance, ed. Biological Diversification in the Tropics. Columbia University Press, New York.

Myers, N. 1983. Conservation of rain forests for scientific research, for wildlife conservation, and for recreation and tourism. Pp. 325–334 in F. B. Golley, ed. Tropical Rain Forest Ecosystems, Structure and Function. Elsevier, Amsterdam.

NRC (National Research Council). 1980. Research Priorities in Tropical Biology. National Academy of Sciences, Washington, D.C. 116 pp.

Norton, B. J., ed. 1986. The Preservation of Species. Princeton University Press, Princeton, N.J. 305 pp.

Oldfield, M. I. 1984. The Value of Conserving Genetic Resources. U.S. Department of the Interior, National Park Service, Washington, D.C. 360 pp.

Pickett, S. T. A., and P. S. White, eds. 1985. The Ecology of Natural Disturbance and Patch Dynamics. Academic Press, Orlando, Fla. 472 pp.

Raven, P. H. 1977. Perspectives in tropical botany: Concluding remarks. Ann. Mo. Bot. Gard. 64(4):746–748.

Sanford, R. L., Jr., J. Saldarriaga, K. E. Clark, C. Uhl, and R. Herrera. 1985. Amazon rain-forest fires. Science 227:53–55.

Sepkoski, J. J., Jr., and D. M. Raup. 1986. Periodicity in marine extinction events. Pp. 3–36 in D. K. Elliott, ed. Dynamics of Extinction. John Wiley and Sons, New York.

Simberloff, D. 1983. Are We on the Verge of Mass Extinction in Tropical Rain Forests? Unpublished monograph, July 1983.

Simberloff, D. 1986. Are we on the verge of a mass extinction in tropical rain forests? Pp. 165–180 in D. K. Elliott, ed. Dynamics of Extinction. John Wiley and Sons, New York.

Soulé, M. E. 1983. What do we really know about extinctions? Pp. 111–124 in C. M. Schonewald-Cox, S. M. Chambers, B. MacBryde, and W. L. Thomas, eds. Genetics and Conservation. Benjamin/Cummings, London.

Stebbins, G. L. 1979. Strategies for preservation of rare plants and animals. Great Basin Naturalist Memoirs 3:87–93.

Tosi, J. 1980. Life zones, land use, and forest vegetation in the tropical and subtropical regions. Pp. 44–64 in S. Brown, A. E. Lugo, and B. Liegel, eds. The Role of Tropical Forests on the World Carbon Cycle. A Symposium held at the Institute of Tropical Forestry in Rio Piedras, Puerto Rico, on March 19, 1980. CONF-800350, U.S. Department of Energy Carbon Dioxide Program. National Technical Information Service, Springfield, Va.

Tosi, J., and R. F. Voertman. 1964. Some environmental factors in the economic development of the tropics. Econ. Geogr. 40:189–205.

Vermeij, G. J. 1986. The biology of human-caused extinction. Pp. 28–49 in B. G. Norton, ed. The Preservation of Species. Princeton University Press, Princeton, N.J.

Wadsworth, F. H., and R. A. Birdsey. 1982. Un nuevo enfoque de los bosques de Puerto Rico. Pp. 12–27 in Noveno Simposio de Recursos Naturales. Puerto Rico Department of Natural Resources, San Juan, Puerto Rico.

Weaver, P. L. 1986. Hurricane damage and recovery in the montane forests of the Luquillo Mountains of Puerto Rico. Caribb. J. Sci. 22:53–70.

WRI and IIED (World Resources Institute and International Institute for Environment and Development). 1986. World Resources 1986. Basic Books, New York. 353 pp.

CHAPTER

7

CHALLENGES TO BIOLOGICAL DIVERSITY IN URBAN AREAS

DENNIS D. MURPHY

Research Programs Director, Center for Conservation Biology,
Stanford University, Stanford, California

Jaws, claws, an explosion of spray, and a grizzly emerges from the shallows, a salmon in its grasp. Mixed herds of elk, deer, and pronghorn antelope graze rolling, grassy slopes. A cougar surveys from broken chaparral and woodland above.

A scene from the shores of Yellowstone Lake? Perhaps. But it is also a scene from the shores of San Francisco Bay just 150 years ago. Now only deer and cougar remain, but well away from those shores in mountainous habitats above the sprawling metropolitan Bay Area. It seems that only the relatively recent European settlement of the West has spared those species at all. In wooded patches surrounding Milwaukee, the woodland bison, moose, wolverine, black bear, elk, and lynx have been long extinct. Now just a very few forest specialists, such as the raccoon, chipmunk, and white-footed mouse, survive in the region, and those species are gone from all but the very largest woodland patches (Matthiae and Stearns, 1981). In patches of eastern deciduous forest near Washington, D.C., migrant bird species restricted as breeders to forest interiors also survive in only the largest natural habitat remnants. A number of warbler species there show signs of imminent regional extinction (Whitcomb et al., 1981).

These are merely obvious examples of an accelerating decline in the global diversity of living things. The term biological diversity has been used to describe "the variety of life forms, the ecological roles they perform, and the genetic diversity they contain" (Wilcox, 1984, p.640). While scientists argue about the relative enormity of tropical deforestation and its impact on biological diversity, the loss of populations, species, and entire ecological communities in human population

71

centers and their surrounding landscapes is well documented and inarguably immense. In urban areas of the eastern United States, only species with the most general habitat and resource requirements have remained in urban corridors. Moreover, the prospect of further erosion of biological diversity looms. In Great Britain, where the sustained assault on the environment is measured in millennia rather than in centuries, and where most vertebrate species are distant memories, a cascade of invertebrate extinctions is now being observed. For example, 80% of the resident butterfly species have declined in number in at least a major part of their British ranges during the past decade (Thomas, 1984). A number of those survive only on reserves and under rigorous management regimes. An estimated 18% of all European butterfly species are considered to be vulnerable to or imminently faced with extinction (Heath, 1981).

Unfortunately, losses of animal and plant species are restricted neither to temperate zone urban areas nor to the developed world. Urban impacts on biological diversity reach their most devastating in the Third World. Less than 2% of the Atlantic forests of coastal Brazil within the urban reach of Sao Paulo remain, and it has been estimated that thousands of species from this region of high endemism have been driven to extinction, most never having been described by taxonomists.

Although the full extent of this urban environmental degradation is virtually impossible to convey, its underlying causes are comparatively simple to identify. With few exceptions, losses of naturally occurring biological diversity are incidental to human activities. Thus, urban areas are effectively synonymous with ecosystem disruption and the erosion of biological diversity. Natural habitats are replaced directly by houses, condominiums, hotels, and malls, as well as by streets, highways, and utilities that support them. Historically, urban areas were the first regions subjected to local overkill of wildlife for food, fur, and feathers, and through misdirected predator control programs. They were also the first to experience logging and weed eradication programs. The biological diversity of urban areas has also been among the most severely affected by the introduction of animal species, which prey on native animal populations, compete for limited resources, and act as vectors for novel diseases and parasites to which native organisms can be particularly susceptible.

Great effects on biological diversity in urban areas also can result from less direct sources, including many of the air- and water-borne pollutants that imperil human health. Toxic by-products of industrial production, such as polychlorinated biphenyls (PCBs), sulfur dioxide, and oxidants as well as pesticides directed at noxious species, have been found to disrupt natural ecosystems (Ehrlich and Ehrlich, 1981). Airborne pollutants are especially insidious, since they expand the reach of urban blight far beyond city limits. More subtle impacts on biological diversity result from overdrafting local aquifers, dropping water tables, and ground subsidence. These processes are often compounded by changes in natural patterns of groundwater percolation caused by the destruction of wetlands and diversion of runoff.

This wide array of obvious and subtle factors contribute to the disruption of ecosystem function, the decoupling of interactions among species, and the disappearance of populations of organisms from urban locales. Why should that concern us? Because losses of just a few populations can result in a great destabilization of natural ecological communities and, as a consequence, in a decrement in the ability of those communities to provide a wide array of services. Thus, many reasons for protecting diversity in urban areas are often highly utilitarian. Benefits include amelioration of climate, because foliage in cities contribute to the reduction of ambient temperatures. Large trees and shrubs reduce wind velocity and reduce evaporation of soil moisture. Plants are also useful in architecture, erosion control, watershed protection, wastewater management, noise abatement, and air pollution control (Grey and Deneke, 1986).

Nevertheless, the aesthetic reasons for preserving biological diversity are often those that most obviously affect the populace of urban areas. The great parks and natural areas of the world's major cities, such as Central Park and the Gateway National Recreation Area in New York City and Golden Gate Park and the Golden Gate National Recreation Area in San Francisco, are regarded as prized jewels, providing opportunities for recreation and relaxation as well as habitat for a wide variety of species.

The arguments for protecting biological diversity in urban areas seem straight-forward, but the implementation of conservation programs in urban areas is among the most difficult problems faced by environmentalists. Some areas are so disturbed that functioning, naturally occurring ecosystems are no longer identifiable, whereas other urban habitats remain effectively undisturbed. Open spaces in inner cities often support only species that are particularly well adapted to human impact. Such areas are nearly always small and extremely isolated, and their maintenance and enhancement demand extensive and continuous hands-on management. The conservation goals in such areas must usually aim at maximizing biological diversity to the extent possible, rather than preserving all remaining resident species.

Inner city park developers have traditionally introduced plantings of exotic species. Such settings fulfill many of the aesthetic and utilitarian roles that natural habitats offer, but their establishment and maintenance costs tend to be high, since few of the self-regenerating functions of natural ecosystems are available. Yet, although human-induced intervention such as the replacement of ecosystem components can increase the number of species locally over at least the short run, these processes nearly always upset the ecological balance of communities; hence it ultimately exerts a negative impact on naturally occurring biological diversity.

Where larger, intact ecosystems exist within cities, they are often restricted to corridors alongside steep stream canyons, such as Rock Creek Park in Washington, D.C., and Fairmont Park in Philadelphia. But the most extensive expanses of natural habitat in urban areas are those surrounding city limits. In those relatively undisturbed areas, prescriptions for the preservation of biological diversity are quite different from those for maximizing diversity in more disturbed areas. Corridors and surrounding habitats are among the most valuable urban natural areas, providing for extensive biological diversity and reducing the isolation of the largest surviving ecosystems, which may be far from urban centers.

The single greatest threat to the biological diversity of relatively intact natural communities in and around urban areas is the destruction of natural habitats and their conversion to other uses. The paving over of natural habitats as urban activities sprawl outward destroys and fragments remnant functioning ecosystems. The redistribution of water through channelization and impoundment of flowing waters, and the draining of some wetlands and the flooding of others, destroys undeveloped habitat areas. Activities as seemingly benign as the planting of exotic trees and shrubs in parks and along byways or the conversion of open space to golf courses disrupt the distribution of natural components of biological diversity. These activities combine to decrease habitat area and disturb the equilibrium between extinction and immigration among remaining natural habitats, with the frequent result that some species are permanently lost.

Decreases in local biological diversity resulting from losses of habitat area and insularization of habitat remnants are compounded by the more subtle effects of fragmentation. Losses of single, specific microhabitats within an otherwise undisturbed habitat can cause the local extinction of certain species. Disruption of even narrow corridors of natural habitat between large habitat patches can lead to losses of species. The removal of understory foliage in manicured park areas and suburban housing developments can result in the loss of numerous species, most conspicuously species of birds. Vast differences in temperature, humidity, light availability, and wind exposure exist between forest edges and interiors and affect habitat suitability for some species. In addition, losses of certain species due to any one or more causes can affect closely associated species sometimes leading ultimately to secondary extinction events (Wilcox and Murphy, 1985).

In light of these basic ecological facts, conservation of the full range of urban biological diversity necessitates the protection of the largest possible expanses of natural habitat. Yet, that simple prescription is usually impossible to fill in urban areas, where the forces acting to decrease the size of remaining natural habitats are greatest. These conflicting pressures interact to determine urban conservation policy and to force biologists to justify the sizes of biological preserves.

Economic and political considerations in urban areas make preservation particularly difficult. Land costs are high because of high demand, and the vast majority of urban space is private property. The few publicly owned open spaces are subject to intensive, varied uses, many of which are incompatible with preserving biological diversity. Local political institutions usually favor development over preservation, and many agencies concerned with land and resource management, such as the U.S. Forest Service and Bureau of Land Management, have no presence in urban areas. Many conservation organizations with largely urban memberships virtually limit their concern to nonurban environments, and those involved with local issues rarely have the resources available for protracted fights over development.

The Endangered Species Act with its mandate outlawing the "take" of any endangered species is the best tool for protecting biological diversity in urban areas of this country. Although the goal of the Act is protection of individual species of concern, its "purposes . . . are to provide a means whereby the ecosystems upon which endangered species depend may be conserved" (USC, 1983, p. 1, §1531). Its strength resides in its ability to protect species regardless of land ownership.

Efforts to conserve the full extent of biological diversity by using the Endangered Species Act must target species that are most susceptible to habitat loss. The protection of extinction-prone species can be the key to facilitating the conservation of biological diversity in urban areas. Species especially prone to extinction include those high on trophic pyramids, widespread species with low vagility (i.e., with poor dispersal ability), endemic and migratory species, and species with colonial nesting habits (Terbough, 1974). Many such species inhabit urban areas during all or major portions of their lives and can act as umbrellas of sorts, often conferring protection to great numbers of species in the same habitats.

The greatest erosion of extinction-prone species has usually occurred in habitat remnants that survive in those urban areas with the longest histories of settlement. Hence prescriptions for conserving remaining biological diversity differ substantially among urban areas. For example, forest patches support many more bird species than do grassland patches of similar size. All else being equal, therefore, protection of the total remaining biological diversity of oak woodlands surrounding San Francisco will demand more and larger preserves than protection of similar habitats to achieve a similar goal near less biologically diverse Washington D.C. In addition, the sizes of preserves necessary to protect biological diversity within an urban area will vary because the diversity itself varies greatly among different natural communities. Oak woodland preserves near San Francisco are likely to require more area to protect their complement of biological diversity than will native grassland preserves in the same geographic area.

In the urban United States, three groups must interact to assist the Endangered Species Act in protecting biological diversity. Field biologists must aid in the identification and survey of potential umbrella species. Conservation organizations must use that information and citizen petitions to get appropriate umbrella species protected via the endangered list. In response, the Office of Endangered Species will have to reassess listing priorities.

The San Francisco Bay Area exemplifies the challenge of preserving urban biological diversity. Without the grizzly bear, tule elk, and even the Xerces blue butterfly, San Francisco might be viewed as biologically impoverished in a sense, but the urban Bay Area remains an exceptionally rich natural region in the biologically richest state in the union. The ecological communities within a 25-kilometer radius of Berkeley include redwood, Douglas fir, and digger pine forests as well as coastal sage and inland chaparral, annual grasslands, dunes, riparian corridors, freshwater lakes, bay marshlands, and even pelagic marine communities and offshore seabird rookeries, an extraordinary array of ecological communities supporting immense biological diversity. The conservation challenge is great, especially in the shadow of a population growing at more than 3% per year; moreover, that shadow is not cast evenly. Less than 15% of San Francisco Bay marshlands remain, but much inland chaparral remains untouched.

Can this urban biological diversity be protected? In this country, the answer is a qualified yes. In many other countries the outlook is not that sanguine. In Austria, prohibitions against the collection of wildlife and plants are strictly enforced, while the conversion of natural habitats to cultivation is effectively subsidized by the government. In the Federal Republic of Germany, as the Black Forest dies from acidification, powerful lobbies thwart the implementation of speed limits on the autobahns; consequently high levels of pollution continue to prevail. Overpopulation, chronic poverty, and fuel shortages in the Third World create unrelenting pressures to exploit all available local resources. These pressures certainly will become more overwhelming in the future.

Our urban centers can be viewed as bellwethers of our global environmental fate. Our success at meeting the challenges of protecting biological diversity in urban areas is a good measure of our commitment to protect functioning ecosystems worldwide. If we cannot act as responsible stewards in our own backyards, the long-term prospects for biological diversity in the rest of this planet are grim indeed.

REFERENCES

Ehrlich, P. R., and A. H. Ehrlich. 1981. Extinction. The Causes and Consequences of the Disappearance of Species. Random House, New York. 305 pp.

Grey, G. W., and F. J. Deneke. 1986. Urban Forestry. 2nd edition. John Wiley & Sons, New York. 299 pp.

Heath, J. 1981. Threatened Rhopalocera (Butterflies) in Europe. Council of Europe, Strasbourg, France. 157 pp.

Matthiae, P. E., and F. Stearns. 1981. Mammals in forest islands in southeastern Wisconsin. Pp. 55–66 in R. L. Burgess and D. M. Sharpe, eds. Forest Island Dynamics in Man-dominated Landscapes. Springer-Verlag, New York.

Terborgh, J. 1974. Preservation of natural diversity: The problem of extinction prone species. BioScience 24:715–722.

Thomas, J. A. 1984. The conservation of butterflies in temperate countries: Past efforts and lessons for the future. Pp. 333–353 in R. I. Vane-Wright and P. R. Ackery, eds. The Biology of Butterflies. Academic Press, London.

USC (United States Code). 1984. Title 16, Conservation; Section 1531 et seq. Endangered Species Act of 1973. United States Code, 1984 Lawyers Edition. Lawyers Co-operative, Rochester, N.Y.

Whitcomb, R. F., C. S. Robbins, J. F. Lynch, B. L. Whitcomb, M. K. Klimkiewicz, and D. Bystrak. 1981. Effects of forest fragmentation on avifauna of the eastern deciduous forest. Pp. 125–205 in R. L. Burgess and D. M. Sharpe, eds. Forest Island Dynamics in Man-dominated Landscapes. Springer-Verlag, New York.

Wilcox, B. A. 1984. In situ conservation of genetic resources: Determinants of minimum area requirements. Pp. 639–647 in J. A. McNeeley and K. R. Miller, eds. National Parks, Conservation, and Development: The Role of Protected Areas in Sustaining Society. Proceedings of the World Congress on National Parks, Bali, Indonesia, 11–22 October 1982. Smithsonian Institution Press, Washington, D.C.

Wilcox, B. A., and D. D. Murphy. 1985. Conservation strategy: The effects of fragmentation on extinction. Am. Nat. 125(6):879–887.

HUMAN DEPENDENCE

ON BIOLOGICAL DIVERSITY

A young Yanomami Indian woman in the
Amazon rain forest relaxes while preparing an
armadillo for a future meal. A tame trumpeter
bird searches for food in the background.
Photo courtesy of Victor Englebert. © *1982
Time-Life Books B.V. from the* Peoples of the
Wild *series.*

8

DEEP ECOLOGY MEETS THE DEVELOPING WORLD

JAMES D. NATIONS
Director of Research, Center for Human Ecology, Austin, Texas

There is a movement afoot in the United States that environmentalists call deep ecology (Tobias, 1985). In a nutshell, its basic tenet is that all living things have a right to exist—that human beings have no right to bring other creatures to extinction or to play God by deciding which species serve us and should therefore be allowed to live. Deep ecology rejects the anthropocentric view that humankind lies at the center of all that is worthwhile and that other creatures are valuable only as long as they serve us. Deep ecology says, instead, that all living things have an inherent value—animals, plants, bacteria, viruses—and that animals are no more important than plants and that mammals are no more valuable than insects (Blea, 1986). Deep ecology is similar to many Eastern religions in holding that all living things are sacred. As a conservationist, I am attracted to the core philosophy of deep ecology. Like the Buddhists, and Taoists, and supporters of the Earth First! movement, I also believe that all living things are sacred. When human activities drive one of our fellow species to extinction, I consider that a betrayal of our obligation to protect all life on the only planet we have.

Where I run into trouble with the philosophy of deep ecology is in places like rural Central America or on the agricultural frontier in Ecuadorian Amazonia—places where human beings themselves are living on the edge of life. I have never tried to tell a Latin American farmer that he has no right to burn forest for farmland because the trees and wildlife are as inherently valuable as he and his children are. As an anthropologist and as a father, I am not prepared to take on that job. You could call this the dilemma of deep ecology meeting the developing world.

The dilemma is softened somewhat by the realization that the farmer in the developing world probably appreciates the value of forest and wildlife better than we do in our society of microwave ovens and airplanes and plastic money. The Third-World farmer appreciates his dependence on biological diversity because that dependence is so highly visible to him. He knows that his life is based on the living organisms that surround him. From the biological diversity that forms his natural environment he gathers edible fruit, wild animals for protein, fiber for clothing and ropes, incense for religious ceremonies, natural insecticides, fish poisons, wood for houses, furniture, and canoes, and medicinal plants that may cure a toothache or a snakebite.

There are indigenous peoples in some parts of the world who have an appreciation for biological diversity that puts our own conservation theorists to shame. I stayed once in southeastern Mexico with a Maya farmer who expressed his view this way:

"The outsiders come into our forest," he said, "and they cut the mahogany and kill the birds and burn everything. Then they bring in cattle, and the cattle eat the jungle. I think they hate the forest. But I plant my crops and weed them, and I watch the animals, and I watch the forest to know when to plant my corn. As for me, I guard the forest."

Today, that Maya farmer lives in a small remnant of rain forest surrounded by the fields and cattle pastures of 100,000 immigrant colonists. He is subjected to the development plans of a nation hungry for farmland and foreign exchange. The colonists have been forced by population pressure and the need for land reform to colonize a tropical forest they know nothing about. The social and economic realities of a modern global economy are leading them and their national leaders to destroy the very biological resources their lives are based upon.

The colonists are fine people who are quick to invite you to share their meager meal. But if you want to talk with them about protecting the biological diversity that still surrounds them, be prepared to talk about how it will affect them directly. If you look a frontier farmer in the eye and tell him that he must not clear forest or hunt in a wildlife reserve and that the reason he must not do these things is because you are trying to preserve the planet's biological diversity, he will very politely perform the cultural equivalent of rolling his eyes and saying, "Sure."

But he will not believe you. Instead, you should be prepared to demonstrate how he can produce more food and earn more money by protecting the biological resources on his land. The developing world colonist may understand his dependence on biological diversity, but his interest in protecting that diversity lies in how it can improve his life and the lives of his children. Colonists on the agricultural frontier do not have the luxury of debating the finer points of deep ecology.

The same thing can be said for the government planner in the nation where the pioneer farmer lives and the development banker in Washington, D.C. The planner and the banker may appreciate the moral and aesthetic values of biological diversity. They may lament the eradication of wilderness and wildlife. But if you want them to protect a critical area of forest or place their hydroelectric dam outside a protected area, be prepared to talk about the economic value of watersheds, income from tourism, and cost-benefit analysis.

In the developing world, as well as in our overdeveloped world, we are obligated to present economic, utilitarian arguments to preserve the biological diversity that ultimately benefits us all. Deep ecology makes interesting conversation over the seminar table, but it won't fly on the agricultural frontier of the Third World or in the board rooms of the Inter-American Development Bank.

The day may come when ethical considerations about biological diversity become our most important reason for species conservation. But in the meantime, if we want to hold on to our planet's biological diversity, we have to speak the vernacular. And the vernacular is utility, economics, and the well-being of individual human beings.

In the 1980s, the question seems to be, "What has biological diversity done for me lately?" The good news is that the answer to that question is, "Plenty, and more than you realize." Our lives are full of examples of the logic of preserving the plants and animals that we depend upon as a species.

Our food is a good example. Human beings eat a wealth of plants and animals in the home-cooked meals and restaurant dinners that we live on day-to-day. Yet one of the most immediate threats posed by the loss of biodiversity is the shrinkage of plant gene pools available to farmers and agricultural scientists. During the past several decades, we have increased our ability to produce large quantities of food, but we have simultaneously increased our dependence on just a few crops and our dependence on fewer types of those crops. As much as 80% of the world food supply may be based on fewer than two dozen species of plants and animals (CEQ, 1981). We are eroding the genetic diversity of the crops we increasingly depend upon, and we are eradicating the wild ancestors of those crops as we destroy wilderness habitats around the world.

We are dependent on biological diversity in ways less visible than the plants and animals we eat and wear. We also depend on them for raw materials and medicines. We depend on the diversity of plants and animals for industrial fibers, gums, spices, dyes, resins, oils, lumber, cellulose, and wood biomass. We chemically screen wild plants in search of new drugs that may be beneficial to humankind. We import millions of dollars worth of medicinal plants into the United States and use them to produce billions of dollars worth of medicines (OTA, 1984).

We use animals in medical research as well, though sometimes with brutal results. We import tens of thousands of primates for drug safety tests and drug production (OTA, 1984). We use Texas armadillos in research on leprosy. When human activities threaten the survival of these animals and their wild habitats, they threaten human welfare as well.

At the same time, we have to acknowledge that we will never be able to demonstrate an immediate, utilitarian reason for preserving every species on Earth. Some of them may have no use for humankind beyond being part of the great mystery. But who will tell us which species are unimportant? Who can tell us which level of extinction will seriously disrupt the web of life that we depend upon as human beings?

Environmental writer Erik Eckholm says that one of the key tasks facing both scientists and governments is to identify and protect the species whose ecological functions are especially important to human societies. And "in the meantime," Eckholm continues, "prudence dictates giving existing organisms as much benefit of the doubt as possible" (Eckholm, 1978).

One of the important factors in providing those species with the benefit of the doubt they deserve is educating ourselves and our governments' policy makers about our dependence, as human beings, on biological diversity. That education tends to emphasize the utilitarian value of species protection. One of the results is that there is a growing, pragmatic ethic among scientists and conservationists. It is an ethic that centers on the realization that our ability to preserve biological diversity depends on our ability to demonstrate the benefits that diversity brings to human beings (Fisher and Myers, 1986).

On one level, these benefits take the form of immediate economic income through activities like wildlife harvesting, tourism, and maintaining agricultural production. On another level, they focus on unfulfilled potential—new crops, new medicines, new industrial products. Taken together, the benefits of biological diversity provide short-term income to individual people and improve the long-term well-being of our species as a whole.

These two levels of benefits work together in the sense that if we hope to see the long-term benefits of biological diversity, we have to focus first—or least simultaneously—on the immediate, short-term benefits to individual people. Few of the wild gene pools—the raw materials for future medicines, food, and fuels—are likely to survive intact in places where people have to struggle simply to provide their basic, daily needs (Wolf, 1985).

One of our long-term goals as a species is to enjoy the uncounted benefits that our planet's biological diversity can eventually bring us. But in the short term, at a minimum for the next few decades, our basic strategy must concentrate on ensuring that people here and on the frontiers of the developing world receive material incentives that will allow them to prosper by protecting biological diversity rather than by destroying it (Cartwright, 1985). That done, we can return to the ethical and aesthetic arguments of deep ecology with the knowledge that when we look up from our discussion, there will still be biological diversity left to experience and enjoy.

The authors of the three chapters that follow are counted among the most successful and most dedicated of the scientists now working to point out the short-term and long-term benefits of biological diversity—three scientists who are working as quickly as possible to discover the unread books of our planet's genetic diversity and to translate those discoveries into practical advantages for their fellow human beings.

REFERENCES

Blea, C. 1986. Individualism and ecology. Earth First! Journal 6(6):21, 23.

Cartwright, J. 1985. The politics of preserving natural areas in third world states. Environmentalist 5(3):179–186.

CEQ (Council on Environmental Quality). 1981. The Global 2000 Report to the President, Vol. II. Council on Environmental Quality and the U.S. Department of State, Washington, D.C.

Eckholm, E. 1978. Disappearing Species: The Social Challenge. Worldwatch Paper 22. Worldwatch Institute, Washington, D.C. 38 pp.

Fisher, J., and N. Myers. 1986. What we must do to save wildlife. Int. Wild. 16(3):12–15.

OTA (Office of Technology Assessment). 1984. Technologies to Sustain Tropical Forest Resources. OTA-F-214. Office of Technology Assessment, U.S. Congress, Washington, D.C. 344 pp.

Tobias, M., ed. 1985. Deep Ecology. Avant Books, San Diego, Calif. 285 pp.

Wolf, E. C. 1985. Challenges and priorities in conserving biological diversity. Interciencia 10(5):236–242.

SCREENING PLANTS FOR NEW MEDICINES

NORMAN R. FARNSWORTH

Research Professor of Pharmacognosy, Program for Collaborative Research in the
Pharmaceutical Sciences, University of Illinois at Chicago, Chicago, Illinois

The U.S. pharmaceutical industry spent a record $4.1 billion on research and development in 1985, an increase of 11.6% from 1984 (Anonymous, 1986). In the same year, the American consumer purchased in excess of $8 billion in community pharmacies for prescriptions whose active constituents are still extracted from higher plants (Farnsworth and Soejarto, 1985). For the past 25 years, 25% of all prescriptions dispensed from community pharmacies in the United States contained active principles that are still extracted from higher plants, and this percentage has not varied more than 1.0% during that period (Farnsworth and Morris, 1976). Despite these data, not a single pharmaceutical firm in the United States currently has an active research program designed to discover new drugs from higher plants.

THE GLOBAL IMPORTANCE OF PLANT-DERIVED DRUGS

Approximately 119 pure chemical substances extracted from higher plants are used in medicine throughout the world (Farnsworth et al., 1985) (see Table 9-1). At least 46 of these drugs have never been used in the United States. For the most part, the discovery of the drugs stems from knowledge that their extracts are used to treat one or more diseases in humans. The more interesting of the extracts are then subjected to pharmacological and chemical tests to determine the nature of the active components. Therefore, it should be of interest to ascertain just how important plant drugs are throughout the world when used in the form of crude extracts. The World Health Organization estimates that 80% of the people in

TABLE 9-1   Secondary Plant Constituents Used as Drugs Throughout the World, Their Sources and Uses

Compound Name	Therapeutic Category in Medical Science	Plant Sources	Plant Uses in Traditional Medicine	Correlation Between Two Uses[a]
Acetyldigitoxin	Cardiotonic	Digitalis lanata Ehrh. (Grecian foxglove)	Not used	Indirect
Adoniside	Cardiotonic	Adonis vernalis L. (Pheasant's eye)	Heart conditions	Yes
Aescin	Antiinflammatory	Aesculus hippocastanum L. (Horse chestnut)	Inflammations	Yes
Aesculetin	Antidysentery	Fraxinus rhynchophylla Hance (variety of Fraxinus chinensis Roxb.)	Dysentery	Yes
Agrimophol	Anthelmintic	Agrimonia eupatoria L. (Common agrimony)	Anthelmintic	Yes
Ajmalicine	Circulatory stimulant	Rauwolfia serpentina (L.) Benth. ex Kurz (Indian snakeroot)	Tranquilizer	Indirect
Allantoin[b]	Vulnerary	Several plants	Not used	No
Allyl isothiocyanate[b]	Rubefacient	Brassica nigra (L.) Koch (Black mustard)	Rubefacient	Yes
Anabasine	Skeletal muscle relaxant	Anabasis aphylla L. (Tumbleweed)	Not used	No
Andrographolide	Antibacterial	Andrographis paniculata Nees. (Karyat)	Dysentery	Yes
Anisodamine	Anticholinergic	Anisodus tanguticus (Maxim.) Pascher (Zàng qié)	Meningitis symptoms	Yes
Anisodine	Anticholinergic	Anisodus tanguticus (Maxim.) Pascher (Zàng qié)	Meningitis symptoms	Yes
Arecoline	Anthelmintic	Areca catechu L. (Betel-nut palm)	Anthelmintic	Yes
Asiaticoside	Vulnerary	Centella asiatica (L.) Urban (Indian pennywort)	Vulnerary	Yes
Atropine	Anticholinergic	Atropa belladonna L. (Belladonna)	Dilate pupil of eye	Yes
Benzyl benzoate[b]	Scabicide	Several plants	Not used	No
Berberine	Antibacterial	Berberis vulgaris L. (Barberry)	Gastric ailments	Yes
Bergenin	Antitussive	Ardisia japonica Thunb. (Japanese ardisia)	Chronic bronchitis	Yes

Borneol[b]	Antipyretic; analgesic; antiinflammatory	Several plants	Not used	No
Bromelain	Antiinflammatory; proteolytic	Ananas comosus (L.) Merrill (Pineapple)	Not used	Indirect
Caffeine	Central nervous system stimulant	Camellia sinensis (L.) Kuntze (Tea)	Stimulant	Yes
Camphor	Rubefacient	Cinnamomum camphora (L.) Nees & Eberm. (Camphor tree)	Not used	No
(+)-Catechin	Hemostatic	Potentilla fragarioides L. (Cinquefoil)	Hemostatic	Yes
Chymopapain	Proteolytic; mucolytic	Carica papaya L. (Papaya)	Digestant	Yes
Cocaine	Local anesthetic	Erythroxylum coca Lam. (Coca)	Appetite suppressant; stimulant	Yes
Codeine	Analgesic; antitussive	Papaver somniferum L. (Opium poppy)	Analgesic; sedative	Yes
Colchiceine amide	Antitumor agent	Colchicum autumnale L. (Autumn crocus)	Gout	No
Colchicine	Antitumor agent; anti-gout	Colchicum autumnale L. (Autumn crocus)	Gout	Yes
Convallatoxin	Cardiotonic	Convallaria majalis L. (Lily-of-the-valley)	Cardiotonic	Yes
Curcumin	Choleretic	Curcuma longa L. (Turmeric)	Choleretic	Yes
Cynarin	Choleretic	Cynara scolymus L. (Artichoke)	Choleretic	Yes
Danthron (1,8-dihydroxyanthraquinone)[b]	Laxative	Cassia species (Senna)	Laxative	Yes
Demecolcine	Antitumor agent	Colchicum autumnale L. (Autumn crocus)	Gout	No
Deserpidine	Antihypertensive; tranquilizer	Rauvolfia tetraphylla L. (Snakeroot)	Not used	Indirect
Deslanoside	Cardiotonic	Digitalis lanata Ehrh. (Grecian foxglove)	Not used	Indirect
Digitalin	Cardiotonic	Digitalis purpurea L. (Common foxglove)	Cardiotonic	Yes
Digitoxin	Cardiotonic	Digitalis purpurea L. (Common foxglove)	Cardiotonic	Yes
Digoxin	Cardiotonic	Digitalis lanata Ehrh. (Grecian foxglove)	Not used	Indirect

TABLE 9.1 Continued

Compound Name	Therapeutic Category in Medical Science	Plant Sources	Plant Uses in Traditional Medicine	Correlation Between Two Uses[a]
L-Dopa[b]	Antiparkinsonism	Mucuna deeringiana (Bort) Merr. (Velvet bean)	Not used	No
Emetine	Amebicide; emetic	Cephaelis ipecacuanha (Botero) A. Richard (Ipecac)	Amebicide; emetic	Yes
Ephedrine[b]	Sympathomimetic	Ephedra sinica Stapf (Ma-Huang)	Chronic bronchitis	Yes
Etoposide[b]	Antitumor agent	Podophyllum peltatum L. (May apple)	Cancer	Yes
Galanthyamine	Cholinesterase inhibitor	Lycoris squamigera Maxim. (Ressurection lily; magic lily)	Not used	No
Gitalin	Cardiotonic	Digitalis purpurea L. (Common foxglove)	Cardiotonic	Yes
Glaucarubin	Amebicide	Simaruba glauca DC. (Paradise tree)	Amebicide	Yes
Glaucine	Antitussive	Glaucium flavum Crantz (Horned poppy, sea poppy)	Not used	No
Glaziovine	Antidepressant	Ocotea glaziovii Mez (Yellow cinnamon)	Not used	No
Glycyrrhizin (Glycyrrhetic acid)	Sweetener	Glycyrrhiza glabra L. (Licorice)	Sweetener	Yes
Gossypol	Male contraceptive	Gossypium species (Cotton)	Decreased fertility observed	Yes
Hemsleyadin	Antibacterial; antipyretic	Hemsleya amabilis Diels (Luó guǒ dì)	Dysentery	Yes
Hesperidin	Capillary antihemorrhagic	Citrus species (Citrus, e.g., orange, lemon)	Not used	No
Hydrastine	Hemostatic; astringent	Hydrastis canadensis L. (Golden seal)	Astringent	Yes
Hyoscyamine	Anticholinergic	Hyoscyamus niger L. (Henbane)	Sedative	Yes
Kainic acid	Ascaricide	Digenea simplex (Wulf.) Agardh (Red alga)	Anthelmintic	Yes

Kawain[b]	Tranquilizer	Piper methysticum Forst. f. (Kava)	Euphoriant	Yes
Khellin	Bronchodilator	Ammi visnaga (L.) Lamk. (toothpick plant)	Asthma	Yes
Lanatosides A, B, C	Cardiotonic	Digitalis lanata Ehrh. (Grecian foxglove)	Not used	Indirect
Lobeline	Respiratory stimulant	Lobelia inflata L. (Indian tobacco)	Expectorant	Yes
Menthol[b]	Rubefacient	Mentha species (Mint, e.g., peppermint, spearmint)	Carminative	No
Methyl salicylate[b]	Rubefacient	Gaultheria procumbens L. (Wintergreen)	Carminative	No
Monocrotaline	Antitumor agent (topical)	Crotalaria spectabilis Roth (Rattlebox)	Skin cancer	Yes
Morphine	Analgesic	Papaver somniferum L. (Opium poppy)	Analgesic; sedative	Yes
Neoandrographolide	Antibacterial	Andrographis paniculata Nees (Karyat)	Dysentery	Yes
Nicotine	Insecticide	Nicotiana tabacum L. (Tobacco)	Narcotic	No
Nordihydroguaiaretic acid	Antioxidant (lard)	Larrea divaricata Cav. (Creosote bush)	Antitussive	No
Noscapine (narcotine)	Antitussive	Papaver somniferum L. (Opium poppy)	Analgesic; sedative	Yes
Ouabain	Cardiotonic	Strophanthus gratus (Hook.) Baill. (Twisted flower)	Arrow poison	Indirect
Pachycarpine [(+)-sparteine]	Oxytocic	Sophora pachycarpa Schrenk ex C. A. Meyer (Pagoda tree)	Not used	No
Palmatine (fibraurine)	Antipyretic; detoxicant	Coptis japonica Makino (Goldthread)	Not used	No
Papain	Proteolytic; mucolytic	Carica papaya L. (Papaya)	Digestant	Yes
Papaverine[b]	Smooth muscle relaxant	Papaver somniferum L. (Opium poppy)	Sedative; analgesic	No
Phyllodulcin	Sweetener	Hydrangea macrophylla (Thunb.) Seringe (Hydrangea)	Sweetener	Yes
Physostigmine (eserine)	Anticholinesterase	Physostigma venenosum Balf. (Ordeal bean)	Ordeal poison	Indirect

TABLE 9-1 Continued

Compound Name	Therapeutic Category in Medical Science	Plant Sources	Plant Uses in Traditional Medicine	Correlation Between Two Uses[a]
Picrotoxin	Analeptic	Anamirta cocculus (L.) Wright & Arn. (Fish berry)	Fish poison	Indirect
Pilocarpine	Parasympathomimetic	Pilocarpus jaborandi Holmes (Jaborandi)	Poison	Indirect
Pinitol[b]	Expectorant	Several plants	Not used	No
Podophyllotoxin	Escharotic	Podophyllum peltatum L. (May apple)	Cancer	Yes
Protoveratrines A & B	Antihypertensive	Veratrum album L. (False hellebore)	Hypertension	Yes
Pseudoephedrine[b]	Bronchodilator	Ephedra sinica Stapf (Ma-Huang)	Chronic bronchitis	Yes
Pseudoephedrine, nor-[b]	Bronchodilator	Ephedra sinica Stapf (Ma-Huang)	Chronic bronchitis	Yes
Quinidine	Antiarrhythmic	Cinchona ledgeriana Moens ex Trimen (Yellow cinchona)	Malaria	No
Quinine	Antimalarial; antipyretic	Cinchona ledgeriana Moens ex Trimen (Yellow cinchona)	Malaria	Yes
Quisqualic acid	Anthelmintic	Quisqualis indica L. (Rangoon creeper)	Anthelmintic	Yes
Rescinnamine	Antihypertensive; tranquilizer	Rauwolfia serpentina (L.) Benth. ex Kurz (Indian snakeroot)	Tranquilizer	Yes
Reserpine	Antihypertensive; tranquilizer	Rauwolfia serpentina (L.) Benth. ex Kurz (Indian snakeroot)	Tranquilizer	Yes
Rhomitoxin	Antihypertensive; tranquilizer	Rhododendron molle G. Don (Yellow azalea)	Contraindicated in low blood pressure	Yes
Rorifone	Antitussive	Rorippa indica (L.) Hiern (Nasturtium)	Chronic bronchitis	Yes
Rotenone	Piscicide	Lonchocarpus nicou (Aubl.) DC. (Cubé root)	Fish poison	Yes
Rotundine [(+)-tetrahydropalmatine]	Analgesic; sedative; tranquilizer	Stephania sinica Diels (Chinese stephania)	Sedative	Yes

Rutin	Capillary antihemorrhagic	*Citrus* species (Citrus, e.g., orange, lemon)	Not used	No
Salicin	Analgesic	*Salix alba* L. (White willow)	Analgesic	Yes
Sanguinarine	Dental plaque inhibitor	*Sanguinaria canadensis* L. (Bloodroot)	Not used	No
Santonin	Ascaricide	*Artemisia maritima* L. (Levant wormseed)	Anthelmintic	Yes
Scillaren A	Cardiotonic	*Urginea maritima* (L.) Baker (Squill)	Cardiotonic	Yes
Scopolamine	Sedative	*Datura metel* L. (Recurved thornapple)	Sedative	Yes
Sennosides A & B	Laxative	*Senna alexandrina* Miller (Alexandria senna)	Laxative	Yes
Silymarin	Antihepatotoxic	*Silybum marianum* (L.) Gaertn. (St. Mary's blessed, milk, or holy thistle)	Liver disorders	Yes
Sparteine	Oxytocic	*Cytisus scoparius* (L.) Link (Scotch broom)	Not used	No
Stevioside	Sweetener	*Stevia rebaudiana* Hemsley (Sweet herb; Caa-hé-hé)	Sweetener	Yes
Strychnine	Central nervous system stimulant	*Strychnos nux-vomica* L. (Nux vomica)	Toxic stimulant	Yes
Teniposide[c]	Antitumor agent	*Podophyllum peltatum* L. (May apple)	Cancer	Yes
Δ9-Tetrahydrocannabinol	Antiemetic; decrease ocular tension	*Cannabis sativa* L. (Marijuana, hemp)	Euphoriant	No
(±)-Tetrahydropalmatine	Analgesic; sedative; tranquilizer	*Corydalis ambigua* Cham. & Schltdl. (Birthwort)	Sedative	Yes
Tetrandrine	Antihypertensive	*Stephania tetrandra* S. Moore (Fáng jî, turtle twig)	Not used	No
Theobromine	Diuretic; vasodilator	*Theobroma cacao* L. (Cocoa, cacao)	Diuretic	Yes
Theophylline	Diuretic; bronchodilator	*Camellia sinensis* (L.) Kuntze (Tea)	Diuretic; stimulant	Yes
Thymol	Antifungal (topical)	*Thymus vulgaris* L. (Common thyme)	Not used	No

TABLE 9-1 Continued

Compound Name	Therapeutic Category in Medical Science	Plant Sources	Plant Uses in Traditional Medicine	Correlation Between Two Uses[a]
Trichosanthin	Abortifacient	*Trichosanthes, kirilowii* Maxim. (Chinese snake gourd)	Abortifacient	Yes
Tubocurarine	Skeletal muscle relaxant	*Chondodendron tomentosum* R. & P. (Curare)	Arrow poison	Yes
Valepotriates	Sedative	*Valeriana officinalis* L. (Valerian)	Sedative	Yes
Vasicine (peganine)	Oxytocic	*Adhatoda vasica* Nees (Malabar nut)	Expectorant	No
Vincamine	Cerebral stimulant	*Vinca minor* L. (Common periwinkle, running myrtle)	Cardiovascular disorders	Yes
Vinblastine (vincaleukoblastine)	Antitumor agent	*Catharanthus roseus* (L.) G. Don (Madagascar periwinkle)	Not used	No
Vincristine (leurocristine)	Antitumor agent	*Catharanthus roseus* (L.) G. Don (Madagascar periwinkle)	Not used	No
Xathotoxin (ammoidin; 8-methoxypsoralen)	Pigmenting agent	*Ammi majus* L. (Bishop's weed)	Leukoderma; vitiligo	Yes
Yohimbine	Adrenergic blocker; aphrodisiac	*Pausinystalia johimbe* (K. Schum.) (Pierre ex Beille)	Aphrodisiac	Yes
Yuanhuacine	Abortifacient	*Daphne genkwa* Sieb & Zucc. (Pinyin; Yuán huā)	Abortifacient	Yes
Yuanhuadin	Abortifacient	*Daphne genkwa* Sieb. & Zucc. (Pinyin; Yuán huā)	Abortifacient	Yes

[a]Yes indicated a positive correlation between the traditional medical use of the plant and the current therapeutic use of the chemical extracted from the plant.

No indicated that there is no correlation as indicated previously.

[b]Now also synthesized commercially.

[c]A minor synthetic modification over a natural product.

developing countries of the world rely on traditional medicine[1] for their primary health care needs, and about 85% of traditional medicine involves the use of plant extracts. This means that about 3.5 to 4 billion people in the world rely on plants as sources of drugs (Farnsworth et al., 1985). Specific data in support of these estimates are difficult to find, but the few examples that are available are quite revealing.

THE IMPORTANCE OF HERBAL DRUGS

In Hong Kong

In the small British colony Hong Kong (1981 population, 5,664,000), there were at least 346 independent herbalists and 1,477 herbal shops in 1981 (Kong, 1982); that same year, there were 3,362 registered physicians and 375 registered pharmacies. Chinese herbalist unions in Hong Kong claim to have a membership of about 5,000 (Kong, 1982). It is claimed that Hong Kong is the largest herbal market in the world, importing in excess of $190 million (US) per year (Kong, 1982). About 70% of these herbal products are used locally, and 30% are reexported. They fall into three roughly equal categories: ginseng products, crude plant drugs other than ginseng, and over-the-counter drugs and medicated wines (Kong, 1982). By comparison, about $80 million worth of Western-style medicines were imported into Hong Kong during the same period. Kong (1982) calculated that the average Hong Kong resident spends about $25 (US) per year for Chinese medicines.

In Japan

The system of traditional medicine in Japan, known as Kampo, is an adaptation of Chinese traditional medicine. Kampo formulations are essentially multicomponent mixtures of natural products, primarily plant extracts. In 1976 more than 69 kinds of Kampo formulae were introduced into the National Insurance Scheme in Japan, and this number has doubled since that time. The total expenditure for all types of pharmaceutical products in Japan was approximately $8.3 billion (US) in 1976, whereas only about $12.5 million (US) was spent on Kampo medicines. Thus in that year, Kampo medicines in the Japanese health care system amounted to only about 0.15% of total pharmaceutical expenditures. In 1983, total pharmaceutical expenditures in Japan were valued at about $14.6 billion (US) and those for Kampo medicines increased to about $150 million (US). Hence, in 7 years, expenditures for Kampo medicines in the Japanese health care system increased to about 1% of total pharmaceutical expenditures (Terasawa, 1986).

In a survey of 4,000 Japanese clinicians conducted in 1983, 42.7% of the respondents reported that they used Kampo medicines in their daily practices. As with most systems of traditional medicine, the applications of Kampo are most

---

[1]Traditional medicine is a term loosely used to describe ancient and culture-bound health practices that existed before the application of science to health matters in official, modern, scientific medicine or allopathy.

successful in the treatment of chronic diseases, most of which are difficult to treat successfully with Western type medicine. Conditions for which traditional medicine is most frequently used include chronic hepatitis, climacteric disorders, common cold, bronchial asthma, high blood pressure, constipation, autonomic insufficiencies, allergic rhinitis, diabetes mellitus, gastritis, headache, and bowel dysfunction (Terasawa, 1986).

In the People's Republic of China

The People's Republic of China includes one-fourth of the world's population. In 1974 I was privileged to visit that country as a member of the Herbal Pharmacology Delegation—the third of nine scientific exchange delegations set up by former President Nixon when he first visited that country. Since then, I have returned to the PRC in 1980 and again in 1985. It is obvious that the system of Chinese traditional medicine, in which the use of plant extracts to treat disease is extremely important, remains today as an important element in providing adequate primary health care for this populous country. Some of the value of Chinese medicine is most likely its use as a placebo, but I for one am convinced that the vast majority of plants used in this system have constituents that produce real therapeutic effects.

THE SEARCH FOR NEW PLANT DRUGS

There is a great deal of interest in and support for the search for new and useful drugs from higher plants in countries such as the People's Republic of China, Japan, India, and the Federal Republic of Germany. Virtually every country of the world is active in this search to a limited degree. However, in light of its size and resources, the United States must be regarded as an underdeveloped country with regard to productivity and programs designed to study higher plants as sources of new drugs, both in terms of industrial and university-sponsored research.

Estimates of the number of higher plants that have been described on the face of the Earth vary greatly—from about 250,000 to 750,000. How many of these have been studied as a source of new drugs? This is an impossible question to answer for the following reason. The National Cancer Institute in the United States has tested 35,000 species of higher plants for anticancer activity. Many of these have shown reproducible anticancer effects, and the active principles have been extracted from most of these and their structures determined. However, none of these new drugs have yet been found to be safe and effective enough to be used routinely in humans. The question then arises, could any of these 35,000 species of plants contain drugs effective for other disease states, such as arthritis, high blood pressure, acquired immune deficiency syndrome (AIDS), or heart trouble? Of course they could, but they must be subjected to other appropriate tests to determine these effects. In reality, there are only a handful of plants that have been exhaustively studied for their potential value as a source of drugs, i.e., tested for several effects instead of just only one. Thus, it is safe to presume that the entire flora of the world has not been systemically studied to determine if its constituent species contain potentially useful drugs. This is a sad commentary when one considers that interest in plants as a source of drugs started at the beginning of the nineteenth century and that technology and science have grown dramatically since that time.

As shown in Table 9-1, the 119 plant-derived drugs in use throughout the world today are obtained from less than 90 species of plants (Farnsworth et al., 1985). How many more can be reasonably predicted to occur in the more than 250,000 species of plants on Earth?

Use of the NAPRALERT Data Base

It is possible to present certain types of data showing the relative interest in studying natural products as a source of drugs by means of the NAPRALERT data base that we maintain at the University of Illinois at Chicago (Farnsworth et al., 1981, 1983; Loub et al., 1985). This specialized computer data base of information on natural products was derived from a systematic search of the world literature. Data that can be retrieved from the system include folkloric medicinal claims for plants, the chemical constituents contained in plants (and other living organisms), the pharmacological effects of naturally occurring substances, or the pharmacological effects of crude extracts prepared from plants. More than 80,000 articles have been entered into the data base since 1975, and about 6,000 new articles are added each year. The system contains folkloric, chemical, or pharmacological information on about 25,000 species of higher plants alone.

Pharmacological Interest in Natural Products

To give some idea as to the interest (or lack thereof) in studying the pharmacological effects of natural products, we can cite the following data from NAPRALERT. In 1985, approximately 3,500 new chemical structures from natural sources were reported. Of these, 2,618 were obtained from higher plants, 512 from lower plants (lichens, filamentous fungi, and bacteria), and 372 from other sources (marine organisms, protozoa, arthropods, and chordates) (Table 9-2). A significant 56.6% of the new chemicals obtained from lower plants (primarily antibiotics produced in industrial laboratories) were reported to have been tested for biological effects. About 23.9% of those obtained from marine sources, protozoa, arthropods, and chordates were studied for biological effects, but only 9.5% of the new structures obtained from higher plants were tested for pharmacological effects. The probable reasons for the low, 9.5% figure are that a majority of these discoveries were reported from university laboratories where the interest is mainly on chemistry, where there is less interdisciplinary research (i.e., botanists, chemists, and biologists working in collaboration), and where routine testing services for pharmacological activity are not readily available.

Why is there so little interest and activity in plant-derived drug development in the United States? An attempt will be made to answer this question, but first it is important to describe briefly some of the more fruitful approaches to drug discovery from higher plants.

TABLE 9-2 New Chemical Structures of Natural Origin Reported in 1985[a]

Source	Pharmacological Evaluation	
	Tested	Not Tested
Higher Plants		
Gymnosperms	2	48
Dicots	238	2,144
Monocots	10	112
Pteridophytes	0	40
Bryophytes	0	24
	250 (9.5%)	2,368
Lower Plants		
Lichen	0	0
Fungi	74	106
Schizomycetes	216	114
	290 (56.6%)	220
Other		
Marine organisms	82	280
Protozoa	4	0
Arthropods	4	0
Chordates	0	2
	90 (23.9%)	282

[a]From NAPRALERT data base at the University of Illinois at Chicago.

Approaches to Drug Discovery from Plants

There are many approaches to the search for new biologically active principles in higher plants (Farnsworth and Loub, 1983). One can simply look for new chemical constituents and hope to find a biologist who is willing to test each substance with whatever pharmacological test is available. This is not considered to be a very valid approach. A second approach is simply to collect every readily available plant, prepare extracts, and test each extract for one or more types of pharmacological activity. This random collection, broad screening method is a reasonable approach that eventually should produce useful drugs, but it is contingent on the availability of adequate funding and appropriate predictable bioassay systems. The last major useful drugs to have reached the marketplace based on this approach are the so-called vinca alkaloids, vincristine sulfate (leurocristine) and vinblastine sulfate (vincaleukoblastine). Vincristine is the drug of choice for the treatment of childhood leukemia; vinblastine is a secondary drug for the treatment of Hodgkin's disease and other neoplasms.

Vincristine was discovered by Gordon H. Svoboda at the Lilly Research Laboratories. In January 1958, Svoboda submitted an extract of the Madagascan periwinkle plant [Catharanthus roseus (L.) G. Don] to a pharmacological screening program at Lilly (Farnsworth, 1982). This was the fortieth plant that he selected for inclusion in the program. Vincristine was marketed in the United States in 1963, less than 5 years after a crude extract of C. roseus was observed to have antitumor activity. In 1985, total domestic and international sales of vincristine (as Oncovin®) and vinblastine (as Velban®) were approximately $100 million, 88% of which was profit for the company (G. H. Svoboda, personal communication, 1986).

This discovery of new drugs from higher plants is one of the few that has evolved from a random-selection broad pharmacological screening program. For example, in the very expensive research and development effort undertaken by the National Cancer Institute described above, not one useful drug has emerged.

Recently we analyzed information on the 119 known useful plant-derived drugs to determine how many were discovered because of medicinal folkloric information on the plants from which they were isolated. In other words, what correlation, if any, exists between the current medical use of the 119 drugs and the alleged medical uses of the plants from which they were derived? As shown in Table 9-1, 74% of the 119 chemical compounds used as drugs have the same or related use as the plants from which they were derived. This does not mean that 74% of all medical claims for plants are valid, but it surely points out that there is a significance to medicinal folklore that was not previously documented.

Thus, in my opinion, future programs of drug development from higher plants should include a careful evaluation of historical as well as current claims of the effectiveness of plants as drugs from alien cultures. Such information is rapidly disappearing as our own culture and ideas permeate the less developed countries of the world where there remains a heavy dependence on plants as sources of drugs.

LACK OF INTEREST IN NEW DRUG DISCOVERY PROGRAMS FROM PLANTS

Why is there such a reluctance to initiate new programs involving plants as sources of drugs in the United States, where we have the most sophisticated pharmaceutical industry in the world and where expenditures for drug development are staggering? In my conversations with staff from U.S. pharmaceutical companies, the following reasons seem to be consistent:

• To recover the costs of developing such drugs, solid patent protection must be secured. It is generally believed that natural products cannot be patented with the same degree of assurance as can synthetic compounds. This of course cannot be a valid deterrent, since patent protection for vincristine and vinblastine was sufficiently secure that the Eli Lilly Company had exclusive marketing rights to these substances for the full term of patent protection.

• Most promising plants seem to be indigenous to developing countries, many of which do not have stable governments and thus cannot provide assurance that there will be a continued supply of the raw material needed to produce the useful drugs. This of course may be true in a strict sense; however, as history shows, it is rare when a useful plant grows only in one isolated developing country. In the course of developing a full program involving plants as a source of raw material, it would be normal logic to immediately seek sources of the useful plant from a large number of geographic areas. Cultivation programs should also be initiated. In the early stages of development of vincristine and vinblastine, the plant source

C. roseus was collected from many different countries of the world and was also cultivated in eastern European countries and in the United States.

• There is reputed to be biological variation from lot to lot of plant drugs, but scientific documentation for this statement is difficult to find. This does not appear to be a problem affecting any of the plant sources required for production of the 119 drugs listed in Table 9-1.

What really seems to be the problem is that most pharmaceutical firms, as well as decision-making offices in government agencies, lack personnel who have a full understanding and appreciation of the potential payoff in this area of research. For example, new programs in drug development are usually initiated by the presentation of a proposal by a research staff member before a group of peers and research administrators. Following is one possible scenario: Dr. E. Z. Greenleaf prepares his arguments for a new drug development program at the ABC Pharmaceutical Corporation in which he proposes to study plants as a source of new drugs. His approach to the program is to examine written medicinal folklore to obtain information on plants allegedly used by primitive peoples for certain specified diseases. He might even be brave enough to suggest that the ABC Pharmaceutical Corporation hire one or two physicians to travel to Africa, Borneo, New Caledonia, or other exotic areas to live with the people for a year or so. During this period, Drs. U. Canduit and I. M. Reliant would observe the witch doctors treating patients and then would make their own diagnoses of each patient and conduct follow-up observations on outcome. When improvement is noted, they would record which plants had been used to treat the patients. These plants would then be collected and sent to the Research Laboratory of the ABC Pharmaceutical Corporation located in Heartbreak, Colorado, for scientific studies. Total cost of such a 5-year program would be less than the cost of a new jet fighter.

The second scientist from the ABC Pharmaceutical Corporation to make a new program presentation is Dr. Adam N. Molecule. He uses a long sequence of chemical equations to illustrate his theory that he can synthesize a series of chemical analogs based on computer analysis of structure-activity relationships in which his theoretical compounds will react favorably with specific receptor sites. He illustrates his plan with a full color videotape presentation of the computerized sequence of events that he hopes will take place at the molecular level. There is nothing left to the imagination. Molecule's computer produces a flowchart projecting the full costs of each stage of the synthesis at 2-month intervals. Everything is predictable, based on a percentage of projected sales should the end product prove to be a useful drug, and ensuring at least a 75% profit margin.

At the end of the two presentations, management must decide on whether to follow the folkloric line of Dr. E. Z. Greenleaf or the molecular biology-computer graphic-theoretical approach of Dr. Adam N. Molecule. Since Dr. Greenleaf is probably the only person in the room with a background and appreciation for his approach and most of the scientists in attendance are well trained and highly skilled synthetic chemists, biochemists, and molecular biologists, it is not difficult to predict which program will be approved and implemented.

SUMMARY

Higher plants have been described as chemical factories that are capable of synthesizing unlimited numbers of highly complex and unusual chemical substances whose structures could escape the imagination of synthetic chemists forever. Considering that many of these unique gene sources may be lost forever through extinction and that plants have a great potential for producing new drugs of great benefit to mankind, some action should be taken to reverse the current apathy in the United States with respect to this potential.

REFERENCES

Anonymous. 1986. Pharmaceutical R&D Spending by US Industry Hits $4.1 Billion, Setting Record, as do Sales. P. 5 in Chem. Mark. Rep. February 3, 1986.

Farnsworth, N. R. 1982. Rational approaches applicable to the search for and discovery of new drugs from plants. Pp. 27–59 in Memorias del 1er Symposium Latinoamericano y del Caribe de Farmacos Naturales, La Habana, Cuba, 21 al 28 de Junion, 1980. Academia de Ciencias de Cuba y Comisin Nacional de Cuba ante la UNESCO, UNESCO Regional Office, Montevideo, Uruguay.

Farnsworth, N. R., and W. D. Loub. 1983. Information gathering and data bases that are pertinent to the development of plant-derived drugs. Pp. 178–195 in Plants: The Potentials for Extracting Protein, Medicines, and Other Useful Chemicals. Workshop Proceedings. OTA-BP-F–23. U.S. Congress, Office of Technology Assessment, Washington, D.C.

Farnsworth, N. R., and R. W. Morris. 1976. Higher plants—the sleeping giant of drug development. Am. J. Pharm. Educ. 148(Mar.–Apr.):46–52.

Farnsworth, N. R., and J. M. Pezzuto. 1983. Rational approaches to the development of plant-derived drugs. Pp. 35–63 in Proceedings of the Second National Symposium on the Pharmacology and Chemistry of Natural Products, Joao Pessoa, Brazil, November 3–5, 1983. Paraiba University, Joao Pessoa, Brazil.

Farnsworth, N. R., and D. D. Soejarto. 1985. Potential consequences of plant extinction in the United States on the current and future availability of prescription drugs. Econ. Bot. 39(3):231–240.

Farnsworth, N. R., W. D. Loub, D.D. Soejarto, G. A. Cordell, M. L. Quinn, and K. Mulholland. 1981. Computer services for research on plants for fertility regulation. Korean J. Pharmacogn. 12:98–110.

Farnsworth, N. R., O. Akerele, A. S. Bingel, D. D. Soejarto, and Z.-G. Guo. 1985. Medicinal plants in therapy. Bull. WHO 63:965–981.

Kong, Y.-C. 1982. The control of Chinese medicines—a scientific overview. Yearb. Pharm. Soc. Hong Kong 1982:47–51.

Loub, W. D., N. R. Farnsworth, D. D. Soejarto, and M. L. Quinn. 1985. NAPRALERT: Computer handling of natural product research data. J. Chem. Inf. Comput. Sci. 25:99–103.

Terasawa, K. 1986. The present situation of education and research work on Traditional Chinese Medicine in Japan. Presentation at the International Symposium on Integration of Traditional and Modern Medicine, Taichung, Republic of China, May 22, 1986.

SERENDIPITY IN THE EXPLORATION OF BIODIVERSITY
What Good Are Weedy Tomatoes?

HUGH H. ILTIS

Director, University of Wisconsin Herbarium, Madison, Wisconsin

For someone studying natural history, life can never be long enough
(Miriam Rothschild, British entomologist,
television interview on *Nova*, 1986).

Biodiversity is out there in nature, everywhere you look, an enormous cornucopia of wild and cultivated species, diverse in form and function, with beauty and usefulness beyond the wildest imagination. But first we have to find these plants and animals and describe them before we can hope to understand what each of them means in the great biological—and human—scheme of things.

The classification of biodiversity is the job of taxonomists who, born as packrats and inspired by a compulsion to explore and collect the world's biological riches, will risk life and limb to solve the great puzzles of biogeography, ethnobotany, and evolution.

But for taxonomists these are paradoxical times. Were Alexander von Humboldt or Charles Darwin (two of our godfathers) alive today, they would marvel at our knowledge and technology, and the relative ease with which we can now explore the most inaccessible places, enabling us to bring back biological treasures even from the darkest jungles of Africa and the greenest hells of Amazonia. That's the good news.

The bad news is that the same roads that allow us to drive jeeps into the rain forests or up the highest tundras of the Andes, and the very technologies that land helicopters on the mist-shrouded mesas of Mt. Roraima in Venezuela's Lost World, also bring in a flood of land-hungry squatters, ambitious cattle ranchers, and greedy corporations, often under the auspices of international promoters of development such as the world's multilateral development banks. All are recklessly destructive of nature and in an orgy of environmental brutality, clearcut the forests, burn the trees, and plow up the land to grow more food or graze more cattle, even before any scientist has had a chance to find out what lives there. In the name of growth, progress, and development, and with a colossal self-confidence, we humans are now messing up even the last wild lands and damming the last wild rivers, oblivious of the irreplaceable biological treasures that are being destroyed.

In short, our twentieth-century civilization still pretty much reflects the short-sighted seventeenth-century pragmatism of Cotton Mather (1663–1728), the witch hunter of Salem, Massachusetts, who proclaimed: "What is not useful, is vicious." But who is to say what is useful and what is not, especially about species not yet discovered that, unknown and unstudied, fall prey to plow or cow? And who can predict the value of a monkey, a butterfly, or a flower? Or of intact ecosystems, to which we are inseparably linked, whether we acknowledge this or not?

Mankind depends on plants for food, fiber, drugs—and a livable world. But more than that, our children will want nature to experience while growing up— to explore, love, and enjoy its beauty and diversity. Corn and cows, concrete and cars are not enough to sustain and empower a human psyche that until only a few generations ago lived in daily contact with a variety of plants and animals, a psyche that, winnowed and sifted by natural selection, is genetically programmed to respond positively to nature and its patterns (Iltis et al., 1970; Wilson, 1984). By destroying so much of the natural environment, we humans are now destroying crucial parts of our own psychological as well as physical habitat. For those in the know, it is a gloomy picture indeed.

Like most taxonomists, I am by nature a born collector, first of stamps, then of plants—a botanical adventurer excited by the prospects of finding species no one has ever seen before. Unlike some botanists, I have never had a compelling interest in increasing the world's food supply. After all, is it not now obvious that *the world hunger problem cannot be solved by growing more food, but only by growing fewer people,* and that more food will always result in still more people, who in turn will devastate ever more nature, inevitably exterminate ever more plant and animal species, and in the long run, make life for themselves and their children ever more difficult? It is then quite ironical that by hunting for the evolutionary origin of potatoes and maize, I was involved in the discovery of two new species of agricultural significance, both splendid examples of wild biodiversity directly useful to humans.

THE DISCOVERY OF A NEW TOMATO

In December 1962, Don Ugent (now a botany professor at Southern Illinois University in Carbondale) and I were collecting wild and weedy potatoes and associated plants in the Peruvian Andes for the University of Wisconsin Herbarium at Madison (Iltis, 1982).

For a month we had studied potato populations in the mountains east of Lima to determine how the modern cultigen might have evolved. In fact, its exact origin is still an ethnobotanical mystery (Ugent, 1970). Was it in Peru, Bolivia, or Chile that people first collected the bitter wild tubers and selected edible potatoes?

We traveled to the Peruvian city of Cuzco by way of a gravelly back road, which crossed the Andes east of Pisco and then traversed above Puquio the vast and unending altiplano, an arid tundra-like grassland called the *puna*. A cold 3,500 to 4,500 meters above sea level, and therefore often higher than Pike's Peak, the *puna* is covered with a fantastic collection of cushion plants, including white fuzzy cacti that look like sleeping sheep, all adapted to withstand grazing by domesticated llamas and alpacas and the rare, wild vicuñas.

On the eastern slope of these gigantic mountains, within sight of snow-covered peaks, this so-called road (at times not much more than a footpath) dipped dizzily down from 4,260 meters to 1,800 meters at Abancay in only 25 kilometers, then crossed the Apurimac River below Curahuasi (where once stood The Bridge of San Luis Rey of Thornton Wilder's novel), and eventually wound its way up again to the altiplano and on to Cuzco, the capital both of Inca kings and of wild and cultivated potato diversity.

A rest in Abancay was welcome after freezing nights tenting above timberline and being miserable with *siroche* (mountain or altitude sickness). On December 21, the early morning was spent packing some 1,500 dried herbarium specimens of the 296 different species collected the week before and getting ready for the push to reach Cuzco in time for Christmas and a hot bath. Then off we drove to the Hacienca Casinchihua in the Rio Pachachaca valley to look for a rare, wild potato species cited by Correll (1962) in a monograph published a short time before.

It was the beginning of the rainy season, and this deep valley was now bursting into bloom. Most memorable were pendent, 4-inch-long, orange trumpet flowers of a *Mutisia*, a gorgeous daisy named by Linnaeus's son for the eighteenth-century Spanish botanist Don José Celestino Mutis.

Above the hacienda, our jeep was soon stopped by a landslide. There was nothing to do but hike along that old Inca road until high above the river we stopped to eat our lunch of avocados, oranges, cheese, and small, boiled Peruvian potatoes, yellow and rich in protein.

All around us was a floristic wonderland, full of rare and beautiful plants. In fact, these arid inter-Andean valleys are veritable biogeographic islands, each with many endemic (i.e., unique) species and isolated from other such valleys by wet tropical forests below and cold Andean tundras above, a situation favoring speciation and, hence, biodiversity. In a nearby gully, iridescent green and blue hummingbirds hovered and flitted about, piercing with their bills the cardinal-red flower tubes of a bushy sage, *Salvia oppositifolia*, one of several hundred (!) Andean species of this prolific genus.

So here we spent the rest of the day, always collecting five specimens of each plant—one each for the University of Wisconsin, the University of San Marcos in Lima, and the U.S. National Herbarium in Washington, D.C., and one or two for botanists specializing in that particular plant family, who would tell us exactly what we had collected. This must be done, for there are no accurate, usable books on the 30,000 species of Peruvian plants, a flora so rich it staggers the imagination.

(The northeastern United States is much larger than Peru but has only about 5,000 species of plants; yet here we have *many* up-to-date botanical compendia called floras by which plant species may be identified.)

Presently we noticed a tangled, yellow-flowered, sticky-leaved, ratty-looking wild tomato, not much different from the weedy tomatillo (*Lycopersicon peruvianum*) so widespread in Peru. Nevertheless, we took immediate notice of it, for tomatoes belong to the potato family and this was a relative of a cultivated species. And wild or weedy tomatoes must always be taken seriously!

Not only did we collect herbarium specimens of this weedy tomato, describing it in our notebook under the serial number 832 (i.e., the 832nd collection of this expedition), but we also gathered two dozen of its green-and-white striped berries, which are smaller than cherries. We smashed the berries between newspapers to dry their seeds, and weeks later, we mailed them together with other tomato seed samples to Charles Rick, tomato geneticist at the University of California, Davis, who, we had heard, would want to grow them in his experimental plots.

This is an old story, of course, and illustrates the network nature of the study of natural history. Taxonomists do this sort of thing for each other all the time, unasked and as a matter of course, whether they know each other personally or not. "I will collect seeds for you of your special plant group, if you will collect seeds for me of mine."

Back at the University of Wisconsin, a thank-you note from Prof. Rick informed us that our No. 832 was most unusual and perhaps useful in plant breeding. Not until 1976, however, after 14 years of research, did Rick (1976) publish this as a new species, naming it *Lycopersicon chmielewskii* in honor of the late Tadeusz Chmielewski, a Polish tomato geneticist and Rick's associate. Another of our tomato collections, obtained below Curahuasi, he described as yet another new and local species, *Lycopersicon parviflorum*. That certainly made us feel good: to have been involved in the discovery of *two* new species in this small though important genus. Previously, taxonomists recognized only seven species of wild tomato, and now there were nine! Our story could have ended here, of course, and still be a good one, what with us showing off the type of specimens housed in the University of Wisconsin Herbarium to interested students and telling tall expedition tales of haciendas and vicuñas, potatoes and tomatoes. But there was more to come.

HOW MUCH IS A WILD TOMATO WORTH?

In July 1980, a letter from Dr. Rick told the following story. When 17 years before, he had received our seeds numbered 832, he crossed their progeny with a commercial tomato variety to improve the latter's characteristics. After nearly 10 (!) generations of back-crossing the first-generation (F_1) hybrids, and with subsequent selection, Rick was able to produce several new tomato strains with larger fruit and a marked increase in fruit pigmentation. But most importantly, they had greatly increased the content of soluble solids, mainly fructose, glucose, and other sugars, all attributes of prime importance to the tomato industry. While the usual type of tomato contains between 4.5 and 6.2% soluble solids, the genes from our

No. 832 increased the content in the new hybrids to 6.6 to 8.6%. In a paper published in 1974, Rick summarized this work as follows:

> An attempt was made to combine the high soluble-solids content of ripe fruits of the small, green-fruited *Lycopersicon* [*chmielewskii*] with the horticulturally desirable characteristics of a standard *L. esculentum* cultivar. By backcrossing from the former to the latter, and by subsequent pedigree selection, pure-breeding lines in which soluble-solids content was elevated to 7–7.5 percent—at least 2 percentage points above that of the recurrent parent—were synthesized (Rick, 1974, Abstract).

Parts of Rick's letter to us are worth reproducing:

> In our assays of [*Iltis and Ugent No. 832* from Hacienda Casinchihua], we discovered that its fruits have a very high sugar content [to 11.5%] as assayed by refractometer readings. Since this species is readily hybridized with the cultivated tomato and the crosses yield relatively fertile hybrids, we initiated a program to introgress the genes responsible for high soluble solids from No. 832 to horticultural lines of [tomatoes. Thus] it was possible to transfer at least some of the genes for this character to produce large, red-fruited lines with significantly elevated sugar content. These derived lines have been widely distributed to tomato workers, some of whom have been exploiting them with the aim of improving sugar content of new tomato cultivars.
>
> The concentration of soluble solids in raw tomatoes is a matter of great economic importance to the processing industry. A number of years ago an expert estimated that each 0.5% increase in soluble solids would be worth about a million dollars. Greatly improved flavor is another benefit. I thought you might be interested in this use of your valuable collection and want to thank you again for your trouble and foresight in sharing it with us.

To make a long story short, and adjusting for inflation to 1987 U.S. dollars, the value to the tomato industry of the genes found in collection No. 832 could, if widely incorporated, be worth about $8 million dollars a year or, to bask in the glory of larger numbers, about $80 million over a decade!

The yearlong expedition (including the jeep) and 3 years of follow-up research cost the National Science Foundation only $21,000, a small amount in the great scheme of things, and yielded more than 1,000 different numbered herbarium collections, a total of 8,000 specimens now scattered in many major herbaria of the world. In other words, each of our collection *numbers* cost the U.S. government on the average only $21 (1962 value), including *Iltis and Ugent 832* and any of the other species previously unknown to science.

In fact, perhaps the most significant values stemming from our expedition are yet to come, possibly from the high-protein potatoes we collected or from the hundreds of bits and pieces of botanical information we passed along to colleagues, graduate students, and others. But as in the case of our tomato, collected in 1962, commercially utilized a decade later, and not described as a new species until 1976, the practical value of an organism can often not be recognized except after years of work, even for plant groups with known economic use that have been well studied by teams of specialists (which does not apply to most taxonomic groups because of lack of funds to support such large efforts).

For this reason, among others, I have no patience with the phony requests of developers, economists, and humanitarians who want us biologists to "prove" with hard evidence, right here and now, the "value" of biodiversity and the "harm" of tropical deforestation. Rather, it should be for them, the sponsors of reckless destruction, to prove to the world that a plant or animal species, or an exotic ecosystem, is *not* useful and *not* ecologically significant before being permitted by society to destroy it. And such proof, of course, neither they nor anybody else can offer!

The benefits of even the most unimportant research are often quite unexpected. Who could have predicted that these tiny, slimy seeds of a useless, ugly weed, stuck to an old newspaper and costing no more than a few dollars and 30 minutes of our time, might enrich the U.S. economy by tens of millions of dollars—in other words (using 1986 dollars in the calculation), a potential $8 million-a-year gain on a one-time $42 investment? Not bad for a government agency (the National Science Foundation) sometimes maligned for supporting such old-fashioned research. Pretty good for a band of field biologists not even wearing white lab coats.

Finally, this discovery is not exceptional. As Rick pointed out, "the literature is replete with examples of the transfer from the [wild species] to acceptable [tomato] cultivars of desirable new traits—mostly resistance to diseases and other pests— often of enormous economic value" (Rick, 1974, p.493). Sweet indeed are the uses of biodiversity!

A NEW SPECIES OF WILD MAIZE

The sensational story of *Zea diploperennis*, a wild species of maize (*teosinte*) recently discovered in the Mexican state of Jalisco, has often been told (Iltis et al., 1979; Vietmeyer, 1979). We cannot here even begin to outline the unlikely events that led a Mexican undergraduate to find this, the fourth species of the genus *Zea*, which includes maize (*Zea mays*)—the world's third most important crop with the enormous 1986 global value of more than $50 billion. But although our new tomato was collected through pure serendipity, the diploperennial *teosinte* owes its discovery to many people, all of whom shared a consuming interest in the Mexican flora and in the mysterious origin of maize (Iltis, 1983).

This story has a very happy ending. Because of the spectacular beauty of the 10,000-foot (2,886-meter)-high Sierra de Manantlán and the potential agricultural value of this rare, perennial, virus-resistant *teosinte*, which grows here on only 6 hectares and nowhere else on Earth, Mexican botanists and others have worked tirelessly, and successfully, to establish the Las Joyas biological field station of the Universidad de Guadalajara and a 135,000-hectare (350,000-acre) *Reserva Biosfera de la Sierra de Manantlán* under State of Jalisco and UNESCO Man in the Biosphere (MAB) auspices. The dedication of this enormous reserve on March 5, 1987, by Mexico's President Miguel de la Madrid H. was very gratifying, because the new reserve will now protect the whole, vast, *intact* biodiversity of that mountain chain, not only the world's only wild populations of this *teosinte* but also the parrots and jaguars, the orchids and ocelots, the crested guans and giant magnolias, and 10,000 lesser species. Moreover, it will allow all these organisms, including the flagship species *Zea diploperennis*, to survive in the very environment to which they are

The author, Hugh Iltis, standing in a field of *teosinte*.
Photo by Michael Nee, New York Botanical Garden.

evolutionarily adapted. In the long run, such in situ (in place) preservation of whole ecosystems in very large nature reserves is really the only effective way these, or any other species, can be assured survival.

THE CONTINUING IMPORTANCE OF BOTANICAL EXPLORATION

The new species of tomato and wild maize are just two examples of valuable plants saved in the very nick of time before their minuscule populations faced extinction. And there are tens of thousands of unknown species yet to be discovered! Biologists, therefore, must insist that the days of exploration are far from over and that the study of biological diversity, and its preservation in situ, is one of their primary scientific responsibilities.

In the final analysis, the necessary investments in nature preserves and in such noncommercial activities as biological expeditions, herbaria, zoological museums, and training of field biologists (especially in the tropics), inexpensive as these are, are far wiser uses of tax dollars than the billions that are so readily spent on space flights or Star Wars. The Moon and the planets will be out there forever, but the Earth's biological diversity is being exterminated now. It is therefore imperative that we study and carefully preserve nature on *this* planet *now*, for this will be our last chance to ensure that biodiversity will survive for future generations. Protection of biodiversity needs to receive top priority in national and international planning. But if nature preservation is to be effective and long-lasting, it must become codified into law and incorporated into ethics and organized religion. Not only biologists and agriculturists, but every thinking citizen, every responsible politician and religious leader, has here an indispensable role.

REFERENCES

Correll, D. S. 1962. The Potato and Its Wild Relative. Texas Research Foundation, Renner, Texas. 606 pp.

Iltis, H. H. 1982. Discovery of No. 832: An essay in defense of the National Science Foundation. Desert Plants 3:175–192.

Iltis, H. H. 1983. From teosinte to maize: The catastrophic sexual transmutation. Science 222:886–894.

Iltis, H. H., O. L. Loucks, and P. Andrews. 1970. Criteria for an optimum human environment. Bull. At. Sci. 26:2–6.

Iltis, H. H., J. F. Doebley, R. Guzmán M., and B. Pazy. 1979. Zea diploperennis (Gramineae): A new teosinte from Mexico. Science 203:186–188.

Rick, C. M. 1974. High soluble-solids content in large-fruited tomato lines derived from a wild green-fruited species. Hilgardia 42(15):492–510.

Rick, C. M. 1976. Genetic and biosystematic studies on two new sibling species of Lycopersicon from interandean Peru. Theor. Appl. Genet. 47:55–68.

Ugent, D. 1970. The potato: What is the botanical origin of this important crop plant, and how did it first become domesticated? Science 170:1161–1166.

Vietmeyer, N. D. 1979. A wild relative may give corn perennial genes. Smithsonian 10:68–79.

Wilson, E. O. 1984. Biophilia, the Human Bond with Other Species. Harvard University Press, Cambridge, Mass. 157 pp.

THE OUTLOOK FOR NEW AGRICULTURAL AND INDUSTRIAL PRODUCTS FROM THE TROPICS

MARK J. PLOTKIN

Director, Plant Conservation, World Wildlife Fund–U.S., Washington, D.C.,
and Research Associate, Harvard Botanical Museum, Cambridge, Massachusetts

Many of the initial international wildlife conservation efforts focused on attractive species of endangered mammals—the so-called charismatic megafauna. Although a number of these programs have proven to be extremely successful, the modus operandi was clearly not entirely applicable to the conservation of all organisms: "Save the Sedges!" is just not as stirring a battle cry as "Save the Tiger!" We cannot save the pandas, however, unless we save the bamboos on which they feed. Furthermore, human existence is much more dependent on the plant kingdom than on animals. Plants are indeed the roots of life.

Because of the sheer diversity of plant life—especially in the tropics—many conservationists in the recent past have had some difficulty trying to decide where to begin. Faced with an area like the Amazon, home to tens of thousands of species of plants, many of which have yet to be discovered by modern scientists, it is clearly impractical to evaluate the conservation status and potential utility of each species on an individual basis. Consequently, there has been a perceptible shift in emphasis toward plants that are either useful or potentially useful to people. The concept of protecting a plant because it shows promise for aiding human well-being seems to have a much wider appeal than preserving a species for purely aesthetic or academic purposes.

Conservationists generally divide useful plants into three categories: medicinal, agricultural, and industrial. Of these three groupings, medicinal plants tend to attract the most attention from the media. There is no denying the appeal of the modern ethnobotanist's ventures into the jungle to work with witch doctors to find healing herbs. Due to a variety of factors—factors that are expected to change in the near future—there has been relatively little development of new wonder drugs from tropical plants during the last decade (Tyler, 1986; see also Farnsworth, Chapter 9 of this volume). The recent history and predicted future for new agriculture and industrial plant products from the tropics have been very positive (Balick, 1985; Schultes, 1979).

AGRICULTURE

The greatest service which can be rendered any country is to add a useful plant to its culture. From Thomas Jefferson, 1821.

A hungry people listen not to reason, nor care for justice, nor are bent by prayers. From Seneca, ca. 60 A.D.

A hungry mob is an angry mob. From Bob Marley, 1979.

Tropical forest plants can be of use to modern agriculture in three different ways: as sources of new crops that can be brought into cultivation; as source material for breeding improved plant varieties; and as sources of new biodegradable pesticides.

NEW CROPS

Only a very small proportion of the world's plants have ever been used as a food source on a large scale. Of the several thousand species known to be edible, only about 150 have ever become important enough to enter into world commerce (R.E. Schultes, Harvard Botanical Museum, personal communication, 1986). In the movement toward a global economy, there has been a trend to concentrate on fewer and fewer species. Today, less than 20 plant species produce most of the world's food (Vietmeyer, 1986b). Furthermore, the four major carbohydrate crop species—wheat, corn, rice, and potatoes—feed more people than the next 26 most important crops combined (Witt, 1985).

The obvious place to turn for new crops to reduce our heavy reliance on such a relatively small number of species is the tropics. North America north of Mexico has contributed relatively little to the storehouse of economically important crop plants. If we had to live on plants that originated in the United States, our diet would consist of pecans, sunflower seeds, cranberries, blueberries, grapes, wild rice, pumpkins, squashes, and jerusalem artichokes. Caufield (1982) estimated that 98% of U.S. crop production is based on species that originated outside our borders. Of our common foodstuffs, corn, rice, potatoes, sweet potatoes, sugar, citrus fruit, bananas, tomatoes, coconuts, peanuts, red pepper, black pepper, nutmeg, mace, pineapples, chocolate, coffee, and vanilla all originated in tropical countries. A typical American breakfast of cornflakes, bananas, sugar, coffee, orange juice, hot chocolate, and hash brown potatoes is based entirely on tropical plant products.

Few people realize how much of our diet today has been determined by exploitation patterns developed when tropical countries were colonies of Europe. In many cases, the advantage that some current crop staples have over other, less-exploited tropical species is the disproportionate amount of research to which they have been subjected. Under the colonial system, only a few key species were chosen for export, and the establishment of a market for these species determined future cultivation

The author collecting medicinal plants with local Indian guide, in southern Suriname.

and research priorities, which excluded lesser-known species. This overreliance on a few species was maintained even after independence, since developing countries had to depend on preexisting markets and technicians trained in temperate countries (NRC, 1975).

Many currently underexploited tropical species will become common sights in the produce sections of our supermarkets during the next decade. Because those species are often best known to aboriginal or peasant peoples, they have often been stigmatized as slave foods in their country of origin. This has impeded the development of these crops, which often tend to be robust, productive, self-reliant, free of indigestible compounds with relatively high nutritive value, and suitable for growing in some sort of agricultural system.

The demand for tropical cuisine continues to grow in this country. The Los Angeles area is said to have more than 200 Thai restaurants, and Mexican fast-food outlets have become a $1.6 billion industry (Vietmeyer, 1986a). Even a short walk down M Street in Washington, D.C., will take you past Chinese, Vietnamese, Thai, Filipino, Mexican, Central American, and South American restaurants. Kiwi fruit from China was not introduced into this country until 1962, yet last year they were purchased by more than 10 million Americans. Furthermore, domestic demographic trends will add to the demand for tropical produce: the U.S. population has increased 17% since 1970, whereas the Hispanic population has risen 87% and the Asian population, 127% (Vietmeyer, 1986a).

The more promising species include the following:

• The uvilla (*Pourouma cecropiaefolia*; family Moraceae). The uvilla is a medium-size tree native to the western Amazon. Both harvested from the wild and cultivated by local Indians as a doorstep crop, it yields fruit in only 3 years time. The tasty fruits can be eaten raw or made into a wine (Balick, 1985; Prance, 1982).

• The lulo (*Solanum quitoense*; family Solanaceae). The lulo, or naranjilla, is one of the most highly prized fruits in Colombia and Ecuador. It is a shrubby perennial bearing pubescent, yellow-orange fruits. The greenish flesh is made into an exceptionally delicious drink. The lulo has already been introduced in Panama, Costa Rica, and Guatemala, where it is being marketed as a frozen concentrate (Heiser, 1985).

• The pupunha (*Bactris gasipaes*; family Palmae). Native to the northwest Amazon, the pupunha, or peach palm, is a 20-meter-tall palm widely cultivated in both South and Central America. Each year this palm can yield up to 13 bunches of fruit, which contains carbohydrates, protein, oil, minerals, and vitamins in nearly perfect proportions for the human diet. Under cultivation, the tree will produce more carbohydrate and protein per hectare than does corn (Balick, 1985; NRC, 1975; Vietmeyer, 1986b).

• The amaranths (*Amaranthus* spp.; family Amaranthaceae). The three major species of amaranths (*Amaranthus caudatus*, *A. cruentus*, and *A. hypochondriachus*) are rapidly growing cereal-like plants that have been cultivated in Central and South America since Pre-Columbian times. The ancient Aztecs considered amaranth a sacred plant and consumed cakes made of ground amaranth seeds and human blood. Because of this religious practice, the Spanish severely suppressed the cultivation of this plant. Amaranth seeds have extremely high levels of total protein and of the nutritionally essential amino acid lysine, which is usually lacking in plant protein (NRC, 1984; Vietmeyer, 1986b). Amaranth is currently being marketed in this country as breakfast cereal and is now being sold in many health food stores.

• The guanabana (*Annona muricata*; family Annonaceae). The guanabana, or soursop, is a medium-size tree native to tropical America. Throughout the year the tree produces fruit whose delicious white flesh has a unique smell and a texture that can be best described as a sort of fibrous pineapple custard. Already popular in China, Australia, Africa, and the Philippines, guanabana can be eaten raw or made into a delicious drink or yogurt (NRC, 1975).

• The buriti palm (*Mauritia flexuosa*; family Palmae). A veritable tree of life to many Amazonian Indians, the buriti palm produces a fruit said to be as rich as citrus in vitamin C content. Its pulp oil is believed to contain as much vitamin A as carrots and spinach. A starch extracted from the pith is used to make bread. An edible palm heart can be extracted from the shoots. The Indians also make wine from its fruit, sap, and inflorescences (NRC, 1975). A strong fiber is obtained from the young leaves, and a useful cork-like material is extracted from the petioles.

The wood of the trunk is used in light construction. Furthermore, the buriti palm thrives in Amazonian swamps of little use for intensive agriculture (Schultes, 1979).

IMPROVEMENT OF CROP SPECIES THROUGH CROSS-BREEDING

Relatives of commercial species must continuously be crossbred with these species to improve crop yield, nutritional quality, durability, responsiveness to different soils and climates, and resistance to pests and diseases (IUCN, 1980). Since many of the world's most important crop species originated in the tropics, we must look to the equatorial regions for wild or semidomesticated relatives of commercial species to maintain or improve our crops.

A barley plant from Ethiopia has already provided a gene that protects a $160-million barley crop in California from the lethal yellow dwarf virus. A wild relative discovered by Iltis in the Peruvian Andes has increased the sugar content of the domestic tomato which has resulted in an increased commercial value estimated at $5 to $8 million per year (Witt, 1985; see also Iltis, Chapter 10 of this volume). In fact, tomatoes are one of the world's most important crops, yet they could not be grown commercially in the United States without the genes provided by wild relatives (Harlan, 1984). Rice grown in Asia is protected from the four main rice diseases by genes provided by a single wild species from India. In both Africa and India, yields of cassava—one of the most important crops throughout the tropics—have been increased up to 18 times because of the disease resistance provided by genes from wild Brazilian cassava. Disease resistance provided by wild Asian species of sugarcane have saved the sugarcane industry in the southeastern United States from total collapse (Prescott-Allen and Prescott-Allen, 1983). Perennial corn, discovered by Guzmán in Mexico in 1977, has proven to be immune or resistant to the seven major diseases of domesticated corn (Witt, 1985).

Although the use of wild and semidomesticated relatives is already extensive, it will undoubtedly increase in the near future because of the wider availability of these plants and the growing documentation of their potential utility (Frankel, 1983; Prescott-Allen and Prescott-Allen, 1983). Rapid advances in genetic engineering will also provide greater access to certain gene pools, which can now only be taken advantage of with special techniques (Frankel, 1983).

Following are some good examples of the types of plants that may prove useful for future breeding purposes:

- Coffee (*Coffea* spp.; family Rubiaceae) is a mainstay of the economy of several tropical countries, yet it is rather susceptible to certain fungal diseases. Although Africa is home to most commercial species (particularly *C. arabica* from Ethiopia), the island of Madagascar has approximately 50 wild species of *Coffea*. Some of these species may prove important for commercial breeding not only for their potential resistance to fungal infections but also because they produce beans with little or no caffeine (Guillaumet, 1984; Plotkin et al., 1985).

- Two wild species of potatoes (*Solanum* spp.; family Solanaceae) have leaves that produce a sticky substance that traps predatory insects, which subsequently die of starvation. This type of self-defense could conceivably reduce or negate the need for using pesticides on cultivated potatoes (Gibson, 1979; Harlan, 1984).

• Formerly one of the most important timber trees of western coastal Ecuador, *Persea theobromifolia*, also called *Caryodaphnopsis theobromifolia* (family Lauraceae) has been pushed to the very brink of extinction by overexploitation. When it was finally described in 1979, it was found to be a relative of the common avocado (*Persea americana*) and might one day prove useful as rot-resistant root graft stock for the cultivated species (Gentry and Wettach, 1986).

NATURAL PESTICIDES

Many tropical plants have developed chemical defenses to deter predation by herbivorous animals. Tropical people possess a sophisticated knowledge of these plants, often using them as medicines or poisons. The calabar bean (*Physostigma venenosum*) was traditionally used as an ordeal poison in West Africa, and studies of the active principle of this species led to the development of methyl carbamate insecticides. World trade in daisy flowers (*Chrysanthemum cinerariaefolium*), the source of insecticidal pyrethrum extracts, is a multimillion dollar business (Oldfield, 1984). This plant was first discovered because of its use by African tribal peoples to control insect pests.

South American Indians use *Lonchocarpus*, a forest vine, as a poison to stun fish. Today we import the roots of this plant as a source of rotenone, a biodegradable pesticide. Other plants used by tribal people as fish poisons have yet to be evaluated for their potential as pesticides. Plants used to make arrow poisons or curares also bear looking into, since one such species, *Chondrodendron tomentosum* already provides us with *d*-tubocurarine—an anesthetic administered during abdominal surgery. Not only do we need to investigate the individual components used in the manufacture of the many different types of curare but we must also study the interactions among different species that are sometimes used together. In the northeast Amazon, the preparation of an arrow poison may involve the mixing of seven different species, and the Indians insist that each plant changes and amplifies the toxicity of the poison.

Yet another category of potentially useful natural pesticides are allelochemicals. These are chemicals produced by plants that inhibit the growth of other plants and of soil microorganisms. Allelochemicals include a number of different types of chemicals and may one day be used directly or serve as models for seminatural or wholly synthetic compounds (Balandrin et al., 1985).

Species that might prove useful as sources of biodegradable pesticides in the future include the following:

• Piquiá (*Caryocar* spp.; family Caryocaraceae). One Amazonian species of *Caryocar* produces a compound that seems to be toxic to the dreaded leaf-cutter ant (*Atta* spp.). This insect is the scourge of South American agriculture, causing millions of dollars of damage each year.

• Guaraná (*Paullinia cupana*; family Sapindaceae). This woody vine is native to central Brazil. It is grown on plantations near Manaus for use in Brazilian soft drinks. Guaraná contains three times as much caffeine as does coffee, and recent tests at Harvard have shown that caffeine and some synthetic analogs can kill or inhibit the growth of mosquitoes and other insects (J. Nathanson, Harvard, personal communication, 1986). Should further testing prove caffeine to be an effective insecticide, guaraná could become a major crop throughout the tropics.

INDUSTRY

> Development of native indigenous plants, particularly with reference to tropical and subtropical soils, will be beneficial at a variety of economies of scale. In some instances, their development will be small and amenable to utilization by individual farmers or farming groups. On the other hand, there will be instances where development will be large scale and have international implications. From McKell, 1980.

During the Arab oil embargo of 1973, the U.S. community was faced not only with the loss of a major energy source but also with the loss of its most important raw material for the manufacture of innumerable synthetic products. Few realize how many of our everyday products are made from petroleum and petroleum by-products, such as plastics, fertilizers, lubricants, and adhesives, to name only a few. It has recently been estimated that almost one-fifth of the petroleum used in this country is devoted to industrial nonfuel purposes (White, 1979). Between 1973 and 1976, the annual use of petroleum-based chemicals in the United States was more than 100 billion pounds (Princen, 1977), yet the majority of these substances can now be synthesized from plant products (Wang and Huffman, 1981). These so-called botanochemicals are destined to become increasingly important as raw materials for industry.

Until 1985, the reasons for reducing our dependence on fossil fuels were obvious. At present, the price of oil has dropped sharply, and there are those who believe that the heyday of the OPEC cartel is over. Nonetheless, experts disagree sharply about predictions of future price trends for petroleum. Since oil is a nonrenewable resource, and since the largest reserves lie in one of the most politically unstable regions of the world, we should try to reduce our dependence on petroleum whenever it is economically feasible.

FATS AND OILS

Approximately 3 million tons of vegetable fats and oils are used each year in the manufacture of coatings, lubricants, plasticizers, and many other products (Prescott-Allen and Prescott-Allen, 1982). In the past, industrial usage of these vegetable products has suffered from competition with cheap synthetic petroleum products (Wang and Huffman, 1981), but this trend is expected to change due to the uncertainty about the future of the petroleum market. Between 1973 and 1981, the price of petrochemicals increased more than 700%, whereas that of vegetable oils rose less than 100% (Prescott-Allen and Prescott-Allen, 1982). Even in the industrialized world, commercial demand for oils for use as a food and in industry continues to grow, and demand often exceeds supply (Schultes, 1979).

The supply of edible oils is seriously inadequate to meet human nutrition requirements, especially in underdeveloped tropical regions. However, several tropical forest species have been used by tribal peoples as sources of edible oils for thousands of years. These oils contain vitamins and minerals and are necessary for cooking in areas where butter or lard are either unavailable or in short supply. There has been little attempt to domesticate some of these species, although ambitious efforts are under way in Brazil, Colombia, and Malaysia (M. Balick, New York Botanical Garden, personal communication, 1986). Domestication would increase yield, lower production costs, and reduce or eliminate characteristics that might inhibit harvesting on a commercial scale while providing a steady supply of edible and/or industrial oils. Some of these plants, e.g., bacabá (*Oenocarpus bacaba*) and patauá (*Jessenia bataua*), can grow in both the forest and on semiforested plantations and thus seem to be potential crop species of great importance in the tropics. Some of the more promising tropical oil plants include the following:

- The patauá palm (*Jessenia bataua*; family Palmae). The patauá palm grows to a height of 20 meters and is found in the lowlands of tropical South America. The oil of the fruit is almost identical to olive oil in its chemical and physical properties, and the biological value of its protein is almost 40% higher than that of soybean protein (Balick, 1985; Balick and Gershoff, 1981).
- The babassú palm (*Orbignya* spp.; family Palmae). The South American babassú palm may reach 60 meters in height. A single tree may produce up to a half ton of a fruit that resembles the coconut, although babassú has a higher oil content. This oil can be refined into an edible oil or used to make plastics, detergents, soap, margarine, and shortening. The seedcake is 27% protein and is an excellent fertilizer and animal feed. Its ability to colonize and thrive in deforested areas makes it an ideal species for turning degraded areas into productive lands (Balick, 1985; Schultes, 1979).
- The vine (*Fevillea*; family Cucurbitaceae). Seeds of the fruits of these vines have a higher oil content than that of any other dicotyledenous plant. Gentry and Wettach (1986) theorized that if naturally occurring lianas in a rain forest were cut and replaced by *Fevillea*, a per-acre oil yield comparable to those obtained in the most productive plantations might be obtained without felling a single tree.

FIBERS

Fiber plants are second only to food plants in terms of their usefulness to humans and their influence on the advancement of civilization. Tropical people use plant fibers for housing, clothing, hammocks, nets, baskets, fishing lines, and bowstrings. Even in our industrialized society, we use a wide variety of natural plant fibers: for ropes, brooms, brushes, and baskets. In fact the so-called synthetic fibers now providing much of our clothing are only reconstituted cellulose of plant origin. [Cellulose is produced in far greater quantities by the world's plants than any other organic compound—up to 3 billion tons a day, according to R. E. Schultes of the Harvard Botanical Museum (personal communication, 1986)]. Several trees in tropical South America could be exploited for the fiber that they produce; their commercial potential is, at this point, unrealized. Promising species include:

- The tucúm palm (*Astrocaryum tucuma*; family Palmae). The tucum palm reaches a height of 20 meters and is native to the western Amazon. Its fiber is considered to be among the finest and most durable of the plant kingdom and is highly valued by Amazonian Indians. Furthermore, the tucúm produces an edible palm heart and a fruit that contains three times more vitamin A than do carrots (Balick, 1985; Schultes, 1977).
- Rattans (*Demoncus* spp.; family Palmae). Rattans are climbing palms native to the Asian tropics. Trade in rattan end products amounts to more than $1 billion a year. Unable to afford imported rattan, Peruvian peasants have begun to use *Demoncus*, a local climbing palm that has proven to be a very satisfactory substitute (A. Gentry, Missouri Botanical Garden, personal communication, 1986).

THE ROLE OF THE ETHNOBOTANIST

Tropical forest peoples are the key to understanding, utilizing, and protecting tropical plant diversity. Virtually every plant mentioned in this paper—not only the lesser-known species like the tucum palm and the buriti but also the well-known ones like corn and chocolate—were first discovered and utilized by indigenous peoples. Although it may come as a surprise to many that modern botanists are learning about useful plants from primitive peoples (the science known as ethnobotany), we are in fact just getting started. A single tribe of Amazonian Indians may use more than 100 different species of plants for medicinal purposes alone, yet very few tribal populations have been subjected to a complete ethnobotanical analysis and the need to do so becomes more urgent with each passing year. As we struggle to protect the dwindling tropical rain forest and find new and useful plant species for the benefit of modern human beings, the people who best understand these forests are dying out. More than 90 different Amazonian tribes are said to have disappeared since the turn of the century (G. Prance, New York Botanical Garden, personal communication, 1986). Through extinction and tribal acculturation, true forest peoples are dying out, and their oral traditions are disappearing with them. Each time a medicine man dies, it is as if a library has burned down.

Conservationists often talk about the problem of disappearing species, but the knowledge of how to use these species is disappearing much faster than the species themselves. In order to collect this information, we need to expand ethnobotanical field research. Organizations like the World Wildlife Fund and the National Geographic Society, together with leading botanical institutes like the Harvard Botanical Museum, the New York Botanical Garden, and the Missouri Botanical Garden, are working to document ethnobotanical lore (Figure 11-1). The results of this type of research are not only lists of useful species but also data on potentially useful wild and cultivated varieties as well as ecological information on how to best utilize tropical ecosystems in a sustainable manner. The collection of this type of information, combined with expanded programs bringing some of the more promising species into cultivation, will eventually enrich our diets, and reduce our overdependence on current crop species and nonrenewable industrial materials.

REFERENCES

Balandrin, M., J. Klocke, F. E. Wurtele, and W. Bollinger. 1985. Natural plant chemicals: Sources of industrial and medicinal materials. Science 228:1154–1160.

Balick, M. 1985. Useful plants of Amazonia: A resource of global importance. Pp. 339–368 in G. Prance and R. Lovejoy, eds. Key Environments—Amazonia. Pergamon Press, Oxford.

Balick, M., and S. Gershoff. 1981. Nutritional evaluation of *Jessenia bataua*: Source of high quality protein and oil from Tropical America. Econ. Bot. 35(3):261–271.

Caufield, C. 1982. Tropical Moist Forests: The Resource, the People, the Threat. International Institute for Environment and Development, London. 67 pp.

Frankel, O. 1983. Genetic principles of *in-situ* preservation of plant resources. Pp. 55–65 in S. Jain and K. Mehra, eds. Conservation of Tropical Plant Resources. Proceedings of the Regional Workshop on Conservation of Tropical Plant Resources in South-East Asia, New Delhi, March 8–12, 1982. Botanical Survey of India, Howrah.

Gentry, A., and R. Wettach. 1986. Fevillea—A new oil seed from Amazonian Peru. Econ. Bot. 40(2):177–185.

Gibson, R. 1979. The geographical distribution, inheritance and pest-resisting properties of stick-tipped foliar hairs on potato species. Potato Res. 22:223–236.

Guillaumet, J. L. 1984. The vegetation: An extraordinary diversity. Pp. 27–54 in A. Jolly, P. Ogberle, and R. Albignac, eds. Key Environments—Madagascar. Pergamon Press, Oxford.

Harlan, J. 1984. Evaluation of wild relatives of crop plants. Pp. 212–222 in J. Holden and J. Williams, eds. Crop Genetic Resources: Conservation and Evaluation. Allen and Anwin, London.

Heiser, C. 1985. Ethnobotany of the naranjilla (*Solanum quitoense*) and its relatives. Econ. Bot. 39(1):4–11.

IUCN (International Union for the Conservation of Nature). 1980. World Conservation Strategy. International Union for the Conservation of Nature, Gland, Switzerland. 55 pp.

McKell, C. M. 1980. Native plants: An innovative approach to increasing tropical food production. Pp. 349–382 in Background Papers for Innovative Biological Technologies for Lesser Developed Countries. Office of Technology Assessment, Washington, D.C.

NRC (National Research Council). 1975. Underexploited Tropical Plants with Promising Economic Value. National Academy of Sciences, Washington, D.C. 187 pp.

NRC (National Research Council). 1984. Amaranth: Modern Prospects for an Ancient Crop. National Academy Press, Washington, D.C. 76 pp.

Oldfield, M. 1984. The Value of Conserving Genetic Resources. U.S. Department of the Interior, National Park Service, Washington, D.C. 360 pp.

Plotkin, M., V. Randrianasolo, L. Sussman, and N. Marshall. 1985. Ethnobotany in Madagascar. Report submitted to the International Union for the Conservation of Nature, Morges, Switzerland. 657 pp.

Prance, G. 1982. The increased importance of ethnobotany and underexploited plants in a changing Amazon. Pp. 129–136 in J. Hemming, ed. Change in the Amazon Basin. Vol. I: Man's Impact on Forests and Rivers. Manchester Press, Manchester.

Prescott-Allen, R., and C. Prescott-Allen. 1982. What's Wildlife Worth? Economic Contributions of Wild Plants and Animals to Developing Countries. International Institute for Environment and Development, London. 92 pp.

Prescott-Allen, R., and C. Prescott-Allen. 1983. Genes from the Wild. Using Genetic Resources for Food and Raw Materials. International Institute for Environment and Development, London. 101 pp.

Princen, L. 1977. Potential wealth in new crops: Research and development. Pp. 134–148 in D. Siegler, ed. Crop Resources. Proceedings of the 17th Annual Meeting of the Society for Economic Botany, the University of Illinois, Urbana, June 13–17, 1976. Academic Press, New York.

Schultes, R. E., 1977. Promising structural fiber palms of the Colombian Amazon. Principes 21(2):72–82.

Schultes, R. E. 1979. The Amazonia as a source of new economic plants. Econ. Bot. 33(3):259–266.

Tyler, V. 1986. Plant drugs in the twenty-first century. Econ. Bot. 40(3):279–288.

Vietmeyer, N. 1986a. Exotic edibles are altering America's diet and agriculture. Smithsonian 16(9):34–43.

Vietmeyer, N. 1986b. Lesser-known plants of potential use in agriculture and forestry. Science 232:1379–1384.

Wang, S., and J. B. Huffman. 1981. Botanochemicals: Supplements to petrochemicals. Econ. Bot. 35(4):369–382.

White, J. 1979. The growing dependency of wood products on adhesives and other chemicals. For. Prod. J. 29:14–20.

Witt, S. 1985. Biotechnology and Genetic Diversity. California Agricultural Lands Project, San Francisco. 145 pp.

DIVERSITY AT RISK:

TROPICAL FORESTS

An example of slash-and-burn agriculture, one of the major mechanisms used to clear forests. *Photo courtesy of the Missouri Botanical Garden.*

OUR DIMINISHING TROPICAL FORESTS

PETER H. RAVEN

Director, Missouri Botanical Garden, St. Louis, Missouri

Ⅰn any discussion of biological diversity, tropical forests must occupy center stage. Broadly defined, these forests are home to at least two-thirds of the world's organisms, a number that amounts to no fewer than 3 million species, and could be 10 or more times greater than that amount. Striking, however, is the fact that only about 500,000 species from the tropical and subtropical regions of the world have been given names and been cataloged in the scientific literature. This means, very simply, that where one might expect to identify the great majority of any collection of insects or other arthropods made within the boundaries of Europe or temperate North America, only a very few of those in any reasonably diverse sample of tropical organisms—at least among relatively small and inconspicuous groups—could be located in the world's collections, or are mentioned in the world's literature.

Even among those very few, only a tiny fraction would be known from more than one or several specimens, a few short lines of technical description, and a locality. In short, identifying them would not provide much help concerning their ecology, their evolutionary relationships, their behavior—or any of the components that might have been involved in their history, or that might contribute to their chances of survival. Such matters must be considered seriously as we learn more about the diversity of organisms itself.

Regardless of whether there are 2.5 million more tropical organisms to be named or 25 million, the task facing us is enormous. All the activities of all those concerned with cataloging organisms over the past centuries in all types of ecosystems throughout the world have resulted in the naming of only about 1.5 million of them, and a task at least twice, and perhaps many times, that large confronts us now. All the scientific and societal gains that depend on an increased knowledge of these organisms (we must know that they exist before we can understand or use them) depend on the degree to which that task can be completed. Since all of human society depends directly or indirectly on our ability to manage plants, animals, and microorganisms effectively, the task is one of enormous importance.

In light of the rapid destruction of tropical forests, it is an especially urgent matter to catalog the organisms in those regions and to establish well-considered priorities for this undertaking. It is clear that most tropical forests will have been destroyed or severely damaged within the next 25 years, because of the size of the human population in the tropics and subtropics, already constituting a majority of the world's people and growing explosively; the extensive poverty there, which afflicts well over a third of the people; and our collective ignorance of effective ways to manage tropical ecosystems so that they will be productive on a sustainable basis. By 2010, the only large blocks of undamaged forest remaining will be those in the western and northern Brazilian Amazon, the interior of the Guyanas, and the central Zaire (Congo) basin in Africa. All the forests in other parts of the tropics and subtropics (those in Mexico, Central America, the West Indies, Andean South America, the eastern and southern portions of the Amazon), all the forests of Africa outside the central Zaire basin, and all the forests of tropical and subtropical Asia will have been devastated by that time.

There will of course be exceptional preserved areas within these regions, their sizes depending on the effectiveness of local conservation programs and on the nature of the soils underlying particular pockets of vegetation. Some areas will simply be too steep to cultivate, others too rocky, and still others too wet. In these pockets of vegetation, populations of organisms will survive; however, they will be reduced to relatively few individuals in most cases, subjected to the effects of light and heat penetrating from the edges of the fragmented patches of forest in which they are surviving, and assaulted by human activities related to their greater accessibility. For example, the hunting of primates and other animals (discussed by Mittermeier in Chapter 16) is often greatly intensified when the surviving patches of vegetation are small. Because of the nature of small populations—they are unlikely to persist long owing to chance alone—and the increasing strains on these pockets of vegetation, many of the species that initially survive locally are likely to become extinct within a very few years.

The question arises as to whether large, important preserves such as Manu Park in Peru or the Tai Forest in the Ivory Coast can survive until the projected stabilization of the human population in the second half of the next century. As in other tropical and subtropical countries, human pressures in Peru and the Ivory Coast are incredible, and resources tend to be consumed in meeting the needs of rapidly growing populations with high proportions of poor, often malnourished people. Over the next few years, the confrontation between human needs and forest preservation, already evident in many areas, will become more acute. The protection of such major reserves is conceivable, however, if there is a genuine willingness to share resources on a global level—to provide major support from industrialized countries not merely for the protection of parks and reserves but for the creation of conditions in which all people can live with a measure of human dignity. The decisive factors will be social, political, and economic; they will not be limited simply to a willingness to conserve.

Putting these relationships in another context, and assuming that two-thirds of the world's 4 to 5 million species are located in the tropics and subtropics, nearly half the world's species of plants, animals, and microorganisms will be destroyed or severely threatened over the next quarter century—well within the expected life span of most people living today. If half these organisms become extinct during the next several decades—surely a conservative estimate—the world will experience a major episode of extinction. This episode could amount to the loss of perhaps 10% of the world's species by the end of the century and to more than a 25% loss within the next couple of decades. These estimates are compatible with the predictions of extinction rates for primates and other relatively well-studied groups of organisms and with the closely coupled nature of the biological relationships involved. To find a comparable rate of extinction, one needs to go back more than 65 million years to the end of the Cretaceous period, when the dinosaurs disappeared along with a major loss of other life on Earth. In fact, the rate of extinction that will be characteristic of most of the remaining lifetimes of those now living is estimated to be at least 1,000 times the normal rate. Since there are now many more species than there were 65 million years ago, the absolute loss in number of species will be much greater.

For a more concrete example, consider the flowering plants. We obtain 85% of our food directly or indirectly from just 20 kinds of plants, and about two-thirds from just three: maize (corn), wheat, and rice. The 20 species were brought into cultivation thousands of years ago, largely because they were easy to grow; they were not selected because of their ability to contribute to the needs of a modern industrial civilization. Despite that, they are precious. Widespread starvation in the tropics and subtropics, however, reminds us that temperate-zone agriculture is not suitable everywhere, and suggests that an enhanced ability to cultivate some of the other 250,000 species of flowering plants might offer rewards by providing food crops that can be cultivated successfully in areas where the cultivation of present food plants is now difficult or impossible.

In evaluating our future opportunities to use the lesser known plants, however, consider the significance of the extinction projections reviewed above. Some 25,000 species of plants—about 5 species a day—are expected to disappear between now and the end of the century, and then perhaps 10 species a day will become extinct over the following couple of decades. Clearly, many of the 50,000 species of plants expected to vanish forever during our lives hold exceptional promise for producing food, fodder, wood, medicine—all the factors that increase the quality and stability of human existence on Earth. Given our record numbers, and the extreme pressure with which we are assaulting the global ecosystem, it seems absolutely mandatory that we redouble our efforts to survey, classify, preserve, and understand these plants, as well as members of other groups of organisms, while they still exist.

The consequences of the destruction of tropical and subtropical forests are grave; basically, our collective actions are denying to our children and grandchildren the ability to play the game of survival with the tools that we have at our disposal today. In effect, we are, by our passivity, making the effort to survive through the creation and maintenance of stable, productive ecosystems more difficult for them than it is for us. The kind of restoration ecology described so eloquently by Janzen for the dry forest in Chapter 14 will undoubtedly come to be practiced widely as the human population stabilizes and our relationships with the global ecosystem become more realistic. The preservation of individual species of plants, animals, and microorganisms *now* offers the best chance of achieving the most complete success in this complex area during the twenty-first century. Analogous is the need to preserve genes for future use in the developing field of genetic engineering. It will be decades or centuries before it is possible to synthesize genes that can confer desirable traits on recipient organisms in any but the most simplistic ways; yet living organisms contain an enormous library of such genes, already tested by nature and available for use until the organisms themselves become extinct. Through our endless preoccupation with immediate, seemingly pressing domestic problems, we are seriously damaging our prospects for the very near future by losing scientific, societally relevant, and aesthetic possibilities beyond imagining.

Nonetheless, as we confront this grim spectacle, we must remember that the opportunities for studying and preserving biological diversity are greater today than they will ever be in the future. Of critical importance will be our ability to abandon our passivity and face the situation as it is, devoting increased resources to the exploration of diversity and using the information that we gain for our common benefit. In this effort, the importance of the kinds of studies described in this section is evident; they provide models of the variety of activities that should be intensified and multiplied while the opportunities are as great as those we enjoy now.

THE TROPICAL FOREST CANOPY
The Heart of Biotic Diversity

TERRY L. ERWIN

Curator, Department of Entomology, National Museum of Natural History,
Smithsonian Institution, Washington, D.C.

A few years ago in a short paper in the *Coleopterists Bulletin*, I hypothesized that instead of the current estimate of 1.5 million species on Earth, there were 30 million species of insects alone (Erwin, 1982). This hypothesis was based on collections of beetles from tropical forest canopy samples in Panama (Erwin and Scott, 1980), rather than on the catalog counts of taxonomic names used in all the earlier estimates. I used simple arithmetic based on actual numbers of beetle species in my samples, estimated numbers of tropical forest tree species given me by the leading botanists, and a conservative estimate of the host specificity of tropical forest canopy insects. Host specificity in this sense means that a species in some way is tied to the host tree species and cannot exist without it.

This reestimation of the magnitude of life on Earth got a lot more coverage than I anticipated and began the usual controversy of right or wrong. Those engaged in the controversy, most of whom never read this obscure paper in the *Coleopterists Bulletin*, in a way actually missed the point of the paper. Consequently, I now want to take the opportunity to clarify the situation.

Science, at least in natural history, proceeds from casual observations, usually in the field or on museum specimens, to the erecting of hypotheses and finally to the testing of those hypotheses. Repeated failure to prove a hypothesis false lends support to the possibility that it may be true. For the 30 million species of insects hypothesis, which was based on a brand new set of observations never before available to scientists, I suggested that testing must begin by refining of our knowledge about host specificity of insects in tropical forests.

In a subsequent paper, analyzing data from the canopies of four different forests in the central Amazon around Manaus, Brazil, I showed that 83% of the beetle species in the samples were found in only the samples of one of the types of forest, 14% of the species were shared between two, and only 1% of the species of beetles was found in all four forest types (Erwin, 1983a). This added fuel to the "numbers" controversy, because of the numerous types of forest known to exist in the Amazon Basin alone and the fact that the analysis was based on more than 1,000 species of beetles, a fairly substantial data base. At this point, I turned my attention to the now well-refined sampling techniques of insecticidal fogging of forest canopies at the Tambopata Reserved Zone in the southeast corner of Amazonian Peru. I developed these techniques for the purpose of testing the main hypothesis regarding biological diversity in tropical forests and the subhypothesis that host specificity is a main feature of the lifestyle of tropical canopy insects. The following paragraphs provide some glimpses of the Tambopata Canopy Project, some preliminary observations on the fauna itself, and what I believe to be the status of the 30 million species hypothesis. With a data base of a million specimens (we'll get to the number of species later), it will take a long time to complete the data analysis from just 1 year of collecting.

THE PROBLEM

It has been predicted that in 25 to 30 years, much of the humid tropical forest could be gone or severely converted (see Raven, this volume, Chapter 12). Between 25 and 40% has already been lost to misguided human exploitation. The best estimate is that an area the size of Honduras is being lost or converted each year, and by the year 2000 some popular accounts have predicted that a million species will become extinct. Although I regard such guesses as a bit low, a point discussed later in this chapter, they mean that in our generation we, the only species on Earth with the mental capacity to reason, will see the virtual disappearance of contiguous tropical forests and probably the extermination of more than 20% of the diversity of life on Earth, and we humans will have caused it.

THE HISTORY

The Amazon basin (Figure 13-1) has the richest biota on Earth. There are several factors involved, not the least of which is the sheer size of the basin. We must start the historical analysis with the Amazon basin as it was on the western portion of the megacontinent Gonwanaland some 100 million years ago. The biota of today is a result of many events that occurred after two supercontinents, South America and Africa, rifted, and South America drifted in a westerly direction. As this occurred, the uplift of the Andes began. This wonderful mountain chain, extending from Venezuela down into Chile, became a dike that reversed the western flow of all the rivers of Gonwana, turning them around and beginning their flow to the east. In the last 40 million years, this event has caused a mosaic of habitats, the fine-grained resolution of which we have no comprehension at this time. As I am discovering in some of my work in Peru, the fine-grainness of habitats is far, far greater than what the botanical classifications have led us to believe. We need from the botanists a better picture of tree species distribution and habitats, and of the small communities made up by these tree species microdistributions.

During this 40 million years of Andean orogeny, there were three uplifts of crystalline rock across the Amazon, represented by the red arches in Figure 13-1. The two gray areas in the north and south are bedrock, the Guyana and Brazilian Shields. From this perspective, we now see the development of this mosaic of habitats, defined by the meandering river systems of the Amazon basin itself. The study of these rivers and the areas between them offers an interpretation of the events of the past (Erwin and Adis, 1982).

A mosaic component that extends throughout the Amazon basin is the oxbow lake, a lake formed when a loop of a river becomes isolated from the river as a result of sedimentation. The formation of an oxbow lake is the first stage in succession that culminates in forest. This small "island" of aquatic life will soon become an island of grassy life, which will then become an island of palm tree life and so on until it returns to climax inundation or upland forest of some type. During succession, it may be crosscut by another twist of the river or another small river, which will then subdivide it into four successional stages each with a different time differential. This kind of successional evolution on a massive 6-million-square-kilometer area is but one of the features that has provided the evolutionary pathway for Amazonia's fantastic diversity.

What we see today from the air is a forest canopy that extends more or less unbroken across those 6-million-square kilometers, except for the rivers, the hy-

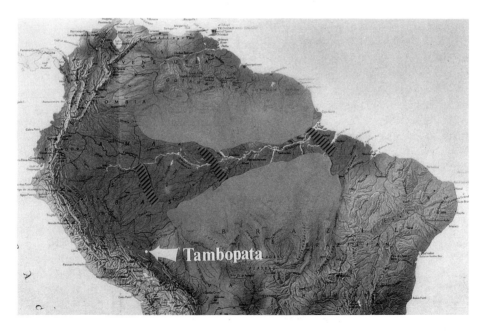

FIGURE 13-1   The South American land mass. The Guyana and Brazilian Shields are shown in gray; the hatched areas represent the three arches of crystalline rock.

droelectric projects, the Rondonia project, and various other development projects that are starting to break up that vast expanse of forest. From the air, one can detect even finer and finer mosaics. It is very easy to pick out the trees in blossom, the trees with tough dark-green leaves, trees that lost their leaves during the dry season and are now getting a new flush of very light pale leaves (the ones the insects like to eat the most), and vines that reach up into the canopy to spread their leaves over the tree leaves or intermingle them with the leaves of the canopy trees. All 150 or more species of canopy trees or vines per hectare contribute to the mosaic. There is an intermingling of leaves between two species of trees, between the vines and the trees, and between one tree overshadowing the other, resulting in the creation of microenvironments for the little creatures that are so important in providing the richness of the world's biotic diversity.

Depending on forest type, the tops of the trees range from 15 meters to as high as 55 meters. Tambopata was chosen for my preliminary studies because logistically it is very difficult to get equipment and people into a virgin rain forest, keep them there for long periods, and get the material back to the museum to study it under the microscope. The average length of the beetles in the canopy is about 2 to 3 millimeters, so one needs pretty good facilities to make detailed studies. Tambopata served the logistic purposes as well as another purpose—approximately 11 different types of forests are found within walking distance. That seemed like too much to handle during 1 year, so only five were selected for intense collecting. In each of these five forests, we selected three 12-meter-square plots (Erwin, 1983b).

All 15 plots were sampled in the early rainy, late rainy, early dry, and late dry seasons. The data collected included tree canopy sizes, species of trees, and exact location of the collecting trays. All this information has been computerized and allows museum specimens to be traced back to the actual square meter of rain forest where they were collected. This gives us the opportunity to return in subsequent years and resample in order to see what the canopy, or what the forest in general, is doing over long periods. Long-term cycles have been largely overlooked, except by a few researchers for only a few species. My research team is now beginning to computerize the canopy in three dimensions so that we can describe exactly where these insect species reside in the canopy.

Beyond this data set, we also have the branching patterns, the leaf structure, and other details of microhabitats. It has taken a long time to develop our data collections, because we have paid attention to the finest details. I am trying to look at the canopy habitat through the eyes of these 2- to 3-millimeter-long beetles.

To date, we have analyzed about 3,000 species of beetles from only five plots. When we complete our analysis, we will have a large data set. A comparison of the tree composition of the different kinds of forest has shown that the forest in Manaus and two of our upland terra firma forests contain entirely different tree families. There are more big trees in the Peruvian sites than in the Manaus sites. Perhaps that accounts in part for the larger size of the insects in the canopy in Peru than in Manaus.

Only 2.6% of the species are shared between Manaus and Tambopata (Figure 13-2). This seems reasonable, because the two sites are 1,500 kilometers apart.

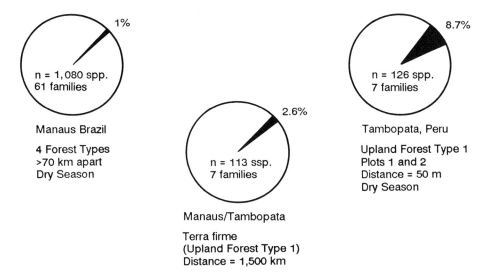

FIGURE 13-2 Pie diagrams of shared beetle species among forests in Peru and Brazil in percent of fauna.

But we found that of the 1,080 species analyzed, there was only a 1% overlap of species in all four forest types in Manaus (Erwin, 1983a).

Data collected during three seasons for two forest plots in the same type of forest 50 meters apart in Tambopata indicate that only 8.7% species are shared. When we add the fourth season data (which will come in shortly), we predict that the percentage of shared species will drop.

Figure 13-3 is a cumulation species curve, which shows the increase in the number of species as we increase the samples. After this figure was made, some more samples were analyzed and the curve became much steeper. These data are just from Plot 1 in Upland Forest Type 1 (Erwin, 1985). The 3,000 species already analyzed amount to more than all the samples from Brazil.

A canopy beetle is shown in Figure 13-4. In fully describing the distribution of these insects in time and space in the tropics, we should think in terms of more than 30 million, or perhaps 50 million or more, species of insects on Earth. A large number of species are tied only to certain forest types that are found on very small patches of soil deposited differentially through time by the vast and meandering Amazon River system. The extermination of 50% or more of the fauna and flora would mean that our generation will participate in an extinction process involving perhaps 20 to 30 million species. We are not talking about a few endangered species listed in the Red Data books, or the few forbish louseworts and snail darters that garner so much media attention. No matter what the number we are talking about, whether 1 million or 20 million, it is massive destruction of the biological richness of Earth.

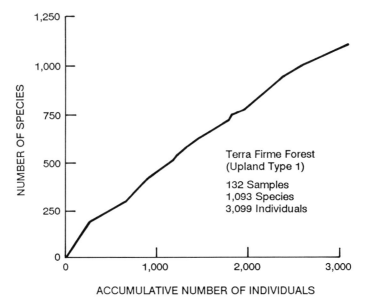

FIGURE 13-3   Numbers of species accumulated per square meter sample in 12-meter-square plot (119 square meters sampled) in Upland Forest Type I at Tambopata Reserved Zone, Peru.

FIGURE 13-4   *Agra arrowi* Liebke, a member of the top predatory carabid beetle group in tropical forest canopies.

We are rapidly acquiring a new picture of Earth, and it is crammed with millions upon millions of nature's species on the verge of being replaced by billions upon billions of hungry people, asphalt, brick, glass, and useless eroded red clay baked by a harsh tropical sun. Many driving forces of evolution have affected carabid beetles and much of the other life on this planet. Very late in the scale of geologic time, a new driving force, humans, appeared. There is little question in my mind that Isaac Asimov (1974) in his wonderful *Foundation Trilogy* may have been particularly visionary when he described the planet Trantor, a sphere of steel and concrete; a hollow joke of its former self. Could Trantor be future Earth? Perhaps; perhaps not. Perhaps the biocrisis can be avoided. Human beings are starting to pay attention to the problem, and we're a very resilient species and have a lot of good ideas. But do we have the resolve to rise above profit and greed?

REFERENCES

Asimov, I. 1974. Foundation Trilogy. Avon, New York. 684 pp.

Erwin, T. L. 1982. Tropical forests: Their richness in Coleoptera and other Arthropod species. Coleopt. Bull. 36(1):74–75.

Erwin, T. L. 1983a. Beetles and other Arthropods of the tropical forest canopies at Manaus, Brasil, sampled with insecticidal fogging techniques. Pp. 59–75 in S. L. Sutton, T. C. Whitmore, and A. C. Chadwick, eds. Tropical Rain Forests: Ecology and Management. Blackwell Scientific Publications, Oxford.

Erwin, T. L. 1983b. Tropical forest canopies, the last biotic frontier. Bull. Entomol. Soc. Am. 29(1):14–19.

Erwin, T. L. 1985. Tambopata Reserved Zone, Madre de Dios, Peru: History and description of the Reserve. Rev. Peru. Entomol. 27:1–8.

Erwin, T. L., and J. Adis. 1982. Amazon inundation forests: Their role as short-term refuges and generators of species richness and taxon pulses. Pp. 358–371 in G. Prance, ed. Biological Diversification in the Tropics. Columbia University Press, New York.

Erwin, T. L., and J. C. Scott. 1980. Seasonal and size patterns, trophic structure, and richness of Coleoptera in the tropical arboreal ecosystem: The fauna of the tree *Luehea seemannii* Triana and Planch in the Canal Zone of Panama. Coleopt. Bull. 34(3):305–322.

TROPICAL DRY FORESTS
The Most Endangered Major Tropical Ecosystem

DANIEL H. JANZEN

Professor of Biology, University of Pennsylvania, Philadelphia, Pennsylvania

The rain forest is not the most threatened of the major tropical forest types. The tropical dry forests hold this honor. When the Spaniards arrived in the Western Hemisphere, there were 550,000 square kilometers of dry forest (approximately five times the size of Guatemala, or the size of France) on the Pacific coast of Mesoamerica (an area extending north from Panama to western Mexico). Today, only 0.09% of that region (approximately 480 square kilometers) has official conservation status, and less than 2% is sufficiently intact to attract the attention of the traditional conservationist. If there is to be a conserved neotropical (i.e., Western Hemisphere) dry forest wildland large enough to maintain the organisms and the habitats that were present when the Spaniards arrived, and if it is to be large enough to be easily maintained and thus a project willingly undertaken and managed into the indefinite future by the society in which it is imbedded, then we will have to grow it (Janzen, 1986a).

The story is the same for the dry tropical regions of Australia, Southeast Asia, Africa, and major parts of South America. The cause of the severe habitat loss is straightforward. Dry forest is easily cleared with fire, and woody regeneration in fields or pastures is easily suppressed with fire. Furthermore, fire does not stay where you put it; many areas are unintentionally cleared. The farmer is also aided by the severe dry season; it suppresses pest and weed populations, facilitates the use of fire as a tool to clean up pastures and fields, and slows soil degradation caused by continuous rain and farming. Many tropical dry forest regions even have good soils; they are downwind of volcanic mountain ranges or are situated on alluvia.

What are the conditions in a tropical dry forest (cf. Hartshorn, 1983; Janzen, 1986a; Murphy and Lugo, 1986)? Its 4- to 7-month rain-free dry season is sufficiently harsh that many species of trees, vines, and herbs are deciduous for 2 to 6 months. Its rainy season, during which 1 to 3 meters of rain can fall, is as wet, if not wetter, than that of a rain forest. In the dry season, the sun penetrates to the forest floor, the leaf litter becomes very dry (and virtually ceases to decompose), watercourses dry up or greatly diminish in flow, and daytime relative humidity ranges from 20 to 60%. The dry forest may appear uniformly green during the rainy season, but during the dry season this homogeneity changes into a complex mosaic of tens of habitat types distinguished by the differential drying rates of different soils and exposures, different ages of succession, and different vegetation types. Many animals migrate to moist refugia (hollow logs and caves, moist riparian sites, north-facing slopes protected from the wind, and sites close to rain forests). During the dry season, most plants cease their vegetative activities, but many species of woody plants flower, mature their fruits, and disperse their seeds. Some species of animals feed on dry season fruits, seeds, and flowers; for them, the dry season is the bountiful time of year and the rainy season, inimical.

DIVERSITY IN THE DRY FOREST

What is the level of diversity in tropical dry forests? A lowland dry forest adjacent to a lowland rain forest (such as a portion of the Pacific dry and Atlantic wet sides of northern Costa Rica) sustains a fauna and flora about 50 to 100% as species-rich as does the neighboring rain forest (Janzen, 1986a). Floras are the least similar in richness of species, largely because the dry forest epiphytes and trees are substantially less rich in species. The greatest similarity in species richness is represented by mammals and major insect groups such as butterflies and moths (Lepidoptera) and Hymenoptera such as bees, wasps, and ants. Species overlap between the two areas, ranging from less than 5% (e.g., epiphytes, amphibians) to as high as 80% (e.g., sphingid moths, mammals). In the 11,000-hectare dry forest of Santa Rosa National Park in Costa Rica, I estimate that there are 13,000 species of insects, and fairly accurate counts indicate that there are 175 breeding species of birds (Stiles, 1983). There are also 115 species of nonmarine mammals (Wilson, 1983) and about 75 species of reptiles and amphibians (Savage and Villa, 1986). All the species of small herbs and grasses have not yet been collected, but the final list of angiosperms (which include vascular plants such as orchids and trees) will probably not exceed 700 species (Janzen and Liesner, 1980). When such a dry forest habitat is replaced by fencerows, ditchsides, unkempt pastures, and woodlots, the species richness of the breeding fauna and flora is reduced by 90 to 95%.

It is true that long lists of species have been used as criteria for identifying tropical habitats worthy of conservation. However, an approach that merely considers the number of species present is incomplete. This can be seen by examining the tropical dry forest, which is less rich than the rain forest in total species but is much richer in its variety of species' activities. It contains many species that remain dormant in inclement (wet or dry) weather, and species that magically find enough water to develop flowers, fruits, and leaves at the height of the dry season (e.g., Janzen, 1967, 1982a,b). Its parasitoids and grazers range from absolutely monophagous to extremely polyphagous; that is, some subsist only on one type of food, whereas others feed on a variety of organisms. Where else can you find white-tailed deer eating fruits dropped by spider monkeys and coyotes foraging side by side with jaguars? Moreover, the Santa Rosa dry forest has the only wind-pollinated legume (Janzen, in press a), the fiercest ant-plant mutualism (Janzen, 1966), and an enormous seasonal pulse of caterpillars that changes to a nearly total absence of caterpillars while the host plants are still in full leaf (Janzen, in press b). If you want a plantation timber tree that will grow throughout the year in a tropical rain forest, yet withstand the droughts produced by agricultural clearing, look to dry forest trees rather than to rain forest trees. The dry forest is also the parental climate for many major tropical crops and food animals such as cebu cattle, chickens, cotton, rice, corn, beans, sweet potatoes, sorghum, and pasture grasses (Janzen, 1986b).

Rather than focusing just on lists of species, tropical conservationists should also concentrate on saving interactions among species. What good are such interactions? Interactions make wildlands interesting, and they provide the raw materials used by the natural historian to construct the stories and the visions that are the real value of the natural world to humanity (Janzen, 1986b).

The conservation world has by and large failed to exploit the real enticements of the areas it conserves. Tropical forests can be likened to libraries and books. The value of a book is not measured by the number of words it contains or even by the number of kinds of words it contains. Put most simply, how long will the public continue to support a library whose goal is to have enormous holdings but no card catalog, no librarians, and thus no readers. Such a library is doomed to fail during the next paper shortage or governmental budget cut. Books are also comparable to the species in the dry forest in that they have little meaning except in context. The context of the dry forest is unique because of its species; it was once widespread and certainly has as much potential for biocultural development as does any type of tropical vegetation. It is this context that we must save.

The likelihood of long-term survival of a conserved wildland area is directly and strongly proportional to the economic health and stability of the society in which that wildland is imbedded. The farm- and ranchland once occupied by dry forest often sustains economically strong regional subcultures, which can thrive without the need to exploit the conserved area. Land blessed with conservation status in such a subculture has a much higher chance of survival than do the more abundant wildlands in frontierlike subcultures, many of which are also based on marginal farmland.

Overemphasis on the length of species lists is also potentially misleading, because the extraordinary peaks in species richness encountered at certain sites can be highly atypical. They are very interesting ecologically but are not representative of the tropics as a whole. They may not even be representative of the site itself, since many of the species-rich sites bear accumulations of strays at the overlap between several less species-rich habitats. Furthermore, the remaining sites of very great endemism and extraordinarily extensive species richness are often sites with peculiar physical characteristics that render them less likely to be occupied by humans at present and therefore easier to conserve. The apparently successful conservation of such areas gives a feeling of accomplishment that makes it easier to accept conservation failures or inactivity with less species-rich, and therefore more ordinary, tropical sites. In the headlong rush to conserve diversity, we risk leaving the next generation with a handful of pretty baubles rather than the substance of the tropics. Saving a habitat with 300 species of endemic orchids on an Andean mountaintop may not have the same long-term ecological, intellectual, or economic value as does saving remnants of once widespread and less species-rich lowland forest.

A MANY-FACED THREAT

The threats to the tropical dry forest are multiple and complex. The concerned observer is correct to be no longer stirred to action by the simplistic chant, "The beef cow is responsible for the demise of the tropics"; the music is more daunting, more complicated, and more site-specific.

There are almost no large blocks of dry forest still standing that can be destroyed and thus cause concern among the public and academic world. Equally important, there are few opportunities to recognize the biocultural deprivation of the ranching and farming cultures that have been sustained for hundreds to thousands of years by the soils that once supported dry forest. As tropical conservation has swung into high gear during the past three decades, it has become comfortable to focus largely on the remaining rain forest and not to worry about scraps of other scattered vegetation types such as the dry forest. A traditional conservation battle for tropical dry forests would have to have been fought in 1900. Today, restoration ecology and habitat management (e.g., Janzen, in press c) are the only answers.

The acquisition and restoration of dry forest wildlands conflict with traditional conservationist protocol in numerous ways:

• Land apparently used for agricultural production, or land that has produced something, is being set aside. This is expensive land, and its acquisition is often accompanied by a last-minute harvest of the few remaining trees and other resources by sellers who can do very serious damage to the anticipated restoration of the site.

• The sellers—poor to very wealthy—are likely to be involved in neighborhood functions and politics far into the future; they cannot be bought out and left as resentful recipients of a bad deal.

• The frontier is gone. The audience is local. The power is local. Within a few decades, if it hasn't happened already, almost all members of tropical societies situated on dry forest soils will be settled on firmly titled land and will be leading a real or vicarious urban life with amenities such as good roads and schools. Survival of a wildland will depend on regional policy decisions by government institutions and planning commissions, and those decisions will be made by or in conjunction with the local community.

• Many dry forest species are relatively robust, largely due to their evolutionary history of exposure to seasonal changes. Thus even tiny population fragments and severely altered populations can be ecologically reworked into viable, interacting populations and complex habitats that are replicas or facsimiles of what once was on the site. Habitats that appear well beyond recovery can be restored if the seed sources are present.

• Intensive cleaning up of the Mesoamerican dry forest agroscape during the last two to three decades is leading to the final extirpation of species and habitats that survived the first wave of megafaunal extinction by hunters 9,000 years ago, the extensive agriculture that began 5,000 years ago, and the ever more intensive agriculture that began 500 years ago. Either we act very soon or we will witness the elimination of many species that have persisted through many seemingly more severe perturbations than the contemporary, innocuous-appearing clearing of the last fencerows.

• Conserved areas of dry forest wildlands will be rich in plant and animal opportunistic species and may well be the only places where most of today's weeds survive. Weeds may be the most information-rich carriers of the genetic information for environmental toughness—information of obvious value in genetic engineering for crop species in harsh habitats.

• Fires and invasion by grasses are the most serious contemporary ecological threats to the restoration and maintenance of dry forest wildlands. Properly manipulated, domestic animals may be the best tools for managing these threats, and they may even pay for their own maintenance: they mow the competing grass, they eat the fuel for the next dry season's grass fires, and they disperse tree seeds far into pastures.

• Dry forest conservation requires not only restoration but also explicit efforts to eliminate the various species initially introduced for agricultural or ranching purposes.

• The biggest and a perpetual problem in dry forest management lies in deciding which areas of the wildland will be managed in what manner and to what end. Yes, it can be returned to a natural state, given the availability of seed sources. But which natural state do you want? On a time scale of at least thousands of years, the state to which it returns or in which it remains depends on many factors, such as the initial condition of the site, the species that arrived, and the order in which they arrived.

What means can be used to restore a tropical dry forest habitat?

• Initiate and maintain a heavy flow of biological information, both biocultural and economic, from the site into the neighboring social system. The process of restoration, and the biology and interactions of the organisms being restored, must become as familiar to the region as are its irrigation projects, school development, and health programs. This task is both more difficult and more sustainable if the conservation effort is focused on habitats, interactions, and caterpillars rather than on redwoods, lions, and condors.

• Stop man-made fires, hunting, cattle ranching, and other free-ranging perturbations. That is to say, give the site back to the remaining or adjacent dry forest organisms to recolonize by their own means. However, while this multihectare regeneration appears to be natural, such megacolonization of pastures and fields does not occur anywhere in nature. It is also not risk-free; neighboring blocks of pristine vegetation are likely to be severely altered by invasions of secondary successional organisms from the large and ever-growing areas under restoration.

• If some organisms are to be reintroduced from elsewhere, how far back in time shall we reach? Do we return the Pleistocene horse to Central American dry forest wildlands? How do we do that without adding its predators? If we put the tapir and white-lipped peccary back into El Salvador, do we also bring back their food plants? There are no correct choices per se, but it is clear that certain major dry forest areas must be set aside purely for agriculture and that no effort must be spared to maintain certain other (much smaller) areas as wildlands.

• If the goal of restoration ecology is to conserve a maximum number of species, the management plan would be quite different than if the goal is to conserve habitats and interactions with whatever species they normally contain. Management directed at the conservation of a maximum number of species leads at least to the fragmentation of the wildland into a mosaic of successional types and ages and the introduction of species from other areas. If the goal is to conserve interactions as well as species, then the wildland manager must predict rather than simply react. The interactions that are saved will depend on the management steps taken, which depend on the interactions that are desired (and there will be errors and surprise outcomes). Conservation abruptly graduates from the art of patrolling a boundary against poachers to a variety of technical activities, such as the research-based studies of ecological succession, evolutionary biology, species packing, competitive exclusion, and epidemiology.

• Rain forest wildlands must be conserved within migratory reach of the dry forest areas that are subject to restoration. A dry forest does not exist unto itself and neither does a rain forest. In Central America, the rain forest and the dry forest are the mutual recipients of each other's migrants—migrants that are important parts of the interactive structure that holds tropical habitats together. Birds migrating from Wisconsin to Costa Rica are not the only ecological link over large agroscapes.

• The dry forest is not only a collection of many kinds of habitats, each rich in unique species, but the members of a given habitat can be important interactants in adjacent habitats. If only certain (usually species-rich) dry forest habitats are slated for conservation, one quickly discovers that a substantial portion of the species in those habitats spend critical parts of their lives in other nearby habitats. To put it another way, the conversion of highly deciduous forest on dry ridges to pasture may have a severely depressing effect on the species-richness of organisms in the very species-rich adjacent alluvial bottomlands.

The dry tropics contain adult remnants of a once thriving forest, juveniles from gradually dwindling seed reservoirs, and waifs from as yet intact wildlands. These organisms now stand on a trashed agroscape and will die without replacement. They are the living dead—all physiologically alive but can be regarded as dead if they were already lying in the litter (Janzen, 1986b). If they flower, they fail to set seed (lack of pollinators). If they set seed, the seeds do not disperse (lack of seed dispersers). If seeds disperse, they do not develop as new members of their population (lack of adequate conditions for growth and development). If they develop, they do not thrive in a sustained population (the caprice of agroecosystem development eradicates conditions needed for population maintenance within a generation or two). These organisms are usually included in the lists of species in a region and are often used to demonstrate that a species is not threatened with extinction—even though it is (Janzen, 1972). Agroscapes, seemingly still supporting long lists of widespread species, are primed for massive extinction as individuals of these species senesce or are killed through intensification of contemporary agriculture. If all these species were to be physically removed as soon as they have no future, the catastrophe would be much more noticeable and would therefore arouse the sentiments normally associated with massive extinction.

The dry forest is more prone to these less visible catastrophes than is the rain forest, in which scraps of vegetation left when the forest is cleared die more quickly than do those of the dry forest. Thus the threatened plant species in dry forest are available for a longer time and can be used as basic stock in restoration projects (though the price paid is that they dilute the visual impact of a largely demolished forest).

A Central American tropical dry forest wildland that is large enough to be visited and used by humans is substantially larger than a wildland that is to exist without human intrusion. Tropical habitats are very rich in behaviorally sensitive species and species that exist in low-density populations. Moreover, dry forests contain many small habitats (e.g., springs, dry ridge tops, marshes, edaphic outcrops, temporary streams, and pickets of forest sheltered from the wind). Visitors (tourists researchers, seed collectors, and habitat managers) will perturb ecological interactions substantially more and effect them more permanently in dry forests than in most extratropical or rain forest habitats. This calls for strict zoning for habitat use and replicate habitats, both of which can be compatible with conservation management only if a large acreage is set aside.

FUTURE PROSPECTS

Ignore the voice that demands that a monetary value be placed on a wildland or a species before it can be conserved. Is that what you would do to determine the need for a public library, a public hospital, a public school? Can you tell me the dollar value of these institutions to your children? Are we to continue to be led by commercial interests to sanctify the production of material goods? The great majority of tropical humans live as draft animals; they are sold to the highest bidders along with the habitats that maintain them, and the purchasers are not generally benevolent. Through the swirl of changing market values, there will eventually come a day when the living organisms in a tropical wildland would be as doomed as would be libraries, if books were valued only for their paper pulp and the price of paper pulp were to rise.

Many organisms we believe to be safe are really endangered, and those we call endangered are in reality extinct. Guards will not save tropical wildlands. The world's dry tropics are already way beyond their capacity for accommodating human activity. Thus a contract between managers of wildlands and society is mandatory. And the scientific community must aggressively participate in writing and executing the contract. Without this participation, tropical biology will be nothing but low-grade and gradually diminishing restoration ecology. The conservation community has valiantly propped up the fortress walls, but they are too few. The future lies in the children, but we cannot wait for a well-educated cohort to replace its parents. The tropical dry forest is a living classroom, and its students are its neighbors. The collective power to turn the game around resides with policy makers. We cannot force the world to conserve tropical nature; we must seduce it, and the bait is intellectual mutualism—not the dollar value of a caterpillar.

ACKNOWLEDGMENTS

The people of Costa Rica have inspired me to believe there is still a chance. U.S. tax dollars through the National Science Foundation have financed the acquisition of the knowledge to see the chance, and the academic community has given the peer approval to know that this is the right direction. I thank P. Raven, W. Hallwachs, P. May, and A. Ugalde for help with the manuscript.

REFERENCES

Hartshorn, G. S. 1983. Plants. Pp. 118–157 in D. H. Janzen, ed. Costa Rican Natural History. University of Chicago Press, Chicago.

Janzen, D. H. 1966. Coevolution of mutualism between ants and acacias in Central America. Evolution 20:249–275.

Janzen, D. H. 1967. Synchronization of sexual reproduction of trees with the dry season in Central America. Evolution 21:620–637.

Janzen, D. H. 1972. The uncertain future of the tropics. Nat. Hist. 81:80–89.

Janzen, D. H. 1982a. Cenizero tree (Leguminosae: *Pithecellobium saman*) delayed fruit development in Costa Rican deciduous forests. Amer. J. Bot. 69:1269–1276.

Janzen, D. H. 1982b. Variation in average seed size and fruit seediness in a fruit crop of a guanacaste tree (Leguminosae: *Enterolobium cyclocarpum*). Amer. J. Bot. 69:1169–1178.

Janzen, D. H. 1986a. Guanacaste National Park: Tropical Ecological and Cultural Restoration. Editorial Universidad Estatal a Distancia, San Jose, Costa Rica. 103 pp.

Janzen, D. H. 1986b. The future of tropical ecology. Ann. Rev. Ecol. Syst. 17:305–324.

Janzen, D. H. In press a. Natural history of a wind-pollinated Central American dry forest legume tree (*Ateleia herbert-smithii* Pittier). In C. H. Stirton and J. L. Zarucchi, eds. Advances in Legume Biology. Mo. Bot. Gard. Monogr. Syst. Bot.

Janzen, D. H. In press b. Ecological characterization of a Costa Rican dry forest caterpillar fauna. Biotropica.

Janzen, D. H. In press c. Management of habitat fragments in a tropical dry forest: Growth. Ann. Mo. Bot. Gard.

Janzen, D. H., and R. Liesner. 1980. Annotated check-list of plants of lowland Guanacaste Province, Costa Rica, exclusive of grasses and non-vascular cryptogams. Brenesia 18:15–90.

Murphy, P. G., and A. E. Lugo. 1986. Ecology of tropical dry forest. Ann. Rev. Ecol. Syst. 17:67–88.

Savage, J. M., and J. Villa. 1986. Introduction to the Herpetofauna of Costa Rica. Society for the Study of Amphibians and Reptiles, Athens, Ohio. 207 pp.

Stiles, F. G. 1983. Checklist of birds. Pp. 530–544 in D. H. Janzen, ed. Costa Rican Natural History. University of Chicago Press, Chicago.

Wilson, D. E. 1983. Checklist of mammals. Pp. 443–447 in D. H. Janzen, ed. Costa Rican Natural History. University of Chicago Press, Chicago.

DEFORESTATION AND INDIANS IN
BRAZILIAN AMAZONIA

KENNETH I. TAYLOR

Executive Director, Survival International (USA), Washington, D.C.

Deforestation of tropical forests affects not only the plants and animals of these regions but also their human inhabitants. The Indian populations of Amazonia are successful managers of the forest. Long ago, they discovered the secrets of sustainable use of its resources. In this chapter I discuss the knowledge and management of the forest environment exhibited by the Yanomami and Kayapo Indians of Brazilian Amazonia and the importance that their knowledge and their presence as part of the forest ecosystem has for us all. Not only is this forest ecosystem now being destroyed at a rapid rate, but we (the non-Indians) do not yet know how to care for and make use of whatever areas of forest will be left when this process of destruction is brought to a halt.

THE YANOMAMI OF NORTHERN BRAZIL

The use and management of natural resources by the Yanomami include hunting, fishing, and collecting faunal resources, gathering and collecting floral resources, and shifting cultivation of bananas, plantains, manioc, several varieties of potatolike tubers, and a number of lesser crops. Their population is small and widely dispersed, resulting in an extremely low population density of 777 hectares per person. For the standard of living to which they are adapted, the forest provides them with an abundance of everything they need for a well-fed, healthy, and gratifying life. To date, there is no satisfactory evidence that they ever overused their resources or in any way degraded their environment. In fact, there are a number of indications that they vitalize and rejuvenate the forest, adding to its diversity and the size of its faunal and floral populations.

I have lived for more than 2 years among the Yanomami. The following account of Yanomami life in the forest is, for the most part, based on my own observations from that period (Taylor, 1974, 1983).

A Yanomami settlement is a clearing in the forest containing one or more of the several types of houses used by the different subgroups of the Yanomami. Directly associated with the site is a year-round source of water at a nearby stream or river. Radiating out from the settlement are numerous trails leading to the fields currently in use, to abandoned fields, to hunting, fishing, and gathering locations, to campsites in the forest, and to other settlements. The several fields actively cultivated by the families of the community are generally cut in primary forest, though occasionally in secondary forest, and usually no more than a 2-hour walk from the settlement. In more distant fields, a second family house is built for temporary stays of 1 to 2 weeks during the dry season. Several hours away from the settlement are a number of campsites used during dry-season fishing expeditions, long-term hunting trips, and journeys to other communities. The forest around the settlement is also criss-crossed by a number of minor trails used on hunting or gathering trips for food and raw materials of all kinds. These trails link together a series of regularly used locations, such as stands of fruit trees where game birds feed; streams where fish, crabs, or frogs can be found at certain times of the year; and places where different species of terrestrial game feed at times. And in all directions a number of major trails extend into the territories and lead to the settlement sites of other communities. These trails are more or less frequently and regularly used as friendships and alliances between communities come and go over the years.

This complex and ever-changing network of trails and the sites that they link together are not, of course, evenly spread out over a uniform and homogeneous circle of forest. In the Yanomami area, when you travel through the forest for more than even a few minutes, one of the most striking things you notice is its extraordinary diversity.

The use of the various hunting zones, and therefore the various biotopes around a Yanomami settlement, varies according to the type of hunting practiced: dawn/dusk, day, or festival hunting. In some parts of Yanomami territory there may be totally unused areas that for years at a time function as game preserves (Gross, 1975; Harner, 1972; Hickerson, 1970).

A Yanomami community certainly needs access to a relatively large, ecologically heterogeneous territory that is contiguous with the territories of a number of neighboring communities, but we may wonder if so much land as 777 hectares per person is really needed. To the best of my knowledge, however, land use in Amazonia with non-Indian techniques, which involve clearing large areas of their protective forest cover for introduced, and inappropriate, crops and livestock, is leading to an ultimate degradation of the environment and is not self-sustaining on a permanent basis. The apparent exceptions of the riverine *caboclos* (forest-dwellers) (Frechione et al., 1985) and the rubber tappers of, for example, the State of Acre (Allegretti, 1985) are, in fact, land uses by settlers of long standing who have learned from the Indians a number of the basic requirements of a self-sustaining life-style in Amazonia. Of primary importance among these, and first to be ignored by government plans for colonization of the region, is the essential feature of low—extremely low—population density.

THE KAYAPO OF CENTRAL BRAZIL

The Kayapo Indians of central Brazil live far to the south of the Yanomami in the watershed of the Xingu River, which is one of the major right-bank tributaries of the Amazon. Their territory is near the southern limit of the tropical forests of Amazonia and includes terra firme and gallery forests interspersed with areas of more or less open cerrado (similar to savannah). Their knowledge, management, and use of the floral and faunal resources of the forests in their territory are astonishingly subtle and complex. It is unlikely that the Kayapo are unique—they are simply, and by far, the best studied of the many Indian groups of Amazonia, with regard to this aspect of their way of life.

Like almost all the Indian groups in Amazonia, the Kayapo hunt, fish, and gather a great many species of the fauna and flora of the forests and practice shifting cultivation. They also concentrate native plants by growing them in resource islands, forest fields, forest openings, tuber gardens, agricultural plots, and old fields, and beside their trails through the forest. They select and transplant a number of semidomesticated native plants and manipulate some species of animals (birds, fish, bees, and mammals) used as food or food sources. The nests of two species of bees, for example, are brought from the forest and mounted on housetops until the honey is ready to be harvested. Forest patches (*apete*) are created from open cerrado in areas prepared with crumbled termite and ant nests and mulch (Posey, 1983, 1985; Posey et al., 1984).

The Kayapo Indians are probably not unique. More likely they are typical of indigenous societies in tropical forests. They not only live a healthy and well-fed life as the human component of a thriving tropical forest ecosystem but they also beneficially manage, manipulate, and modify the flora and fauna of their territory. As a result of their presence and remarkable way of life, the plant and animal resources of their area are more diverse, more locally concentrated, of greater population size and density, and more youthful and vigorous than would be found in a forest empty of these Indian resource managers.

Perhaps the most surprising and significant of their many resource management techniques is the creation of the *apete* forest patches. Posey became aware that these isolated patches of forest were man-made only in the seventh year of his research among the Kayapo (D. A. Posey, Museu Emilio Goeldi, Belém, Brazil, personal communication, 1986). As he pointed out, "Perhaps the most exciting aspect of these new data is the implication for reforestation. The Indian example not only provides new ideas about how to build forests 'from scratch,' but also how to successfully manage what has been considered to be infertile campo/cerrado" (Posey, 1985, p. 144).

THE RIGHTS AND WRONGS OF
SHIFTING CULTIVATION

Shifting cultivation involves the felling or cutting of the vegetation in an area selected for a field or garden and the burning of the felled trees, bushes, underbrush, or grasses. It is a widely used technique that has been around for a long, long time. Conklin (1961) gives a definitive overview of the long history (since the Neolithic period) and distribution (worldwide, especially in the tropics) of this form of agriculture and discusses the various forms it can take, in terms of whether primary or secondary forest or grasslands are being used, crop-fallow time ratios, types of crop, dispersal relative to human settlements, concomitant presence of livestock, and tools and techniques used. It is unlikely that a form of agriculture so time-honored and widespread would be inefficient or destructive of the environment; yet many people regard it as just that—as one way in which the remaining tropical forests are being destroyed.

Between 1968 and 1976, I had the opportunity to fly in light aircraft and helicopter over most of Yanomami territory in the Ajarani, Catrimani, Mucajai, Parima, and Auaris river basins in the territory of Roraima, Brazil. The vitality, the exuberance, and the seeming endlessness of the dense carpet of forest cover made a lasting impression on me. Yet this is where most of the Yanomami lead their lives. Whatever else the Yanomami may or may not be doing, they are most certainly not destroying the forest. As discussed above, Posey and collaborators describe the Kayapo as Indians who are greatly enhancing the vitality of the forests of their region.

But aren't the Yanomami and the Kayapo what some people call slash-and-burn agriculturalists? And isn't it by cutting and burning that all those thousands of acres of tropical forest are being destroyed in Amazonia and around the world? The answer to both these questions is yes, but obviously there is a difference. The difference, of course, is one of scale.

The Yanomami and the Kayapo Indians live in the forest and are part of the forest. If they destroy it, they destroy themselves. They therefore make their modest-size fields and plant crops sufficient only for their needs. It is the non-Indian agriculturalists (or investors) who order the destruction of a forest they may never have seen in order to install quite inappropriate plantations or cattle ranches. This, of course, they must do on as large a scale as they can afford to ensure that their profit margin is to their liking. The enormous clearings that result are far beyond the ability of even the most healthy forest to regenerate. One example of such extensive destruction is the notorious case of the Volkswagen ranch whose burning became public knowledge only when seen from space by the Skylab satellite (Bourne, 1978).

In contrast, the forest itself begins reclaiming the relatively tiny Indian fields cut in its midst by supporting the growth of pioneer species (weeds, some would call them) even before the Indians have taken the two or three harvests they find practical before returning the field to the forest and its regenerative process. In many cases, in fact, the Indians stop using a field not so much because the productiveness of the tropical soils decline so rapidly but because clearing the field of weeds is just more trouble than it is worth.

Done the right way, the Indian way, shifting cultivation rejuvenates the forest. It is the use of the technique on too large a scale by the non-Indian that is destructive.

INDIAN PRODUCTIVITY IN THE TROPICAL FOREST

The Indians of Amazonia have what we would consider an extremely low standard of living. Living in relative isolation from the national societies of the countries within whose boundaries they live, their economic production, whether from their agricultural practices or their use of game, fish, and natural forest resources, is strictly for their own subsistence. As a result, they are commonly stereotyped as poor and lazy with no potential as producers of anything for the regional, national, or international markets. Quite the opposite is true. A now-famous example is the successful production and marketing of highly marketable Brazil nuts by the Gavioes of Para State, Brazil—an activity they began on their own initiative in the mid-1970s (Ferraz, 1982; Ramos, 1980). Almost overnight, the Gavioes became not only quite well-to-do by local (non-Indian) standards but also transformed themselves, in the eyes of their non-Indian neighbors, from lazy good-for-nothings to productive members of society. In 1975 I knew of one Yanomami community that after exhausting the supply of bananas in its own fields, began a new, additional plantation so that it could continue to sell bananas to the tin miners who worked in Yanomami territory for a time in 1975 and 1976. As long ago as 1930, Curt Nimuendaju (1974) spoke of how the Ramkokamekra Canela Indians could have produced a marketable surplus of manioc flour but explained that they never did develop this potential because they had no way to transport the flour to market. Another example is the production and marketing of natural rubber in a recent community development project undertaken by one subgroup of the Nambiquara.

These are only a few examples. There has been considerable discussion of the possibly marketable products that can be grown in a properly and sustainably managed tropical forest (see, for example, Goodland, 1980). It is not yet known quite how productive the tropical forest can be or how large (or small) a population of resident producers it can support. The point here is simply that the Indians who live in the forest and know its ecology so well have long ago demonstrated their ability to function as valuable and effective producers of its marketable resources.

THE IMPACT OF DEFORESTATION ON INDIAN LIFE

The destruction of the tropical forests has both a direct and an indirect impact on the resources and livelihood of Indian populations. In some cases, deforestation is occurring inside recognized Indian areas. Even when the deforestation occurs outside these areas, however, the impact on the animal and plant resources, the water supply, and the rivers, which serve as avenues of transportation, in and near Indian areas can be devastating.

Deforestation takes place along with traditional frontier expansion and with the implementation of large-scale development projects. In either case, there is an influx of outsiders into regions previously but sparsely inhabited. Among the inhabitants of these regions there are often relatively isolated Indian populations. This isolation is broken overnight as land-hungry settlers begin invading Indian lands. In Brazil, colonists are aware of this well-known technique even before they move to Amazonia. If a settler can establish an illegal smallholding on Indian lands (among the least protected by the authorities of all land categories in Brazil's interior) then, either as a matter of squatter's rights or in compensation for being evicted, he will have improved his chances of acquiring a plot of land through the government colonization program.

In all such cases, the Indians' control of their own natural resources is eroded and the supply of these resources declines. One of the first results of these processes is the impoverishment, if not the dispossession, of the Indian populations, leading to their migration to towns and cities where the best they can hope for is to swell the ranks of the urban under- and unemployed and a life of disease, prostitution, homelessness, and begging.

INDIAN MANAGERS OF THE RAIN FORESTS

The indigenous inhabitants of the tropical rain forests are valuable participants in these ecosystems. Their relationship and interaction with their forest environment not only affords them a sustainable and nondestructive livelihood but also enhances the vigor, diversity, and population size of the forest's flora and fauna. Deforestation, both within and outside Indian lands, can so devastate the natural resources on which the Indian way of life depends that it becomes impossible for the Indian population to remain in the area and lead any semblance of its traditional way of life. But given the opportunity, Indian groups can rapidly adapt to a productive relationship with national and international society. They can produce a wide range of natural and cultivated products for the marketplace while still pursuing their way of life in and as part of the forest.

When we speak of the preservation of the tropical forests we must make clear, explicitly and emphatically, that we mean the preservation of the forests' flora and their fauna *and* their indigenous human inhabitants. These indigenous peoples are our representatives of choice who can bring the forests, little by little, to their full productive potential and keep them healthy and well and with their magnificent biological diversity intact for the benefit of us all.

REFERENCES

Alegretti, M. H. 1985. The Rubber Tappers and Environmental Issues in the Amazon Region. Message sent to the UN World Commission on Environment and Development. Instituto de Estudos Socio-Economicos (INESC), Brasilia.

Bourne, R. 1978. Assault on the Amazon. Victor Gollancz, London. 320 pp.

Conklin, H. C. 1961. The study of shifting cultivation. Curr. Anthropol. 2(1):27–61.

Ferraz, I. 1982. Os Indios Gavioes: Observacoes Sobre uma Situacao Critica. Unpublished manuscript. Rio de Janeiro. 31 pp.

Frechione, J., D. A. Posey, and L. Francelino da Silva. 1985. The Perception of Ecological Zones and Natural Resources in the Brazilian Amazon: An Ethnoecology of Lake Coari. Paper presented at the annual meeting of the American Anthropological Association, Washington, D.C.

Goodland, R. J. A. 1980. Environmental Ranking of Amazonian Development. Pp. 1–20 in F. Barbira-Scazzocchio, ed. Land, People and Planning in Contemporary Amazonia. Centre of Latin American Studies, Occasional Publication No. 3. Cambridge University, Cambridge.

Gross, D. R. 1975. Protein capture and cultural development in the Amazon Basin. Am. Anthropol. 77(3):526–549.

Harner, M. J. 1972. The Jivaro: People of the Sacred Waterfalls. Doubleday/Natural History Press, Garden City, N.Y. 233 pp.

Hickerson, H. 1970. The Chippewa and Their Neighbors: A Study in Ethnohistory. Holt, Rinehart and Winston, New York. 133 pp.

Nimuendaju, C. 1974. Farming among the Eastern Timbira. Pp. 111–119 in P. J. Lyons, ed. Native South Americans. Little, Brown, Boston.

Posey, D. A. 1983. Indigenous knowledge and development: An ideological bridge to the future. Cienc. Cult. 35(7):877–894 .

Posey, D. A. 1985. Indigenous management of tropical forest ecosystems: The case of the Kayapo Indians of the Brazilian Amazon. Agroforestry Sys. 3:139–158.

Posey, D. A., J. Frechione, J. Eddins, and L. F. da Silva. 1984. Ethnoecology as applied anthropology in Amazonian development. Hum. Org. 43(2):95–107.

Ramos, A. R. 1980. Development, integration and the ethnic integrity of Brazilian Indians. Pp. 222–229 in F. Barbira-Scazzocchio, ed. Land, People and Planning in Contemporary Amazonia. Centre of Latin American Studies, Occasional Publication No. 3. Cambridge University, Cambridge.

Taylor, K. I. 1974. Sanuma Fauna, Prohibitions and Classifications. Monograph No. 18. Fundacion La Salle de Ciencias Naturales, Instituto Caribe de Antropologia y Sociologia, Caracas, Venezuela. 138 pp.

Taylor, K. I. 1983. Las necesidades de tierra de los Yanomami. (Abstract in English) America Indigena XLIII(3):629–654.

PRIMATE DIVERSITY AND THE TROPICAL FOREST
Case Studies from Brazil and Madagascar
and the Importance of the Megadiversity Countries

RUSSELL A. MITTERMEIER

Vice-President for Science,
World Wildlife Fund/The Conservation Foundation, Washington, D.C.

Much of the early interest in wildlife conservation grew out of a desire to save some of the world's most spectacular mammals, and to some extent, these so-called charismatic megavertebrates are still the best vehicles for conveying the entire issue of conservation to the public. They are really our flagship species, both here in the United States and in the developing countries, and primates in particular are perhaps the best flagships for tropical forest conservation. Nonhuman primates are of particular interest in this context for three basic reasons: they are of great importance to our own species; they are largely a tropical order, roughly 90% of all primate species being restricted to the tropical forest regions of Asia, Africa, and the Neotropics; and they are members of the elite group called the charismatic megavertebrates.

The threats to primates and their tropical forest habitats can be seen by examining two tropical forest regions: Brazil, particularly the Atlantic forest region of eastern Brazil, and the island of Madagascar. These are clearly two of the most important countries for primate conservation, and they are among the world's richest countries for living organisms in general—countries that I call the *megadiversity countries* and that are critical to the survival of the majority of the world's biological diversity.

Most people are aware of the importance of the Order Primates, which of course includes our own species, *Homo sapiens*. However, few realize how diverse the Order of Primates actually is, including as it does some 200 species that range from the tiny mouse lemur (*Microcebus murinus*) of Madagascar and the tarsiers (*Tarsius* spp.) of Southeast Asia to the great apes, which include our closest living relatives, the chimpanzee (*Pan troglodytes*) and the pygmy chimpanzee (*Pan paniscus*). Our nonhuman primate relatives are valuable to us in many ways, and the rapid growth of the science of primatology over the past 25 years has reflected this. Studies of these animals have taught us a great deal about the intricacies of our own behavior, they have clarified questions about our evolution and our origins, and they have played a significant role in biomedical research. Furthermore, the importance of primates as key elements of the tropical forest (e.g., seed dispersers) is only starting to be understood.

Unfortunately, wild populations of most nonhuman primates are decreasing all over the world. Many spectacular species like the mountain gorilla (*Gorilla gorilla beringei*) from Rwanda, Uganda, and Zaire, the golden lion tamarin (*Leontopithecus rosalia*) and the muriqui (*Brachyteles arachnoides*) from Brazil, and the indri (*Indri indri*) and the aye-aye (*Daubentonia madagascariensis*) from Madagascar are already endangered, and many others are headed in the same direction.

Without a doubt, the major cause of the decline of primate populations is destruction of their tropical forest habitat, which is occurring at a rate of some 10 to 20 million hectares per year (OTA, 1984), the latter figure being equivalent to a loss of an area the size of California every 2 years.

Another very important factor in the decline of these populations is hunting of primates, mainly as a source of food, but these animals are also hunted for their supposed medicinal value, for the ornamental value of their skins and other body parts, and for their use as bait for other animals, or to eliminate them from agricultural areas where they have become crop raiders. The effects of hunting vary greatly from region to region and from species to species, but hunting of primates as food is known to be a very serious threat in at least three parts of the world— the Amazonian region of South America, West Africa, and Central Africa. Many thousands of primates are killed every year in these regions for culinary purposes, and such overhunting has already resulted in the elimination of certain species from large areas of otherwise suitable forest habitat (e.g., the elimination of woolly monkeys and spider monkeys in Amazonia) (Mittermeier, 1987; Mittermeier et al., 1986).

Live trapping of primates, either for export or for local use, plays an important role as well. Live primates are used in biomedical research and testing, or they may be sold as pets or exhibits, both internationally and within their countries of origin. For the most part, this is a less important factor than habitat destruction or hunting, but for certain endangered and vulnerable species that happen to be in heavy demand, it can be quite serious. Species that have been hurt by the trade in live primates include the chimpanzee and the cotton-top tamarin (*Saguinus oedipus*), both of which were important biomedical research models, and the woolly monkeys (*Lagothrix* spp.), which were and still are very popular as pets for local people in Amazonia.

All these factors have combined to bring about a worldwide decline in primate populations. According to the International Union for Conservation of Nature (IUCN), one out of every three of the world's 200 primate species is already in some danger and one in seven is highly endangered and could be extinct by the turn of the century or even sooner if something isn't done quickly. These are minimum estimates. Very often when specialists go into the field to investigate the status of poorly known species, they find it necessary to add to the endangered list.

To prevent the extinction of the world's nonhuman primates, the Primate Specialist Group of IUCN's Species Survival Commission put together a Global Strategy for Primate Conservation in 1977. This document (Mittermeier, 1977) was the first effort to take a worldwide view of primate conservation problems, and its purpose was to make the Primate Specialist Group's goal of maintaining the current diversity of the Order of Primates a reality. It placed dual emphasis on ensuring the survival of endangered species wherever they occur and on providing effective protection for large numbers of primates in areas of high primate diversity or abundance. This original Global Strategy, which is now out of date, is being updated by a series of new regional plans for Africa (Oates, 1986), Asia (Eudey, 1987), Madagascar, and the Neotropical region, which will guide primate conservation activities for the remainder of this decade.

To find the financial support for the activities identified in the original Global Strategy, the World Wildlife Fund established a special Primate Program in 1979. Since that time, the program has funded and helped implement more than 150 projects, large and small, in 31 different countries, and it continues to grow. In addition, the program produces a wide variety of educational materials and publishes *Primate Conservation*, the newsletter and journal of the IUCN/SSC Primate Specialist Group, which is the major means of communication among the world's primate conservationists.

A number of other organizations have also helped to support projects identified by the Primate Specialist Group, among them the New York Zoological Society, the Wildlife Preservation Trust International, the African Wildlife Foundation, the Fauna and Flora Preservation Society, the Brookfield Zoo, and the Frankfurt Zoological Society, to name just a few. Although the combined efforts of the World Wildlife Fund Primate Program and the other organizations have achieved a great deal on behalf of primates over the past decade, it is clear that much more will have to be done over the next few years to ensure that all of the world's 200 primate species are still with us as we enter the next century.

Two tropical countries, Brazil and Madagascar, are particularly important in efforts to conserve primate diversity, since they alone are home to 40% of the world's living primate species. Brazil, with 357 million hectares of tropical forest, is by far the richest country in the world for this biome, containing more than three times more forest than the next country on the list, which is Indonesia, and 30% of all the tropical forest on our planet (Table 16-1). Not surprisingly, Brazil is also home to far more primates than any other country; its 53 species account for about 27%, or one in every four, primates in the world (Table 16-2).

Although one usually hears much more about Amazonia, the highest priority area within Brazil is its Atlantic forest region, which is the most developed and most devastated part of the country. The Atlantic forest is a unique series of ecosystems quite distinct from the much more extensive Amazonian forests to the northwest. At one time, it stretched pretty much continuously from the state of Rio Grande do Norte at the easternmost tip of South America out as far as Rio Grande do Sul, the southernmost state in Brazil, and it included some of the

TABLE 16-1 Countries of the World Containing
the Largest Areas of Closed Tropical Forest[a]

Country	Areas of Closed Forest (hectares)
Brazil	357,480,000
Indonesia	113,895,000
Zaire	105,750,000
Peru	69,680,000
India	51,841,000
Colombia	46,400,000
Mexico	46,250,000
Bolivia	44,010,000
Papua New Guinea	34,230,000
Burma	31,941,000
Venezuela	31,870,000
Congo	21,340,000
Malaysia	20,995,000
Gabon	20,500,000
Guyana	18,475,000
Cameroon	17,920,000
Suriname	14,830,000
Ecuador	14,250,000
Madagascar	10,300,000

[a]From OTA, 1984, and Mittermeier and Oates, 1985.

richest, tallest, and most beautiful forest on Earth. In its primeval state, the Atlantic forest complex covered over 1 million square kilometers in 14 states or about 12% of Brazil, and its length from north to south extended a greater distance than the entire Atlantic seaboard of the United States from northern Maine to the Florida Keys. However, this region was the first part of Brazil to be colonized, it has developed into the agricultural and industrial center of the country, and it has within its borders two of the three largest cities in all of South America—Rio de Janeiro and São Paulo, which is now one of the largest cities on Earth.

The result has been large-scale forest destruction, especially in the last two decades of rapid economic development, to obtain lumber and charcoal and to make way for plantations, cattle pasture, and industry—to the point that only 1 to 5% of the original forest remains in this region. As might be expected, the animals and plants native to the Atlantic forest are not doing very well under such circumstances. Many of these are endemic (including 40% of all the small, non-volant, i.e., nonflying, mammals, 54% of the trees, and 64% of the palms), and increasing numbers are being added to the endangered species list. The best example is probably the effect on the primates, 80% of which are endemic to the Atlantic forest. Twenty-one species and subspecies of monkeys are found in this region, and the studies that have been carried out with World Wildlife Fund support since 1979 indicate that fully 14 of these are endangered and that several are literally on the verge of extinction. Of these 14 endangered species, 13 are found nowhere else in the world (Mittermeier et al., 1986).

Two of the Atlantic forest primate species stand out among the rest: the muriqui (*Brachyteles arachnoides*), which is the largest and most apelike of the South American monkeys, and the golden lion tamarin (*Leontopithecus rosalia*), which is surely one of the most beautiful of all mammals. These animals are representatives of the two highly endangered genera that are endemic to the Atlantic forest, and they have been subjects of two major public awareness campaigns that have been under way for the past 5 years (Dietz, 1985; Mittermeier et al., 1985). They have really become the flagship species for the entire region, and the campaigns using them as symbols are excellent examples of the way in which key groups of animals can be used to sell the whole issue of conservation, both in the tropical countries and in the developed world.

The campaigns for the muriqui and the golden lion tamarin have been multifaceted, including ecological research, survey work, development of museum exhibits, production of films, and distribution of a wide variety of educational and promotional materials, including posters, stickers, T-shirts, and various publications. The result is that these two species, which were virtually unknown to the general public in Brazil 5 years ago, are now so popular that they appear on the cover of phone books, on postage stamps, as themes of parades and theater presentations, and as subjects of numerous magazine and newspaper articles. All this, and of course a broad spectrum of some 50 other conservation projects being supported by the World Wildlife Fund in this region, has led to a general increase in conservation awareness, which we hope will be instrumental in helping to save what remains of the Atlantic forest and its spectacular fauna and flora.

The situation in Madagascar is even more critical than in the Atlantic forest region of eastern Brazil. Madagascar is a unique evolutionary experiment and a living laboratory that is unlike anyplace else on Earth. The island has been separated

TABLE 16-2   Countries of the World Containing the Greatest Primate Diversity[a]

Country	No. of Species	No. of Genera
Brazil	52	16
Indonesia	33–35	9
Zaire	29–32	13–15
Madagascar	28	13
Cameroon	28–29	14
Peru	27	12
Colombia	27	12
Nigeria	23	13
Congo	22	14
Equatorial Guinea	21–22	12
Central African Republic	19–20	11–12
Gabon	19	11
Uganda	19	11
Bolivia	17–18	11–12
Angola	18–19	10–11

[a]Modified from Mittermeier and Oates, 1985.

TABLE 16-3   Primate Endemism in the 15 Countries With the Greatest Primate Diversity

Country	Endemic Species (%)	Endemic Genera (%)
Madagascar	93	92
Indonesia	44–50	12.5
Brazil	35	12.5
Colombia	11	0
Peru	7	0
Zaire	6–7	0
Nigeria	4	0
Cameroon	0	0
Congo	0	0
Equatorial Guinea	0	0
Central African Republic	0	0
Gabon	0	0
Uganda	0	0
Bolivia	0	0
Angola	0	0

from the African mainland for perhaps as long as 200 million years, if in fact it was ever connected, and most of the plant and animal species found there have evolved in isolation and are unique to the island.

The most striking and conspicuous animals on Madagascar are the primates, which consist entirely of lemurs. Among these lemurs are some of the most unusual primates on Earth, ranging from the mouse lemur, which is the smallest living primate, to the indri, which is the largest living prosimian, and the aye-aye, which is the strangest of all primates and the only representative of an entire primate family, the Daubentoniidae.

This lemur radiation on Madagascar is one of the most diverse primate faunas anywhere, its 29 species placing it fourth on the world list of primate diversity behind Brazil, Zaire, and Indonesia (even though it is only 7% the size of Brazil, Table 16-2). When endemism is considered, Madagascar's primate fauna seems even more impressive, since 93% of all its species are restricted to that country—a figure not even approached by any other country (Table 16-3). Furthermore, the two lemur species found outside Madagascar reside only on the nearby Comoros Islands and are probably recent introductions by humans.

The situation is much the same for most other groups of organisms in Madagascar. Seven of the eight species of carnivores found there are endemic, as are 29 of the 30 tenrecs, 106 of the 250 birds, 233 of the 245 reptiles, 142 of the 144 frogs, 110 of the 112 species of palms, and 80% of its nearly 8,000 angiosperm plants. It is not just endemism that is impressive on Madagascar, however, but total diversity as well. Although Madagascar is only about 40% again as large as the state of California and accounts for less than 2% of the African region, its 8,000 angiosperm plants represent 25% of all angiosperms in Africa (P. Lowry, personal communication, 1987), it has more orchids than the entire African mainland, and its 13 living primate genera approach the 14 to 15 mainland genera in total diversity.

Unfortunately, most of Madagascar's spectacular fauna and flora is endangered, mainly, once again, because of forest destruction. Although human beings arrived on Madagascar only some 1,500 to 2,000 years ago, human activity has resulted in the loss of some 80% of Madagascar's forests, and the major remaining forest formations are being chipped away for firewood and charcoal and for slash-and-burn agriculture. Hunting is a problem as well, especially with the breakdown of local cultures, which formerly included many taboos against the hunting of primates and other wildlife.

Lest anyone believe that extinctions are a figment of the conservationist's imagination, he or she need only look at what has already been lost on Madagascar over the past 2,000 years. Among the species that have disappeared are the elephant birds (*Aepyornis* spp.), which were the largest birds that ever lived, a pygmy hippopotamus, an aardvark, and fully six genera of lemurs, representing one-third of all known Malagasy lemur species. Included among the species lost are animals like *Megaladapis* (Figure 16-1), which moved like a huge koala and grew to be as large as a female gorilla (Sussman et al., 1985).

Almost all the species that have already disappeared were diurnal and larger than the surviving species. If this trend continues, the next in line would be the indri, which is the largest, and the sifakas (*Propithecus* spp.), which are next in size. In fact, several of these are already endangered. One, the black sifaka (*Propithecus diadema perrieri*) from northeastern Madagascar, is now down to only about 100 individuals and must be considered on the verge of extinction.

FIGURE 16-1 Above: The extinct giant lemur *Megaladapis* from Madagascar, as reconstructed by Stephen D. Nash. Left: An extant ring-tailed lemur in Madagascar.

At present, about 40% of Malagasy lemurs are considered endangered and many more are likely to enter the endangered category as we learn more about them. And what is happening to lemurs is happening to the rest of Madagascar's fauna and flora as well.

Despite the many problems, there is cause for optimism in Madagascar. In November 1985, a special National Conservation Strategy Conference held there attracted representatives from many international organizations, including IUCN, the World Wildlife Fund, the United Nations Environment Program, the Food and Agriculture Organization, the World Bank, and a number of bilateral aid organizations, including the U.S. Agency for International Development. This conference generated a great deal of enthusiasm for conservation among the Malagasy themselves and should serve as an important take-off point for future conservation activities. Several projects supported by the World Wildlife Fund are also serving as models for community involvement in conservation, and are attracting international attention to the need for conservation in this all-important country. Of particular importance in this respect is the Beza-Mahafaly project in southwestern Madagascar, which is being conducted by researchers from the University of Madagascar, Yale University, Washington University, and the Missouri Botanical Garden (Sussman et al., 1985).

To be sure, a great deal still needs to be done in Madagascar to ensure that the country's amazing biological diversity is maintained for future generations. Nevertheless, the time appears to be ripe to accomplish something of major proportions there and in effect to change the course of conservation history in this unique country.

As indicated in Table 16-2, there is a very disproportionate distribution of primate diversity in the world. Just four countries, Brazil, Madagascar, Zaire, and Indonesia, by themselves account for approximately 75% of all the world's primate species. If we are going to maintain global primate diversity, we must pay special attention to these countries over the next few decades, not to the exclusion of others but certainly more than we have in the past.

Needless to say, these *megadiversity* countries are not just important for primates. Although we are still in the process of compiling data, it appears that approximately 50 to 80% of the world's total biological diversity will be found in some 6 to 12 tropical countries. The first 6 of these to have emerged from the preliminary analysis are Brazil, Colombia, Mexico, Zaire, Madagascar, and Indonesia (see Figure 16-2). Not only do these countries have a major portion of the world's biological diversity, they have an even higher percentage of the world's diversity at risk—the very diversity that is in danger of disappearing over the next decade and that is of so much concern to conservation biologists. All these countries are undergoing rapid environmental change, are facing severe economic problems, and in general, lack the resources to develop the broad-based conservation programs needed to conserve biological diversity on their own. This means that people of the developed world are going to have to work in much closer collaboration with colleagues in these countries in the years to come and that the developed countries will have to provide far more resources for conservation than ever before.

FIGURE 16-2 Megadiversity countries identified by the World Wildlife Fund.

I do not believe in a gloom-and-doom approach to conservation, which can be quite detrimental to our efforts. On a more upbeat note, I believe that much of our planet's biological diversity can be maintained and that conservation in general has to be considered the art of the possible. The example of Brazil, which may be the single most diverse country in the world, is most encouraging. One hears a great deal about destruction and the many environmental problems faced by Brazil but very little about the successes. Nonetheless, the successes are there, and for those of us who have been working in Brazil for two decades, the advances in conservation in that country seem little short of phenomenal. They lead me to believe that a very large proportion of Brazil's biological diversity can be maintained. With the proper input of resources from both the developed world and the developing countries themselves, there is no reason why these successes cannot be repeated on a global basis.

REFERENCES

Dietz, L. A. 1985. Captive-born lion tamarins released into the wild: A report from the field. Primate Conserv. 6:21–27.

Eudey, A. A. 1987. Action Plan for Asian Primate Conservation. International Union for the Conservation of Nature and Natural Resources/Species Survival Commission Primate Specialist Group, World Wildlife Fund, and United Nations Environment Programme, Washington, D.C. 70 pp.

Mittermeier, R. A. 1977. A Global Strategy for Primate Conservation. International Union for the Conservation of Nature and Natural Resources/Species Survival Commission Primate Specialist Group, Cambridge, Mass. 325 pp.

Mittermeier, R. A. 1987. The effects of hunting on rain forest primates. Pp. 109–146 in C. Marsh and R. A. Mittermeier, eds. Primate Conservation in the Tropical Rain Forest. Alan R. Liss, New York.

Mittermeier, R. A., and J. F. Oates. 1985. Primate diversity: The world's top countries. Primate Conserv. 5:41–48.

Mittermeier, R. A., C. Valle, I. B. Santos, C. Alves, C. A. Machado Pinto, and A. F. Coimbra-Filho. 1985. Update on the muriqui. Primate Conserv. 5:28–30.

Mittermeier, R. A., J. F. Oates, A. A. Eudey, and J. Thornback. 1986. Primate conservation. Pp. 3–72 in G. Mitchell and J. Erwin, eds. Comparative Primate Biology, Vol. 2A, Behavior, Conservation and Ecology. Alan R. Liss, New York.

Oates, J. F. 1986. Action Plan for African Primate Conservation. International Union for the Conservation of Nature and Natural Resources/Species Survival Commission Primate Specialist Group, World Wildlife Fund and United Nations Environment Programme, Washington, D.C. 41 pp.

OTA (Office of Technology Assessment). 1984. Technologies to Sustain Tropical Forest Resources. Congress of the United States, Office of Technology Assessment, Washington, D.C. 344 pp.

Sussman, R. W., A. F. Richard, and G. Ravelojaona. 1985. Madagascar: Current projects and problems in conservation. Primate Conserv. 5:53–59.

PART

4

DIVERSITY AT RISK:

THE GLOBAL PERSPECTIVE

Aerial view of a coral reef in the Capricorn Group at the southern end of the Great Barrier Reef, Australia. *Photo courtesy of G. Carleton Ray.*

17

LESSONS FROM MEDITERRANEAN-CLIMATE REGIONS

HAROLD A. MOONEY

Professor of Biological Sciences, Department of Biological Sciences,
Stanford University, Stanford, California

Discussions on the loss of biological diversity are correctly focused on tropical regions because of the massive, rather recent alterations in the structure of these extensive biotic communities. The consequences of these alterations are many. There are of course no landscapes on Earth that have not been modified to some extent by the human species. Many of these landscapes have been totally altered from their prehuman configuration and functioning, and others appear less affected; however, none are protected from the types of global changes that are resulting from human-induced alterations of the Earth's atmosphere.

This section focuses on the nature and some of the consequences of alterations of nontropical biogeographic regions. The discussions are selective, concentrating on selected processes and organisms within a few systems. In Chapter 18, Franklin deals with temperate and boreal forests, which occupy 16% of Earth's land surface—an area equivalent to that covered by tropical forests (Waring and Schlesinger, 1985)—and which have provided to a large degree the timber and in part the fuel to support the growing human population. In the next chapter, Risser discusses the impact of humans on biological diversity in grasslands, the biome that has largely provided, either directly or indirectly, the food for the world's human population. Finally, in Chapter 20, Vitousek details the kinds of biotic changes that have resulted from human settlement on Hawaii and on oceanic islands in general—systems that have proven to be particularly susceptible to losses and additions of species.

Each chapter emphasizes somewhat different points. Franklin focuses on the consequences of structural diversity loss in forest ecosystems, drawing examples from the magnificent coniferous forests of the Pacific Northwest. Risser notes the low loss of species in the high-impact North American grasslands and the potentially high loss of ecotypes. He also discusses the variable consequences of different land-use patterns on species diversity. Vitousek relates the apparent devastating effects of species invaders on the endemics of the Hawaiian Islands, noting that although species diversity has actually increased, ecosystem types have been lost.

As an introduction to these chapters on threats to diversity in nontropical systems, I first compare the community diversity of tropical systems with those of temperate regions, providing plants as examples. I then focus more specifically on Mediterranean-climate (cool wet winter, dry summer climate) regions to balance the presentations on forests and grasslands. Mediterranean-climate regions, of which there are five in the world, are of special interest for two reasons: they rival tropical regions for their biological richness, and because they have had very different histories of human settlement, they serve as interesting comparison areas in studies to determine the human impact on biotic diversity.

COMMUNITY DIVERSITY IN TROPICAL AND TEMPERATE REGIONS

The fact that tropical regions are biologically richer than temperate regions has been stated repeatedly: for example, Raven (1976) has noted that 65% of the world's 250,000 flowering plants are found there. Until recently, the tropics, particularly the lowland wet tropics, have remained one of the last areas that has not been subjected to extensive human exploitation. In temperate regions of the world, many of the natural ecosystems have been massively altered by human settlement and activities. By looking at some of these disturbed regions, we can assess the consequences of human activities on biological diversity and, to some extent, learn what we should expect in the tropics in the future. If we were to pick only one biome type to serve as a model of comparison, it should be the Mediterranean-climate regions of the world. These regions are remarkably diverse by any measure.

Gentry (1979) reported that the number of plant species he encountered in 0.1-hectare plots increased as he moved from dry tropical to wet tropical forests (Table 17-1). In his most diverse sites in Panama he encountered more than 150 species of woody plants thicker than 1 inch in diameter at breast height. In contrast, only 21 woody species were found in a temperate forest in Missouri. Data on total species counts in tropical forests have not been available. However, Whitmore (1986) reported the results of a survey in which 236 species of vascular plants were counted in a 0.01-hectare plot in Costa Rica; he estimated that "one man decade would be required to enumerate one hectare" (Whitmore, 1986). Counts of all the vascular plants in sample plots in other climatic regions are available for comparison.

In the Mediterranean-climate region of Israel, Naveh and Whittaker (1979) found sites that included as many total species as woody species found by Gentry in Panama. The richest sites were those with some degree of current disturbance. Mediterranean-climate sites of the same size in other parts of the world also have relatively high species counts in comparison to counts of temperate-zone vegetation (Whittaker, 1977).

The bases for the high diversities among the different Mediterranean-type vegetations differ. In Israel, the diversity is accounted for mostly by herbaceous species, principally annuals, and is the result of human-driven "relatively rapid evolution under stress by drought, fire, grazing and cutting" (Naveh and Whittaker, 1979). In contrast, the high diversity of the South African fynbos (Mediterranean-climate scrubland) vegetation consists of woody species, of which there are few annuals. This type of vegetation has not been subject to a long history of human disturbance.

The data thus indicate that tropical systems are probably among the world's richest in terms of local, or alpha, diversity, but that the vegetation of Mediter-

TABLE 17-1   Mean Numbers of Species per 0.1-Hectare Sample Area (Non-Mediterranean Sites Include Only Data for Woody Plants over 1 Inch in Diameter at Breast Height)

Sample Area	Mean No. of Species
Dry Tropical Forest[a]	
Costa Rica upland, Guanacaste	41
Costa Rica riparian, Guanacaste	64
Venezuelan Llanos, Calabozo	41
Venezuelan coastal, Boca de Uchire	67
Moist Tropical Forest[a]	
Panama Canal Zone, Curundu	88
Brazil, Manaus	91
Panama Canal Zone, Madden Forest	125
Wet Tropical Forest[a]	
Panama Canal Zone, pipeline road	151
Ecuador, Rio Palenque	118
Costa Rica, near La Selva[b]	236
Temperate Zone[a]	
Missouri, Babler State Park	21
Temperate Zone[c]	
Australia, forests and woodlands	48
Tennessee, Great Smoky Mountains	25
Oregon, Siskiyou Mountains	26
Arizona, Santa Catalina Mountains	21
Colorado, Rocky Mountain National Park	32
Mediterranean Zone[d]	
Israel, grazed woodlands	136
Israel, open shrubland	139
Israel, closed shrubland	35
California, grazed woodlands	64
California, closed shrubland	24
Chile, open shrubland	108
Australia, heath	65
South Africa, fynbos	75

[a]Data from Gentry, 1979.
[b]Data from Whitmore, 1986, for a 0.01-hectare plot.
[c]Data from Whittaker, 1977.
[d]Data from Naveh and Whittaker, 1979.

ranean-climate regions is also quite rich. In Mediterranean-climate regions the basis for the localized diversity can differ with the pattern of disturbance. In some systems with a long history of association with human activities, diversity has actually increased (Naveh and Whittaker, 1979).

Data on diversity at a given site indicate its structural dynamics as related to both evolutionary history and pattern of disturbance. We are just now beginning to appreciate the role of both natural disturbances and the impacts of humans in controlling community structure, including its diversity (Bazzaz, 1983). Such knowledge is essential for understanding and hence managing a given level of diversity.

MEDITERRANEAN-CLIMATE FLORISTIC DIVERSITY

Data on local diversity are an indication of disturbance pattern and evolutionary history leading to niche diversification. Another view of the biotic richness of an area is the degree of endemism of the biota. Data on species numbers and degree of endemism for Mediterranean-climate regions form the basis for identifying them as critical sites for conservation. An indication of the diversity and uniqueness of Mediterranean-climate plant life is given below for South Africa, California, and the Mediterranean basin—areas that share unusually high biotic diversities but have dissimilar histories of human impact. For example, South Africa has large tracts of land dominated by the original species-rich shrubland, and the Mediterranean basin contains predominantly herb or shrub degradation forms of the original vegetation. The diversity of South Africa is threatened by development and the invasion of alien species; the Mediterranean basin diversity, by changes in land-use patterns.

South Africa

The Mediterranean-climate region (fynbos biome) of South Africa covers 75,000 square kilometers. This area includes 8,550 vascular plants (Macdonald and Jarman, 1984), three-quarters of which are endemic (Jarman, 1986). According to estimates by Hall (1978), the flora indigenous to the South African Cape, which is found in an area of 46,000 square kilometers, contains at least 6,000 higher plant species—a species richness three times that found in tropical regions of similar areas. This subregion has been considered one of the world's six distinctive floristic regions.

In the fynbos biome, 1,585 plant species are considered rare and threatened (Macdonald and Jarman, 1984), and 39 have recently become extinct (Jarman, 1986). Although the fynbos region occupies less than 1% of southern Africa, it contains 65% of the threatened plant species (Hall, 1979).

Much of the vegetation in this region has been destroyed by human activities, but not to the extent it has occurred in other Mediterranean-climate areas. In the lowland regions, only about 30% of the original vegetation remains, whereas in the mountains, approximately 80% of the vegetation remains intact. Overall, about 67% of the natural fynbos vegetation remains (Jarman, 1986). One threat to the native flora is the presence of alien, generally woody species, which have invaded about one-fourth of the native vegetation (Jarman, 1986). Of 70 critically threat-ened or recently extinct taxa, 23% are threatened by invading acacias, 8% by pines, and 2% by hakeas (Hall, 1979).

In summary, the South African Mediterranean-climate vegetation is as rich as any found on Earth. This richness is being threatened by human development, as everywhere, but also by a rather remarkable invasion of woody plants that are altering the basic functioning of these systems (Macdonald and Jarman, 1984).

California

There is rather complete information describing the biotic richness of the State of California, most of which falls within a Mediterranean-type climate. Although not as rich as South Africa in plant species, it certainly is one of the world's most biotically diverse areas. In an area of 411,000 square kilometers, there are more than 5,046 native vascular plant species, 30% of which are endemic. (In com-parison, there are about 20,000 vascular plant species in the continental United States.) About one-tenth of the flora in these regions of California has recently become extinct or endangered. This represents 25% of all the extinct and endan-gered species of the United States as a whole (Raven and Axelrod, 1978).

California has suffered great losses of natural communities through human de-velopment of agriculture, industry, and housing, especially in coastal and valley regions. Entire ecosystems have evidently been irrevocably lost. One of the most spectacular examples of this is the native perennial grassland of the Central Valley and north coastal regions, which has been replaced by an annual grassland dom-inated by species mostly inadvertently introduced from the Mediterranean basin (Burcham, 1957). Raven and Axelrod (1978) estimate that more than 10% of the flora in these regions is now composed of naturalized aliens. Thus California, like other Mediterranean-climate regions, has an unusually diverse biota that is being threatened by human activities. But to a greater extent than in other regions, substantial areas of the state have been set aside as parks and preserves.

The Mediterranean Basin

The entire Mediterranean basin encompasses more than 2 million square kilo-meters and may include as many as 25,000 higher plant species, about half of which are endemic (Quezel, 1985). Of 2,879 species endemic to individual Med-iterranean countries (excluding Syria, Lebanon, Turkey, and the Atlantic islands), 1,529 are rare (1,262) or threatened, and 300 are not categorized. If the Atlantic islands (Azores, Madeira, and the Canaries) are included, these figures increase to 3,583 endemics and 1,968 rare or threatened plant species (Leon et al., 1985)

In contrast to California and South Africa, where large areas of climax vegetation remain, much of the Mediterranean basin has been completely transformed from its native state. Naveh and Dan (1973, p. 387) reported that the region as a whole "is composed of innumerable variants of different degradation and regeneration stages." Since the impact of humans in this region has been so extensive for a long time, it is believed that the Mediterranean endemic has evolved under conditions of frequent disturbance or in depauperate microsites, such as rock outcrops (Gomez-Campo, 1985). Greuter (1979, p. 90) observed that "the rare threatened taxa are seldom members of the characteristic vegetation units as defined by the plant sociologists: they are marginal creatures living on the borderline of biota. . . ." This general viewpoint has led to the following conclusion of Ruiz de la Torre (1985, p. 197): "Unlike the tropical rain forest, where most of the indigenous species can be conserved with climax formations under conditions of maximum stability, the Mediterranean region has been severely influenced by man and various other factors and is still very rich in species. Very few of these species are known to be part of Mediterranean climax vegetation. Most of them correspond to successional stages affected by either natural or artificial exploitation, and they should be conserved under the prevailing conditions of relative instability."

INCREASING BIOTIC DIVERSITY—THE INVADERS

As indicated above, plant diversity in Mediterranean-climate regions is among the world's richest in terms of numbers of species, but there have been losses of species and continuing threats of extinction to many others. However, there have also been additions of new species to these and other regions of the world. As shown in Table 17-2, the floras of certain islands, ranging from subarctic to tropical, have been enriched half again by species from other biographic regions. In mainland Mediterranean-climate regions such as California, and even to a greater extent in South Australia, there are also substantial numbers of invading species that have become naturalized, many maintaining large and dominating populations. In these regions, as elsewhere, these invading species are not distributed uniformly in the landscape but are generally associated with ecosystems that have experienced human impact. Organisms other than plants are also being enriched by the addition of species in these climates. In California, for example, 49 species have been added to the 132 indigenous inland fishes (Moyle, 1976).

Thus in some cases, human disturbance can actually enrich biotic diversity. However, species counts in a given area give us little understanding of ecosystem functioning and how the invasions affect it. Some invaders may become the dominant species in the host-region ecosystem. Examples of this include a species of oat (*Avena fatua*) in the grasslands of California (Burcham, 1957) and brome grass (*Bromus tectorum*) in the intermountain West (Mack, 1986). Many of the invaders are pest species of one sort or another and may cause economic havoc. These species of course receive considerable attention, and their biology and community role is generally well known. However, we generally know little about the effects of most invaders on the ecosystem or, for that matter, the effects of most species on natural communities.

Are these invaders enriching biotic diversity? They are when considered in absolute numbers of species. In many cases, however, they are impoverishing the biota by leading to species exclusions (Race, 1982) or even to extinctions. The invaders are generally symptoms of an abused landscape, one that has been disturbed and has generally lost some of its original productive capacity. The successful introduction of exotic mammals has often resulted in greatly perturbed ecosystem function and losses of indigenous species. In general, new community types are being added to the original ones that in turn are being reduced in extent. The

TABLE 17-2 The Plant Invaders

Region	Area (thousands of square kilometers)	Indigenous Species	Endemic Species	Naturalized Species	Type of Plant[a]	Reference
Non-Mediterranean-Climate Islands						
Seychelles	0.3	153	69	165	F	Proctor, 1984. Status of 130 species unknown
Faeroe Islands	1.4	370	0	30	F	Hansen and Johansen, 1982
New Zealand	268.0	1,996	1,618	500	V	Godley, 1975
Hawaiian Islands	16.7	ca. 1,250	ca. 1,180	600	F	Wagner et al., 1985; Smith, 1985
Mediterranean-Climate Regions						
South Australia	984.4	2,380	NR[b]	654	F	Specht, 1972
California	411.0	5,046	1,517	674	V	Raven and Axelrod, 1978
Canary Islands	7.3	1,050	550	700	V	Kunkel, 1976

[a]V, vascular plants; F, flowering plants.
[b]Not reported.

landscapes are becoming more complex. Yet, when viewed on a more global scale, the biota is becoming less interesting because of homogenization. For example, geographically separate and distinctive biological regions are often invaded by the very same weedy species. As a result, regions such as parts of California and Chile, which once had only a few plant species in common, now share hundreds.

The maintenance of a diverse landscape, rich in community types and species, requires knowledge of the dynamics of ecosystems as well as the ecology of individual species. Since this information is generally lacking, attempts to conserve individual species or populations are still filled with surprises, even in preserves.

REFERENCES

Bazzaz, F. A. 1983. Characteristics of populations in relation to disturbance in natural and man-modified ecosystems. Pp. 259–275 in H. A. Mooney and M. Godron, eds. Disturbance and Ecosystems. Springer-Verlag, New York.

Burcham, L. T. 1957. California Rangeland. California Division of Forestry, Sacramento, Calif. 261 pp.

Gentry, A. 1979. Extinction and conservation of plant species in Tropical America: A phytogeographical perspective. Pp. 110–126 in I. Hedberg, ed. Systematic Botany, Plant Utilization and Biosphere Conservation. Almqvist and Wiksell International, Stockholm.

Godley, E. J. 1975. Flora and vegetation. Pp. 177–229 in G. Kuschel, ed. Biogeography and Ecology in New Zealand. Monographiae Biologicae, Vol. 27. W. Junk, The Hague, the Netherlands.

Gomez-Campo, C. 1985. The conservation of Mediterranean plants: Principles and problems. Pp. 3–8 in C. Gomez-Campo, ed. Plant Conservation in the Mediterranean Area. W. Junk, Dordrecht, the Netherlands.

Greuter, W. 1979. Mediterranean conservation as viewed by a plant taxonomist. Webbia 34:87–99.

Hall, A. V. 1978. Endangered species in a rising tide of human population growth. Trans. R. Soc. S. Afr. 43:37–49.

Hall, A. V. 1979. Invasive weeds. Pp. 133–147 in Fynbos Ecology: A Preliminary Synthesis. South African National Scientific Programmes Report No. 40. Cooperative Scientific Programmes, Council for Scientific and Industrial Research, Pretoria, South Africa. 166 pp.

Hansen, K., and J. Johansen. 1982. Flora and vegetation of the Faeroe Islands. Pp. 35–52 in G. F. Rutherford, ed. The Physical Environment of the Faeroe Islands. Monographiae Biologicae, Vol. 46. W. Junk, The Hague, the Netherlands.

Jarman, M. L. 1986. Conservation Priorities in the Lowland Regions of the Fynbos Biome. South African National Scientific Programmes Report No. 87. Cooperative Scientific Programmes, Council for Scientific and Industrial Research, Pretoria, South Africa. 55 pp.

Kunkel, G. 1976. Notes on the introduced elements in the Canary Islands flora. Pp. 249–266 in G. Kunkel, ed. Biogeography and Ecology in the Canary Islands. Monographiae Biologicae, Vol. 30. W. Junk, The Hague, the Netherlands.

Leon, C., G. Lucas, and H. Synge. 1985. The value of information in saving threatened Mediterranean plants. Pp. 177–196 in C. Gomez-Campo, ed. Plant Conservation in the Mediterranean Area. W. Junk, Dordrecht, the Netherlands.

Macdonald, I. A. W., and M. L. Jarman. 1984. Invasive Alien Organisms in the Terrestrial Ecosystems of the Fynbos Biome, South Africa. South African National Scientific Programmes Report No. 85. CSIR Foundation for Research Development, Council for Scientific and Industrial Research, Pretoria, South Africa. 72 pp.

Mack, R. 1986. Alien plant invasion into the Intermountain West: A case history. Pp. 191–213 in H. A. Mooney and J. Drake, eds. The Ecology of Biological Invasions into North America and Hawaii. Springer-Verlag, New York.

Moyle, P. B. 1976. Inland Fishes of California. University of California Press, Berkeley. 405 pp.

Naveh, Z., and J. Dan. 1973. The human degradation of Mediterranean landscapes in Israel. Pp. 373–390 in F. de Castri and H. A. Mooney, eds. Mediterranean Type Ecosystems: Origin and Structure. Springer-Verlag, Berlin.

Naveh, Z., and R. H. Whittaker. 1979. Structural and floristic diversity of shrublands and woodlands in northern Israel and other Mediterranean areas. Vegetatio 41:171–190.

Procter, J. 1984. Floristics of the granitic islands of the Seychelles. Pp. 209–220 in D. R. Stoddart, ed. Biogeography and Ecology of the Seychelles Islands. Monographiae Biologicae, Vol. 55. W. Junk, The Hague, the Netherlands.

Quezel, P. 1985. Definition of the Mediterranean region and the origin of its flora. Pp. 9–24 in C. Gomez-Campo, ed. Plant Conservation in the Mediterranean Area. W. Junk, Dordrecht, the Netherlands.

Race, M. S. 1982. Competitive displacement and predation between introduced and native mud snails. Oecologia 54:337–347.

Raven, P. H. 1976. Ethics and attitudes. Pp. 155–181 in J. B. Simmons, R. I. Beyer, P. E. Brandham, G. Lucas, and V. T. H. Parry, eds. Conservation of Threatened Plants. Plenum, New York.

Raven, P. H., and D. I. Axelrod. 1978. Origin and Relationships of the California Flora. Univ. Calif. Pub. Bot. 72:1–134.

Ruiz de la Torre, J. 1985. Conservation of plant species within their native ecosystems. Pp. 197–218 in C. Gomez-Campo, ed. Plant Conservation in the Mediterranean Area. W. Junk, Dordrecht, the Netherlands.

Smith, C. W. 1985. Impact of alien plants on Hawai'i's native biota. Pp. 180–250 in C. P. Stone and J. M. Scott, eds. Hawai'i's Terrestrial Ecosystems: Preservation and Management. Cooperative National Park Resources Studies Unit, University of Hawaii, Honolulu, Hawaii.

Specht, R. L. 1972. The Vegetation of South Australia. A. B. James, Adelaide, Australia. 328 pp.

Wagner, W. L., D. R. Herbst, and R. S. N. Yee. 1985. Status of the native flowering plants of the Hawaiian Islands. Pp. 23–74 in C. P. Stone and J. M. Scott, eds. Hawai'i's Terrestrial Ecosystems: Preservation and Management. Cooperative National Park Resources Studies Unit, University of Hawaii, Honolulu, Hawaii.

Waring, R., and W. H. Schlesinger. 1985. Forest Ecosystems: Concepts and Management. Academic Press, New York. 340 pp.

Whitmore, T. C. 1986. Total species count on a small area of lowland tropical rain forest in Costa Rica. Bull. Br. Ecol. Soc. 17:147–149.

Whittaker, R. H. 1977. Evolution of species diversity in land communities. Evol. Biol. 10:1–67.

STRUCTURAL AND FUNCTIONAL DIVERSITY IN TEMPERATE FORESTS

JERRY F. FRANKLIN

Chief Plant Ecologist, USDA Forest Service, U.S. Department of Agriculture, and
Bloedel Professor of Ecosystem Analysis, College of Forest Resources,
University of Washington, Seattle, Washington

Temperate zones, including their Mediterranean subzones, are the regions of the world most uniformly and extensively altered by human activities. Settlement and development of these productive and hospitable regions have a long history and have had dramatic impacts on biological diversity. Many ecosystems and organisms have been entirely eliminated, and most remaining examples of natural ecosystems are fragmented and highly modified. Intensive human activities, including the relatively recent addition of environmental pollutants, provide continuing threats to biota.

Preserving biotic diversity in temperate zones therefore represents a major challenge. Restoring some of the lost biodiversity is an element of this challenge as is protecting what remains. Positive factors in preservation include the general resilience of temperate forests, the relatively high level of relevant knowledge, and the wealth and educational level of temperate-zone nations and inhabitants. A resurgence of temperate forests on abandoned agricultural and cutover forest lands, such as in the northeastern United States, also contributes to the potential for restoration of biodiversity.

This chapter contains my views on some major needs in preserving and enhancing biotic diversity in temperate forest regions. These needs are to maintain, or, where absent, to create a complete array of forest successional stages, including old-growth forest conditions; to maintain structural and functional diversity throughout the forest landscape, e.g., by retaining standing dead trees and fallen logs; to protect aquatic diversity in the streams, lakes, and rivers associated with temperate forests; and to develop effective stewardship programs that can maintain (and create, when necessary) natural area preserves within intensively utilized landscapes. There is also a critical need to integrate biodiversity objectives into management of all our landscapes because preservation of selected tracts of land, even at the largest scale possible, will not by itself achieve the desired goal of maintaining Earth's biodiversity.

MAINTAINING SUCCESSIONAL STATES

Preserving biodiversity in temperate regions requires the maintenance of all successional stages. Since early successional stages are typically well represented, a major concern is preserving or recreating old-growth forests. Such old-growth forests typically contrast sharply with early successional stages in composition, structure, and function.

Most forests in the temperate zone are secondary forests that developed after logging of primeval forests or abandonment of agricultural lands. In the United States, these forests are typically young, having originated during the last 100 to 150 years. The composition and structure of these forests are different—often drastically different—from those they have replaced. We see, for example, forests of birch (*Betula* spp.) and aspen (*Populus* spp.) in the Great Lakes states, where the forests were originally dominated by long-lived pioneer species, such as red and eastern white pine (*Pinus resinosa* and *P. strobus*), and late successional species of hardwood.

Old-growth temperate forests dominated by coniferous species still cover substantial acreages in the western United States; research in these forests is clarifying the contrasts between young- (e.g., <100 year) and old-growth (e.g., >200 year) forests (see, e.g., Franklin et al., 1981). For example, old-growth forests of Douglas fir and western hemlock (*Pseudotsuga menziedii* and *Tsuga heterophylla*) (Figure 18-1) provide essential habitats for a set of highly specialized vertebrate species, including the northern spotted owl (*Strix occidentalis*). Research presently under way will provide a definitive list of old-growth-dependent species within these temperate conifer forests. This list may include several other birds, several mammals (bat species may be notable), and several amphibians (particularly salamanders). Such forests are also very rich in mosses, lichens, and liverworts, of which at least one species—a lichen—is strongly related to old-growth forests. That species, *Lobaria oregana*, is an important nitrogen-fixing foliose lichen that grows in the crowns of old-growth Douglas-fir trees. Research will almost certainly show that some of the rich invertebrate community is also old-growth-dependent; more than 1,000 species have been identified within a single old-growth stand, the upper bole and crown providing particularly rich habitat. The old-growth forests obviously have a high genetic content and are far from the biological deserts that some game biologists and foresters once suggested.

Functional differences between old-growth and younger forests are often qualitative rather than quantitative. That is, forests at all stages fix and cycle energy or carbon, regulate hydrologic flows, and conserve nutrients. Some stages carry out these activities more efficiently than others, however. Old-growth forests in the Douglas-fir region are particularly effective at regulating water flows and re-

FIGURE 18-1 Old-growth forests are an important successional stage that needs to be protected in any overall scheme for protection of temperate zone biodiversity; 500-year-old *Pseudotsuga menziesii-Tsuga heterophylla* forest on the H. J. Andrews Experimental Forest in the central Oregon Cascade Range. Courtesy Glen Hawk.

ducing nutrient losses. Nutrient losses from old-growth watersheds in the Pacific Northwest are, for example, extremely low (Franklin et al., 1981), although this is not always true in other regions (see, e.g., Martin, 1979). Old-growth forests may contrast with younger forests in their influence on some important hydrologic processes. Old-growth coniferous forests present a very large crown surface and occupy an extensive volume of space, because dominant trees are commonly taller than 75 meters. Such forests are particularly effective at gleaning moisture from clouds and fog, which can substantially increase precipitation (Harr, 1982). These forests may also influence the amount and spatial distribution of snowfall thereby minimizing the potential for the damaging rain-on-snow floods that are charac-teristic of the Pacific Northwest. In addition, the old-growth Douglas-fir forests provide several important sites for nitrogen fixation (e.g., epiphytic lichens and rotting wood), which are more limited or absent in earlier stages of succession.

Old-growth coniferous forests contrast most visibly with earlier successional stages in their structure (Franklin et al., 1981). Old-growth stands obviously have a greater range of tree sizes and conditions than do younger stands and generally have a more heterogeneous forest understory. Large live trees, large standing dead trees (or snags), and large fallen logs are the most conspicuous structures that distinguish old-growth forests. Furthermore, these structures are often the key to the unique compositional and functional attributes of the forest, such as habitat for the northern spotted owl and its prey. Early successional forests developing after natural catastrophes, such as wildfires or hurricanes, often contain large standing dead trees and fallen logs because most catastrophes kill trees but do not consume the wood structures. Young forests developing after timber cutting or agricultural abandonment do not have snags and woody debris, however, because the boles are removed.

Although these examples are all drawn from the temperate coniferous forests of the Pacific Northwest, old-growth forests in other temperate regions probably exhibit similar distinctions of composition, structure, and function. Ecological investigations of old-growth forests in northeastern North America are just beginning, but differences between early and late successional stages in composition and structure are already apparent. Old-growth-dependent wildlife species have not yet been identified, but some of them may already have been eliminated; at present, no investigations of lower plants or invertebrates have been undertaken. Ongoing investigations of remnant primeval forests in northeastern North America, China, South America, New Zealand, and Europe should clarify the distinctive characteristics of old-growth forests throughout the temperate zones.

Old-growth forests and the organisms and processes that they represent are an essential aspect of the global biodiversity at risk. Thus, preserving or recreating old-growth temperate forests should be a key objective of any conservation program. Such efforts would be timely, since there are still opportunities to retain examples of old-growth ecosystems in northwestern North America and eastern Asia and to allow areas of maturing woodlands in northeastern North America to develop into old-growth forests. Additional research on the characteristics of old-growth hardwood and hardwood-conifer forests is critical as a basis for conservation efforts.

MAINTAINING STRUCTURAL AND FUNCTIONAL DIVERSITY

We tend to be intent on preserving genetic diversity as represented by species, but ecosystem simplification and loss of biodiversity is proceeding rapidly in other ways. Maintaining structural and functional diversity in temperate regions is an important need, particularly in intensively managed landscapes. Unfortunately, such efforts run contrary to our cultural tendencies to simplify ecosystems, even when such simplification is not essential to our objectives. Large snags and fallen logs are examples of structural diversity (Figure 18-2). Retaining nitrogen-fixing organisms exemplifies a functional aspect of biotic diversity within an ecosystem or landscape.

FIGURE 18-2 Coarse woody debris, including standing dead trees and downed boles, are an important structural component of forests. An important goal in preserving ecological diversity is to maintain such structures within managed forest ecosystems. This rotting log is serving as habitat for a large variety of heterotrophic and autotrophic organisms. (Goar Marsh Research Natural Area, Giffort Pinchot National Forest, Washington.) Courtesy U.S. Forest Service.

Standing dead trees and fallen logs are essential to many organisms and biological processes within forest ecosystems (Harmon et al., 1986); yet, such structures have rarely been retained within managed forests. For example, Thomas (1979), in his compilation of the wildlife of northeastern Oregon forests, found that 178 vertebrates—14 amphibians and reptiles, 115 birds, and 49 mammals—used fallen logs as habitats. Elton (1966, p. 279) recognized the broad importance of dead wood structures for biotic diversity: "When one walks through the rather dull and tidy woodlands [of England] that result from modern forestry practices, it is difficult to believe that dying and dead wood provides one of the two or three greatest resources for animal species in a natural forest, and that if fallen timber and slightly decayed trees are removed the whole system is gravely impoverished of perhaps more than a fifth of its total fauna." In addition to its role as a habitat for land animals, woody debris also provides habitats, structure, energy, and nutrients for aquatic ecosystems (Harmon et al., 1986). Furthermore, it provides sites for nitrogen fixation, sources of soil organic matter, and sites for the establishment of other higher plants, including tree seedlings (Harmon et al., 1986). Maintaining dead-wood structures should be a regular objective of silvicultural activities within the forests of the temperate zone and other zones, quite apart from any program for maintaining old-growth-forest conditions.

Maintaining nitrogen-fixing organisms within our forest landscapes is an example of maintaining functional diversity. Many nitrogen-fixing species of plants, such as ceanothus (*Ceanothus* spp.) and alder (*Alnus* spp.), are associated with early stages of succession. Others, such as the lichen mentioned earlier, are associated with old growth; still others (microbial) are associated with woody debris. Forest management activities have tended to eliminate these sources to minimize competition from noncrop species and speed development of a closed canopy of crop trees.

Efforts to conserve structural and functional diversity are often linked; for example, by maintaining woody debris, one of the sites for nitrogen fixation is retained within the ecosystem. Another example is maintaining large volume, complex crown structures that are especially effective at scavenging moisture and particulate materials from the atmosphere.

Obviously, maintaining structural and functional diversity is an objective that is broadly applicable to temperate landscapes and not just to forests. For example, continuous efforts are under way to convert complex shrub-steppes or savannas to grasslands or even monocultures of seeded grasses by eliminating woody plants such as sagebrush (*Artemisia* spp.) or junipers (*Juniperus* spp.). Such programs are capable of causing great damage to structural, functional, and genetic diversity over large areas.

PROTECTING AQUATIC DIVERSITY

Protecting aquatic diversity, including that of the riparian zones, is one of the most difficult tasks within the temperate zone. Streams and rivers have been dammed, diverted, and polluted. Organisms have been extirpated and many new organisms introduced, either purposely or accidentally. Control of large land areas (watersheds) is required to provide complete protection for many bodies of water (Figure 18-3). Legal problems are often overwhelming in view of the large number of jurisdictions involved and, at least in the United States, the peculiarities of water rights and law.

The risk to aquatic biodiversity within temperate regions is great and has not received much effective attention, despite the attention given waterfowl and fisheries and the recognized importance of wetlands. Loss of diversity in river ecosystems may be particularly serious and certainly affects invertebrates (e.g., insects and molluscs) as well as vertebrates (e.g., fish). One need only be reminded of the loss of anadromous fish from many river systems after dams were built to realize that these changes involve loss of other important compositional, structural, and functional features from these ecosystems as well.

Developing effective programs to protect aquatic biodiversity is a priority of the highest order. Even the initial step—an adequate analysis of the problem—will require additional research as well as syntheses of existing information. Creative new approaches to conservation will be required, such as acquisition of water rights and licenses for dam construction. The Nature Conservancy has pioneered development of such creative approaches in their recent wetlands initiative.

Protecting aquatic biodiversity is a problem in all segments of the temperate zone—from forests to deserts. The most critical problems in protecting aquatic

FIGURE 18-3 Maintaining examples of natural river and stream ecosystems is one of the most challenging tasks facing society in temperate as well as other biotic zones. (San Juan Mountains, Colorado.) Courtesy U.S. Forest Service.

biodiversity are probably associated with bodies of water in arid regions where they are a critical and often overallocated resource.

DEVELOPING EFFECTIVE STEWARDSHIP PROGRAMS

Maintaining biodiversity is a continuing and multifaceted task. It cannot be permanently accomplished by a single action, such as establishing a national park or biological preserve. Indeed, we often forget that establishing a preserve is only the first step in the infinite responsibility that we have assumed for keeping many organisms and ecosystems afloat (Figure 18-4).

Fulfilling our stewardship responsibility will require a great deal more attention than it has been receiving. Maintaining a viable biological preserve in the densely settled and intensively used temperate zones requires sophistication and dedication. Large amounts of information about the ecology of the target ecosystems and organisms and about environmental conditions in and around these preserves will be required. This means intensive research and monitoring programs, often of long

FIGURE 18-4 Maintaining ecological reserves in the heavily settled temperate zone will require extensive knowledge and sophisticated technology. Prescribed burning is one of the methodologies already commonly utilized in both prairie and forest reserves in North America. (Konza Prairie Biosphere Reserve, Kansas.) Courtesy U.S. Forest Service.

duration. Trained personnel will have to develop and implement complicated management programs. To meet all these needs will require large and stable financial support and the development of professional cadres trained and experienced in stewardship.

The key to such a large and long-term commitment can ultimately come only from society at large. Resolving the risks to biodiversity in the temperate zones and developing the philosophy and technology of stewardship can provide an essential example for tropical regions.

INCORPORATING BIODIVERSITY OBJECTIVES INTO MANAGEMENT

We cannot accomplish our objectives simply by creating preserves; the objectives of maintaining biodiversity must be incorporated into intensively managed temperate landscapes. The bulk of the temperate landscape will be used for production of commodities and for human habitation. We must therefore develop management strategies for forestry, agriculture, water development, and fisheries that incorporate the broader diversity. Most intensive management strategies currently do not take biological diversity into consideration; rather, they emphasize simplifying and sub-

FIGURE 18-5   It is essential that the objective of preserving ecological diversity be incorporated into management programs on lands used for production of commodities; reserves or "set-asides" on the public lands will not adequately accomplish the essential goals. This will have to include considerations of landscape ecology, such as the effects of patch patterns on biota. (Dispersed patch clearcutting on the Gifford Pinchot National Forest, Washington.) Courtesy U.S. Forest Service.

sidizing ecosystems, i.e., organismal, structural, successional, and landscape homogenization (Franklin et al., 1986).

In forestry practices, we can see this emphasis on simplification from the level of the tree, where great efforts are being expended to create genetically uniform material, through the geometrically arranged stand to the landscape, where multiple age classes of conifer monocultures are sometimes cited as evidence of commitment to biological diversity. We must modify our treatments of forest stands and arrangements of forest landscapes to incorporate the objective of protecting biodiversity (Figure 18-5). This can be done with very little reduction in the production of commodities. Failure to do so will result in immense losses of genes and processes within the temperate zone.

Biodiversity is abundant in the temperate zone, and it, too, is worth saving.

REFERENCES

Elton, C. S. 1966. Dying and dead wood. Pp. 379–305 in The Pattern of Animal Communities. Wiley, New York.

Franklin, J. F., K. Cromack, Jr., W. Denison, A. McKee, C. Maser, J. Sedell, F. Swanson, and G. Juday. 1981. Ecological characteristics of old-growth Douglas-fir forests. USDA Forest Service General Technical Report PNW-118. Forest Service, Pacific Northwest Forest and Range Experiment Station, U.S. Department of Agriculture, Portland, Oreg. 48 pp.

Franklin, J. F., T. Spies, D. Perry, M. Harmon, and A. McKee. 1986. Modifying Douglas-fir management regimes for nontimber objectives. Pp. 373–379 in C. D. Oliver, D. P. Hanley, and J. A. Johnson, eds. Douglas-Fir Stand Management for the Future. College of Forest Resources, University of Washington, Seattle.

Harmon, M. E., J. F. Franklin, F. J. Swanson, P. Sollins, S. V. Gregory, J. D. Lattin, N. H. Anderson, S. P. Cline, N. G. Aumen, J. R. Sedell, G. W. Lienkaemper, K. Cromack, Jr., and K. W. Cummins. 1986. Ecology of coarse woody debris in temperate ecosystems. Pp. 133–302 in Advances in Ecological Research, Vol. 15. Academic Press, New York.

Harr, R. D. 1982. Fog drip in the Bull Run municipal watershed, Oreg. Water Res. Bull. 18(5):785–789.

Martin, C. W. 1979. Precipitation and streamwater chemistry in an undisturbed forested watershed in New Hampshire. Ecology 60(1):36–42.

Thomas, J. W. 1979. Wildlife habitats in managed forests: The Blue Mountains of Oregon and Washington. USDA Agricultural Handbook 553. U.S. Department of Agriculture, Washington, D.C. 512 pp.

19

DIVERSITY IN AND AMONG GRASSLANDS

PAUL G. RISSER

Vice President for Research, University of New Mexico, Albuquerque, New Mexico

Grasslands cover broad areas of both temperate and tropical regions, but they occur primarily in climatic zones with a pronounced dry season (Axelrod, 1985). They are characteristically found in regions where there is insufficient soil water to support an arboreal canopy yet adequate moisture to permit the existence of a grass-dominated canopy rather than desert vegetation. Technically, grasslands can be described as types of vegetation that are subjected to periodic drought, that have a canopy dominated by grass and grasslike species, and that grow where there are fewer than 10 to 15 trees per hectare. The number of grass species in these areas, however, is frequently lower than the number of forbs, e.g., composites such as daisies and sunflowers (Curtis, 1959). Grasslands are found in such diverse locations as the steppes of the Soviet Union, the Serengeti of Africa, the dry grasslands of Australia, the pampas of Argentina, and the Central Plains of the United States. Given this wide range of variations, it is not surprising that the grasslands of the world contain a large amount of native biodiversity.

GRAZING AND AGRICULTURE CONVERSIONS

All grasslands support an array of native herbivores. In terms of energy consumption, the impact of herbivores is usually quite low but differs among the various grassland types. However, the greatest impacts on most grassland ecosystems are caused by domestic herbivores that have been introduced by human societies. Grasslands can withstand moderate grazing, especially when weather conditions are favorable, but overgrazing frequently causes important changes in the composition of the plant and animal population. A common response is that the grassland becomes converted to a relatively sparse shrubland composed of less-palatable herbaceous or woody species. This type of conversion can be found, for example, along the India-Pakistan borders in the Sind-Kutch region, throughout much of the Sahelian area of Africa, and in parts of the southwestern United States (Brown, 1950; Buffington and Herbel, 1965; Howard-Clinton, 1984). An obvious consequence of this impact is a loss of the native biodiversity of grasslands throughout the world.

Many grasslands, especially those in relatively humid environments, produce large amounts of below-ground growth consisting primarily of roots and rhizomes. As these plant parts naturally die, the organic matter is incorporated into the soil. These enriched soils are prime agricultural soils, and as a result, the grasslands are quite vulnerable to conversion to croplands. An obvious example is the prairie peninsula (Transeau, 1935) in the east central part of the United States—an area that is now almost completely cropland rather than the original tall-grass prairie with its deep, organic-rich melanized soils.

Well over 100 species of native plants commonly grow in prairie remnants smaller than 2 hectares. Within the Central Plains and tall-grass prairie, between 250 and 300 species are usually found in remnants with areas of approximately 250 hectares (Steiger, 1930). Although the loss of grassland habitat has been calculated for selected states and specific grassland types (Risser, 1986), there are no general figures on the loss of grassland species.

MORE SUBTLE IMPACTS

Although overgrazing and conversion to croplands represent the most obvious impacts on the native biodiversity of grasslands, a true diagnosis requires a more refined analysis. For example, relatively recent widespread overgrazing and resultant major changes in species composition of grassland habitats has occurred in developing countries where there is and has been enormous pressure to produce food. Conversion of grassland to agricultural cropland has taken place primarily in humid grasslands. Thus, in the United States, most dry western grasslands, or steppes, can now be adequately managed to remain as perpetuating rangelands. And although there have been misguided efforts to plow the rangelands and some cases of grassland abuse by overstocking with grazing animals, most of these western grasslands now remain intact. In the eastern prairie region, most of the grasslands have been converted to cropland, but important preservation and restoration efforts are now under way (Risser, 1986).

Prairie fires have been a persistent characteristic of grasslands that produce enough fuel for them (Daubenmire, 1968). In fact, in the tall-grass prairie, periodic burning increases the species diversity above that found on an unburned prairie, especially one that is not grazed. Burning is routinely used intentionally in grassland management to reduce invasion by woody shrub species and to encourage native perennial species. However, these burning practices are frequently not beneficial to insects (Cancaledo and Yonke, 1970) or to the small mammals and birds. Thus, to attain optimum biological diversity, either the scheduling of prescribed burning must be compromised or alternative treatments must be administered to patches within the grassland.

Light to moderate levels of grazing usually result in a richer diversity of plant species than do heavy levels of grazing or no grazing at all, especially in the more humid grasslands such as the tall-grass prairie in the United States (Risser et al., 1981). Presumably, this increased diversity is caused by opening the vegetation canopy and allowing more species to compete successfully. Thus, species diversity is maximized by light to moderate grazing intensities—no grazing by domestic herbivores reduces diversity because of the thick vegetation canopy and heavy litter layer, and heavy grazing reduces species diversity by selectively eliminating the more palatable species.

The southwestern grasslands of the United States are dominated by warm-season perennial species. Since these species mature later in the growing season, many ranching operations include pastures of cool-season species, such as crested wheat-grass (Agropyron cristatum) to serve as livestock forage early in the year. These crested wheatgrass fields contain very little species diversity.

In the eastern tall-grass prairie, where remaining grasslands are relegated primarily to small patches in an otherwise agricultural landscape, there are several threats to biodiversity. One is the intrusion of several aggressive alien plant species, which are invading and replacing native species. In Illinois, more than a dozen species have invaded prairies to such a degree that the prairies themselves are now threatened.

The small, isolated prairie remnants are unable to support the normal complement of either native flora or native fauna. In Missouri, only 0.5% of the original tall-grass prairie remains, mostly in isolated prairie islands within a matrix of improved pastures and croplands. Sampson (1980) compared prairie sizes with the presence and absence of the greater prairie chicken (Tympanucus cupido) and found that grasslands without prairie chickens averaged 172 hectares, but those without prairie chickens averaged only 33 hectares. Furthermore, those grassland remnants without prairie chickens were isolated from other grasslands by 81 kilometers, whereas those with prairie chickens were, on average, only 14 kilometers from other grasslands. Sampson concluded that Missouri grasslands capable of supporting the greater prairie chicken should be about 300 hectares larger and within 20 kilometers of other grasslands.

Sampson (1980) also computed the probability that a given habitat size will annually contain a breeding population of selected native grassland bird species. The minimum habitat sizes were calculated as follows: for the eastern meadowlark (Sturnella magna), less than 1 hectare; for the horned lark (Eremophila alpestris) and grasshopper sparrow (Ammodramus sarannarum), more than 1 hectare; for Henslow's sparrow (A. henslowii), the upland sandpiper (Bartramia longicanda), and the greater prairie chicken, more than 10 hectares. In general, the size and not the habitat heterogeneity had a significant influence on the number of breeding prairie bird species. In Illinois, Graber and Graber (1976) also found that the size of the grassland had a major influence on the number of bird species and that in small patches over a 20-year period, the number of bird species decreased at a much faster rate than the simple reduction in the total area of grassland.

In their study of grassland invertebrates, Whitcomb et al. (1986) found that more than 100 dominant grass species, and perhaps an equal number of forbs, are important contributors to the diversity of sap-sucking insects in North American grasslands. They reported that perennial and dominant (but not annual or sub-dominant) grass and forb species tended to have specific assemblages of cicadellids (leafhoppers) in a given geographic region but that the species composition of these assemblages varied geographically. Patch size and structure of the host vegetation stands were of considerable importance, and the rarity of these leafhoppers was directly attributable to the rarity of the host plant species. Even in host patches of sufficient size to support reasonably large numbers of cicadellids, insect populations were reduced by such disturbances as fire, drought, floods, predators, and, especially, parasites.

The origin of North American grasslands is relatively recent—they formed approximately 12,000 years ago (Dort and Jones, 1970). There is a low rate of vertebrate and plant endemism in these areas, and the origins of their flora and fauna are diverse. Therefore, despite the massive loss of grasslands in the United States and elsewhere, there are fewer than 15 true grassland species listed or proposed as federally threatened or endangered. However, as has been recognized for decades, grassland plant species have undergone a significant amount of ecotypic variation (Olmsted, 1945), and the reduction in grasslands has resulted in a reduction of genetic diversity—diversity losses that are not apparent in simple measures of species diversity.

Thus biodiversity in and among grasslands is complicated because of the rather subtle nature of the grassland ecosystem. Major, obvious impacts such as widespread overgrazing and conversion to agricultural croplands have significantly reduced the native biodiversity (Weaver, 1954). Among the more subtle impacts are the effects of reduced habitat size, the lack of endemic species, which are so easily recognized, and the highly developed ecotypic differentiation in grasslands, which is not detected in conventional measures of biodiversity.

REFERENCES

Axelrod, D. I. 1985. Rise of the grassland biome, central North America. Bot. Rev. 51:163–201.

Brown, A. L. 1950. Shrub invasion of southern Arizona desert grassland. J. Range Manage. 3:172–177.

Buffington, L. C., and C. H. Herbel. 1965. Vegetational changes in a semidesert grassland range from 1858 to 1963. Ecol. Monogr. 35:139–164.

Cancaledo, C. S., and T. R. Yonke. 1970. Effect of prairie burning on insect populations. J. Kans. Entomol. Soc. 43:274–281.

Curtis, J. T. 1959. Vegetation of Wisconsin. University of Wisconsin Press, Madison. 657 pp.

Daubenmire, R. F. 1968. The ecology of fire in grasslands. Adv. Ecol. Res. 5:209–266.

Dort, S. W., Jr., and J. K. Jones, Jr., eds. 1970. Pleistocene and Recent Environments of the Central Great Plains. University Press of Kansas, Lawrence. 433 pp.

Graber, J. W., and R. R. Graber. 1976. Environmental evaluations using birds and their habitats. Biological Notes No. 297. Illinois Natural History Survey, Champaign. 39 pp.

Howard-Clinton, E. G. 1984. The emerging concepts of environmental issues in Africa. Environ. Manage. 8:187–190.

Olmsted, C. E. 1945. Growth and development of range grasses. V. Photoperiodic responses of clonal divisions of three latitudinal strains of side-oats grama. Bot. Gaz. 106:382–401.

Risser, P. G. 1986. Preservation status of true prairie grasslands and ecological concepts relevant to management of prairie preserves. Pp. 339–344 in D. L. Kulhavy and R. N. Connor, eds. Wilderness and Natural Areas in the Eastern United States: A Management Challenge. Papers presented at a symposium: Wilderness and Natural Areas in the East, held in Nacogdoches, Texas, on May 13–15, 1985. Center for Applied Studies, Stephen F. Austin State University, Nacogdoches, Tex. 416 pp.

Risser, P. G., E. C. Birney, H. D. Blocker, S. W. May, W. J. Parton, and J. A. Wiens. 1981. The True Prairie Ecosystem. Hutchinson Ross, Stroudsburg, Penn.

Steiger, T. L. 1930. Structure of prairie vegetation. Ecology 11:170–217.

Sampson, F. B. 1980. Island biogeography and the conservation of prairie birds. Pp. 293–299 in C. L. Kurera, ed. Seventh North American Prairie Conference, Proceedings. Southwest Missouri State University, Springfield, Mo.

Transeau, E. N. 1935. The prairie peninsula. Ecology 16:423–437.

Weaver, J. E. 1954. North American Prairie. Johnson Publishing Company, Lincoln, Nebr. 348 pp.

Whitcomb, R. F., J. Kramer, M. D. Coan, and A. L. Hicks. In press. Ecology and evolution of leafhopper-grass host relationships in North American prairies, savanna and ecotonal biomes. Curr. Top. Vector Res. 3.

CHAPTER

20

DIVERSITY AND BIOLOGICAL INVASIONS OF OCEANIC ISLANDS

PETER M. VITOUSEK

Associate Professor, Department of Biological Sciences, Stanford University,
Stanford, California

To date, human-caused species extinctions are more an island-based than a continental phenomenon. Of the 94 species of birds known to have become extinct worldwide since contact with Europeans, only 9 were continental (Gorman, 1979). Currently, more endemic Hawaiian bird species are officially listed as endangered or threatened than are listed for the entire continental United States. Where information is available on other groups of animals, it indicates that human-caused extinctions are invariably more frequent on islands.

Heywood (1979) summarized the causes of extinction on islands as deforestation and fire, the introduction of grazing mammals, cultivation, and the introduction of weedy plants. All these factors can be important on continents as well, but species introductions (deliberate or accidental) are disproportionately important on islands (Elton, 1958). Isolated islands and archipelagos often lack major elements of the biota of continents, and their native species often lack defenses against grazing or predations.

Biological invasions are not the only factor leading to elevated extinction rates for island species. Extinction rates are also higher on islands because island species generally have small populations, restricted genetic diversity, and narrow ranges prior to human colonization, and because human alterations of land through use destroy an already-limited critical habitat. The plant and animal hitchhikers and fellow travelers who accompany humans to isolated islands interact with these other causes of extinction, however, and biological invaders endanger native species in reserves and other protected lands.

The fact that biological invasions decrease diversity on islands is paradoxical, because, as pointed out by Lugo in Chapter 6, the introduction of alien species generally increases the total number of species on an island, often spectacularly. However, most of the introduced species are cosmopolitans that are in no danger of global extinction, whereas most species on isolated islands are endemic. Biological invasions can therefore cause a net loss of species worldwide and a homogenization of the biota of Earth (Mooney and Drake, 1986).

SCOPE OF THE PROBLEM

The disproportionate effects of human colonization and attendant biological invasions on island ecosystems are well known (Carlquist, 1974; Darwin, 1859; Elton, 1958; Wallace, 1880); they can be demonstrated even on large islands such as Madagascar and Australia (Carlquist, 1974). The most severe consequences are experienced on old, isolated, mountainous, tropical, or subtropical islands or archipelagos. Islands located near continents receive organisms from those continents and rarely develop unique species. Truly oceanic islands have rates of evolution and speciation greater than those of immigration; hence, their biota contains many endemic species. Low islands (such as atolls) lack the range of environments that permits evolutionary radiation, while islands at high latitudes are subjected to strong climatic fluctuations (Bramwell, 1979), which prevent radiation.

Together these factors suggest that the Hawaiian Islands, the most isolated archipelago in the world, should have a large number of exotic species and a large potential for loss of endemic species as a consequence of biological invasions. The very large number of endemic species on these islands is well documented (Carlquist, 1974); the importance of biological invasions can also be demonstrated. For example, a survey of exotic plants on National Park Service lands (Loope, in press a) shows that island parks have a much larger proportion of alien species in their flora than do continental parks (Table 20-1). Moreover, in most continental parks alien species are largely confined to roadsides and areas occupied by humans before the park was established. In contrast, Channel Islands National Park in California, Everglades National Park (an island of tropical vegetation at the tip of the Florida peninsula), and the Hawaiian parks contain alien species that establish themselves in otherwise undisturbed native ecosystems and change the nature of the sites they occupy (Ewel, 1986; Stone and Scott, 1985; Stone et al., in press a).

The problems in the Hawaiian parks reflect in part the overall abundance of exotic species in Hawaii. As many as 1,765 native species of vascular plants (probably fewer as taxonomic revisions take hold) existed in the islands when the Polynesians arrived, and 94 to 98% of them were endemic (Kepler and Scott, 1985). Polynesians brought additional species, perhaps 30 of them (Nagata, 1985), when they colonized Hawaii and journeyed among the Pacific islands. The advent of more rapid transportation from distant areas and especially the occupation of Hawaii by people from diverse western and eastern cultures, each with its distinctive food, medicinal, and ornamental plants, greatly increased the number of species present. More than 4,600 species of introduced vascular plants are now known to

TABLE 20-1 Proportion of Alien Plants in the
Vascular Flora of Selected U.S. National Parks[a]

National Park	Alien Species (% of total)
Sequoia-Kings Canyon	6–9
Rocky Mountain	7–8
Yellowstone	11–12
Mount Ranier	12–14
Acadia	21–27
Great Smoky Mountain	17–21
Shenandoah	19–24
Channel Islands	16–19
Everglades	15–20
Haleakala	47
Hawaii Volcanoes	64

[a]From Loope, in press a.

grow in Hawaii, and at least 700 of these are reproducing successfully and maintaining populations in the field (Smith, 1985; Wester, in press). At the same time, more than 200 endemic species are believed to be extinct, and another 800 are endangered (Jacobi and Scott, in press). Most sites below 500 meters elevation, and many higher ones, are entirely dominated by alien species (Moulton and Pimm, 1986).

Similar patterns of introduction of alien insects, mammals, reptiles and amphibians, and birds have been described (Carson, in press; Moulton and Pimm, 1986). The birds are probably the best documented (Moulton and Pimm, 1986; Olson and James, 1982), although mammals are the most spectacular (from 1 native bat to at least 18 species of alien mammals). At least 86 species of land birds are known to have been present in Hawaii 2,000 years ago, and at least 68 of them were endemic passerines. Forty-five species, including 30 passerines, disappeared around the time of Polynesian colonization; another 11 have disappeared since Europeans arrived; and several more are on the verge of extinction (Moulton and Pimm, 1986; Stone, 1985). In contrast, at least 50 species of alien passerines have become established since 1780. Even casual observers of lower-elevation birds in Hawaii have noted a kaleidoscope of shifting dominance by different species of alien birds over the past 30 years; the one constant has been the near absence of natives.

This pattern of successful invasion by cosmopolitan species and the decline of certain native species is not unique to Hawaii. A similar conversion of native-dominated to alien-dominated ecosystems occurs on isolated islands in all the oceans—from the Galapagos to New Zealand to Diego Garcia to Tristan da Cunha and St. Helena (Bramwell, 1979; Carlquist, 1974; King, 1984; Wace and Ollier, 1982). In many cases, the successful invaders are identical—goats (*Capra hircus*) and guava (*Psidium guajava* and *P. cattleianum*) are problems in Hawaii, the Galapagos, and the Rodrigues Islands in the Indian Ocean.

WHY ARE ISLANDS SUSCEPTIBLE?

The reasons why biological invasions are disproportionately successful on islands, and why island species seem more likely to become extinct, have long been debated. Loope (in press b) summarized this discussion with seven possible explanations for the observed patterns:

- Reduced competitive ability due to repeated "founder effects," i.e., chance events during colonization by small initial populations
- Disharmony of functional groups and relative lack of diversity
- Small populations and genetic variability; restrictive specialization
- Relative lack of adaptability to change; loss of resistance to consumers and disease
- Loss of essential co-evolved organisma
- Relative lack of natural disturbance, especially fire, in the evolutionary history of many island biotas
- Intensive exploitation by humans

He also pointed out that the apparent lack of vigor of island species can be overstated, sometimes with negative consequences. For example, Lyon (1909) interpreted a decline of native óhiá (*Metrosideros polymorpha*) in Hawaii as reflecting that species' inability to survive in the modern world, and spearheaded the introduction of many alien species to replace it. In fact, periodic diebacks of natural populations of *Metrosideros* are a natural feature of forest dynamics in Hawaii and elsewhere in the Pacific (Mueller-Dombois, 1983), and *Metrosideros* naturally recolonizes most of these areas. More generally, many native island species maintain themselves quite successfully in mixed native/exotic ecosystems (Mueller-Dombois et al., 1981).

At the other extreme, it has been argued that alien species are merely temporary components of island ecosystems, certain to be replaced by natives in the course of ecological succession (Allan, 1936; Egler, 1942). In fact, some aliens invade intact native ecosystems, whereas others alter the course of succession in already disturbed sites (Smith, 1985) and seem capable of persisting in those altered sites.

Although biological invasions clearly have contributed to the extinction of native species on islands, the importance of direct competition between native and exotic species in causing these extinctions is uncertain. Habitat destruction by humans and feral animals, alterations in basic ecosystem properties caused by newly introduced species, grazing and predation pressure from introduced consumers, and exotic animal diseases (such as avian pox and malaria) appear to be at least equally important.

The importance of grazing and predation by alien animals deserves special emphasis. Most isolated oceanic islands originally lacked whole groups of organisms; mammals were especially sparse. Even ants were nearly or entirely absent on some islands, including Hawaii (Medeiros et al., 1986).

The introduction of mammals has had enormous effects on island ecosystems throughout the world. Comparisons of islands with introduced ungulates and those without such animals in widely separated Pacific island groups (the Hawaiian

Islands, the Cook Islands, and the Kermadec Islands) demonstrate that native communities often hold their own in the absence of mammals but that invasions by plants are much more common and disruptive of native communities on heavily grazed islands (Merlin and Juvik, in press).

WHAT CAN BE DONE?

Biological invasions of oceanic islands appear to be an immense and largely unmanageable problem. Of the approximately 4,600 species of alien plants on Hawaii, more than 700 reproduce in the wild and 86 are considered serious threats to native ecosystems (Smith, 1985). At present, there are neither the resources nor the will to attack a problem of this magnitude. Moreover, while interception and quarantine systems can slow the further introduction of additional exotic species and stop a few indefinitely, the sheer volume and pace of transport by jet aircraft may overwhelm most controls. Finally, any inspection system detailed enough to be broadly effective would necessarily hinder and annoy tourists that are the major economic support of many oceanic islands. Moreover, many island residents have strong reasons for importing or protecting introduced species as agricultural, timber, or forage crops, medicinal or ornamental plants, watershed protection, domestic livestock, pets, agents of biological control, or targets of sport or commercial hunting or fishing. These economic or cultural attachments to alien species mean that there is little chance of developing broad-based, politically effective support for controlling alien species that are not regarded as weeds in the classical (economic) sense.

There are nevertheless several steps that can be taken to reduce the effects of biological invasions and protect some of the native biological diversity on isolated oceanic islands:

- identification of the aliens most likely to threaten native ecosystems and concentration of control efforts on those species;
- selection of critical habitat areas from which most or all species of aliens are excluded;
- protection of areas from further habitat destruction; and
- study of biological invasion and species extinction on islands to learn how these same processes may affect continents.

IDENTIFICATION OF PROBLEM SPECIES

Identification of the invading species most likely to disrupt native ecosystems requires some understanding of the biology of both the invader and the invaded community. Research designed to obtain that information is now being conducted, and its results are being used in management decisions on many islands. The most disruptive species (not necessarily in order of importance) include herbivorous mammals, vertebrate and invertebrate predators, species that can alter ecosystem-level characteristics of invaded areas, and species that can invade otherwise undisturbed native ecosystems.

Herbivorous Mammals

Grazing and browsing mammals effect islands in such pervasive ways that it is difficult to see how native ecosystems can be protected unless they are eliminated. Studies of whole islands and of exclosures have clearly demonstrated that ungulate populations affect erosion, soil fertility, and the success of invasions by alien plants (Loope and Scowcroft, 1985; Merlin and Juvik, in press; Mueller-Dombois and Spatz, 1975; Vitousek, 1986). Island plants often lack defenses, such as thorns and toxic chemicals, against herbivores, and herbivority reduces total plant cover and selects for better defended alien plants. Moreover, feral pigs (which are widespread on many oceanic islands) directly disrupt soil structure in the course of their feeding. Efforts to eliminate mammals are expensive and difficult, but they have been highly successful in a number of areas (Bramwell, 1979; Stone et al., in press b). In many cases, the removal of grazing animals has been followed by the recovery of native plants and even by the discovery of entirely new species of native plants (Bramwell, 1979; Mueller-Dombois and Spatz, 1975).

Predators

Alien vertebrate and invertebrate predators can have significant effects on island ecosystems both directly, by eliminating natives, and indirectly, by altering community structure. For example, rats and feral cats affect the breeding success of ground-nesting birds in many areas (Clark, 1981; King, 1984; Wace, 1986). Alien ants altered invertebrate communities in the Hawaiian lowlands years ago, and other ant species are now threatening to do so at high elevations (Medeiros et al., 1986). Invertebrate predators are particularly problematic in that they may eliminate important native pollinators from island faunas.

Ecosystem-Level Effects

Any alien species that alters ecosystem-level characteristics (such as primary productivity, nutrient availability, hydrological cycles, and erosion) of the area it invades alters the living conditions for all organisms in that area (Vitousek, 1986). It may also alter the kind or quality of the services that natural ecosystems provide to human societies (Ehrlich and Mooney, 1983). Alien animals clearly alter ecosystem properties in a number of ways (as described above), and it is becoming clear that alien plants can do so as well. In Hawaii, for example, the exotic nitrogen-fixing fire tree (*Myrica faya*) increases the availability of the soil nitrogen in nitrogen-limited volcanic ash deposits (Vitousek, in press). Similarly, the alien grasses *Andropogon virginicus* and *A. glomeratus* provide fuel for fires and also sprout rapidly following fires, thereby greatly increasing both their abundance and the overall frequency of fires to the detriment of native species not adapted to fire resistance (Smith, 1985).

Invasion of Intact Native Ecosystems

Alien animals are frequently (not invariably) able to invade intact native ecosystems, but plants species that can do so are not common. Most often, alien plants invade undisturbed native ecosystems in association with alien animals. In Hawaii, alien birds and mammals consume and disseminate the fruit of the aggressive alien plants strawberry guava (*Psidium cattleianum*) and banana poka (*Passiflora mollissima*) throughout native forest areas. Interactions between feral pigs and these invading plants are particularly severe: pigs disseminate seeds of these fleshy-fruited aliens, mix them with organic fertilizer, and deposit them into seedbeds, which are cleared by the pigs' rooting activity. The pigs' descendants then use fruit of the daughter plants as a major food source (Smith, 1985; Stone, 1985). Similar interactions between cattle and common guava (*Psidium guajava*) occur in the Galapagos (Bramwell, 1979). These interactions between alien plants and animals further illustrate why control of alien animals is fundamental to protecting the native ecosystems of islands.

IDENTIFICATION OF CRITICAL HABITATS

A second strategy for limiting the effects of biological invaders is to control manageable alien species in selected critical habitats. This process is expensive and time-consuming, but it does lead to the maintenance of areas as close to their natural state as possible (although birds, flying insects, and microorganisms are of course difficult or impossible to control). Management in "Special Ecological Areas" of Hawaii Volcanoes National Park has been designed to protect areas that represent the major ecosystems in the park by minimizing the influence of alien species. These areas can then act as refugia for threatened native biota and as areas for ecological study and education (Stone et al., in press a; Tunison et al., 1986).

HABITAT DESTRUCTION

Control over habitat destruction is also essential to protecting biological diversity on oceanic islands. Land clearing or fire in native systems can both destroy individuals of threatened native species and lead to the establishment of alien-dominated successional ecosystems. Conflicts in achieving this objective are inevitable; most islands are neither museums nor biological preserves, and one person's "habitat destruction" will certainly be another's source of food or income. Destruction of critical habitat on islands is perhaps most severe on Madagascar, but it is not a problem confined to developing countries. Nearly half of Hawaii's largest native-dominated lowland rain forest was cleared during 1984 and 1985 in a subsidized endeavor to generate electricity from wood chips.

ECOLOGICAL RESEARCH ON ISLANDS

Controlling the effects of biological invasions on islands is paramount, but there is also a great deal to be gained from studying their effects carefully. The relative simplicity of the biota of many islands perhaps enables invading species to have greater effects on native communities than they would in continental areas; it certainly facilitates a much more complete evaluation of those effects. Better understanding of biological invasions and their consequences for biological diversity on islands will contribute to the development and testing of basic ecological theory on all levels of biological organization. Few of the effects of biological invasions described here are unique to islands; they are only more highly developed and occur most rapidly there, as demonstrated by the invasion of European wild boars into Great Smoky Mountains National Park (Singer et al., 1984). An understanding of the effects of invasions on biological diversity in rapidly responding island ecosystems may give us the time and the tools needed to deal with similar problems on continents; it may even contribute to the prediction and evaluation of the effects of environmental releases of genetically altered organisms.

REFERENCES

Allan, H. H. 1936. Indigene versus alien in the New Zealand plant world. Ecology 17:187–193.

Bramwell, D., ed. 1979. Plants and Islands. Academic Press, London. 459 pp.

Carlquist, S. 1974. Island Biology. Columbia University Press, New York. 660 pp.

Carson, H. L. In press. Colonization and speciation. In A. H. Gray, M. Crawley, and P. J. Edwards, eds. Colonization and Succession. Blackwell Scientific, Oxford.

Clark, D. A. 1981. Foraging patterns of black rats across a desert-montane forest gradient in the Galapagos Islands. Biotropica 13:182–194.

Darwin, C. R. 1859. On the Origin of Species by Means of Natural Selection, or the Preservation of Favored Races in the Struggle for Life. 1st edition. John Murray, London. 502 pp. (Facsimile of the first edition published by the Harvard University Press, Cambridge, 1964.)

Egler, F. E. 1942. Indigene vs. alien in the development of arid Hawaiian vegetation. Ecology 23:14–23.

Ehrlich, P. R., and H. A. Mooney. 1983. Extinction, substitution, and ecosystem services. BioScience 33:248–253.

Elton, C. S. 1958. The Ecology of Invasions by Animals and Plants. Methuen Co., London. 181 pp.

Ewel, J. J. 1986. Invasability: Lessons from south Florida. Pp. 214–230 in H. A. Mooney and J. Drake, eds. The Ecology of Biological Invasions of North America and Hawaii. Springer-Verlag, New York.

Gorman, M. 1979. Island Ecology. Chapman and Hall, London. 79 pp.

Heywood, V. H. 1979. The future of island floras. Pp. 431–441 in D. Bramwell, ed. Plants and Islands. Academic Press, London.

Jacobi, J., and J. M. Scott. In press. An assessment of the current status of native upland habitats and associated endangered species on the Island of Hawai'i. Pp. 3–22 in C. P. Stone, C. W. Smith, and J. T. Tunison, eds. Alien Plant Invasions in Hawaii: Management and Research in Near-Native Ecosystems. Cooperative National Park Resources Studies Unit, University of Hawaii, Honolulu.

Kepler, C. B., and J. M. Scott. 1985. Conservation of island ecosystems. ICBP Tech. Pub. 3:255–271.

King, C. M. 1984. Immigrant Killers: Introduced Predators and the Conservation of Birds in New Zealand. Oxford University Press, Aukland. 224 pp.

Loope, L. L. In press a. An overview of problems with introduced plant species in national parks and reserves of the United States. In C. P. Stone, C. W. Smith, and J. T. Tunison, eds. Alien Plant Invasions in Hawaii: Management and Research in Near-Native Ecosystems. Cooperative National Park Resources Studies Unit, University of Hawaii, Honolulu.

Loope, L. L. In press b. Haleakala National Park and the "island syndrome." In L. K. Thomas, ed. Proceedings of a Symposium on Ecology and Management of Exotic Species. Conference on Science in the National Parks, Ft. Collins, Colorado, July 1986. U.S. National Park Service and The George Wright Society, Washington, D.C.

Loope, L. L., and P. G. Scowcroft. 1985. Vegetation response within exclosures in Hawai'i: A review. Pp. 377–402 in C. P. Stone and J. M. Scott, eds. Hawai'i's Terrestrial Ecosystem: Preservation and Management. Cooperative National Park Resources Study Unit, University of Hawaii, Honolulu.

Lyon, H. L. 1909. The forest disease on Maui, Hawaii. Plant. Rec. 1:151–159.

Medeiros, A. C., L. L. Loope, and F. R. Cole. 1986. Distribution of ants and their effects on endemic biota of Haleakala and Hawaii Volcanoes National Parks: A preliminary assessment. Pp. 39–

51 in Proceedings of the Sixth Conference in Natural Sciences, Hawaii Volcanoes National Park, June 1986. Cooperative National Park Resources Studies Unit, University of Hawaii, Honolulu.

Merlin, M. D., and J. O. Juvik. In press. Relationships between native and alien plants on oceanic islands with and without wild ungulates. In C. P. Stone, C. W. Smith, and J. T. Tunison, eds. Alien Plant Invasions in Hawaii: Management and Research in Near-Native Ecosystems. Cooperative National Park Resources Studies Unit, University of Hawaii, Honolulu.

Mooney, H. A., and J. Drake, eds. 1986. The Ecology of Biological Invasions of North America and Hawaii. Springer-Verlag, New York. 321 pp.

Moulton, M. P., and S. L. Pimm. 1986. Species introductions to Hawaii. Pp. 231–249 in H. A. Mooney and J. Drake, eds. The Ecology of Biological Invasions of North America and Hawaii. Springer-Verlag, New York.

Mueller-Dombois, D. 1983. Canopy dieback and successful processes in Pacific forests. Pac. Sci. 37:317–325.

Mueller-Dombois, D., and G. Spatz. 1975. The influence of feral goats on the lowland vegetation of Hawaii Volcanoes National Park. Phytocoenologia 3:1–29.

Mueller-Dombois, D., K. W. Bridges, and H. L. Carson, eds. 1981. Island Ecosystems: Biological Organization in Selected Hawaiian Communities. Hutchinson-Ross, Stroudsburg, Pa. 583 pp.

Nagata, K. M. 1985. Early plant introductions in Hawai'i. Hawaii. J. Hist. 19:35–61.

Olson, S. L., and H. F. James. 1982. Fossil birds from the Hawaiian Islands: Evidence for wholesale extinction by man before western contact. Science 217:633–635.

Singer, F. J., W. T. Swank, and E. E. C. Clebsch. 1984. Effects of wild pig rooting in a deciduous forest. J. Wildl. Manage. 48:464–473.

Smith, C. W. 1985. Impact of alien plants on Hawai'i's native biota. Pp. 180–250 in C. P. Stone and J. M. Scott, eds. Hawai'i's Terrestrial Ecosystems: Preservation and Management. Cooperative National Park Resources Studies Unit, University of Hawaii, Honolulu.

Stone, C. P., 1985. Alien animals in Hawai'i's native ecosystems: Towards controlling the adverse effects of introduced vertebrates. Pp. 251–297 in C. P. Stone and J. M. Scott, eds. Hawai'i's Terrestrial Ecosystems: Preservation and Management. Cooperative National Park Resources Studies Unit, University of Hawaii, Honolulu.

Stone, C. P., and J. M. Scott, eds. 1985. Hawai'i's Terrestrial Ecosystems: Preservation and Management. Cooperative National Park Resources Studies Unit, University of Hawaii, Honolulu. 584 pp.

Stone, C. P., C. W. Smith, and J. T. Tunison, eds. In press a. Alien Plant Invasions in Hawaii: Management and Research in Near-Native Ecosystems. Cooperative National Park Resources Studies Unit, University of Hawaii, Honolulu.

Stone, C. P., P. K. Higashino, J. T. Tunison, L. W. Cuddihy, S. J. Anderson, J. D. Jacobi, T. J. Ohashi, and L. L. Loope. In press b. Success of alien plants after feral goat and pig removal. In C. P. Stone, C. W. Smith, and J. T. Tunison, eds. Alien Plant Invasions in Hawaii: Management and Research in Near-Native Ecosystems. Cooperative National Park Resources Studies Unit, University of Hawaii, Honolulu.

Tunison, J. T., C. P. Stone, and L. W. Cuddihy. 1986. SEAs provide ecosystem focus for management and research. Park Sci. 6(3):10–13.

Vitousek, P. M. 1986. Biological invasions and ecosystem properties: Can species make a difference? Pp. 163–176 in H. A. Mooney and J. Drake, eds. The Ecology of Biological Invasions of North America and Hawaii. Springer-Verlag, New York.

Vitousek, P. M. In press. Effects of alien plants on native ecosystems. In C. P. Stone, C. W. Smith, and J. T. Tunison, eds. Alien Plant Invasions in Hawaii: Management and Research in Near-Native Ecosystems. Cooperative National Park Resources Studies Unit, University of Hawaii, Honolulu.

Wace, N. M. 1986. Control of rats on islands—research is needed. Oryx 20:79–86.

Wace, N. M., and C. D. Ollier. 1982. Biogeography and geomorphology of South Atlantic Islands. Pp. 733–758 in National Geographic Society Research Reports. National Geographic Society, Washington, D.C.

Wallace, A. R. 1880. Island Life. Macmillan, London. 526 pp.

Wester, L. L. In press. Alien plants and their status in Hawaii. In C. P. Stone, C. W. Smith, and J. T. Tunison, eds. Alien Plant Invasions in Hawaii: Management and Research in Near-Native Ecosystems. Cooperative National Park Studies Unit, University of Hawaii, Honolulu.

THE VALUE OF BIODIVERSITY

Native varieties of peppers and corn at a
village marketplace near Cuzco, Peru, high in
the Andes mountains. *Photo courtesy of Noel
D. Vietmeyer.*

ECONOMICS AND THE PRESERVATION OF BIODIVERSITY

W. MICHAEL HANEMANN

Associate Professor, Department of Agricultural and Resource Economics,
University of California, Berkeley, California

Any analysis of the value of preserving biodiversity requires the attention of many disciplines. The chapters that follow in this section define the role of economics in this endeavor and assess its contribution. In this chapter, I offer a brief overview of some of the issues involved in the economics of biodiversity.

There are many different questions that economists ask in connection with biodiversity. Is economic growth harmful to biological diversity? What are the reasons why it may turn out to be harmful? How can harmful impacts be avoided? What institutions are required to ensure a better outcome? At a more specific level, the questions tend to focus on project analysis and specific environmental policy decisions. For example, what are the benefits and costs of a particular investment project or conservation program? Should they be undertaken?

These questions involve a mixture of normative and positive analyses. For the normative issues, economists have something to contribute—the criterion of economic efficiency—but we recognize the fundamental role of equity considerations and value judgments. Our theories reserve a place for the equity criterion and are capable of tracing its implications for private and public decisions, but as economists, we do not have a theory to explain what those value judgments should be. The positive issues are different, however; these fall squarely within our bailiwick and within those of the other social sciences that seek to explain and predict human behavior. As resource economists, it is one of our direct obligations to measure, explain, and predict how individuals and institutions manage natural resource systems, value biological diversity, and make decisions affecting its preservation.

We may disagree on techniques of measurement and theories of behavior, and some of our measurements and theories may be wrong, but as a discipline, we have standing in this area.

In explaining why anyone may rationally choose to deplete natural resources and destroy ecosystems, mainstream economists adduce a variety of arguments. The most important is intertemporal preferences and discounting. In contrast to conventional commodities, a distinctive feature of natural resources is that they are not instantly renewable; they can be restocked, if at all, only with time and subject to the constraints of biological processes. Consequently, harvesting these resources—whether for commercial gain or otherwise—involves a trade-off between present benefits and future costs that depends on how the latter are discounted relative to the former.

Economists have theories about how the interest rate level, the nature of the net benefit function and its movement over time, and the dynamics of the resource's natural growth process combine to determine the optimal intertemporal path of exploitation—whether or not it is desirable to aim for a steady state and if so, what that steady state should be. Other things being equal, the higher the interest rate at which future consequences are discounted, the more it is optimal to deplete the resource now. We have theories about what interest rates will pertain if resource management decisions are determined by market forces side-by-side with commercial investment decisions, and we also have arguments about why market-based interest rates may be inappropriate for natural resource management (as well as other social investment policies).

With regard to the latter, Sen (1967) and Marglin (1963) in the 1960s put forth the following two arguments. First, a person acting in a public role, e.g., voting as a citizen on an issue of social policy, may place a different weight on the welfare of future generations than he or she does in making private market decisions. Second, even if each person has a single set of preferences, members of the present generation may be willing to join in a collective contract calling for more savings but all while being unwilling to save more in isolation. In effect, the act of sacrificing present consumption opportunities to benefit future generations is a collective good. Both arguments lead to the conclusion that the intertemporal allocations resulting from a decentralized market system may in fact be undesirable to the present generation. This is separate from the commonly voiced argument that the present generation may place too little weight, in terms of some external ethical criterion, on the welfare of future generations.

Not only may the act of conserving natural resources be a collective good, but the resources themselves frequently are collective goods. This is the second major reason why excessive depletion may occur. Either there may not be any well-defined property rights, or the property rights may not create adequate incentives to private individuals to engage in conservation activities. This can result, for example, in a prisoners' dilemma situation: conservation may be the optimal strategy collectively, but it is not a dominant strategy for each individual privately.

The third distinctive feature of many natural systems is the often considerable degree of uncertainty concerning the future consequences of present conservation/depletion actions. This is partly a consequence of the intertemporal aspects of resource management—the further into the future the consequences, the harder to predict them—but it is also due to the enormous variability that is inherent in many natural systems.

There is a variety of economic theories about how uncertainty should be factored into decisions made by private persons or public policy makers. These theories focus on the implications of various forms of risk-averting behavior for the optimal pattern of resource conservation over time. To implement the theories, however, one needs empirical measurements of the type of risk preferences possessed by the decision makers, in the case of a positive analysis, or value judgments about the type of approach to decisions about risk that should be adopted, in the case of a normative analysis.

As with discounting, there are also economic arguments about why private decision makers may be more or less averse to risk than would be appropriate for making social decisions. For example, the presence of financial constraints may make individuals more averse to risk than society would wish to be. Moreover, as Arrow and Lind (1970) have pointed out, society may have opportunities to pool risks that are not available to individual decision makers; it could therefore be appropriate for society to treat small risks in a risk-neutral manner. An important exception, noted by Fisher (1973), is when collective risks cannot be pooled because one person's assumption of the risk does not reduce the risk shouldered by others. Many environmental hazards, including the destruction of biodiversity, are likely to be categorized as collective risk and thus require the application of a risk premium even in public decision making. Why private decision makers might be insufficiently averse to risk has been discussed in some of the other social sciences, particularly in psychology, and by Kahneman et al. (1982), who have charted systematic patterns of bias in individual perceptions of risk. While economics is predisposed to assume rationality in human behavior, it is important to leave some room for other varieties of conduct—short-sightedness, wishful thinking, self-deception, and (occasionally) stupidity—as potent explanations of failures in resource management.

To the extent that the outcomes of some decisions may be irreversible, such as those leading to species extinction, there is an additional twist to the way in which uncertainty and time combine to influence decisions involving the concept of option value. Because the passing of time brings information about the consequences of present actions, there is a premium on actions that preserve the flexibility to exploit this information. If a current decision is physically or economically irreversible, that flexibility is abandoned. To the extent that decision makers disregard the potential value of future information, they will systematically undervalue policies, such as conservation programs, that maintain flexibility and preserve options for future action.

To these explanations of economic decisions that may lead to destruction of biodiversity, Norgaard adds another—the force of specialization on the basis of competitive advantage (see Chapter 23). Whereas an isolated region must produce all the natural resources that its members wish to consume, once the region enters into trade relations with other areas it can satisfy its wants by specializing in the production of a small set of goods and exchanging its surplus production for the other needed commodities. Indeed, it is economically efficient to do this: given differences in relative production efficiencies among trading partners, everybody can gain from the exchange.

There are some qualifications to this conclusion. First, the argument applies in principle to trade between any areas not yet developed, and developing countries. Second, trade between developed and developing countries has historically involved elements of coercion that are extraneous to the argument based on efficiency. Third, the calculation of relative production efficiencies and the estimate of gains from trade involve trade-offs between gains to the present generation and potential losses to future generations through the depletion of nonrenewable resources and may be biased by the use of an inappropriate discount rate for the reasons mentioned above.

All these arguments apply to the management of natural resources in general as well as to the conservation of biological diversity in particular. Insofar as they offer explanations about why decentralized private behavior may result in decisions that are socially undesirable, they also suggest possible remedies, e.g., changes in institutions or specific government policies such as wilderness protection acts, procedures for reviewing proposed development projects, or interest rate subsidies for conservation programs. They are all, of course, economic explanations and, therefore, capture only part of the picture. Ecologists have their own explanations of species extinction, such as the perils of K-selection in a suddenly unstable environment, the requirement of a specialized habitat, or specialized feeding needs, for example. As long as the arguments of economists are regarded as explanations of human behavior rather than as an apologia, they have a legitimate claim to the attention of ecologists and other natural scientists concerned with the preservation of biodiversity.

So far I have focused on the larger questions that economists pose about biodiversity. The more specific questions (e.g., What are the benefits of preserving some particular ecosystem?) may be closer to what a lay audience expects of economists and may explain why it has doubts about their role. In this context, it is crucial to distinguish between positive questions (What value do people place on the preservation of the ecosystem?) and normative ones (What value ought they place on its preservation?) Economists deal professionally with the former; the latter is an ethical question about which they may have feeling but no particular expertise to provide an answer.

With respect to the positive analysis of ecosystem values, it is useful to distinguish between the environment as a marketed good, or an input to the production of marketed goods, and the environment as a nonmarketed good of concern to people in its own right. The first situation is certainly easier to deal with and represents, I believe, the stereotype of what economic analysis is about. The economist figures out the market price of alligator handbags, say, and multiplies that by the reduction in the quantity of alligator handbags resulting from the destruction of alligator habitat. Although that certainly is part of economic analysis, it has not been at the cutting edge of research in environmental economics for the past decade or more.

Environmental economists are interested in markets not because they want to use market prices to multiply something but because they are interested in measuring the preferences of individuals and ascertaining their trade-offs between environmental resources and money or conventional market commodities. Organized markets are one forum in which people reveal their preferences through the choices that they make—but markets are not the only forum, and they are not essential to the enterprise of environmental valuation. Instead, economists have come to rely quite extensively on simulated markets, or their analogs, in which individuals reveal their preferences through interviews or experimental games involving trade-offs between money and environmental outcomes. Moreover, when they do analyze actual markets, economists are interested not in the market prices per se but, rather, in the patterns of selection and the types of preferences that these imply.

Both direct and indirect techniques for eliciting or inferring the preferences of individuals have been greatly refined in recent years. Rather than attempting to summarize them, I will mention several potential limitations of general concern.

First, at what level of aggregation should the valuation be conducted? Should one analyze each species separately or the ecosystem as a whole? In principle, this is an empirical issue, and the solution depends on two sets of factors—the way in which individuals perceive and care for natural environments (What are the aspects that matter to them?) and the way in which the ecosystem functions (What are the biological linkages?) In practice, fashioning a sensible set of units of analysis requires an interdisciplinary approach—a dialog between the natural and social scientists.

The second pertains to the complex types of preferences involved when one is dealing with uncertain and intertemporal outcomes. There is conflicting empirical evidence on how people approach these issues and the types of decision rules they use. As noted above, we have a variety of axiomatic systems and theories about how uncertainty and time could be factored into decision making but much less empirical information on what people do in practice. Moreover, there is some evidence from psychologists that raises doubts about whether individuals have consistent risk or intertemporal preferences at all. At the very least, there is evidence that preferences depend on the type of choice available. To the extent that this is so, the problems of identifying the appropriate choice and appropriate measurement technique, or of extrapolating preferences revealed by one type of choice to the valuation of another, are indeed challenging.

Third, there is the problem of aggregating preferences or values across individuals. Preferences vary among people: some may even dislike biological diversity and prefer concrete parking lots to natural wilderness. In any case, a resource management program is likely to create both winners and losers. How should one sum the gains and losses and compare them with one another? Is everyone weighted equally (which, in principle, enshrines the existing income distribution), or do some count more than others? The question cannot be answered by economists alone: it must be resolved by reference to some philosophical or ethical system.

Fourth, the sheer difficulty of dealing with futurity in resource management and preservation issues can scarcely be overemphasized. So many consequences involve future events, which are extremely difficult to predict—long-run ecosystem impacts, economic variables (such as future energy prices), and even the preferences of future generations, who are not around now for one to observe their behavior. Any economic analysis of the benefits and costs of biodiversity preservation involves predictions (i.e., guesses), some of which will inevitably be wrong. One can attempt to counteract this by choosing sophisticated analytical techniques and decision criteria that recognize the uncertainties and the potential for error, but the feeling of unease cannot be avoided.

Lastly, the postulate underlying the entire enterprise of positive analysis (the legitimacy of attempting to establish what value people place on biodiversity) is itself a value judgment and one that can be questioned. Consumer sovereignty, and the notion that people may have consistent and stable preferences, can be challenged. People may be ignorant, ill-informed, or fickle in their attitudes. Why should we care about what they think? A justification comes from the utilitarian ethical system that mainstream economics embraces. The issue is discussed eloquently by Randall in Chapter 25. The point I want to emphasize here is that this question defines both the strengths and the weaknesses of economic analysis. In their positive analysis, economists are essentially holding a mirror to society. If the picture is unattractive, that is, if people individually and collectively place a low value on the preservation of diversity (which I actually think is not the case), that is not the fault of the economist. It is critical to distinguish the legitimacy of the homocentric, instrumentalist, and utilitarian ethical framework from the degree of success with which economists measure human values. Moreover, there are the fundamental difficulties in making decisions on broad issues of resource preservation. Economics is certainly not immune to them, but neither is it uniquely susceptible to them. These difficulties must be confronted in any type of discourse about human affairs.

In concluding I will return to the special question of biodiversity in Third World countries. Within the economics profession, there has been too little communication between those who specialize in economic development and those who focus on natural resource economics. Except for minerals and energy resources, development economics has paid relatively little attention to biological and environmental implications of economic growth. For its part, environmental economics in countries like the United States, Britain, France, and Scandinavia has its intellectual origins in public finance and cost-benefit analysis. That is to say, it has a domestic focus. Like the doctors in George Bernard Shaw's play, we resource economists have tended to specialize in the diseases of the rich. I hope that this book will help to alter this state of affairs.

In this context I should point out that when economists from the United States and other developed countries urge developing countries to preserve their biological resources, there is a certain awkwardness. It is like an aging rake urging chastity on a young man: the advice is certainly based on a wealth of experience, but it may not be entirely persuasive. It would carry more weight if it were backed up by financial incentives. That is to say, if we want developing countries to protect their biological resources we should be willing to pay them to do so. This topic deserves immediate attention from policy makers as well as from the academic community.

REFERENCES

Arrow, K. J., and R. C. Lind. 1970. Uncertainty and the evaluation of public investment decisions. Am. Econ. Rev. 60(3):364–378.

Fisher, A. C. 1973. Environmental externalities and the Arrow-Lind public investment theorem. Am. Econ. Rev. 63(4):722–725.

Kahneman, D., P. Slovic, and A. Tversky, eds. 1982. Judgment Under Uncertainty: Heuristics and Biases. Cambridge University Press, New York. 555 pp.

Marglin, S. 1963. The social rate of discount and the optimal rate of investment. Q. J. Econ. 77(1):95–111.

Sen, A. K. 1967. Isolation, assurance, and the social rate of discount. Q. J. Econ. 81:112–124.

COMMODITY, AMENITY, AND MORALITY
The Limits of Quantification in Valuing Biodiversity

BRYAN NORTON

Professor of Philosophy, Division of Humanities, New College of the
University of South Florida, Sarasota, Florida

What is the value of the biological diversity of the planet? That question reminds me of a game we used to play at ice cream socials and church picnics when I was growing up in the Midwest. Someone on the entertainment committee would count an assortment of screws and gimcracks, or nuts and bolts, and put them into a mason jar. At the Christmas party, it was pecans, walnuts, and hickory nuts. Everybody else had to guess: How many whatchamacallits are in the jar?

Pretend we're having an ice cream social on an improved version of the space shuttle. Someone looks down and says, "What's the value of the life on that planet down there?" The closest guess wins a door prize.

But our question is tougher than nuts and bolts. Recently, scientists discovered bones from a dinosaur they have called seismosaurus. That animal was 18 feet tall, more than 100 feet long, and weighed 80 tons. The diversity in size between a seismosaurus and the smallest microbe is staggering. And I used to be thrown off when they put washers of two different sizes in the mason jar! Given the diversity in size among species, not to mention the fact that many species live inside others, it is not surprising that scientists have left themselves some latitude in their guesses as to how many species there are: they estimate that there are between 5 and 30 million species.

That's O.K. I never did very well at the guessing game myself. One time I guessed that a jar contained 452 nuts and bolts. The correct answer was more than 2,000. I won the booby prize for being the farthest off; my prize was the jar and its contents.

But again, I can't help mentioning how much more difficult our current task is. We would hardly have begun to place a value on biodiversity if we had known how many species there are. We're supposed to put a *value* on them. In what terms?

When I looked into a jar, I was always a bit overwhelmed at first, so let's not give up yet. Eventually, I'd decide to be systematic about my guess. I'd divide the jar into somewhat equal sections and try to do a rough count for one of them. Then, I'd multiply by the number of sections. Despite my lack of success, it's a reasonable approach; we can call it the divide-and-conquer method. Economists and other policy analysts have adopted a similar method for valuing biotic resources. They usually try to estimate, however roughly, a value for one species (Fisher, 1981; Fisher and Hanemann, 1985). If they could assign a value to a few species, such as the snail darter, the Furbish lousewort, and the California condor, then we might average the values of those species and then multiply that average value by the number of species there are, if we only knew how many species there are.

All this averaging and multiplying will require that we use numerical values, so we might as well follow economists in trying to use present dollars as the unit of value. Before introducing the technical terms used by economists, let's start with some ordinary concepts: species can have value as commodities and as amenities, and they can have moral value.

We'll say that a species has *commodity value* if it can be made into a product that can be bought or sold in the marketplace. In this category, alligators have potential value in the manufacture of shoes, but they may also have indirect commodity value if it turns out that vinyl shoes stamped in an alligator pattern sell for more than plain vinyl shoes. Indirect value of this sort is especially important in the pharmaceutical industry, since many of our most valuable medicines are synthetic copies of biologically produced chemicals (Lewis and Elvin-Lewis, 1977; Myers, 1983).

A species has *amenity value* if its existence improves our lives in some nonmaterial way, e.g., when we experience joy at sighting a hummingbird or when we enjoy walks in the forest more when we sight a ladyslipper. Hiking, fishing, hunting, bird-watching, and other pursuits have a huge market value as recreation, and wild species contribute, as amenities, to these activities. Bald eagles, for example, have not only inspired the production of millions of dollars worth of Americana, but they also generate aesthetic excitement through a whole area that is blessed with a nesting pair of them.

Finally, species have *moral value*. Here, we begin to encounter controversy. Some philosophers would say that species have moral value on their own. They are, according to this view, valuable in themselves, and their value is not dependent on any uses to which we put them (Regan, 1981; Taylor, 1986). We will not be able to settle this issue. Suffice it to say that species have moral value even if that moral value depends on us. Here, Thoreau comes to mind. He believed that his careful observation of other species helped him to live a better life (Thoreau, 1942). I believe this also. So there are at least two people, and perhaps many others, who believe that species have value as a moral resource to humans, as a chance for humans to form, re-form, and improve their own value systems (Norton, 1984; Norton, in press).

Moral values that people attach to species are quite high. Responses to questionnaires have indicated that people place a surprisingly high value on just the knowledge that a thing exists independent of any use (Randall, 1986, and Chapter 25 of this book). Economists, using a method called *contingent valuation*, create shadow markets in which they can ask people how much they would be willing to pay to protect a species, quite independent of any use of the species (see Chapter 25). If existence values can be thought of as a rough indicator of moral values for present purposes, we can say that species also have considerable moral value, measurable in dollars.

So, we can say with some confidence that some species have considerable commodity, amenity, and moral value. The problem that economists have encountered is that these values are distributed very unevenly among species, at least given our current knowledge. For example, Hanemann and Fisher (1985) have surmised that under certain assumptions, a wild grass recently discovered in Mexico, a perennial related to corn, may prove to have a value of $6.82 billion annually, and they calculated its value for only one possible use—the creation of a perennial hybrid of corn (Fisher and Hanemann, 1985; see also Chapters 10 and 11 of this volume).

At present, however, we do not have sufficient knowledge to calculate the value of most species. Consequently, in addition to the known values that economists note with respect to some small number of species, they also calculate an *option value* for species of unknown worth, i.e., the value we should place on the possibility that a future discovery will make useful a species that we currently think useless (Fisher and Hanemann, 1985; see also Chapter 25 of this book). If we extinguish a species now, such discoveries are precluded. Fisher and Hanemann therefore define option value as the present benefit of holding open the possibility that some species we might eradicate today may prove valuable in the future. They would ask people how much they are willing to pay to retain the option of saving the species, given the possibility that new knowledge indicating its value may be discovered in the future.

One important aspect of option value is that it applies equally to commodity, amenity, and morality. As time passes, we gain knowledge in all of these areas, and new knowledge may lead to new commodity uses for a species or to a new level of aesthetic appreciation, or our moral values may change and some species will, in the future, prove to have moral value that we cannot now recognize.

If placing a dollar figure on these option values seems a daunting task, the situation is actually far worse than it first seems. Calculations of option value can only be begun after we identify a species, guess what uses that species might have, place some dollar value on those uses, and estimate the likelihood of such discoveries occurring at any future date (so that we can discount the values across time). Once we've done all that, we can try to figure out how to translate those future, possible values into present dollars. I think it is safe to say that despite the great theoretical interest in assigning use and option values to species, and some impressive strides in modeling these formally, it may be a long time before the total value of even one species can be stated in terms of present dollars (Norton, in press).

It is worth stepping back to look at the most difficult problems faced by the divide-and-conquer method. First, there is the problem of irreversibility. In general, economists have trouble with decisions where one of the options cannot be reversed. This is an especially important problem for biodiversity. If we decide to have a dam and give up a species, blowing up the dam won't bring the species back.

Second, we are forced to make present decisions under conditions of uncertainty—another problem for assigning present values. Our ignorance of species is mind-boggling. Suppose you're walking on a hillside in Mexico. Your eyes fall on a few tufts of nondescript grass. Would you guess that grass is worth $6.82 billion annually? Only if you knew that it was a member of the corn family, that it is a perennial, that . . . , and so on. Scientists believe that they have identified and named approximately 15% of the species on Earth (Myers, 1979), and we have rudimentary knowledge of the life characteristics of only a few of them. It is an understatement to refer to this level of ignorance as mere "uncertainty."

A third problem with the divide-and-conquer method derives from ecological knowledge. Species do not exist independently; they have coevolved in ecosystems on which they depend. This means that each individual species depends on some set of other species for its continued existence. A species may depend on just one other species for food, or it may depend on an entire complex of interrelated species. This seems to imply that if we now take actions that cause the extinction of any species, then the loss in future benefits should include losses accruing if any other dependent species succumbs as well. Species on which others depend therefore have contributory value in addition to their direct uses (Norton, in press). To extinguish a species on which two other species depend is to extinguish three species. Thus to get the full value of a species, we would somehow have to determine the values of all the other species that depend on it.

It also appears that some species are keystones in their ecosystems. For example, when the Florida alligator populations dipped dangerously low about 15 years ago, wildlife biologists noticed that many populations of other species also declined. During the dry winters in the Florida Everglades, other species depended on alligator wallows as their source of water (Taylor, 1986). Must we say then, that the value of the alligator includes the value of most of the wildlife in the Everglades?

In principle, these ecological facts add no complication. We need only factor in the ecological information regarding the interdependencies among species in ecosystems. Then, we could tally the direct uses and option values of a species and add to this the uses and option values of all dependent species, and so forth. But, of all the areas of biology and ecology, few are less understood than interspecific dependencies. Ecologists cannot even identify all the interdependencies in the systems they understand best. There is no hope that sufficient information will become available for us to determine the interdependencies in tropical forest ecosystems before the forests are destroyed.

Aside from all these problems, the divide-and-conquer method is not even asking the right question. The value of biological diversity is more than the sum of its parts. Even if we could place a value on the biological diversity represented by all species, we would be only part way to an answer to the question, "What is the value of biodiversity?" To answer that question, we would have to include also the genetic variation within species across populations and the variety of interrelationships in which species exist in different ecosystems.

The reason my guesses on nuts and bolts were often very far off, even with my divide-and-conquer method, was that I never completed my calculations before an answer was required. I was always overcome by the uncertainties involved. Did the little area I counted represent one-twentieth or one-twenty-fifth of the jar? Is it representative? In order to answer that, I'd shake the jar, only to discover all the small washers were at the bottom. So, I'd have to count again and recalculate. "Time's up. Turn in the scrap of paper with your name and number. The game's over." I'd end up writing down a random number and suffering the embarrassment attendant thereto. As species become extinct at an ever-increasing rate, resulting in the loss of a fifth or a fourth of all species in the next two decades, according to various estimates, I fear economists and biologists are in a similar situation.

Rather than continuing my attempt to answer this difficult question on the value of diversity, it may make more sense to take a careful look at the question itself and why we are trying to answer it. The question says a lot about us, the questioners. It is a measure of our unique arrogance that we are the only species that calls symposia and writes books to address that question. The sense of arrogance is hardly diminished when we note our usual reasons for asking it. Why are some people so insistent that we put dollar values on species diversity? Because, we are told, important decisions are being made that may extinguish other species. These decisions must be based on some kind of analytic framework (which means each species must be given value in our economic system). If we do not put some dollar value on a species, it will get left out altogether. In other words, they want us to put dollar values on species so they can compare these to the value of real estate around reservoirs and to kilowatt-hours of hydroelectric power.

Suddenly, the fun goes out of our guessing game. A new analogy seems more apt: I have been in a terrible accident, and I wake up in a hospital bed on a life-support system. The hospital is short on funds, and the hospital administrators are having a meeting at my bedside. They say they have examined all the other methods to raise the necessary money, and they are proposing to sell a few spare parts from my life-support system at a yard sale. One of them says, "This equipment is so complicated, a few parts won't be missed." "How much do you think this part is worth?" asks another, pointing toward a piece of shiny metal. I try to see what the part is connected to, but it is screwed into a big metal box that looks important. "Or that one over there; it looks like it's just cosmetic," another of them suggests. I almost agree, and then I notice that a main power line passes through it. "Stop! Not that one," I say. Just in time.

It is one thing to treat the valuation of biodiversity as a guessing game or as a set of very interesting theoretical problems in welfare economics. It is quite another thing to suggest that the guesses we make are to be the basis of decision making that will affect the functioning of the ecosystems on which we and our children will depend for life.

If we are not taken seriously unless we quantify our answer, I would like to suggest some new units of measurement. An *oops* is the smallest unit of chagrin that we would feel if we willfully extinguish a species we need later on. A *boggle* is the amount of ignorance encountered when an economist asks a biologist a question about species and ecosystems, and the biologist answers: "I don't know, and I'm so far from knowing, it boggles the mind." If I understand what the economists are saying, irreversible oopses and boggles of uncertainty are the main factors in decisions affecting biodiversity. In the passion to express the values of a species in dollar figures, it will be unfortunate if we forget to count oopses and boggles as well.

I believe that we should abandon the divide-and-conquer approach. I suggest we use the big picture method instead. Now, the question is easier. The value of biodiversity is the value of everything there is. It is the summed value of all the GNPs of all countries from now until the end of the world. We know that, because our very lives and our economies are dependent upon biodiversity. If biodiversity is reduced sufficiently, and we do not know the disaster point, there will no longer be any conscious beings. With them will go all value—economic and otherwise.

I am afraid this answer will not be useful to those who want to know the value lost when they act to extinguish a species, but it seems a better answer than a guess, even a guess that counts oopses and boggles as well as dollars.

One thing we know: if we lose enough species, we will be sorry. The guessing game is really Russian roulette. Each species lost without serious consequences has been a blank in the chamber. But how can we know before we pull the trigger? That is the question we should be asking (Ehrlich and Ehrlich, 1981).

REFERENCES

Ehrlich, P. R., and A. Ehrlich. 1981. Extinction. The Causes and Consequences of the Disappearance of Species. Random House, New York. 305 pp.

Fisher, A. C. 1981. Economic Analysis and the Extinction of Species. Report No. ERG-WP-81-4. Energy and Resources Group, Berkeley, Calif. 19 pp.

Fisher, A. C., and W. M. Hanemann. 1985. Option Value and the Extinction of Species. California Agricultural Experiment Station, Berkeley. 35 pp.

Lewis, W. H., and M. P. F. Elvin-Lewis. 1977. Medical Botany. John Wiley & Sons, New York. 515 pp.

Myers, N. 1979. The Sinking Ark. Pergamon, Oxford. 307 pp.

Myers, N. 1983. A Wealth of Wild Species: Storehouse for Human Welfare. Westview Press, Boulder, Colo. 274 pp.

Norton, B. G. 1984. Environmental ethics and weak anthropocentrism. Environ. Ethics 6:131–148.

Norton, B. G. In press. Why Save Natural Variety? Princeton University Press, Princeton, N.J.

Randall, A. 1986, Human preferences, economics, and the preservation of species. Pp. 79–109 in B. G. Norton, ed. The Preservation of Species. Princeton University Press, Princeton, N.J.

Regan, T. 1981. The nature and possibility of an environmental ethic. Environ. Ethics 3:19–34.

Taylor, P. W. 1986. Respect for Nature. Princeton University Press, Princeton, N.J. 323 pp.

Thoreau, H. D. 1942. Walden. New American Library, New York. 221 pp.

THE RISE OF THE GLOBAL EXCHANGE ECONOMY AND THE LOSS OF BIOLOGICAL DIVERSITY

RICHARD B. NORGAARD

Associate Professor of Agricultural and Resource Economics, University of California,
Berkeley, California

T he global loss of biological diversity has been described as a product of two phenomena. First, population levels have forced the transformation of heretofore relatively undisturbed areas into lands used for agriculture. Second, both industrial and agricultural pollutants have applied a new and narrowly uniform selective pressure on species. Population growth and technological changes have a multiple, rather than simply additive, impact on biological diversity. There has, however, been a third and probably equally important change that factors into the explanation. During this past century, world agriculture has been transformed from a patchwork quilt of nearly independent regions to a global exchange economy. This change in social organization also contributes to the loss of diversity.

While historians and anthropologists maintain a wealth of knowledge about our past, most people—including developmental economists, planners, and agricultural scientists—have little conceptual understanding of the development process prior to modernization. The past was traditional instead of modern, preindustrial instead of industrial, earlier on the road of progress, a void relative to the present. Neither neoclassical nor Marxist economic theory explains how the human population doubled eight times between the agricultural revolution and the industrial revolution without a proportionate accumulation of capital and use of materials and energy (Norgaard, 1984). A richer vision of the past might help us understand the present.

The world before the industrial revolution can be envisioned as a mosaic of coevolving social and ecological systems. Within each area of the mosaic, species were selected for characteristics according to how well they fit the evolving values, knowledge, social organization, and technologies of the local people. At the same time, each of these components of the social system was also evolving under the selective pressure of how well it fit the evolving ecological system and the other social components. Local knowledge, embedded in myths and traditions, was correct, for it had proven fit and through selective evolutionary pressure, had become consistent with the components of social and ecological systems it explained (Norgaard, 1984).

Within the coevolving mosaic, the boundaries of each area were not distinct or fixed. Myths, values, social organization, technologies, and species spilled over the boundaries of the areas of the mosaic within which they initially coevolved to become exotics in other areas. Some of these exotics were preadapted and thrived; some coevolved; and some died out. But to some extent they all influenced the further coevolution of system characteristics in their new areas. Because of the many combinations of spillovers, the pattern of coevolving species, myths, organization, and technology remained patchy and constantly changing.

Tattered remnants of coevolutionary agricultural development remain today to give us clues to the past. A few agricultural scientists during the past decade have followed the path of anthropologists and discovered a wide array of traditional agroecosystems (Altieri and Letourneau, 1982; Chacon and Gliessman, 1982; Gliessman et al., 1981). In nearly all these systems, farmers deliberately intermix many crop and noncrop species and occasionally animal species. These agroecosystems coevolved with the values, beliefs about nature, technologies, and social organization of indigenous peoples over centuries, sometimes millennia. Farmers selected for adequate and stable rates of food production through as much of the growing season as possible. A dependable food supply was achieved in part by planting many different crops in different places at different times such that average production from year to year varied little because of the law of large numbers (Richards, 1985).

The increased interest in agroecology coincides with an increased recognition of people as biological participants. Whereas natural historians have consistently portrayed the influence of humans as destructive of natural systems, we are now beginning to learn how traditional people at low population densities were less destructive and under some circumstances contributed to the growth of genetic diversity (Alcorn, 1984; Altieri and Merrick, 1987; Brush, 1982). There traditional people created environments within which plants and microorganisms coevolved under selective pressures that were different from those that occur in environments only marginally disturbed by people. Environmental uniformity was not imposed; farmers developed different approaches to agriculture for different microenvironments, adding to variation in selective pressure (Richards, 1985).

Recent development has been distinctly different from the coevolving mosaic of the past. The mechanistic grid of universal truths developed by Western science has boldly overlaid and simplified most of the elaborate coevolutionary mosaic. The global adoption of Western knowledge and technologies has set disparate cultures on convergent paths. And the environment has not been immune to this globally unifying process. Environments are also merging due to the common selective pressure from the cropping, fertilization, and pest control practices of modern agriculture. Global markets, global values, global social organizations, and global technologies have resulted in global criteria for environmental fitness. Diversity of all kinds has been lost.

The economic way of thinking sustains the global exchange economy. The concepts of comparative advantage, specialization, and the gains from exchange are central to the neoclassical economic model. Comparative advantage stems from differences in the productivities of people, tools, and land in various economic activities. It immediately follows that total output can be increased through specialization of people, tools, and land in those activities for which they have a comparative advantage. Specialization in particular activities leaves each producer with lots of one product. Producers then exchange with each other until they have a mix of goods, which makes each of them as happy as possible given the willingness of others to exchange. Comparative advantage, the efficiency of specialization, and the gains through exchange are basic to our understanding of economic systems and to our understanding of the development process.

The gain from trade arguments underlies many development policies and justifies many specific projects. Road construction, much of it financed by international lending agencies, has encouraged traditional farmers to switch to cash crop agriculture, specializing in only a few crops according to market prices rather than to criteria of sustainable environmental management. Farmers who once planted diverse crops for subsistence thus have become connected with the global exchange. Other subsistence farmers were simply bought out or moved out by larger commercial agricultural ventures. Since labor with specific skills as well as capital equipment can be purchased in the market, the pattern of agriculture tends to be determined by the physical environment. For this reason, large, physically homogeneous regions now specialize in only a few crops.

The reduction in the number of crop species grown results in an even larger reduction in the number of supporting species. The locally specific nitrogen-fixing bacteria, fungi that facilitate nutrient intake through mycorrhizal association, predators of pests, pollinators and seed dispersers, and other species that coevolved over centuries to provide environmental services to traditional agroecosystems have become extinct or their genetic base has been dramatically narrowed. Deprived of the flora with which they coevolved, soil microbes disappear. Specialization, exchange, and the consequent regional homogeneity of crop species have reduced biological diversity.

Participation in the global exchange economy also transforms local agroecosystems because it forces farmers to stay competitive with other farmers who have been put in the same bind. This encourages use of inputs common to modern agriculture worldwide—fertilizers, pesticides, and high-yielding seed varieties—thereby eliminating many of the remaining regional differences in selective pressure (Ehrlich and Ehrlich, 1981). The adoption of modern technologies, however, must be understood in context of the complementary change in social organization (de Janvry, 1981).

The global exchange economy also induces temporal variation for which species have not evolved the strategies needed to cope. Crop failures, new technologies, changing tastes, variations in interest rates, changes in the strength of cartels, and variations in trade barriers—all these redefine comparative advantage. This re-definition is accommodated, at least in theory, by a shift in the specialization of people, tools, and land to different lines of production and by a new pattern of exchange. Economists assume that factors involved in production are mobile, i.e., labor, capital, and land can shift between lines of production in a way that optimizes benefit to all.

Other things being equal, these adjustments to exogenous change lead to eco-nomic well-being. With all producers adjusting to compensate for a change in the best possible way, the overall impact of the change is minimized. The adjustments keep aggregate well-being as close to the undisturbed maximum as possible and hence more stable than it would be if the adjustments did not take place. But this stabilizing process for humanity as a whole increases the amount of change for individuals in terms of who does what with which tools and land. Variation in aggregate economic welfare is reduced by increasing the variation for the individual components in the economic system.

The economic model is used for designing exchange policies based on the implicit assumption that land can move between uses much like people and tools. But environmental services cannot freely shift from the support of rice to the support of cotton, to suburban lawns, to concrete, to alfalfa, to marsh habitat for waterfowl, and back to rice much the same as a reasonably adaptive person might shift from being a farmer to an urban gardener, to a game warden, and back to being a farmer.

There are many similarities between economic and ecological models (Rapport and Turner, 1977). Economic models have people with different capabilities filling different niches much like different species fill different niches. But the two models differ dramatically with respect to how the systems are presumed to adjust to exogenous change. Biological species evolve to fill their niches. The recent shift from thinking of each species as having evolved individually in response to a changing physical environment to thinking that species coevolved has led to a new understanding of evolutionary dynamics (Lewin, 1986). The new emphasis also stresses how the coevolutionary process defines the niches themselves. Ecol-ogists do not assume that predefined species sort themselves into predefined niches according to their comparative advantages, resulting in what is best for all given the exogenous influences at the time. The differences between economic and ecological understanding help explain why the global exchange economy has led to extinction.

May (1973) hypothesized that biological diversity is greater in the tropics than in the arctic because the climatic constancy facilitated the evolution of greater niche specialization. This conjecture matches theory with evidence very nicely and has considerable appeal. Climates with little variation lead to the coevolution of highly specialized, interdependent species dependent on particular conditions. Con-versely, in a tropical rain forest ecosystem, a small change from the conditions in which species coevolved is more likely to lead to extinction than a change of comparable magnitude in an arctic tundra system. This explains why the tropical rain forests, with their great species diversity and the complex relationships among them, have proven so vulnerable to changes wrought by modern technologies.

There would not be a problem if the species that supply the environmental services appropriate to particular crops could coevolve to fill their supporting niches as fast as the global exchange economy leads farmers to shift crops. In this sense, the mismatch between economic and ecological models can be reduced to differences in the speed of their adjustment. Adjustment rates are important even within economics. Doctors can switch from treating a flu epidemic in March to advising on hay fever in April. The auto industry, however, cannot shift from the production of low-fuel- to high-fuel-efficiency cars and back again as fast as the Organization of Petroleum Exporting Countries coalesced and raised the price of oil and then collapsed and let the price fall. The oil price perturbations and differences in adjustment rates have resulted in the extinction of the specialized auto companies and many firms that supported the auto industry. Fluctuations in agricultural prices have similarly put farmers out of business. This understanding simply needs to be extended to biological species.

The tractable, formal models of economics dominate development policy and generate the arguments for exchange. But by linking with the global exchange, all participants in the economy are forced to respond to each other. The difficulties of adjustment imposed upon capitalists, entrepreneurs, and laborers are not a part of the tractable model. But the hardships are very well acknowledged informally and in practice. Every reasonably developed economy has additional mechanisms—unemployment insurance, the expending of moving costs, and capital loss write-offs—to cushion and reduce the hardships of adjustment. Our informal acknowledgment of the hardship, however, has not been extended to biological species. As a minimal remedial measure, we should protect biological species from the hardships of adjustment to the exchange economy much like we protect people.

People and the economic decisions they make are an integral part of the ecological system. To think of them separately is one of the unfortunate consequences of the idea of objective knowledge. The diversity of the ecological system is intimately linked to the diversity of economic decisions people make. There was considerable economic diversity in the past due to cultural diversity. How people interact with ecological systems today is heavily influenced by the signals—common over large areas, yet unstable over time—of the global exchange economy.

Economists heretofore have used economic reasoning to determine when extinction is economically rational, to estimate the economic value of species, and to suggest measures to correct the economic system to compensate for the absence of markets for diversity. In these efforts, the neoclassical economic model is used as the starting point. In this chapter, I have shown how this economic view of the world leads to extinction because of the presumption that gains will result from the shifting of production factors into activities for which they have a comparative advantage. Since biological species are generally less able than people to shift between lines of activity, the implications are clear. Economists should give more attention to the basic assumptions of their model and interpret the conclusions of economic arguments in the context of these limitations. The global economy is constantly being fine-tuned by the actions of each country as they amend their trade, aid, and lending and borrowing policies. These amendments should be made with more attention to ecological considerations.

REFERENCES

Alcorn, J. B. 1984. Development policy, forests, and peasant farms: Reflections on Huastec-managed forests' contributions to commercial production and resource conservation. Econ. Bot. 38:389–406.

Altieri, M. A., and D. K. Letourneau. 1982. Vegetation management and biological control in agroecosystems. Crop Protect. 1:405–430.

Altieri, M. A., and L. C. Merrick. 1987. In situ conservation of crop genetic resources through maintenance of traditional farming systems. Econ. Bot. 41:86–96.

Brush, S. B. 1982. The natural and human environment of the central Andes. Mt. Res. Dev. 2:14–38.

Chacon, J. C., and S. R. Gliessman. 1982. Use of the "non-weed" concept in traditional agro-ecosystems of south-eastern Mexico. Agro-Ecosystems 8:1–11.

de Janvry, A. 1981. The Agrarian Question and Reformism in Latin America. Johns Hopkins University Press, Baltimore. 311 pp.

Ehrlich, P. R., and A. Ehrlich. 1981. Extinction. The Causes and Consequences of the Disappearance of Species. Random House, New York. 305 pp.

Gliessman, S. R., E. R. Garcia, and A. M. Amador. 1981. The ecological basis for the application of traditional agricultural technology in the management of tropical agro-ecosystems. Agro-Ecosystems 7:173–185.

Lewin, R. 1986. Punctuated equilibrium is now old hat. Science 231(Feb. 14):672–673.

May, R. M. 1973. Stability and Complexity of Model Ecosystems. Princeton University Press, Princeton, N.J. 235 pp.

Norgaard, R. B. 1984. Coevolutionary development potential. Land Econ. 60(2):160–173.

Rapport, D. J., and J. E. Turner. 1977. Economic models and ecology. Science 195(Jan. 28):367–373.

Richards, P. 1985. Indigenous Agricultural Revolution: Ecology and Food Production in West Africa. Hutchinson, London; Westview Press, Boulder, Colo. 192 pp.

WHY PUT A VALUE ON BIODIVERSITY?

DAVID EHRENFELD

Professor of Biology, Cook College, Rutgers University, New Brunswick, New Jersey

In this chapter, I express a point of view in absolute terms to make it more vivid and understandable. There are exceptions to what I have written, but I will let others find them.

That it was considered necessary to have a section in this volume devoted to the value of biological diversity tells us a great deal about why biological diversity is in trouble. Two to three decades ago, the topic would not have been thought worth discussing, because few scientists and fewer laymen believed that biological diversity was—or could be—endangered in its totality. Three or four decades before that, a discussion of the value of biological diversity would probably have been scorned for a different reason. In the early part of this century, that value would have been taken for granted; the diversity of life was considered an integral part of life, and one of the nicest parts at that. Valuing diversity would, I suspect, have been thought both presumptuous and a terrible waste of time.

Now, in the last part of the twentieth century, we have meetings, papers, and entire books devoted to the subject of the value of biological diversity. It has become a kind of academic cottage industry, with dozens of us sitting at home at our word processors churning out economic, philosophical, and scientific reasons for or against keeping diversity. Why?

There are probably many explanations of why we feel compelled to place a value on diversity. One, for example, is that our ability to destroy diversity appears to place us on a plane above it, obliging us to judge and evaluate that which is in our power. A more straightforward explanation is that the dominant economic realities of our time—technological development, consumerism, the increasing size of governmental, industrial, and agricultural enterprises, and the growth of human populations—are responsible for most of the loss of biological diversity. Our lives and futures are dominated by the economic manifestations of these often hidden processes, and survival itself is viewed as a matter of economics (we speak of tax shelters and safety nets), so it is hardly surprising that even we conservationists have begun to justify our efforts on behalf of diversity in economic terms.

It does not occur to us that nothing forces us to confront the process of destruction by using its own uncouth and self-destructive premises and terminology. It does not occur to us that by assigning value to diversity we merely legitimize the process that is wiping it out, the process that says, "The first thing that matters in any important decision is the tangible magnitude of the dollar costs and benefits." People are afraid that if they do not express their fears and concerns in this language they will be laughed at, they will not be listened to. This may be true (although having philosophies that differ from the established ones is not necessarily inconsistent with political power). But true or not, it is certain that if we persist in this crusade to determine value where value ought to be evident, we will be left with nothing but our greed when the dust finally settles. I should make it clear that I am referring not just to the effort to put an actual price on biological diversity but also to the attempt to rephrase the price in terms of a nebulous survival value.

Two concrete examples that call into question this evaluating process come immediately to mind. The first is one that I first noticed a number of years ago: it was a paper written in the *Journal of Political Economy* by Clark (1973)—an applied mathematician at the University of British Columbia. That paper, which everyone who seeks to put a dollar value on biological diversity ought to read, is about the economics of killing blue whales. The question was whether it was economically advisable to halt the Japanese whaling of this species in order to give blue whales time to recover to the point where they could become a sustained economic resource. Clark demonstrated that in fact it was economically preferable to kill every blue whale left in the oceans as fast as possible and reinvest the profits in growth industries rather than to wait for the species to recover to the point where it could sustain an annual catch. He was not recommending this course— just pointing out a danger of relying heavily on economic justifications for conservation in that case.

Another example concerns the pharmaceutical industry. It used to be said, and to some extent still is, that the myriad plants and animals of the world's remaining tropical moist forests may well contain a great many chemical compounds of potential benefit to human health—everything from safe contraceptives to cures for cancer. I think this is true, and for all I know, the pharmaceutical companies think it is true also, but the point is that this has become irrelevant. Pharmaceutical researchers now believe, rightly or wrongly, that they can get new drugs faster and cheaper by computer modeling of the molecular structures they find promising on theoretical grounds, followed by organic synthesis in the laboratory using a host of new technologies, including genetic engineering. There is no need, they claim, to waste time and money slogging around in the jungle. In a few short years, this so-called value of the tropical rain forest has fallen to the level of used computer printout.

In the long run, basing our conservation strategy on the economic value of diversity will only make things worse, because it keeps us from coping with the root cause of the loss of diversity. It makes us accept as givens the technological/ socioeconomic premises that make biological impoverishment of the world inevitable. If I were one of the many exploiters and destroyers of biological diversity, I would like nothing better than for my opponents, the conservationists, to be bogged down over the issue of valuing. As shown by the example of the faltering search for new drugs in the tropics, economic criteria of value are shifting, fluid, and utterly opportunistic in their practical application. This is the opposite of the value system needed to conserve biological diversity over the course of decades and centuries.

Value is an intrinsic part of diversity; it does not depend on the properties of the species in question, the uses to which particular species may or may not be put, or their alleged role in the balance of global ecosystems. For biological diversity, value *is*. Nothing more and nothing less. No cottage industry of expert evaluators is needed to assess this kind of value.

Having said this, I should stop, but I won't, because I would like to say it in a different way.

There are two practical problems with assigning value to biological diversity. The first is a problem for economists: it is not possible to figure out the true economic value of any piece of biological diversity, let alone the value of diversity in the aggregate. We do not know enough about any gene, species, or ecosystem to be able to calculate its ecological and economic worth in the larger scheme of things. Even in relatively closed systems (or in systems that they pretend are closed), economists are poor at describing what is happening and terrible at making even short-term predictions based on available data. How then should ecologists and economists, dealing with huge, open systems, decide on the net present or future worth of any part of diversity? There is not even a way to assign numbers to many of the admittedly most important sources of value in the calculation. For example, we can figure out, more or less, the value of lost revenue in terms of lost fisherman-days when trout streams are destroyed by acid mine drainage, but what sort of value do we assign to the loss to the community when a whole generation of its children can never experience the streams in their environment as amenities or can never experience home as a place where one would like to stay, even after it becomes possible to leave.

Moreover, how do we deal with values of organisms whose very existence escapes our notice? Before we fully appreciated the vital role that mycorrhizal symbiosis plays in the lives of many plants, what kind of value would we have assigned to the tiny, threadlike fungi in the soil that make those relationships possible? Given these realities of life on this infinitely complex planet, it is no wonder that contemporary efforts to assign value to a species or ecosystem so often appear like clumsy rewrites of "The Emperor's New Clothes."

The second practical problem with assigning value to biological diversity is one for conservationists. In a chapter called "The Conservation Dilemma" in my book *The Arrogance of Humanism*, I discussed the problem of what I call nonresources (Ehrenfeld, 1981). The sad fact that few conservationists care to face is that many species, perhaps most, do not seem to have any conventional value at all, even hidden conventional value. True, we can not be sure which particular species fall into this category, but it is hard to deny that there must be a great many of them. And unfortunately, the species whose members are the fewest in number, the rarest, the most narrowly distributed—in short, the ones most likely to become extinct—are obviously the ones least likely to be missed by the biosphere. Many of these species were never common or ecologically influential; by no stretch of the imagination can we make them out to be vital cogs in the ecological machine. If the California condor disappears forever from the California hills, it will be a tragedy: but don't expect the chaparral to die, the redwoods to wither, the San Andreas fault to open up, or even the California tourist industry to suffer—they won't.

So it is with plants (Ehrenfeld, 1986). We do not know how many species are needed to keep the planet green and healthy, but it seems very unlikely to be anywhere near the more than quarter of a million we have now. Even a mighty dominant like the American chestnut, extending over half a continent, all but disappeared without bringing the eastern deciduous forest down with it. And if we turn to the invertebrates, the source of nearly all biological diversity, what biologist is willing to find a value-conventional or ecological— for all 600,000-plus species of beetles?

I am not trying to deny the very real ecological dangers the world is facing; rather, I am pointing out that the danger of declining diversity is in great measure a separate danger, a danger in its own right. Nor am I trying to undermine conservation; in fact, I would like to see it find a sound footing outside the slick terrain of the economists and their philosophical allies.

If conservation is to succeed, the public must come to understand the inherent wrongness of the destruction of biological diversity. This notion of wrongness is a powerful argument with great breadth of appeal to all manner of personal philosophies.

Those who do not believe in God, for example, can still accept the fact that it is wrong to destroy biological diversity. The very existence of diversity is its own warrant for survival. As in law, long-established existence confers a powerful right to a continued existence. And if more human-centered values are still deemed necessary, there are plenty available—for example, the value of the wonder, excitement, and challenge of so many species arising from a few dozen elements of the periodic table.

And to countenance the destruction of diversity is equally wrong for those who believe in God, because it was God who, by whatever mechanism, caused this diversity to appear here in the first place. Diversity is God's property, and we, who bear the relationship to it of strangers and sojourners, have no right to destroy it (Berry, 1981; Lamm, 1971). There is a much-told story (Hutchinson, 1959) about the great biologist, J. B. S. Haldane, who was not exactly an apostle of religion. Haldane was asked what his years of studying biology had taught him about the Creator. His rather snide reply was that God seems to have an "inordinate fondness for beetles." Well why not? As God answered Job from the whirlwind in the section of the Bible that is perhaps most relevant to biological diversity, "Where were you when I laid the foundations of the earth?" (Job 38:4). Assigning value to that which we do not own and whose purpose we can not understand except in the most superficial ways is the ultimate in presumptuous folly.

The great biochemist Erwin Chargaff, one of the founders of modern molecular biology, remarked not too many years ago, "I cannot help thinking of the deplorable fact that when the child has found out how its mechanical toy operates, there is no mechanical toy left" (Chargaff, 1978, p. 121). He was referring to the direction taken by modern scientific research, but the problem is a general one, and we can apply it to conservation as well. I cannot help thinking that when we finish assigning values to biological diversity, we will find that we don't have very much biological diversity left.

REFERENCES

Berry, W. 1981. The gift of good land. Pp. 267–281 in The Gift of Good Land. North Point Press, San Francisco.

Chargaff, E. 1978. Heraclitean Fire: Sketches from a Life Before Nature. Rockefeller University Press, New York. 252 pp.

Clark, C. W. 1973. Profit maximization and the extinction of animal species. J. Pol. Econ. 81:950–961.

Ehrenfeld, D. 1981. The Arrogance of Humanism. Oxford University Press, New York. 286 pp.

Ehrenfeld, D. 1986. Thirty million cheers for diversity. New Sci. 110:38–43.

Hutchinson, G. E. 1959. Homage to Santa Rosalia, or Why are there so many kinds of animals? Am. Nat. 93:145–159.

Lamm, N. 1971. Ecology in Jewish law and theology. Pp. 161–185 in Faith and Doubt. Ktav Publishing House, New York.

25

WHAT MAINSTREAM ECONOMISTS HAVE TO SAY ABOUT THE VALUE OF BIODIVERSITY

ALAN RANDALL

Professor of Agricultural Economics, Department of Agricultural Economics and
Rural Sociology, Ohio State University, Columbus, Ohio

A wide variety of methodological and ideological perspectives has informed and directed economic inquiry. Nevertheless, in each of the topical areas where economists specialize, it seems that one or, at most, a few approaches are now recognized as mainstream. For evaluating proposed policies to influence the way resources are allocated, the welfare change measurement approach (which includes benefit-cost analysis, BCA) currently enjoys mainstream status. My purpose here is to explain what this approach can contribute to understanding the value of biodiversity. I will distill the basic message into a few simple propositions, stating them one by one and offering a few paragraphs of elaboration on each.

WELFARE CHANGE MEASUREMENT IMPLEMENTS AN EXPLICIT ETHICAL FRAMEWORK

Each human being is assumed to have a well-defined set of preferences. While the way these preferences are ordered should satisfy certain logical requirements, preferences may be *about* literally anything in the range of human concerns. Mainstream economists argue that preferences are seldom whimsical or capricious. Rather, people come by their preferences consciously, in a process that involves learning, acquisition of information, and introspection. The mainstream economic approach is doggedly nonjudgmental about people's preferences: what the individual wants is presumed to be good for that individual.

The ethical framework built on this foundation is *utilitarian*, *anthropocentric*, and *instrumentalist* in the way that it treats biodiversity. It is utilitarian, in that things

217

count to the extent that people want them; anthropocentric, in that humans are assigning the values; and instrumentalist, in that biota is regarded as an instrument for human satisfaction.

There may be other views of the role of nonhuman life forms. For example, animals and plants may be seen as having a good of their own, possessing rights, or being the beneficiaries of duties and obligations arising from ethical principles incumbent on humans. Some people, including some economists, may subscribe to some of these views. Nevertheless, my purpose here is to confine myself to one particular instrumental, utilitarian, and anthropocentric formulation, exploring its implications for valuation. Implications of other approaches will, on their own merits, provide perspectives in addition to those offered here.

Having established preferences as a basis for valuation, any utilitarian formulation must come to grips with two additional issues: resource scarcity and interpersonal conflicts. The mainstream economic approach recognizes the role of ethical presumptions in resolving these conflicts and asserts two explicit ethical propositions. First, at the level of the individual, value emerges from the process in which each person maximizes satisfaction by choosing, on the bases of preference and relative cost, within a set of opportunities bounded by his or her own endowments (i.e., income, wealth, and rights). Thus, individuals with more expansive endowments have more to say about what is valued by society. Second, societal valuations are determined by simple algebraic summation of individual valuations. This means that from society's perspective, a harm to one person is cancelled by an equal-size benefit to someone else. By way of comparison with the ethics of welfare change measurement, note that individualism, as an ethic, accepts the first of these propositions, but explicitly rejects the second and instead, argues for protections against individual harm for the benefit of society as a whole. The classical market, in which all exchange is voluntary, institutionalizes (in principle) the individualist ethic.

Many economists are to some extent uncomfortable with the propositions that underlie welfare change measurement—and they are sympathetic with the discomfort of noneconomists—but these propositions have the virtue of explicitness: at least, one knows where mainstream economics stands.

THE ECONOMIC APPROACH IS NOT LIMITED TO THE COMMERCIAL DOMAIN

The explicit ethical framework of mainstream economics leads to the following definitions of value. To the individual, the value of gain (i.e., a change to a preferred state) is the amount he or she is willing to pay (WTP) for it, and the value of a loss is the amount he or she would be willing to accept (WTA) as sufficient compensation for the loss. For society, the net value of a proposed change in resource allocation is the interpersonal sum of WTP for those who stand to gain minus the interpersonal sum of WTA for those who stand to lose as a result of the change.

Because most laypersons have encountered the ideas that economics is concerned with markets and that since Adam Smith economists have believed that an invisible hand drives market behavior in socially useful directions, it is important for me to be precise about the relationship between economic values (WTP and WTA) and market prices. If everything people care about were private (in technical terms, rival and exclusive) and exchanged in small quantities in competitive markets, prices would reveal WTP and WTA for small changes. Conversely, prices are uninformative or positively misleading where any of the following is true: where people are concerned about goods and amenities that are in some sense public (i.e., nonexclusive or nonrival); where impediments to competitive markets are imposed (by governments or by private cartels and monopolies); and where the proposed change involves a big chunk rather than a marginal nibble of some good, amenity, or resource. The point is that market prices reveal value (in the mainstream economic sense of that term) not in general but only in a rather special and limiting case.

Most issues involved with biodiversity violate the special case where market price is a valid indicator of economic value. Nevertheless, the general theory of economic value encompasses these broader concerns. Here lies the distinction between economic values and commercial values; the essential premises for economic valuation are utility, function, and scarcity; organized markets are essential only to commerce. It is a fundamental mistake to assume that economics is concerned only with the commercial.

THERE IS AN (ALMOST) ADEQUATE CONCEPTUAL BASIS FOR ECONOMIC VALUATION OF BIODIVERSITY

The total value of a proposed reduction in biodiversity is the interpersonal sum of WTA. This total value has components that arise from *current use*, *expected future use*, and *existence*. Use values derive from any form of use, commercial or noncommercial, and including use as a source of raw materials, medicinal products, scientific and educational materials, aesthetic satisfaction, and adventure, personally experienced or vicarious. Future use values must take into account the aversion of humans to risks (e.g., the risk that the resource may no longer be available when some future demand arises) and the asymmetry between preservation and some kinds of uses (preservation now permits later conversion to other uses, whereas conversion now eliminates preservation as a later option). The concerns have encouraged the conceptualization of various kinds of option values, which are adjustments to total value to account for risk aversion and the irreversibility of some forms of development.

To keep the value of existence separate and distinct from the value of use, existence value must emerge independently of any kind of use, even vicarious use. That is a stringent requirement. Nevertheless, valid existence values can arise from human preference for the proper scheme of things. If some people derive satisfaction from just knowing that some particular ecosystem exists in a relatively undisturbed state, the resultant value of its existence is just as real as any other economic value.

For evaluating proposals that would have long-term effects, it is a fairly standard practice in economics to calculate present values by discounting future gains and losses. This procedure seems reasonable when evaluating alternative investments expected to last no more than one generation. When it is applied to potential disasters in the more distant future, it makes many people, including quite a few economists, uneasy. By discounting at standard rates, the inevitable collapse of the living systems on this planet several hundred years from now could be counterbalanced by relatively trivial economic gains in the immediate future. This unresolved issue of how to deal with long-range future impacts is what led me to insert the caveat "almost" in the heading of this section.

TECHNIQUES FOR EMPIRICAL VALUATION EXIST AND ARE APPLICABLE TO MANY BIODIVERSITY ISSUES, BUT LACK OF INFORMATION CAN BE DAUNTING

When price information is available and is informative about value, the analytics are relatively simple and familiar to most economists. The challenges in valuation arise where direct price information is unavailable and when price is not a valid indicator of value. For those situations, the valuation methods that have been developed and are considered reputable by economists fall into two broad classes: implicit pricing methods and contingent valuation.

The *implicit pricing methods* are applicable when the unpriced amenity of interest can be purchased as a complement to, or a characteristic of, some ordinary marketed goods. For example, travel services are purchased as a complement to outdoor recreation amenities, which permits valuation of outdoor recreation amenities by the travel cost method (Clawson and Knetsch, 1966). Hedonic analysis of the housing market may be used, for example, to estimate the value of such nonmarketed amenities as access that housing provides to open space or to a shoreline (Brown and Pollakowski, 1977).

Contingent valuation methods are implemented in survey or experimental situations (Cummings et al., 1986). Alternative policy scenarios are introduced and the choices made by citizen participants reveal WTP or WTA, directly or indirectly. Like other survey or experimental methods, the results may be sensitive to the design and conduct of the research. Nevertheless, there is growing theoretical and empirical evidence that contingent valuation yields results that are replicable and accurate within broad limits. The major advantage of this type of valuation is its broad applicability: it can determine WTP or WTA for any plausible scenario that can be effectively communicated to the sample of citizens. For estimating existence values, for instance, it may be the only feasible method.

With respect to biodiversity, the experts (i.e., ecologists and paleontologists) often have little confidence in their estimates of the impacts of ecosystem encroachment or disturbance. All too often the experts disagree. In these areas, contingent valuation cannot compensate for ignorance. If the experts cannot construct credible scenarios describing the effects of alternative policies on biodiversity, the WTP or WTA of citizens reacting to these scenarios will reflect that uncertainty and misinformation as well as any additional uncertainty they may have about their own preferences concerning biodiversity. More generally, the accuracy of any measure of value based on the preferences of ordinary citizens is limited by the reliability of citizen knowledge about the consequences of alternative actions for biodiversity. Some may regard this as an argument that policy should be based on the judgments of experts rather than of citizens. I disagree. It seems that public opinion quite rapidly reflects expert opinion when the latter is confidently held and expressed with convincing argument. On the other hand, confusion, ignorance, and apathy among the laity typically reflect incomplete and dissonant signals from the specialists.

POLICY DECISION CRITERIA HAVE BEEN PROPOSED

Mainstream economists have proposed two alternative criteria for deciding preservation issues. The *modified BCA* (benefit-cost analysis) approach attempts to implement the conceptual framework of welfare change measurement by identifying and measuring (insofar as possible) the benefits and costs of the alternative courses of action. This approach requires major efforts to measure the noncommercial components of economic value, including amenity, option, and existence values. The benefit-cost decision criterion itself is modified, however, by assigning any benefits of doubt to the preservation side of the ledger. The logic for this is that more is often known and can be documented about the benefits obtainable from commercial uses than is known about the benefits of preservation.

In another approach, the *safe minimum standard* (SMS) is defined as the level of preservation that ensures survival. Proponents of the SMS approach argue that although measuring the benefits of diversity in every instance is a daunting task, there is ample evidence that biodiversity is (in broad and general terms) massively beneficial to humanity.

Whereas the modified BCA approach starts each case with a clean slate and painstakingly builds from the ground up a body of evidence about the benefits and costs of preservation, the SMS approach starts with a presumption that the maintenance of the SMS for any species is a positive good. The empirical economic question is, "Can we afford it?" Or, more technically, "How high are the opportunity costs of satisfying the SMS?" The SMS decision rule is to maintain the SMS unless the opportunity costs of so doing are intolerably high. In other words, the SMS approach asks, how much will we lose in other domains of human concern by achieving a safe minimum standard of biodiversity? The burden of proof is assigned to the case against maintaining the SMS.

The SMS approach avoids some of the pitfalls of formal BCA, e.g., the treatment of gross uncertainty as mere risk, the false appearance of precision in benefit estimation, and the problem of discounting. In contrast to the procedure of discounting, the SMS approach simply accepts that the costs of preservation may fall disproportionately on present generations and the benefits on future generations. Its weakness is that it redefines the question rather than providing the answers. Nevertheless, an appealing argument can be made that "can we afford it?", with a presumption in favor of the SMS unless the answer is a resounding NO, is the proper question.

THE EMPIRICAL CUPBOARD IS NOT BARE

It is customary to draw attention to the scarcity of hard information about the economic value of biodiversity. But for each of the valuation methods discussed above, there has been a smattering of apparently successful empirical application. Fisher and Hanemann (1984) have used ordinary market data to estimate the potential value of the plant breeding that recently resulted in the discovery of perennial grass related to corn (see Iltis, Chapter 10 of this book). Literally dozens of economists have used implicit pricing methods to estimate the values of various environmental amenities. Stoll and Johnson (1984) used the contingent valuation method (CVM) to estimate the existence values for whooping cranes. Bishop (in press) used CVM to estimate the existence values for Wisconsin's bald eagles and striped shiners (a rather obscure freshwater fish). Bennett (1984) used CVM to estimate the existence value of a unique ecosystem that survives in a remote part of the coastline of southeastern Australia. Bishop (1980) has also completed some empirical analyses based on the SMS criterion. For several cases in the United States (the California condor, snail darter, and leopard lizard) and for mountain gorillas in low-income tropical Ruanda, he found the opportunity costs of preservation to be reasonably low. In such cases, preservation decisions are not difficult.

Clearly, the empirical evidence is spotty at this stage, but these examples serve to counter the impression that high-quality empirical work on the value of diversity is not feasible.

FURTHER COMMENTS ON THE MAINSTREAM ECONOMICS APPROACH

The mainstream economic approach has a built-in tendency to express the issues in terms of trade-offs. In that respect, it has much in common with the common law notion of balancing the interests. This makes the mainstream economic approach potentially helpful in the resolution of conflicts. Perhaps it also makes the economic approach anathema to those who would brook no compromise.

Important problems in making decisions concerning biodiversity are seldom of the all-or-none variety. It is easy to provide the mainstream economic answer to the question, What is the value of all the nonhuman biota on the planet Earth? Its value is infinite based on the following logic: elimination of all nonhuman biota would lead to the elimination of human life, and a life-loving human would not voluntarily accept any finite amount of compensation for having his or her own life terminated. Earth's human population surely includes at least one such person. Thus, across the total population, the sum of WTA for elimination of all nonhuman biota is clearly infinite. Nonetheless, the question posed is not very useful. The meaningful questions concern the value lost by the disappearance of a chip of biodiversity here and a chunk there. For this smaller question, it is often possible to provide an economic answer that is useful and reasonably reliable.

The goal of the mainstream economic approach is to complete a particular form of utilitarian calculation. This calculation is expressed in money values and includes (in raw or modified form) the commercial values that are expressed in markets.

However, it expands the account to include things that enter human preference structures but are not exchanged in organized markets. This extension and completion of a utilitarian account, where preservation of biodiversity is at issue, is useful because it shows that commercial interests do not always prevail over economic arguments.

The claim that it is useful to complete this utilitarian account does not depend on any prior claim that the utilitarian framework is itself the preferred ethical system. Environmental goals that may be served by arguments that the biota has rights that should be considered, or that it is the beneficiary of duties and obligations deriving from ethical principles incumbent on humans, may also be served by completing a utilitarian account that demonstrates the value implications of human preferences that extend beyond commercial goods to include biodiversity. Some people would argue that a complete discussion of the value of biodiversity should extend beyond utilitarian concerns. Even these people would, presumably, prefer a reasonably complete and balanced utilitarian analysis to the truncated and distorted utilitarian analysis that emerges from commercial accounts.

REFERENCES

Bennett, J. 1984. Using direct questioning to value the existence benefits of preserved national areas. Aust. J. Agric. Econ. 28:136–152.

Bishop, R. 1980. Endangered species: An economic perspective. Trans. North Am. Wildl. Nat. Resour. Conf. 45:208–218.

Bishop, R. In press. Uncertainty and resource valuation: Theoretical principles for empirical research. In G. Peterson and C. Sorg., eds. Toward the Measurement of Total Value. USDA Forest Service, Rocky Mountain Forest and Range Experiment Station General Technical Report, Fort Collins, Colo.

Brown, G., and H. Pollakowski. 1977. Economic value of shoreline. Rev. Econ. Stat. 69:273–278.

Clawson, M., and J. Knetsch. 1966. Economics of Outdoor Recreation. Johns Hopkins University Press, Baltimore. 327 pp.

Cummings, R., D. Brookshire, and W. Schulze. 1986. Valuing Environmental Goods: A State of the Art Assessment of the Contingent Valuation Method. Rowman and Allenheld, Totowa, N.J. 270 pp.

Fisher, A., and M. Hanemann. 1984. Option Values and the Extinction of Species. Working Paper No. 269. Giannini Foundation of Agricultural Economics, Berkeley, Calif. 39 pp.

Stoll, J., and L. Johnson. 1984. Concepts of value, nonmarket valuation, and the case of the whooping crane. Trans. North Am. Wildl. Nat. Resour. Conf. 49:382–393.

HOW IS BIODIVERSITY MONITORED AND PROTECTED?

Product of a butterfly farm near Wewak,
Papua New Guinea. Bird-wing butterflies are
grown on vegetation planted by a farmer on
steep slopes near his village. *Photo courtesy of
Noel D. Vietmeyer.*

MONITORING BIOLOGICAL DIVERSITY FOR SETTING PRIORITIES IN CONSERVATION

F. WILLIAM BURLEY

Senior Associate, World Resources Institute, Washington, D.C.

Identifying the elements of biological diversity and monitoring their changes through time is a daunting task. Biologists have long recognized that the full array of biological diversity will never be known completely—that is, not all species and ecosystems will be identified, named, cataloged, and studied in any detail before many of them are lost. For example, it is likely that there are many more than 10 million species living today. Only 1.4 million of these have been described and named, and a tiny fraction of them have been studied thoroughly for potential use by humans. Ecosystems also vary greatly in size, composition, complexity, and distribution, and it is not uncommon for ecologists to differ in describing and defining them. For example, despite many studies by vegetation ecologists and biogeographers in the United States, today there is no single, agreed-upon vegetation classification that can be used by the federal land-management agencies or by the many state and private organizations that could productively use a national classification scheme.

All this makes the work of systematically conserving species and ecosystems more difficult. It presents a real problem when we try to determine how well various ecosystems are protected or represented in the global, national, and state systems of protected areas (Harrison et al., 1984).

THE GAP ANALYSIS CONCEPT

To tackle this problem, conservation biologists for years have intentionally or unwittingly used the process called gap analysis to establish short-term and longer-term conservation priorities. The concept is deceptively simple, if not simplistic:

227

within a particular country or region, first identify and classify the various elements of biological diversity in several ways. Then examine the existing and proposed systems of protected areas and other land-management units that help conserve biological diversity. Finally, using various classifications, determine which elements (e.g., major ecosystems, vegetation types, habitat types, species) are unrepresented or poorly represented in the existing system of conservation areas. Once this is known with reasonable precision, priorities for the next set of conservation actions can be established. The process continues indefinitely, and the conservation system is refined as land use changes and as better information about the distribution and status of species and ecosystems is obtained.

In practice, this process usually entails comparing and analyzing many different sets of information by using maps or computers to identify, for example, the gaps in coverage. Many countries have begun this process, but unfortunately, very few have attempted it in a thorough, systematic fashion with defined conservation objectives. Notable exceptions include Great Britain, Peru, Australia, and South Africa (Specht et al., 1974).

The example from Australia illustrates the process well. By the mid-1970s, there was an adequate description of the major vegetation types found throughout Australia. A comprehensive review of the various park and reserve systems was begun to determine which vegetation types were already represented and seemingly adequately protected. The thoroughness of this effort varied considerably from state to state, but by the late 1970s and early 1980s, it was becoming possible to make more objective statements about which vegetation types were poorly represented in the national and state systems and therefore which ones needed conservation attention first. The state of Queensland has taken this broad-scale analysis one step further (Sattler, 1986). The Queensland National Parks and Wildlife Service recently completed mapping the vegetation of Queensland's 90 national parks, environmental parks, fauna reserves, and scientific reserves larger than 1,000 hectares, and it now is analyzing gaps in the representation of vegetation types by protected areas. As these are identified, steps will be taken to protect or otherwise conserve good representative examples of the highest-priority vegetation types (Sattler, 1986).

AN UNDERLYING CONCEPT

An important concept underlies the gap analysis process: by ensuring that all vegetation types are well represented in a system of conservation areas, it is assumed that much if not most of the biological diversity (species and ecosystems) will be protected. Systems in practice verify this, e.g., in much of the United States, Australia, and Europe, but in addition, special efforts must be made to ensure the protection of particularly critical species and ecosystem types.

Debates continue to rage among biologists about the minimum critical sizes of populations and ecosystems that are necessary to conserve the biota over the long term. A very practical question emerges from these debates: should a particular ecosystem type already represented in the system of conservation areas be better represented, or should the next conservation effort be aimed at conserving other ecosystems that are either unrepresented or not adequately represented? All these considerations and unanswered questions in conservation biology do not obviate the need for gap analyses in all regions and countries, however, because inevitably we must have a good information base on which to base better conservation decisions.

Gap analysis exercises similar to the one for Australia mentioned above have been undertaken in other countries. In Chapter 29, Huntley describes progress to date in several countries of southern Africa. A somewhat similar process is also being used to identify global priorities in plant conservation, as described by Williams and Lucas in Chapters 28 and 30, respectively.

In the United States, no such countrywide analysis of ecosystems exists, except for several very preliminary studies using coarse classifications of ecoregions and the most general vegetation types. The federal agencies have never agreed on the methods to be used, but a nongovernmental organization, The Nature Conservancy, has made the most thorough state-by-state investigations using several vegetation classifications and all the species distribution data available. By developing a standardized methodology for all the states, the Conservancy is now able to make regional and preliminary national assessments of the most important gaps in ecosystem and species coverage, thereby establishing conservation priorities in a more systematic fashion than was previously possible. Chapter 27 by Jenkins describes this effort further.

On the global scale, the International Union for Conservation of Nature and Natural Resources (IUCN) and the World Wildlife Fund have worked over the past decade to identify major conservation priorities. A biogeographic classification developed by Dasmann, Udvardy, and others was used to determine which major biomes and biogeographic provinces worldwide are relatively well represented in the global system of protected areas and where there are major gaps in the system (Udvardy, 1975). The analysis itself does not take into account the quality or level of management (and therefore the degree and quality of protection) of the conservation areas, but it has been useful to IUCN and others in helping to determine the allocation of program funds and to design conservation activities on a global scale and in particular regions. IUCN is carrying this global analysis further, and its Commission on National Parks and Protected Areas is coordinating a series of regional analyses to identify high-priority ecosystems and to recommend the establishment of additional protected areas (MacKinnon and MacKinnon, 1987).

The next step in refining this process, however, is to do essentially the same type of analysis at the country and local levels. This is already under way in several Latin American countries. The Conservation Data Center (CDC) in Lima, Peru, for example, recently analyzed gaps in ecosystem coverage by overlaying biogeographic provinces, life zones, selected vegetation types, and existing and planned conservation areas. Although biologists in Peru have known for some time that the Andean cloud forests and coastal vegetation types were being decimated by human impacts and were important ecosystems to be conserved, the gap analysis done by the CDC revealed these priorities in a much more systematic, quantified way and identified particular areas that should be put under some form of protective management.

Unfortunately, however, most developing countries are not this far along in the process. This is ironic, because the cost and time needed to reach this level of data richness are not great, and a small team of biologists could pull together the necessary information in much less than a year. If every country in the tropics, for example, could generate this level of conservation information quickly, it would mean that more objective conservation priorities could be identified and made available for use not only by conservationists but also by land-use planners and development agencies.

REFERENCES

Harrison, J., K. Miller, and J. McNeeley. 1984. The world coverage of protected areas: Development goals and environmental needs. Pp. 24–33 in J. A. McNeeley and K. R. Miller, eds. National Parks, Conservation, and Development. Smithsonian Institution Press, Washington, D.C.

MacKinnon, J., and K. MacKinnon. 1987. Review of the Protected Area Systems of the Afrotropical Realm. International Union for the Conservation of Nature and Natural Resources, Gland, Switzerland. 350 pp.

Sattler, P. S. 1986. Nature conservation in Queensland: Planning the matrix. Proc. R. Soc. Q. 97:1–21.

Specht, R. L., E. M. Roe, and V. H. Boughton, eds. 1974. Conservation of Major Plant Communities in Australia and Papua New Guinea. Aust. J. Bot., Supp. No. 7. Commonwealth Scientific and Industrial Research Organization, East Melbourne, Australia. 667 pp.

Udvardy, M. D. F. 1975. A Classification of the Biogeographical Provinces of the World. IUCN Occasional Paper No. 18. International Union for the Conservation of Nature, Morges, Switzerland. 48 pp.

27

INFORMATION MANAGEMENT FOR THE CONSERVATION OF BIODIVERSITY

ROBERT E. JENKINS, JR.
Vice President, Science Programs, The Nature Conservancy, Arlington, Virginia

Everyone is beginning to recognize that biodiversity at all levels—gene pool, species, and biotic community—is important for many reasons and that it is being rapidly diminished by habitat destruction and other damaging influences resulting from human population growth, pollution, and economic expansion. No one seems to think that we can do anything effective to control the root causes soon enough to provide breathing space for the biota, so biological conservationists devote themselves to the use of techniques that are believed to be helpful in the context of a shrinking natural estate. All these techniques involve forms of triage, increasingly complex interventions, and decreasing margins for error. To conserve biodiversity in less and less space under greater and greater pressure requires that we have more and more knowledge.

We need to know about the existence, identity, characteristics, numbers, condition, status, location, distribution, and ecological relationships between biotic species and biological communities or assemblages; their individual occurrences in the landscape; existing preserves and what they contain; the most important unprotected areas; related land ownerships; and sources of further information, among other things. With such knowledge we can select and design new preserves, improve existing ones, determine what sort of management is needed, establish priorities for ex situ conservation of species, monitor changing status, note restoration opportunities, and maybe even decide where we need to cut our losses and move on.

Of course much of the basic information we would like to have has not even been collected in the field or, if collected, not yet developed into usable form. Kosztarab (1986, p. 23) estimated that "In the United States biologists have described only one-third of the living organisms and their developmental stages."

Erwin (1982) has shown that the extent of our knowledge of the tropics is much more limited. From his collection of insects in tree canopies, he estimates that there may be 30 million species of organisms on Earth. This means that only about 5% have been described and classified at the species level, much less for all their developmental stages. Most of the undescribed organisms are invertebrates, fungi, and nonvascular plants. Kosztarab (Virginia Polytechnic Institute and State University, personal communication, 1986) believes it would take 50 entomologists 40 years just to describe the rest of the insects in the United States, not counting the work needed for additional collecting. It is implausible to think that we can mount a sufficient effort to amass this basic knowledge in time to have much bearing on the conservation of biodiversity. Fortunately, we can do a lot with the basic information already in hand.

Before existing data can be put to use, they must be compiled from scattered secondary sources and repositories and organized into usable form. Accomplishing just this is a very complicated matter, much more so than can be realized without getting deeply into such matters as the details of system design, operational administration, and interinstitutional arrangements. The job would be quite challenging under ideal conditions, but in virtually all instances, this work has had to be carried out under severe financial constraints. Therefore, it is no wonder that many attempts to master the data compilation task have failed. The rest of this chapter describes an effort that seems to be succeeding.

NATURAL HERITAGE DATA CENTERS

The Nature Conservancy was established to conserve biodiversity by establishing natural area preserves. These preserves are selected and designed to protect examples of the widest possible spectrum of native ecosystems and species habitats. It was always clear that identifying the most important areas to be conserved was crucial to making the best use of limited resources. Therefore, gathering and organizing scientific information for conservation became one of the first orders of business. For many years, however, we were overmastered by the complexity of the task, and even the best of our efforts were essentially defeated by methodological flaws, limited duration of effort, underfunding, and similar problems.

In 1974, in cooperation with the South Carolina Department of Wildlife and Marine Resources, The Nature Conservancy initiated the first of what are called State Natural Heritage Inventories. In this program, we finally brought together a sufficiently well-engineered mix of concepts and operational procedures for data gathering and management to successfully function as an efficient and effective tool for planning the conservation of biological diversity. The methodology used in that program has been continuously improved since 1974 and has become the most systematic, comprehensive, and widely used technology in existence for gathering and organizing the information needed for biodiversity conservation. (See Jenkins, 1985, for fuller explanations.)

Natural Heritage Data Centers now exist in nearly every state of the United States, almost always as cooperative ventures between agencies of state government and the Conservancy (see Figure 27-1). Heritage inventory technology is also being

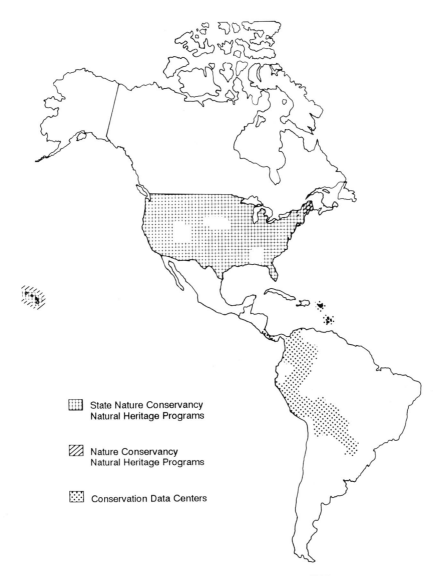

FIGURE 27-1   Distribution of the Natural Heritage Data Centers in 1986.

used to help an increasing number of Latin American institutions to develop their own Centros de Datos para la Conservacion (CDCs), and requests for assistance in starting more such data centers have been coming in from other institutions all over the world. Collectively, the existing programs employ more than 250 biologists and have total annual operating budgets over $10 million.

Each of these inventories used a highly standardized methodology, which provides some important economies in terms of system development and administration. In other words, because of this standardization, the Conservancy has been able to develop, refine, and support systems on a very efficient basis, drawing new experience from all over, incorporating it into the central model, and propagating it throughout the network. Standardization also facilitates data exchange among the individual programs and permits higher order aggregation of data derived from many data centers.

NETWORKING AND CENTRAL DATA BASES

The standardization referred to above does not merely facilitate higher order aggregation of data, it virtually demands it. To optimize the effectiveness of the crucial bottom-up data collection of the individual centers, the Conservancy has gradually accumulated more and more data in a series of national data bases. We needed, for example, to have consistent taxonomic names for species to avoid the confusion brought about by locally used synonyms. We therefore kept central lists of names along with standard "element" codes for all rare and endangered plant and animal species being investigated by any state program. (An "element" is a species, a community type, or some other feature or phenomenon of special interest to conservationists.) We also began to accumulate national Element Manual Files on such species with information added and extracted by individual Heritage programs. We first thought that our central files would only accumulate information on a limited number of rare species. Since everything is rare somewhere in its range, however, we eventually discovered that we had at least some information on nearly every vascular plant and vertebrate animal species in the United States. We therefore went ahead and entered data on the rest of the species in these groups. We also have extensive information on species in other taxonomic groups, and these species- and community-type tracking data bases have become the most comprehensive compilations of such information in existence (Figure 27-2).

Beyond standard taxonomy, there are many other data of common interest to more than one geographic area. To spare each data center the redundant effort of collecting and managing such data, we are dividing the labor by assigning lead responsibility for a given data item to one data center; results are then added to the central data bases, from which the other centers needing that same information can retrieve it.

We have also found it useful to work this process in reverse, by undertaking special projects to add data obtained directly from national or regional sources for downloading to one or more individual data centers. From the beginning we have been compiling data from the National Museum of Natural History, for example, and then sending it back to Minnesota or Arizona where it came from. The central data bases are greatly facilitating this sort of procedure. Another reason for creating central data bases is the increasing need to compare information over wider geographic areas in order to gain a better perspective on the relative significance of conservation objectives and projects and to monitor status changes across a species' or community's entire range.

In addition to the central data bases on North America, the Conservancy's international office has made significant progress through a special Latin American

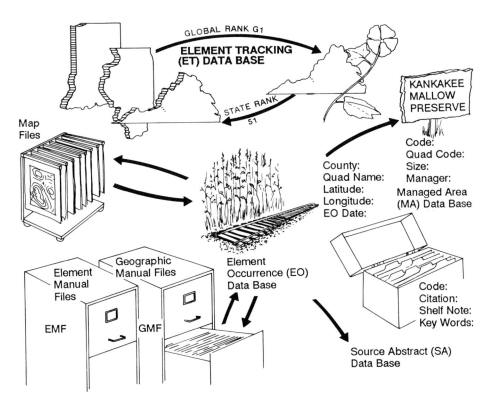

FIGURE 27-2 Data-processing procedure in the Natural Heritage program.

Biogeography Project in assembling parallel data bases on the biota, ecosystems, and conservation areas throughout the rest of the Western Hemisphere. These central data bases are in Spanish and exchange information with the Latin American Conservation Data Centers. Since so many of the Latin American scientific data are located in U.S. repositories, the kind of top-down data repatriation from the central data base referred to above will be particularly helpful to the CDCs.

Altogether, the network of local and central data bases, with its continually improving methods and procedures, has become a sort of machine for collective learning and retention of carefully winnowed facts. This network has been cumulatively amassing an immense body of knowledge with the capability of providing ever better insights about biological conservation needs and priorities. It is already being tapped as an important aid to decision making by a continually widening usership.

APPLICATIONS OF THE DATA

Following are some of the principal uses of these data:

• Facilitating continuing inventory. By organizing data well enough to tell what is and is not known, data needs can be targeted with precision. This is crucial in order to use our limited resources efficiently and to avoid choking the data system with low-value data.

• Conservation priority setting. By its process of continual data enhancement, the Heritage network obtains an ever-better picture of the relative endangerment of species, natural communities, and other natural features. This perspective is encapsulated in Element Priority Ranks. These are assigned to each species and community type on a comparative basis through the use of consistent criteria of rarity, threat, etc. From these, we determine which species to recommend for listing by the Office of Endangered Species, which should be taken off the list, which are candidates for ex situ conservation, which species and communities are most in need of habitat protection, and other categorizations. Specific Element Conservation Plans are developed for many species and communities, and like everything else, are continually revised.

• Preserve selection. The greatest threat to biodiversity comes from habitat destruction. The identification of areas that still sustain the critical elements is one of the most important tasks of a heritage program. The Element Tracking and Site Tracking lists are continually revised to organize the most focused biological land conservation agenda possible.

• Preserve design. Once a site has been selected as a potential preserve, the Heritage or Conservancy staff undertakes intensive field investigations and defines optimal boundaries for a viable, defensible, and manageable preserve for the targeted species or community. Individual property ownership tracts within each site are also identified.

• Land conservation administration. The Site Tracking lists, the Site, Tract, and Managed Area (existing Preserve) data bases, and various other data bases and files are used to organize the actual land protection effort. The systematization of this activity is facilitating the more efficient and sophisticated use of a full spectrum of protection devices from land owner notification to registration, rights-of-first-refusal, cooperative management agreements, leases, easements, agency designation, and land acquisition (see Hoose, 1981). It is also helping in coordination of the activities of many separate agencies and institutions, e.g., in the Tennessee Interagency Protection Planning Committee.

• Element status monitoring. A large fraction of the total heritage data consists of records on Element Occurrences (actual field localities with rare species populations, rare community remnants, or other features). Collectively, the network manages data on more than 150,000 such occurrences, including data on all the rarest and most endangered elements known. This enables any user to make a very quick field reassessment of status, often with the assistance of cooperators from the Heritage Data Center itself or from among its other users. For the rarest biodiversity elements, all necessary efforts to monitor their status, such as establishing permanent plots and conducting annual population censuses, are being undertaken.

• Element management. Special efforts are devoted to the biodiversity elements that require for survival not only protection but also active management. Through Element Stewardship Abstracts and manual files, we attempt to compile pertinent information on their ecological responses to various management options and to put this information to use in the field.

- Site management. Site Stewardship Summaries and Managed Areas Records aid in various administrative and monitoring activities on existing preserves.
- Development planning. Information from the data centers is available to all legitimate users, including corporations and consultants engaged in land development. By supplying information about sensitive sites and project design requirements before decisions and investments have been made, less-damaging alternatives are chosen and much fruitless conflict avoided.
- Environmental impact analysis. Most of the state programs are routinely consulted on all proposed projects requiring special permits or adherence to various environmental regulations. Objective reviews based on real data are highly valued by everyone involved and produce the same kinds of results as above. A typical state program may review more than 2,000 project proposals a year.
- Access to additional information. Information in the system is flagged with citations to published documents, individual experts, museums, agency files, other existing data bases, map series, unpublished data on file, etc. This greatly facilitates follow-up investigations by system users and magnifies the utility of other data repositories.
- Predictive modeling. There is a lot of very definite information in the network, much of it associative, e.g., species with habitats, one species with another, total composition of communities, and habitat-habitat juxtaposition. Some of this lends itself to predictive extrapolation for a variety of purposes. Some of the newer units of the evolving system, such as the Vertebrate Characterization Abstract, facilitate prediction of what is most apt to occur in as yet unsurveyed landscapes.
- Research. Many of the kinds of uses cited above, such as finding field localities of entities or combinations of entities desired for study, can be of assistance to the scientific community. The data bases also constitute one of the largest and most accurate biogeographic resources in existence and may increasingly be used directly as grist for the analysis mill. As a conservation planning device, the Heritage enterprise can also help identify for the growing body of conservation biologists which research problems could produce practical results for applied conservation in such areas as minimum critical habitats, minimum viable population sizes, genetic variation among metapopulations, landscape ecology and patch dynamics, the role of corridors between reserves, management of vegetation succession, nuclear preserves in macrosites, and archipelagos of associated reserves.

In these ways the Heritage data network is put to a tremendous amount of use. Just in conservation or development siting and land management, we estimate that the U.S. programs alone are consulted on more than 50,000 decisions per year. The total amount of practical use is much greater.

INTERACTIONS WITH SCIENTISTS AND THE PUBLIC

The Heritage programs do not accomplish all these things alone—they simply specialize in long-term organization and management of information needed for the conservation of biodiversity. They are not the original source of the vast majority of data, they do not try to be a substitute for the systematic community on species taxonomy, they do not undertake the job of the museums, they do not engage in comprehensive, scientific abstracting or bibliographic work, they are not usually an active conservation or land management entity, and they are not an environmental regulatory agency. Therefore, they depend on cooperative relationships with a broad array of other individuals and institutions as both contributors to the effort and users of it.

If it weren't for John Kartesz' monumental and heroic compilations in botanical taxonomy, for example, I don't know what we would do in this area (Kartesz, 1985; Kartesz and Kartesz, 1980). We are similarly indebted to the Association of Systematic Collections for much of our vertebrate taxonomy. The Center for Plant Conservation is a major user of data on plant status and distribution for their conservation collections for which they will be keeping parallel and mutually accessible data bases. The U.S. Fish and Wildlife Service is our closest cooperator on endangered species and even funds some of our status surveys. Paul Opler of the U.S. Fish and Wildlife Service has compiled Global Element Ranking forms on most of the butterflies in the United States as a labor of nights and weekends. The U.S. Forest Service is cofunding a regional ecologist to work on plant community classification in the Southeast. Stanford's Center for Conservation Biology is cooperating closely with the Conservancy on its studies of minimum viable populations and preserve design. We freely exchange information and efforts with the National Park Service Natural Landmarks program. Various utility companies have provided financial support for Heritage inventories so they can learn whether their proposed power lines are going to run afoul of endangered species populations or other significant elements of biodiversity. Private foundations, including the A. W. Mellon Foundation, the J. N. Pew Charitable Trust, the Hewlett Foundation, the MacArthur Foundation, and many others, have supported this work because they believe it can make a difference. Innumerable individual scientists and natural historians are working with the network. U.S. and Latin American data centers are being used as first and last stops for many field biologists, that is, they use the centers to learn what is known and what needs to be learned in their areas of interest and then report their discoveries to the centers so the data can be put to good use. I cannot begin to cite all the individuals and organizations without whose cooperation this work could not proceed.

The Natural Heritage Data Network is the furthest thing imaginable from a proprietary enterprise. It has been developed with the goal of serving the needs of anyone and everyone involved in the conservation of biodiversity, all of whom are cordially invited to work with the various operating units within the system as much as possible in the furtherance of the system's purposes and of their own.

REFERENCES

Erwin, T. 1982. Tropical forests: Their richness in Coleoptera and other arthropod species. Coleopterists Bull. 36(1):7475.

Hoose, P. M. 1981. Building an ark: Tools for the preservation of natural diversity through land protection. Island Press, Covelo, Calif.

Jenkins, R. E. 1985. Information methods: Why the Heritage programs work. Nature Conservancy News 35(6):21–23.

Kartesz, J. T. 1985. 1985 Draft Revised Kartesz Vascular Plant Checklist. Unpublished review draft distributed by the author, Reno, Nev. 694 pp.

Kartesz, J. T., and R. Kartesz. 1980. A Synonymized Checklist of the Vascular Flora of the United States, Canada, and Greenland. University of North Carolina Press, Chapel Hill, N.C. 498 pp.

Kosztarab, M. 1986. Prefatory comments: Some of the activities leading to this symposium. Pp. 23–27 in K. C. Kim and L. Knutson, eds. Foundations for a National Biological Survey. Association of Systematics Collections, Lawrence, Kans.

The Nature Conservancy. Undated (continually revised). Natural Heritage Program Operations Manual (some parts available on request).

IDENTIFYING AND PROTECTING THE ORIGINS OF OUR FOOD PLANTS

J. TREVOR WILLIAMS

Director, International Board for Plant Genetic Resources,
c/o Food and Agriculture Organization, Rome, Italy

Plants that are used for food present special problems in the monitoring of genetic diversity and its conservation. The diversity of food plants in existence today evolved over the past 10,000 years or so since the origins of agriculture in the Neolithic era. The progenitors of cultivated plants were wild species, which along with countless others were harvested by hunters and gatherers. The genetic diversity of crops increased greatly as the wild species were domesticated and were moved from environment to environment; cultivars adapted to specific environments and were put to different uses. Many cultivars must have been discarded even in early days, but some of them remained in certain areas and others diversified further as they were taken far from their original homes. Hence, the evolution of major crops has been a continuing process—from the wild progenitors to the products of modern plant breeding and genetic manipulation.

By contrast, other species have been used as food but have never been domesticated. Included among these are a multitude of minor fruit and forage plants that are grown in gardens. Included too are many other species that conservationists stress for their potential value in medicine, as food, or for other purposes. The intraspecific variability of such species may be rather insignificant compared with that of major cultigens and is often poorly documented. There has been less opportunity for diversification over millenia of cultivation.

There are thus three types of wild species of interest for conservation: the progenitors of cultivated plants; species used but not domesticated; and those that might be of use in the future. The distinction between these categories has been inadequately considered both by conservationists and by international funding organizations in budgeting. It has led to the misuse of the term "genetic resources" as a justification for almost any conservation activity, and may also lead to confusion and to the unscientific planning and operation of programs. A significant exception remains the work on crop plants guided by the International Board for Plant Genetic Resources (Williams, 1984).

THE INTERNATIONAL BOARD FOR PLANT GENETIC RESOURCES PROGRAM

The International Board for Plant Genetic Resources (IBPGR) was established within the Food and Agriculture Organization because of the loss of substantial intraspecific diversity produced by domestication as fewer and fewer successful modern cultivars were grown over large areas of land. This loss was termed genetic erosion.

When IBPGR was created in 1974, the mandate given to it by the Consultative Group on International Agricultural Research (CGIAR) was to establish a global network of activities to further the collection, conservation, documentation, and use of germplasm for crop species. The network, which has been in existence for some time, is complex and includes gene banks (repositories of seeds, tubers, and other sources of genetic materials) at conservation centers and activities dealing with the collection, characterization and documentation of crop genetic resources, and training. Beginning with a handful of countries in 1974, IBPGR'S work now involves 106 countries, and the number of gene banks has grown from half a dozen in 1974 to more than 100, about 40 of which have agreed to accept responsibility for long-term maintenance of genetic resources. The more than 50 crops in the IBPGR program include cereals, legumes, vegetables, oil seeds, fruits, and a limited number of cash crops of importance to rural farmers. More recently, forage plants have been included.

Emphasis was initially placed on the collection and conservation of local cultivars of crops that had evolved in diverse environments under traditional agricultural practices; these are often known as landraces. Farming methods were also changing rapidly due to the availability of modern cultivars and agricultural techniques. The landraces have provided and continue to provide a rich resource for exploitation, but it is neither practicable to collect and conserve all of them nor is it necessary. Attempts have mainly been directed to securing representative samples from those areas with the richest diversity. These genetic resources have always been regarded by the IBPGR as a heritage of mankind to be freely available to all bona fide users.

Much of the diversity of crop plants of use in plant breeding is of cryptic expression. There are clearly limited numbers of morphological characteristics of the plant, and many of them, with the exception of those controlling such yield determinants as the number and size of ears of corn, may be agronomically irrelevant. Physiological characteristics abound, however, controlling life cycle through photoperiodicity, breeding system, and numerous responses to stress factors such as salinity, drought, and especially pests and diseases. Conserved samples are used primarily as a source of genes for broadening the genetic base of modern cultivars and thereby counteracting the vulnerability of cultivars with too narrow a genetic base to resist new races of pests and diseases.

IBPGR's emphasis began to shift away from the collection of landraces as its work progressed and as plant breeding techniques advanced, and to include the wider gene pools of the crops such as the wild progenitors and close relatives. The materials conserved embrace an evolutionary spectrum of the past 10,000 years and include obsolete and fairly recent cultivars, but breeders' lines rarely need conservation; their diversity is already present elsewhere. This point, which is not widely understood, has led pressure groups to urge the conservation of breeders' lines, which except in a few cases is not justified. In any case, germplasm for major staple crops is available from the International Centres of the CGIAR, the parent body of IBPGR, and from several national agricultural research organizations.

For most of the major crops, germplasm collections provide opportunities for continued introduction of genes with conventional or sophisticated breeding techniques. For many minor crops that have not greatly diversified following domestication, collections may contain cultivars that may be introduced as crops to other countries without further breeding. Hence, the collections of the major crops, e.g., rice, wheat, millet, and groundnut, will be large, whereas those of the minor ones, e.g., okra, many tropical fruits, and forage crops, will be relatively small. As the value of such minor crops becomes appreciated, however, there is likely to be a need for a more diverse germplasm base.

The IBPGR budget is small (a little over U.S. $5 million per year); hence, it does not provide long-term institutional or program maintenance support but, rather, serves as a stimulus or catalyst to work done by national or international organizations. Some of its funds are used to carry out urgent work and to fill gaps in the collection. Linked with the IBPGR program are all crop genetic resources activities in the world, many collections of which were initiated and supported by IBPGR. IBPGR coordinates rather than directs these activities in which all participants are equal partners.

To date, IBPGR has organized more than 500 collecting missions in scores of countries. Germplasm has been acquired in accordance with established priorities (IBPGR, 1981; Williams, 1982) and conserved in long-term storage in 40 centers around the world. All major crops have now been placed in designated gene banks. The IBPGR is currently reviewing the financial support it gives to such gene banks and to centers that are actively involved in ensuring that genetic resources are available for use in crop improvement programs.

CROP ORIGINS AND GERMPLASM USE

Over the past few decades, a great deal of information has been accumulated on the origin and evolution of cultivated species. In general, the wild species tend to have limited patterns of distribution. Although Vavilov (1951) laid the basis for further work in defining centers of origin and centers of diversity based on observed botanical variation, the nuclear centers of the origin of agriculture have been confused with the areas of evolution and diversity of crop plants (Hawkes, 1983). Harlan (1951) identified smaller microcenters that are rich in diversity and not necessarily near civilization nor confined to plains or mountains—features strongly associated with the Vavilovian centers—usually in the areas that would be regarded as broader centers of diversity.

There are therefore different definitions of the regions that comprise centers— from Vavilov's to those of Zhukovsky (1975). Whatever the definitions, it is still clear that there are areas where crop plant diversity is great and others where it is slight. The centers of diversity of several crops coincide and comprise limited numbers of regions in contrast to the 13 centers proposed by Vavilov. Crop germ-plasm is by no means obtained solely from developing countries; centers also cover areas of the developed world. Examples of crops that originated in the developed world include the adzuki bean (*Vigna angularis*) in Japan, oats (*Avena sativa*) and rye (*Secale cereale*) in northern Europe, and the sunflower (*Helianthus annuus*) in the United States. Many old landraces of the developed world are extremely valuable resources being fed back, through breeding, to Third-World agriculture.

In view of the apparent confusion surrounding the definition of the regions of diversity, it is pertinent to question how the material can best be identified and preserved. Over the past decade IBPGR's strategy has been to mobilize scientific information from hundreds of the world's best breeders and crop botanists. Scientists who have grown, used, and evaluated germplasm know far more about its variability than collectors who can only observe phenotypes. This strategy has been highly successful in ensuring the preservation of intraspecific diversity in a very limited time and is still implemented.

For the wider gene pools, where species and species relationships are the principal interest, taxonomy plays a major role in the conservation of genetic resources work, especially experimental taxonomy designed among other things to clarify genomic relationships. But information on the origins and evolution of crop gene pools is obtained not only from taxonomy but also from a synthesis of disciplines, including archaeology, linguistics, ecology, and molecular biology.

For some of the major crops, e.g., wheat and maize, there are still gaps in our knowledge, but in many other cases, the progenitors are well known. Although the ancestor of the faba bean (*Vicia faba*) has not been equivocally identified, several closely related species have frequently been found, the most recent during an IBPGR mission to Syria in 1986. Truly wild forms of cassava are not definitely known either, although Brazilian workers supported by IBPGR may have found one in 1986.

When a theory of origin has been postulated but not verified by taxonomic work, it may not be clear what germplasm should be collected and where. The cowpea (*Vigna unguiculata*) is a good example; taxonomically its ancestral home has been variously proposed as Ethiopia or central/southern Africa. IBPGR ex-amined all herbarium specimens of this species in 1986 and found that the pro-genitor of the cowpea occurs in every country of sub-Saharan Africa. Thus until experimental work on freshly collected material is complete, we will not know the patterns of variation or where to collect samples. Meanwhile, collection of material will be ad hoc and provide useful allelic diversity but not necessarily from the ancestral area of domestication.

In other cases, such as that of citrus crops (*Citrus* spp.), existing classifications are substantially untested in the field. IBPGR has put emphasis on relatives of the lime, lemon, and orange and will be organizing fieldwork to this end—not knowing whether the taxonomy works or not. Obviously, the field strategy may have to be modified to fit new knowledge and adapted as necessary. The collection and conservation work will no doubt greatly add to information on the origins, evolution, and distribution of *Citrus* gene pools. This has already proved true for beans of the genus *Phaseolus* (G. Debouck, IBPGRI/Centro Internacional de Agricultura Tropical, Colombia, personal communication, 1986), coles (*Brassica*)(wild gene pool of the Mediterranean), perennial soya bean (*Glycine*)(IBPGR collaborative work in Australia; A. H. D. Brown, Commonwealth Scientific and Industrial Research Organization, Australia, personal communication, 1986), African eggplants (*Solanum* spp.)(Lester et al., 1986), and cucurbits (Cucurbitaceae) (L. Merrick, University of California, Davis, personal communication, 1986). Numerous other crops will be worked on in a similar way.

Harlan and de Wet (1971) defined primary, secondary, and tertiary gene pools on the basis of the degree to which closely and distantly related species can be cross-bred. Hitherto a gene-pool approach has been taken in defining IBPGR work. This has obviously been the best strategy, because wild progenitors and their related species have increasingly been shown to be valuable sources of genes for crop improvement. Others, such as wheat, are used for transferring entire chromosomes or parts of chromosomes. Both somatic hybridization and conventional crossing are likely to be more successful among closely related species, but embryo rescue techniques are already commonly used where the parents are less closely related, e.g., in wide crosses between wheat or barley and related wild grasses.

Many advances will continue to come from breeding closely related plant species, but where breeding is far advanced, even species of the tertiary gene pool are useful. IBPGR has recognized this for two groups: first, wild grasses of the tribe Triticeae, which includes wheat, barley, rye, forages, and dozens of species in more than 20 genera distributed throughout the world, and second, the East Asian-Pacific-Australian species of soya bean of the tertiary gene pool. In other cases, the wider gene pool has been used very little; hence, conservation priorities have been imprecisely defined. Wild plants related to maize in the genus *Tripasacum* are unlikely to be used in maize breeding before the next century.

EVALUATION OF THE RESOURCES

Integral to crop genetic conservation, but a step subsequent to collection, is description and evaluation of germplasm. Although this activity lags behind, it is an ongoing, long-term process. In autumn 1985, IBPGR held a small strategy workshop to look at the collections and study how they could be used more effectively. It has initiated, in particular, research and development to see how best to work on subsets of large collections, thereby accelerating work. Physiological or otherwise cryptic variation must be thoroughly assessed during this process.

By contrast, other conservation programs lack this user-driven input. It seems to me that scientists must develop aesthetic, moral, and evolutionary arguments for wider conservation practices that are not based on whims of individuals but, rather, are geared to perceived needs such as environmental stabilization and rational land-use planning. Only when studies on biological diversity provide data for these arguments can a clear rationale be presented. It is evident that the environmental data bases needed by my community of scientists are only just being developed. Interestingly, when speaking about wild relatives of crops, a well-known crop botanist recently said that identification of environments was more important than proceeding from past taxonomy and searching large areas with often poor chances of success (G. Ladizinsky, Hebrew University of Jerusalem, Israel, personal communication, 1986).

SCIENTIFIC SUPPORT FOR GENETIC RESOURCES WORK

Conservationists outside the field of crop genetic resources frequently base their plea for conservation of wild species on spurious evidence and unrecognized needs resulting from a lack of adequately trained scientific manpower. The work in crop genetic resources has been successful because of its user orientation, and in fact has been solely directed to this end. There is a clear lesson here for the wider conservation movement and also a reminder of the need to identify gaps in scientific knowledge.

The success of IBPGR would not have been possible without the work of botanists, geneticists, and scientists in related disciplines. Nonetheless, there has been insufficient scientific research pertinent to the conservation of crop genetic resources. This has been partly due to lack of funding and partly to the fact that strategic research is often not as attractive to the academic community as basic research. Accordingly, IBPGR has had to shift its program to accommodate strategic problem-solving research, much of it of an interdisciplinary nature.

Areas of research where such problems exist are:

- the elucidation of patterns of genetic variation using such disciplines as taxonomy, ecology, cytogenetics, molecular biology, and population genetics; and
- the understanding of genetic stability in conservation systems, again as interdisciplinary work involving genetics, tissue culture, molecular biology, and seed physiology.

IBPGR has initiated relevant research that not only will provide practical solutions to specific problems but also will add to scientific knowledge. Traditional disciplines such as taxonomy and cytogenetics have to keep up with exciting new developments in, for instance, molecular biology. One recent development is the proposal that DNA is in a state of flux. Exchange of genetic material between organelles and nuclei may be possible, while bacterial infection and tissue culture, for instance, may induce new genomic changes in response to changed environments (Hohn and Dennis, 1985).

On the whole, conservation biologists have not fully considered the technologies necessary for scientific and practicable solutions and need a synthesis of data from numerous disciplines. One recent initiative, that is supported in part by the government of the United Kingdom and draws on a wide range of expertise, is the International Legume Database Information Service (ILDIS). IBPGR has been associated with this service from its inception and views positively the mobilization of scientists within several institutions. Yet the provision of checklists of taxa continent by continent draws heavily on traditional taxonomy. This venture and the Missouri TROPICOS system and the BIOSIS (BioSciences Information Service) Taxonomic Reference File are to be applauded.

EX SITU AND IN SITU PRESERVATION

Because of their nature, most crop genetic resources are conserved ex situ. Many crop relatives are annuals or weeds associated with disturbed agricultural environments, and most do not lend themselves to in situ conservation. Landraces cannot be conserved by growing them in primitive agricultural conditions; it is neither practical nor can it be justified morally. There are exceptions to what can be conserved in situ, especially perennial species associated with complex ecosystems such as tropical rain forests, e.g., cocoa, oil palm, relatives of *Citrus*, and some root crops. To encourage in situ conservation where it is relevant and to rationalize further collecting work, IBPGR has recently been developing its expertise in the ecogeographic sampling and monitoring of genetic resources. But the opportunities for in situ conservation are limited and should not be overemphasized.

In addition, preservation of species diversity by ecosystem maintenance is not always relevant to the conservation of crop genetic resources. Allelic diversity is rarely considered by ecosystem conservationists, yet it is needed for utilization by crop botanists (Frankel and Soulé, 1981; Ingram and Williams, 1984).

The interface between ecosystem conservation and the special needs to conserve crop genetic resources has been discussed by Ingram and Williams (1984). Many of the scientific points have not yet been addressed by the wider conservation movement and merit further attention if scientific objectivity is to prevail over emotive generalization.

REFERENCES

Frankel, O. H., and M. E. Soulé. 1981. Conservation and Evolution. Cambridge University Press, Cambridge. 327 pp.

Harlan, J. R. 1951. Anatomy of gene centers. Am. Nat. 85:97–103.

Harlan, J. R., and J. M. J. de Wet. 1971. Towards a rational classification of cultivated plants. Taxon 20:509–517.

Hawkes, J. G. 1983. The Diversity of Crop Plants. Harvard University Press, Cambridge, Mass. 184 pp.

Hohn, B., and E. S. Dennis, eds. 1985. Genetic Flux in Plants. Springer-Verlag, New York. 253 pp.

IBPGR (International Board for Plant Genetic Resources). 1981. Revised Priorities Among Crops and Regions. IBPGR Secretariat, Rome. 18 pp.

Ingram, G. B., and J. T. Williams. 1984. *In situ* conservation of wild relatives of crops. Pp. 163–179 in J. H. W. Holden and J. T. Williams, eds. Crop Genetic Resources: Conservation and Evaluation. Allen and Unwin, London.

Lester, R. N., J. J. G. Hakiza, N. Stavropoulos, and M. M. Teixiera. 1986. Variation patterns in the African scarlet eggplant. Pp. 283–308 in B. T. Sytles, ed. Infraspecific Classification of Wild and Cultivated Plants. Clarendon Press, Oxford.

Vavilov, N. I. 1951. The origin, variation, immunity and breeding of cultivated plants. Chronica Bot. Volume 13. 364 pp.

Williams, J. T. 1982. International programs with special reference to genetic conservation and the role of the United States. Pp. 52–55 in Proceedings of the U.S. Strategy Conference on Biological Diversity 1981. Publication No. 9262. U.S. Department of State, Washington, D.C.

Williams, J. T. 1984. A decade of crop genetic resources research. Pp. 1–17 in J. H. W. Holden and J. T. Williams, eds. Crop Genetic Resources: Conservation and Evaluation. Allen and Unwin, London.

Zhukovsky, P. M. 1975. Mega-genecenters and Endemic Micro-genecenters: World Gene Pool of Plants for Breeding. USSR Academy of Sciences, Leningrad. 116 pp.

CONSERVING AND MONITORING BIOTIC DIVERSITY
Some African Examples

BRIAN J. HUNTLEY

Director, National Programme for Ecosystem Research, Council for Scientific and
Industrial Research, Pretoria, South Africa

Africa has excited the imagination of explorers, naturalists, and conservationists more profoundly than any other continent. The writings of Rooseveld, Blixen, and Haggard brought the romance of the African veld into the homes of millions who had never, nor would ever, toil under the African sun. But the romance of the colonial era has now been replaced by the realism of thirst, starvation, and desertification. The wide open spaces are gone, and pressures on productive lands are greater than they can sustain. There is neither money nor trained manpower to implement comprehensive conservation programs in Africa today though the task of the conservationist in Africa is currently much greater and more urgent than it has ever been.

Credibility and realism are needed. But these qualities have not always characterized conservationists in Africa. The impressive heritage of national parks and reserves is a legacy from the colonial era, built on the favorite stamping grounds of repentant white hunters. Fortunately, a new era may be dawning as the result of the World Conservation Strategy (IUCN, 1980)—a program that can be explained in terms that are meaningful to the African husbandman. The concepts of biotic diversity and the sustainability of production and life-support systems carry with them the promise of tangible values and benefits to the community at large, not just to the affluent foreign tourist visiting parks that are national in name alone.

The question addressed in this chapter is quite simple. Given the diversity of African wild plants, animals, and ecosystems and the severe constraints on money, manpower, and time, how should one identify priorities for biotic conservation

248

and monitor progress toward achieving objectives? This question is examined by drawing on experience in several southern African countries against a background of the diversity and dynamics of African ecosystems.

BIOGEOGRAPHIC SETTING

Africa broke away from South America and the rest of the ancient supercontinent of Gondwanaland some 140 million years ago. Dinosaurs and gymnosperms occupied the landscape but were soon replaced by mammals and the flowering plants that we know today. Equatorial rain forests dominated the vegetation for a long period and were only replaced by the savanna systems characteristic of most of Africa today at the onset of the Pleistocene ice age, some 2 million years ago. Throughout the Pleistocene epoch the expansion and shrinking of the polar ice caps were reflected by major changes in the global climate. Cooler, drier periods of approximately 100,000 years alternated with shorter, warmer, and moister spells, and variants of these. This exerted major environmental pressures on the continent's flora and fauna. What we see of African ecosystems today is merely a narrow slice of the diversity that has existed on the continent throughout time. Through much of these massive oscillations of Pleistocene climate and vegetation, humans have exerted a profound influence on the shaping of African environments both as hunter and fire maker. A consequence of these varied influences is the tremendously dynamic character of African ecosystems, in which both speciation and extinction are ongoing processes.

PATTERNS OF SPECIES RICHNESS AND ENDEMISM

Students of plant and animal distributions have divided the world into eight major regions, or biogeographic realms, one of which is Africa south of the Sahara. Known as the Afrotropical realm (Udvardy, 1984), it includes not only sub-Saharan Africa but also the island continent of Madagascar, the islands of the western Indian Ocean, and the southern tip of Arabia. It possesses the richest mammal fauna of any realm and a rich and distinctive fish fauna but relatively impoverished bird, reptile, and amphibian faunas. The flora of tropical Africa is not nearly as rich as that of South America or Southeast Asia, but that of southern Africa is extraordinarily rich.

Of the 20 biogeographic units defined for the African continent, 17 fall within the Afrotropical realm and 7 of these are recognized as centers of endemism, the cradles of floristic evolution and speciation (White, 1983). The species richness and percentage of species restricted to each center of endemism are summarized in Table 29-1. Plants, birds, and fishes provide excellent examples of the differing patterns of biological diversity in Africa and can be used to illustrate the vulnerability of localized centers of species richness to human-induced impacts. The examples chosen also demonstrate that much of Africa's biotic wealth lies outside the tropical savannas, which have enjoyed so much attention from wildlife conservation organizations.

TABLE 29-1 The Seven Centers of Endemism of the Afrotropical Realm, With Estimates of the Numbers of Seed Plant, Mammal (Ungulates and Diurnal Primates), and Passerine Bird Species in Each and the Percentage of These Endemic to the Unit[a]

Biogeographic Unit	Area (1,000 km²)	Plants		Mammals		Birds	
		No. of Species	% Endemic	No. of Species	% Endemic	No. of Species	% Endemic
Guineo-Congolian	2,815	8,000	80	58	45	655	36
Zambezian	3,939	8,500	54	55	4	650	15
Sudanian	3,565	2,750	33	46	2	319	8
Somali-Masai	1,990	2,500	50	50	14	345	32
Cape	90	8,500	80	14	0	187	4
Karoo-Namib	692	3,500	50	13	0	112	9
Afro-montane	647	3,000	75	50	4	220	65

[a]After MacKinnon and MacKinnon, in press.

TABLE 29-2 The Number of Seed Plant Species in Various Regions of the World[a]

Biogeographic Unit	No. of Species	Area (1,000 km²)	Species (per 1,000 km²)
Tropical Africa	30,000	20,000	1.5
West Tropical Africa	7,300	4,500	1.6
Sudan	3,200	2,505	1.28
Southern Africa	22,977	2,573	8.93
Cape Floral Kingdom	8,504	90	94.49
Brazil	40,000	8,456	4.73
Peninsular India	20,000	4,885	4.09
Australia	25,000	7,716	3.24
Eastern N. America	4,425	3,238	1.37

[a]After Gibbs Russell, 1985.

The flora of the Afrotropical realm probably includes some 40,000 species, the richest component of which is found in southern Africa, especially in the Cape Floristic Kingdom, one of the six major floristic divisions of the world's flora defined by Good (1974). The magnitude of the floristic richness of southern Africa is indicated in Table 29-2, which provides data for a wide range of regions. The 8,500 species of the Cape flora are compressed into an extremely small area. The change of species composition from one patch of vegetation to the next is very high—two sides of the same valley may differ by 45%, adjacent large areas may share less than 40% out of more than 2,500 species. Such rapid changes in floristic composition are unknown elsewhere, not even in the Indo-Malayan rain forest (Kruger and Taylor, 1979). The Cape Peninsula, only 470 square kilometers in area, possesses 2,256 indigenous species—more than half the flora of eastern North America. The unique floristic richness of the Cape is unfortunately matched by unusually serious threats to its survival. A detailed 10-year survey of the conservation status of southern Africa's 23,000 species of plants indicates that some 2,373 are threatened. The Cape Floristic Kingdom, occupying less than 4% of southern Africa, accounts for 68% of the threatened species (Hall et al., 1984). Satellite imagery indicates that 34% of the region's natural vegetation has been transformed by agriculture and other human activities. The second most serious threat to the small heathland shrubs characteristic of this region is the aggressive competition exerted by large, introduced woody weeds. In addition, an invasive ant has been found to suppress populations of the native seed-storing ants, thus exposing critical seed sets to predation by rodents or destruction by the intense fires that characterize these Mediterranean-climate shrublands.

The distribution of Afrotropical birds is rather different from that of the patterns of floristic richness. A recent analysis of the distribution of 1,595 species of Afrotropical birds indicated fairly clear correlation between bird diversity and rainfall, vegetation, and other factors in the present environment. But many species of forest birds displayed distribution patterns that could best be interpreted in terms of past climatic and habitat conditions (Crowe and Crowe, 1982). These somewhat anomalous distributions indicate the occurrence of what are known as Pleistocene refugia—sites that would have escaped the dramatic environmental changes that took place during repeated ice ages. They are not only sites of species richness and endemism but are also the habitat islands of rare and threatened species. Of the 168 species listed in *Threatened Birds of Africa and Related Islands, the ICBP/IUCN Red Data Book*, 87 are forest species (Collar and Stuart, 1985).

Many of these birds occur as isolated populations in widely separated montane forests. The most critically threatened group of these rare birds survive in the small patches of forest on Mount Moco in central Angola, more than 2,500 kilometers distant from similar but much larger forests in Cameroon, eastern Zaire, Tanzania, and the South African escarpment (Huntley, 1974). Pressure for timber and fuelwood on this 100-hectare remnant is severe—the rural peoples living on the cold mountain slopes have no alternative resources.

The Great Lakes of Africa are the only massive freshwater bodies in the tropics. They are comparable in size to the North American Great Lakes but are much older. The African lakes harbor the world's richest palustrine fish faunas, one family of which (Cichlidae) provides the supreme example of vertebrate evolution within geographically isolated communities—upwards of 900 species—far more spectacular than that described and immortalized by Darwin in the 13 species of Galapagos finches.

Table 29-3 indicates the levels of species richness and endemism in the three largest African lakes. There are probably more than 1,100 species of indigenous fishes in these lakes, compared with less than 160 species in the much larger North American lakes. Not only are these fishes of immense ecological, evolutionary, and conservation interest, they also support a major traditional fishery. This fishing culture, and its socioeconomic fabric, is now being threatened by the introduction of sophisticated and capital-intensive fisheries based on Nile perch, an introduced species (Coulter et al., 1986). This piscivorous species was introduced into the northern part of Lake Victoria in 1960 in a well-intentioned but shortsighted effort to improve the commercial fishing industry. It has rapidly expanded its range in the lake at the expense of the endemic species. Massive changes in the abundance and distribution of these endemics are occurring, and it is estimated that up to 30 species have already become extinct. This is probably the highest rate of human-induced extinction of vertebrates yet recorded. The tragedy of the situation is that a vast number of the species currently threatened with extinction have not yet been collected, classified, and named.

TABLE 29-3   The Great Lakes of Africa: Area, Fish Species Richness, and Endemism[a]

| Lake | Area (km²) | Cichlid | | Other Families | |
		No. of Species	% Endemic	No. of Species	% Endemic
Victoria	68,000	250	99	38	42
Tanganyika	33,000	165	99	75	70
Malawi	30,800	500	99	45	63

[a]From Coulter et al., 1986.

APPROACHES TO THE CONSERVATION OF BIOTIC DIVERSITY IN AFRICA

The spectacular diversity of landscapes and biota in Africa is not matched by the human and financial resources needed to protect and manage this heritage. It is therefore essential that available skills and funds be directed to sites of the highest conservation priority. An objective system of assessing priorities in relation to clear and unambiguous goals is needed. Within the context of the goals of the World Conservation Strategy, the specific objective of in situ biodiversity conservation in Africa might take the following form:

To establish a minimum set of protected areas that provides for the preservation of the full range of African ecosystems and their biota, including marine and coastal species and systems.

Some of the steps to be taken to reach this objective include:

• the development of a hierarchical series of biotic classification systems (from continental to regional to local scales);
• the assessment of the current level of protection given to each element of these classifications;
• the identification of gaps in the protected area network within a ranked listing of priorities; and
• the mobilization of the funds and manpower needed to incorporate these areas within the network.

These needs will now be examined in the light of African examples.

BIOTIC CLASSIFICATION SYSTEMS

The first need in any review of biotic resources is information on the types, numbers, and distribution of the plants and animals to be found in the area under study. Such information is best synthesized within major vegetation and plant-geographic units. UNESCO has recently published a comprehensive review of the vegetation of Africa (White, 1983) and a detailed map at a scale of 1:5 million, which is of tremendous value in assessing the protected area cover of biota at a continental scale. Numerous regional and national vegetation classifications and maps that are also available in other publications permit analyses at finer resolution. But the complexity of biotic communities and the distribution of plants and animals require that much additional information be made available. The small, but species-rich communities of lakes, wetlands, rivers, estuaries, coastal dunes and mangroves, inselbergs, and escarpments, and many other specialized habitats are seldom included at the scale of even national vegetation maps. Conservation plans that ignore these communities will miss much of a region's biotic richness (Clarke and Bell, 1986).

A more complex problem relates to endemism. Centers of endemism are seldom reflected in distinctive and mappable vegetation types, yet they are of considerable conservation interest. Of greater concern is the fact that centers of endemism can only be revealed through exhaustive floristic analysis, a task that has been accomplished for very few species groups in Africa. Even in South Africa, which has benefited from more than 200 years of biological survey, new centers of endemism of considerable importance are only now being detected.

ASSESSING THE ADEQUACY OF CURRENT PROTECTED AREA COVER

During the past 30 years, the International Union for the Conservation of Nature and Natural Resources has developed a comprehensive data bank on the protected area systems of the world. It has recently completed a detailed listing of all protected areas of the Afrotropical realm, providing exhaustive information on the geographic, faunal, floral, and management attributes of each of some 620 protected areas greater than 50 square kilometers in the realm (IUCN/UNEP, 1987). This data base served as the foundation for the review of Afrotropical protected areas undertaken by MacKinnon and MacKinnon (in press), whose analysis indicates that 4.7% of the realm falls within protected areas totalling 949,500 square kilometers, considerably larger than the State of Texas. By comparing the total area of each mapped vegetation unit with that falling within protected areas, they found that only one of the seven major centers of endemism of the Afrotropical realm has more than 10% of its area protected. The other six centers have between 3.6 and 7.0% of their area within national parks and reserves. Even at the extremely coarse scale of resolution afforded by this analysis, the finding that as little as 3.6% of the biotic resource is protected gives cause for concern. On closer inspection, the situation is even worse—many of the so-called national parks in Africa are little more than yellowing documents in government archives. The Giant Sable Integral Nature Reserve in central Angola is occupied by more than 20,000 peasant farmers, several trading villages, and until guerilla activity prevented it, extensive diamond prospecting. The Reserve has not seen a game ranger in 10 years.

Factors such as the above necessitate a more objective evaluation of the effective protection afforded each biogeographic unit and protected area. In recent years, a variety of scoring systems have been proposed. These systems take account not only of the relative area protected but also of the effectiveness of government action to provide long-term security to the area (Clarke and Bell, 1986; Cumming, 1984). MacKinnon and MacKinnon (in press) developed a scoring system based on the size, protection objectives, and management effectiveness reported for each site. These data were then summarized by biogeographic unit and weighted for the number of distinct habitats and altitudinal ranges represented within the conservation network. Assessment of priority for action was based on the principle that action should be taken where it could have the best effect, not on lost causes— an all too common failing of conservation efforts based on sentiment rather than science.

At a finer scale of resolution, an analysis of the protected area cover of 189 vegetation units in 10 southern African states demonstrated the existence of major deficiencies in many of the 24 major vegetation divisions recognized (Huntley and Ellis, 1984). The most seriously threatened systems included the lowland forests of Angola and Mozambique and the Highveld grasslands, lowland fynbos, and succulent karoo of South Africa. All these systems face rapid reduction due to agricultural development or exploitation of timber resources for foreign exchange or fuelwood.

To assess the adequacy of the protected area cover of 15 communities in the exceptionally rich communities of the lowland fynbos (Cape heathlands), Jarman (1986) made use of 1.25 million maps prepared from satellite imagery. More than 69% of this species-rich vegetation formation had already vanished under urban, industrial, and agricultural development, and 21% of the 8,955 square kilometers still in a seminatural state had been invaded by alien woody plants.

A working group of over 40 researchers, administrators, and land owners participated in a 3-year study of the remnant patches of lowland fynbos. The survey identified 153 sites of conservation value and ranked them according to a formula that incorporated quantified attributes such as the rarity of the vegetation type, habitat diversity, total species richness, and the number of threatened plants found on the site. The rating was weighted in terms of the size and shape of the site and its distance from other protected areas and the degree to which the site had been transformed by introduced woody plants or other forms of disturbance. The conservation merit ratings ranged from 13 to 80 out of a possible 100. Only 5 out of 32 sites with a rating above 50 were currently protected, whereas the majority of the other existing reserves had ratings below 30 and were considered either too small, too greatly disturbed, or too low in biotic richness to merit inclusion in a costly protected area network. The study was probably the most detailed of its kind ever undertaken in Africa; indeed, the variety and quality of data available for the analyses are unlikely to become available elsewhere on the continent for many years. The significance of the results lies in the finding that even in an area of considerable financial and manpower resources, past decisions on the selection of sites for protection have been wholly inadequate to meet biological conservation needs.

IDENTIFYING GAPS IN THE NETWORK

The results of surveys of protected area cover based on vegetation maps can only provide the first step in the process of identifying gaps in the network. Much of the diversity of African ecosystems lies in communities that are too restricted or too narrow or patchy in their distribution to be included in the analytical approaches described in the last section of this chapter. Rivers, wetlands, and coastal ecosystems fall in this category. As a consequence of this, they have been largely ignored by African conservation agencies. During the last 10 years, long overdue attention has been devoted to these ecosystems in southern Africa. Some of the experience gained can be described here.

Wetlands in the form of seasonally waterlogged grasslands (dambos) are a characteristic feature of the vast moist savannas of the central African plateau. Drainage of these dambos provides the only rich agricultural soils over vast areas, and much of these systems have been transformed into agricultural lands, dramatically reducing the habitat available to the vulnerable Wattled crane (*Grus carunculata*) populations, which are dependent on these frequently burnt short grasslands. In South Africa, concern for the future of the Wattled crane has brought new emphasis to the conservation of the wetlands that support the remaining 100 pairs breeding in the country. Furthermore, 48% of the birds listed in the latest *South African Red Data Book—Birds* are grassland and wetland species (Brooke, 1984). Even more urgent, however, is the need to rehabilitate wetlands in the catchments of the country's major rivers, which now carry up to 375 tons per square kilometer per year of soil lost from overgrazed rangelands and cultivated slopes.

Despite the importance of wetlands, conservation efforts have ignored them because of difficulties in defining, identifying, and mapping them. These problems have now been overcome by the use of hydromorphic soils as the key indicator of wetlands (Begg, 1986). Maps of such soils are readily available in most African countries, and their identification on aerial photographs for detailed checking in the field is relatively easy. Because the soils retain their structural characteristics longer than their vegetation cover, it is also possible to obtain a rapid estimate of the rate of change in wetland systems. In the Tugela Basin of Natal, up to 34% of wetland communities have been destroyed in the past 50 years due to cultivation or overgrazing followed by extensive soil erosion.

The most seriously neglected biotic systems in Africa are the tens of thousands of kilometers of streams and rivers that drain the continent. Even those rivers protected within national parks and reserves are subject to severe impacts from developments upstream or downstream of the protected section. Their narrow linear structure and diffuse spread make them difficult to contain within all but the largest protected areas, and detailed information on their biological values, degree of disturbance, and conservation needs are difficult to synthesize. They are, like wetlands, invariably ignored by conservation planners.

During the last 3 years, considerable progress has been made in overcoming the difficulties of analyzing river conservation needs in South Africa. A computer-based expert systems technique developed by O'Keeffe et al. (1986) simulates the logic processes of river ecologists and converts the multivariate probabilities and diffuse intuitions of real-life situations into a simplified expression of river conservation status. The advantage of the system is that the complex calculations of interrelationships are handled by the computer, but the flexibility to take account of unusual situations is retained. A wide range of attributes are included in the system framework, and the users enter the best-available information on each of these, ideally within a workshop situation where the researchers, conservationists, and planners can pool their resources.

Because each river is different and, therefore, all attributes do not always apply in the same way, a number of rules are included in the program to interrelate attributes or modify their effect. The system is designed to assess whole river systems, individual rivers, parts of rivers, or points in a river. The extent of the assessment must be defined by the user beforehand. Obviously, the assessment of a whole river system will be performed at a coarser resolution than that for a small tributary. The flexibility of the system, allowing successive levels of data to be added as knowledge improves, makes it especially useful in Africa, where few river systems are adequately documented. The system has been developed for use on personal computers and is thus within the reach of most government agencies in Africa.

MONITORING THE SYSTEM

The dynamic nature of African ecosystems was mentioned at the beginning of this chapter. Continent-wide changes in the distribution of forests, savannas, grasslands, and deserts have occurred during the last 18,000 years due to major climatic events. San bushmen monopolizing isolated desert waterholes, iron-age communities deforesting coastal woodlands, and honey-gatherers burning moist savannas have induced subtle but significant fluxes in the distribution and abundance of plants and animals over the past 1,000 years. More recently, the changes brought about by both colonial and independent governments have been more extensive and less benign. Superimposed on these latter changes have been oscillating dry and wet rainfall patterns with intervals of 10 to 20 years.

Any attempt to monitor the status of species and ecosystems must be cognizant of such fluxes. Conservation biologists in Africa seldom occupy research posts for more than 10 years, and few remain at a given station for more than 5 years. Their observations are therefore of limited generality and frequently result in misleading rather than accurate predictions of a system's behavior. Nature is often counterintuitive, and the obvious management response to a problem is not always appropriate—it may even produce an effect directly opposite to that intended (Caughley and Walker, 1983). Research is needed to develop a predictive understanding of ecosystem structure and functioning in response to environmental changes and must be linked to monitoring systems that measure the direction and rates of these changes and responses.

There is extensive literature on the philosophy and technology of environmental monitoring. In Africa, much of this is irrelevant. What monitoring is being done varies tremendously in spatial and temporal scales and in duration and precision. Even where detailed long-term studies have been undertaken, few of the results are amenable to statistical analysis and valid interpretation due to faults in their experimental design (O'Connor, 1985). There are simply no reliable sets of data on some of the most critical issues in biotic diversity, such as the reduction of moist forests and the floristic impoverishment of arid lands. The need for a carefully planned international program of biotic diversity analysis and monitoring in Africa is an urgent priority. Without a reliable data base, cost-effective conservation measures cannot be planned. A few examples of successful monitoring activities suggest possible lines of approach.

At continental and regional scales, *Red Data Books* (RDBs) of plants, animals, and habitats are beginning to provide a valuable first approximation to the monitoring of biotic diversity (Anonymous, 1985). The accuracy and detail of information on many species in these lists are inadequate, but the mere publication of these data leads to critical review and improvement. The 1976 RDB on South African birds was cited in more than 150 papers within 8 years of its publication, and the latest edition (Brooke, 1984) includes the reclassification of 37 of the original listing of 101 species and adds 30 more. Similar rapid improvements of the data base on mammals, fishes, and plants have been witnessed in the new editions of these volumes. Although most of these changes in status reflect the inadequacies of the original data rather than real changes in the field situation, the existence of the RDBs triggered an upsurge of interest in monitoring rare species. This activity has been followed by the launch of annual counts of storks and cranes at a southern African scale; monitoring of all RDB bird species is now undertaken within a national bird atlassing project (Hockey and Ferrar, 1985).

Perhaps the most ambitious and detailed monitoring project yet undertaken in Africa is that initiated in the Kruger National Park in South Africa in 1975. Following on 50 years of data collection on large mammal numbers and distribution, rainfall patterns, and fire occurrence, the current program includes a network of climate, vegetation transect, and fixed-point photographic stations plus detailed helicopter and fixed-wing aerial counts of 12 species of large mammals, estimates of forage and water availability, season and extent of controlled burns and wildfires, and other information, with a sampling scale of 4-square-kilometer units within the Park's 19,853 square kilometers. The aerial and ground survey data are integrated within a series of computer programs, which provide a robust data base for the analysis and interpretation of large-scale patterns of change in savanna ecosystems (Joubert, 1983).

Even more detailed monitoring of the dynamics of large mammal populations has been undertaken at Sengwa Wildlife Research Area in Zimbabwe, where the movements and social behavior of the elephant population have been tracked by radiotelemetry for over a decade. The elephant study is supplemented by 20 years of detailed transect surveys of the habitat use of 15 other species of mammal. The Kruger Park and Sengwa projects, along with similar studies elsewhere in Africa, provide the practical experience and theoretical framework for much less sophisticated monitoring systems for implementation in countries with more limited resources. Ironically, the lessons learned in such extended and expensive exercises are seldom noted by expatriate conservation biologists sent to Africa.

LESSONS LEARNED

The review in this chapter is far too brief to provide more than a superficial treatment of the problems and progress in conserving biodiversity in Africa. A few key points arise from experience in this field over the past 20 years.

• African ecosystems are not as fragile and vulnerable as is popularly believed. Throughout their evolution, they have been subjected to enormous environmental pressures—including the hunter and the fire maker through the last few hundred thousand years. But current accelerated rates of change leave little room for complacency regarding the identification of real rather than perceived conservation priorities.

• Biotic diversity is not linked to the distribution of elephants, rhinos, and other so-called charismatic megaherbivores. The massive investment in conservation campaigns directed at these species does more for the souls of the donors and the egos of the elephant experts than it does for biotic diversity, which is centered on less exciting communities of montane forests, Mediterranean heathlands, wetlands, lakes, and rivers.

• The analysis of the level of protection afforded mapped vegetation types is a valuable first approximation to evaluating the effectiveness of existing protected area systems. But attention must also be directed to centers of species richness and endemism and to sites of threatened species, which are seldom reflected in vegetation maps. RDBs of threatened plants, animals, and habitats are of considerable value in stimulating interest at a national level in issues of biodiversity conservation.

• The existence of highly sophisticated environmental monitoring systems in several African states offers experience and expertise to other countries that require simple but effective approaches to monitoring biodiversity.

• Expert systems based on the synthesis of available ecological principles and local knowledge offer tremendous potential in the development of objective decision rules for identifying conservation priorities in areas and on topics with limited expertise or data. With very few exceptions, knowledge of biotic conservation needs and priorities far exceeds the ability of African governments to implement conservation action plans.

REFERENCES

Anonymous. 1985. The plant sites Red Data Book. Pp. 3–24 in Plant Conservation in Africa. A Joint Session with the International Union for the Conservation of Nature. Association for the Taxonomic Study of the Flora of Tropical Africa.

Begg, G. 1986. The Wetlands of Natal. Natal Town and Regional Planning Report. No. 68. Pietermaritzburg, South Africa. 114 pp.

Brooke, R. K. 1984. South African Red Data Book—Birds. South African National Scientific Programmes Report. No. 97. Foundation for Research Development, Council for Scientific and Industrial Research, Pretoria, South Africa. 213 pp.

Caughley, G., and B. Walker. 1983. Working with ecological ideas. Pp. 13–33 in A. A. Ferrar, ed. Guidelines for the Management of Large Mammals in African Conservation Areas. South African National Scientific Programmes Report. No. 69. Foundation for Research Development, Council for Scientific and Industrial Research, Pretoria, South Africa.

Clarke, J. E., and R. H. V. Bell. 1986. Representation of biotic communities in protected areas: A Malawian case study. Biol. Conserv. 35:293–311.

Collar, N. J., and S. N. Stuart. 1985. Threatened Birds of Africa and Related Islands, the ICBP/IUCN Red Data Book. International Union for Conservation of Nature and Natural Resources, Gland, Switzerland. 761 pp.

Coulter, G. W., B. R. Allanson, M. N. Bruton, P. H. Greenwood, R. C. Hart, P. B. N. Jackson, and A. J. Ribbink. 1986. Unique qualities and special problems of the African Great Lakes. Environ. Biol. Fish. 17(3):161–183.

Crowe, T. M., and A. A. Crowe. 1982. Patterns of distribution, diversity and endemism in Afrotropical birds. J. Zool. 198:417–442.

Cumming, D. H. M. 1984. Toward establishing priorities for funding and other international support for protected areas in Africa. Pp. 108–111 in Proceedings of the Twenty-Second Working Session of the Commission on National Parks and Protected Areas, Victoria Falls, Zimbabwe, 22–27 May, 1983. International Union for the Conservation of Nature and Natural Resources, Gland, Switzerland.

Gibbs Russell, G. E. 1985. Analysis of the size and composition of the southern African flora. Bothalia 15:613–629.

Good, R. 1974. The Geography of the Flowering Plants. 4th Edition. Longmans, London. 557 pp.

Hall, A. V., B. de Winter, S. P. Fourie, and T. H. Arnold. 1984. Threatened plants in southern Africa. Biol. Conserv. 28:5–20.

Hockey, P. A. R., and A. A. Ferrar. 1985. Guidelines for the Bird Atlas of Southern Africa.

Ecosystem Programmes Occasional Report. No. 2. Council for Scientific and Industrial Research, Pretoria, South Africa. 55 pp.

Huntley, B. J. 1974. Outlines of wildlife conservation in Angola. J. S. Afr. Wildl. Manage. Assoc. 4:157–166.

Huntley, B. J., and S. Ellis. 1984. Conservation status of terrestrial ecosystems in southern Africa. Pp. 13–22 in Proceedings of the Twenty-Second Working Session of the Commission on National Parks and Protected Areas, Victoria Falls, Zimbabwe, 22–27 May, 1983. International Union for the Conservation of Nature and Natural Resources, Gland, Switzerland.

IUCN (International Union for Conservation of Nature and Natural Resources). 1980. World Conservation Strategy. Living Resource Conservation for Sustainable Development, International Union for Conservation of Nature and Natural Resources, Gland, Switzerland. Four-brochure set.

IUCN/UNEP (International Union for Conservation of Nature and Natural Resources/United Nations Environment Programme). 1987. The IUCN Directory of Afrotropical Protected Areas. International Union for Conservation of Nature and Natural Resources, Gland, Switzerland. 1,034 pp.

Jarman, M. L. 1986. Conservation Priorities in Lowland Regions of the Fynbos Biome. South African National Scientific Programmes Report. No. 87. Foundation for Research Development, Council for Scientific and Industrial Research, Pretoria, South Africa. 53 pp.

Joubert, S. C. J. 1983. A monitoring programme for an extensive national park. Pp. 201–212 in R. N. Owen-Smith, ed. Management of Large Mammals in African Conservation Areas. HAUM Educational Publishers, Pretoria, South Africa.

Kruger, F. J., and H. C. Taylor. 1979. Plant species diversity in Cape Fynbos: Gamma and delta diversity. Vegetatio 41:85–93.

MacKinnon, J., and K. MacKinnon. In press. Protected Areas Systems Review of the Afrotropical Realm. International Union for the Conservation of Nature and Natural Resources, Gland, Switzerland.

O'Connor, T. G. 1985. Synthesis of Field Experiments Concerning the Grasslayer in the Savanna Regions of Southern Africa. South African National Scientific Programmes Report. No. 114. Foundation for Research Development, Council for Scientific and Industrial Research, Pretoria, South Africa. 126 pp.

O'Keeffe, J. H., D. B. Danilewitz, and J. A. Bradshaw. 1986. The River Conservation System, a User's Manual. Ecosystem Programmes Occasional Report. No. 9. Council for Scientific and Industrial Research, Pretoria, South Africa. 17 pp.

Udvardy, M. D. F. 1984. A biogeographical classification system for terrestrial environments. Pp. 34–38 in J. A. McNeely and K. R. Miller, eds. National Parks, Conservation, and Development: The Role of Protected Areas in Sustaining Society. Proceedings of the World Congress on National Parks, Bali, Indonesia, 11–12 October 1982. Smithsonian Institution Press, Washington, D.C.

White, F. 1983. The Vegetation of Africa: A Descriptive Memoir to Accompany the UNESCO/AERFAT/UNSO Vegetation Map of Africa. United Nations Educational Scientific Cultural Organization, Paris. 356 pp.

SCIENCE AND TECHNOLOGY:

HOW CAN THEY HELP?

Sterile culture is one propagation method for
the ex situ preservation of plant diversity.
Photo courtesy of John Einset.

CAN TECHNOLOGY AID SPECIES PRESERVATION?

WILLIAM CONWAY

Director, New York Zoological Society, Bronx, New York

In the preservation of biological diversity, the use of technology is a last resort. When the preservation of ecosystems falters, their fragments may have to be cared for piece by piece.

FOUR OBSTACLES TO SPECIES SURVIVAL

A species of limited distribution faces at least four obstacles. First, there may not be sufficient habitat and the possibility of obtaining numerous large new nature preserves is remote. Even protecting some areas already designated as preserves is not proving possible, and no land whatsoever will be set aside for large numbers of species. Second, many of the preserved habitats will be in pieces too small and too subject to change to sustain unmanaged, genetically and demographically viable populations of the animals and plants they seek to protect (Soulé and Wilcox, 1980). Third, although the majority of wild species must persist outside of wildlife preserves, large land vertebrates and great aggregations that conflict with humans will be mostly confined to refuges and those outside will require continual monitoring, protection, and help. Finally, human populations will continue to grow for some time, inexorably reducing resources available to other species, while human land-use patterns, cultural attitudes, and economic practices will continue to shift and change (Myers, 1979).

PROBLEM AND APPLICATION

Despite the factors mentioned above, the loss of a wild population is not always the result of irreversible habitat change. It can come about for transient economic and cultural reasons, such as overhunting of the American bison, Arabian oryx, white rhinoceros, and American beaver. Or it may happen for correctable environmental reasons, such as introduced pesticides (the peregrine falcon and bald eagle) and introduced predators and competitors (the Hood island giant tortoise and Howe Island wood rail). And the loss may also last for unpredictably long periods, as with the Pere David's deer, Mongolian wild horse, European wisent, perhaps the Siberian tiger, brown-eared pheasant, Mauritius pink pigeon, and Guam rail; it could even be forever. But populations of all these species and many more have been increased in one situation or another by intervention strategies (see Table 30-1).

Where a wild population's ability to survive is lost, especially where the threat and destruction may be temporary, e.g., for the American bison, peregrine falcon

TABLE 30-1 Ex Situ Care and Biotechnology. Each Technique Has Been Utilized with the Species Listed Below It on a Long-Term or Experimental Basis

Intervention Technique	Species
Short-term propagation and reintroduction	Golden lion tamarin, cheetah, wolf, red wolf, American and European bison, Arabian oryx, onager, Andean condor, bald eagle, peregrine, Hawaiian goose, Lord Howe Island wood rail, Guam rail, European eagle owl, Guam kingfisher, Galapagos giant tortoise, Galapagos land iguana, Ash Meadows Amargosa pupfish
Long-term propagation	Lion-tailed macaque, Siberian tiger, Pere David's deer, European bison, Przewalski horse, brown-eared pheasant, Edward's pheasant, Bali myna, white-naped crane, addax, slender-horned gazelle, scimitar-horned oryx, gaur, Gérevy's zebra, Puerto Rican horned toad, Chinese alligator, Mauritius pink pigeon, Madagascar radiated tortoise, Aruba Island rattlesnake
Relocation, transplantation	Koala, mongoose lemur, aye-aye, brown lemur, chimpanzee, gorilla, squirrel monkey, wooly monkey, spider monkey, common marmoset, black rhinoceros, white rhinoceros, red deer, white-tailed deer, mule deer, moose, Tule elk, bighorn sheep, musk-ox, pronghorn antelope, roan antelope, mountain goat, African elephant, more than 400 species of birds, many reptiles and amphibians
Fostering, cross-fostering	Peregrine, bald eagle, whooping crane, masked quail, polar bear (captive), many species of waterfowl, pigeons, cranes (in nature and captivity), and passerine birds in captivity
Artificial incubation	Gharial, Siamese crocodile, Chinese alligator, green turtle, ridley, hooded crane, whooping crane, white-naped crane, and many other birds, reptiles, amphibians, and fishes
Artificial rearing	Hundreds of species of most vertebrate groups
Artificial insemination	Alligator, ocellated turkey, brown-eared pheasant, whooping crane, squirrel monkey, yellow baboon, giant panda, guanaco, Speke's gazelle, gemsbok, bighorn sheep
Embryo transfer	Gaur, bongo, eland, common zebra, Przewalski horse, cottontop marmoset, yellow baboon

(see Cade, Chapter 32), bald eagle, Arabian oryx, or Ash Meadows Amargosa pupfish, we need new scientific comprehension and a responsive technology. We must be able to relocate, sustain, or store a threatened population, to start and stop its propagation, and to reintroduce or remove it.

Because many populations of species in nature are becoming fragmented and isolated from each other, the emigration and immigration necessary for them to find unrelated mates of the right age and gender are becoming impossible. In such instances, intervention technologies will be necessary to effect the required movements. In such small populations, localized catastrophes, disease, sex and age imbalances, and even inbreeding can threaten viability (Schonewald-Cox et al., 1983). In response, technology may make it possible to remove or insert individuals into populations or even embryos or zygotes with needed characteristics *into individuals* (see Dresser, Chapter 34).

Where conservation biologists identify threatened but critical coevolutionary links, especially those between keystone species essential to ecosystem stability and diversity, sustaining these links for a time by scientific management of predators, competitors, even of environmental chemistry, microclimate, and with reintroductions, may be our only option for preservation.

LIMITATIONS OF SCIENCE AND TECHNOLOGY

But if such technological treatments and repairs are possible, why can not science and technology simply save biodiversity? Perhaps, as H. L. Mencken is reported to have said in a different context: "For every complex problem there is a simple answer and it is wrong."

Most losses of biological diversity, to say nothing of lost ecological services, are quite beyond human ability to repair. Too many very intricately interdependent species are being lost too rapidly with too many unpredictable consequences for others. Besides, sustaining species in a freezer, in a captive population, or in small fragmented refuges provides little to the Earth in the way of basic ecological services. However, intensive care and biotechnology can preserve some diversity that would otherwise be lost. But the greatest dimension of such preservation is depressingly slight compared with that which can be or could have been sustained in adequately designed and protected nature preserves and by understanding accommodation outside preserves.

NUMBERS VERSUS TECHNOLOGY

It is the numbers, whether they be those of the great variety of creatures requiring help or those representing the scarcity of biologists and dollars to help them, that discourage prospects for sustaining a sizeable proportion of living creatures solely through technology—despite our most ardent wishes or most arrogant imaginings. There are perhaps 400,000 species of plants. In Chapter 31, Ashton discusses their preservation in gardens and arboreta. But there may be 30 million kinds of invertebrates, mostly insects. Despite their importance, the overwhelming number of specialized invertebrates makes it logistically impossible for technology to contribute to the preservation of a significant representation for restoration programs.

There are only about 41,000 vertebrates. Of these, 19,000 or so are fishes, about 9,000 are birds, 6,000 are reptiles, 3,100 are amphibians, and 4,300 are mammals (E. O. Wilson, personal communication, 1987). For those that have been or will be totally displaced from nature, we have only a few specialized propagation centers and the world's zoos.

Zoos currently house about 540,000 mammals, birds, reptiles, and amphibians—an almost trivial number in relation to original wild populations but significant in the impact on human interest. The number is roughly equal to 1% of the domestic cats in American households, 10% of the cats and dogs euthanized annually in the United States, or about 25% of the deer taken by U.S. sportsmen each year. Zoos are popular but have little room. The spaces for animals in the world's zoos could all fit comfortably within the District of Columbia. Even if half these spaces were suitable for propagation of vanishing animals, the individual numbers of each species necessary to keep viable populations would make it impractical for zoos to sustain more than 900 species very long and probably far fewer in conventional breeding programs (Conway, 1986). But it will not always be necessary to sustain a population for a long time or to do so conventionally. Zoos, revised and improved, can come to have a special role in species preservation, for they represent a unique devotion of local human resources to the care of foreign wildlife.

In the past few years, the world's zoos have bred more than 19% of all the living mammals and more than 9% of the birds. Thus far, criteria of genetic uniqueness as well as the practicality of care have guided long-term propagation programs. In the future, more attention must be given to ecological criteria with an eye toward the future needs of restoration programs, to species that naturally occur at low density, to the great predators and large ungulates, and to the primates. A foundation for such help rests in growing international collaboration and, in the United States, in unequalled programs of coordinated animal data gathering in the International Species Inventory System and species management in the Species Survival Plan of the American Association of Zoological Parks and Aquariums.

Zoos are breeding orangutans and Chinese alligators, Bali mynas, pink pigeons, and Puerto Rican horned frogs, addax, slender-horned gazelles, wattled cranes, and black lemurs. They have pioneered rare embryo transfers between animal species, artificial rearing techniques, cross-fostering between species, and necessary long-term contraceptives for population management—a host of fundamental technological tools essential to the prospects of helping species in extremis.

Even so, extensive scientific and technological advances would be necessary to appreciably expand the space available for the care of species losing their homes. But where cold storage of sperm and embryos is possible, a herd of wild cattle or antelope can be cared for in a space no larger than a soda straw, moved without risk of trauma, and stored indefinitely—if we are satisfied to have our wild cattle and antelopes in soda straws. Yet, we must consider the alternatives and their time scales.

Unfortunately, practically all that we know of animal reproductive physiology has been worked out with a few domestic and laboratory species. The technology of sperm and embryo storage in use with domestic cattle is the product of 20 years of research, millions of dollars, and thousands of specimens—a critical matrix for investigation and discovery. A vast amount of research would have to be undertaken before a technology of embryo and sperm storage and transfer as reliable as that in use with cattle could be available for wild species without domestic analogs. But where is the economic incentive for such research? Where are the animals? The apparently simple techniques of artificial insemination, for example, have been successful with scarcely 20 wild species of mammals. Nonetheless, development of the scientific understanding necessary to long-term propagation is a technological fulcrum for many intensive species care programs. In Chapter 33, Seal discusses some of the challenges.

TECHNOLOGY IS EXPENSIVE

Unhappily, high cost is characteristic of high-tech applications, and whereas the capability and the money to apply advanced technologies to preservation is located mostly in wealthy northern countries, the largest problems of species loss are in poor tropical countries. Money used for high-tech intervention strategies obviously can not be used to preserve habitat. For less-developed countries, habitat preservation is the only realistic strategy, unless help comes from outside.

Whatever the help, no available amounts of money can ensure the protection of many species in nature, even vertebrates, such as the addax and scimitar-horned oryx from the Sahel or Guam's kingfisher and rail. Ex situ care and biotechnology are their only hope. Besides, support from different sources is usually restricted to different purposes. Except in local education, research, and propagation programs in zoos, for example, municipal funds are usually unavailable to international species preservation. In such differentiation of source, competition for funds between preservation options can be diminished. After all, ex situ care and technology are used only after it is evident that conventional conservation efforts could fail.

BUYING TIME

Can technology be used to ensure continuing evolution? Both intensive management and habitat reduction reduce the chances of directional habitat-responsive evolution. And in small unmanaged populations, genetic drift is much more powerful than natural selection. But before worrying about whether species must continue to evolve to survive, please reflect upon the time scale of concern. The profoundly immediate problem is to save as many species as possible through the next 150 years.

It seems inevitable that most large land vertebrates and many plants eventually will survive only as wards of humans, scientifically managed or cared for, even reestablished, at some point. Because of an overall decline of diversity, those species that persist through the feeble efforts of science and technology will become proportionately more important. Saving 200 of 2,000 mammal species seems more important than saving 200 of 4,300. Furthermore, it is the larger forms, among animals if not plants, that will most likely profit from these intervention strategies;

creatures that are not only spiritual symbols for the ongoing promotion of conservation but evolutionarily conservative; more irreplaceable than the merely irreplaceable. Preservation science and technology must become an active branch of conservation biology, because future habitat restoration, if any, will depend upon its progress.

Thus, technology is not a panacea for the disease of extinction. It is a palliative—a topical treatment with which to buy time, to preserve options for a few populations and species judged of special value. In the final analysis, it is no more important than the species it sustains, which would otherwise be lost forever, and no less.

REFERENCES

Conway, W. 1986. The practical difficulties and financial implications of endangered species breeding programmes. Int. Zoo Yearb. 24/25:210–219.

Myers, N. 1979. The Sinking Ark. Pergamon Press, Oxford. 307 pp.

Schonewald-Cox, C. M., S. Chambers, B. MacBryde, and W. L. Thomas, eds. 1983. Genetics and Conservation. Benjamin/Cummings, London. 722 pp.

Soulé, M. E., and B. A. Wilcox, eds. 1980. Conservation Biology: An Evolutionary-Ecological Perspective. Sinauer Associates, Sunderland, Mass. 395 pp.

CONSERVATION OF BIOLOGICAL DIVERSITY IN BOTANICAL GARDENS

PETER S. ASHTON

Director, Arnold Arboretum, Harvard University, Cambridge, Massachusetts

Conservation is already, and very appropriately, recognized as being a major activity for botanical gardens in both their research and educational programs. In this field, arboreta and botanical gardens have a particular and important potential, which I discuss in this chapter.

In nature, plants frequently exist in small populations. Examples include many rare endemics, such as those of mountain peaks and many in the Mediterranean dry sclerophyll scrublands, especially in the Cape Province of South Africa and in Southwest Australia, and those of certain rain forests. Over the relatively short time we realistically have had to work as conservation managers, extremely small stands have been found to persist in nature. Higher plants, being sedentary, are often highly site-specific. This facilitates the development of logical plans for demarcating minimal areas for in situ conservation based on ecological knowledge and principles of island biogeography. On the whole, the most favorable sites are a few environmentally heterogeneous reserves of sufficient size to minimize edge effects (e.g., changes in species composition at the periphery caused by in- and out-migrations from adjacent unprotected lands). Ideally, these would be loosely connected by small stepping stones or corridors to allow for the exchange of genes (Diamond, 1975). Identification and immediate protection of sites of high conservation value must be our highest priority in the absence of even the grossest information upon which to base plans, including basic inventory as well as distributional and ecological data on many of the richest biota. This underlines the vital necessity of increasing inventory and ecological information as a prerequisite to developing any logical plan for conservation. In practice, of course, the luxury of regional planning often does not exist. The conservationist only succeeds in

269

raising awareness when the plant is reduced to endangerment in one or a few isolated localities or, at best, is offered a patchwork of lands for which the farmer and the planner have failed to find other uses. As development proceeds and natural habitats become increasingly fragmented, extinction accelerates (Wilcox and Murphy, 1985). The most endangered floras are those of the arable lands; the current distribution of preserves takes little account of this.

Even when the luxury of time for planning does exist and centers of species richness and endemism can be identified and conserved, many locally endemic plant species refuse to follow the rules and occur in isolated areas where, overall, conservation priorities are low. Even under ideal circumstances, though, decreases in and fragmentation of natural areas is certain to lead, as predicted by the theory of island biogeography, to substantial increases in extinction rates, though it is uncertain whether these rates would be on the scale calculated for large animals (e.g., Schonewald-Cox, 1983; Soulé et al., 1979; Simberloff and Abele, 1984). This is therefore a case for some form of selective program of ex situ conservation, that is, conservation through cultivation.

Too little is known of the ecology of any plant species, let alone those that are rare and endangered, to consider the transfer of species from one natural community to another. But in addition to their immobility, plants have many practical advantages over animals for ex situ conservation.

Plants are generally easy to propagate asexually through a variety of methods, including division of the rootstock, cuttings, and tissue culture. For conservation of heritable character traits, experience with a variety of crop plants has shown that gene cloning is a realistic possibility. Propagants, vegetative or seed, can be collected with minimal disturbance to the wild population and are cheap and easy to transplant.

In comparison to animals, management of ex situ plant populations is relatively simple and inexpensive. Plants do not require caging, and in practice, genotypes can often be maintained for long periods, though probably not permanently, through propagation or forced rejuvenation (e.g., see Rackham, 1976, on the effect of pollarding on trees recorded over half a millennium). Furthermore, plants need less constant care than animals. The bisexuality of the majority of higher plants implies broadly that minimum population sizes for maintenance of heterosis can be half those of populations comprising two sexes. In addition, their modular construction allows considerable phenotypic plasticity. Their habitat requirements can generally be reasonably accommodated ex situ provided that competition is excluded. They do not manifest demanding behavioral traits and are rarely dangerous to humans. Added to these attributes, plants are both attractive and unobtrusive. They are more often scented than smelly, and they are generally perceived by humanity as benign. In short, they are welcomed adornments to the human environment.

Is is certainly true, then, that the rapidly increasing demand for ex situ conservation, occasioned by the inexorable destruction of natural habitats, presents botanical gardens with both a challenge and an opportunity, the likes of which have not arisen for more than a century.

METHODS OF EX SITU CONSERVATION

In comparison to conservation of animals, flowering plants have a disadvantage in the extraordinary diversity of their reproductive systems and, notably, their sexual differentiation. These differences are considered to be major determinants of genetic patterns within populations. The kind of reproductive system influences the minimal viable population sizes needed for conservation (Wilcox and Murphy, 1985) both in situ and ex situ and controlled pollination strategies for stock regeneration. In brief, self-pollinating and vegetatively reproducing species will vary more genetically between than within breeding populations (Allard, 1960). But among outbreeders, especially dioecious species, the reverse will hold true, though differences in gene frequency between reproductively isolated populations will increase over time. In self-pollinating species, representative samples of a wide range of breeding populations should be sampled, but individual samples need be represented by comparatively few individuals. For outbreeders, individual populations should be well sampled, but fewer representatives of different populations will generally be necessary.

Methods of ex situ conservation now available can conveniently be classified according to the part of the plant that is conserved—the whole organism, seed, tissues, or genetic material in culture. When kept in ex situ living collections, whole plants have educational value and can be displayed, and for species that take a long time to reach reproductive maturity, mature specimens on hand are advantageous for research. A relative disadvantage of whole plants is their higher maintenance costs in comparison to other means of ex situ plant conservation. For example, there are high requirements for space, especially for trees. Conversely, annuals require frequent, controlled pollination and reestablishment, unless they are inbreeders or methods of vegetative propagation are available. Whole plants conserved in gardens and plantations often readily hybridize with related taxa. Controlled pollination is therefore obligatory for regeneration from seed among outbreeders.

When grown in single-species plantations, whole plants are more susceptible to communicable diseases than when scattered, as in nature, in a matrix of other species. On the other hand, plants will as a rule prosper outside their natural range in the absence of coevolved pathogens. This explains why crop plants, particularly long-lived tropical species such as *Hevea* rubber, have flourished best in plantations outside their region of origin. Clearly, this biological reality causes political problems, which must be faced and overcome if ex situ conservation is to succeed and its subjects are to be exploited to benefit humans. It also identifies a conflict between the needs for conservation per se and the need for display or education when demonstration of indigenous flora has high priority. In contrast to masses in plantations, however, specimen plants will not incur this danger.

With present technology, the preferred method of ex situ conservation is through storage as seed. The principal advantage of seed banks is their economy of space and the larger sample sizes that then become possible and, in countries with high labor costs, their low labor demands. The principal practical disadvantage is their reliance on a dependable power supply, the need for meticulous monitoring of germinability over time, and the need for periodic regeneration under conditions that minimize selection among the residual seed stock. Added to this is the fact that research in seed storage has overwhelmingly been done with crop plants. Many of these are plants of early succession, a habitat in which many species possess seed dormancy. Many plants in nature will more often and even predominantly possess seeds that lack dormancy and will prove recalcitrant to induction of dormancy with current methods. It will not pass unnoticed that many long-lived species, often large at maturity, also predominate in these same habitats.

Tissue cultures, especially of meristems, can maintain genotypes unaltered over long periods (Henshaw, 1975; Wilkins and Dodds, 1983). They also provide an economic means of suspending, at least temporarily, changes in gene frequency in cultivated populations (in this case represented by cells or tissue in culture). Mutations will continue to occur, particularly in cell and callus cultures, but are less frequent than in whole plants.

Each taxon has it own requirements for successful establishment and regeneration in tissue culture. At present, the basic mechanisms are poorly understood, so that successful techniques must be developed through tedious trial and error. Nevertheless, experience is accumulating. It now seems possible that established techniques are only applicable to half the taxa of higher plants at most, although response can be predicted from evolutionary relationships (Einset, 1985).

Because tissue cultures using current media generally appear to be chromosomally unstable over long periods, they too must be viewed as evolving populations, albeit in a microenvironment in vitro. Nonetheless, tissue cultures provide an invaluable alternative for conservation of multiple lines of taxa not easily conserved ex situ on this scale by other measures.

Currently, then, all these methods of ex situ preservation of live plant material require periodic regeneration and sexual reproduction of the stock. The latter presupposes knowledge of the breeding system and pattern of genetic variability of the species concerned.

Although cryogenic storage of seed may in the future provide a solution for long-term preservation of natural patterns of genetic variation within population samples in vitro, DNA libraries are probably the most stable form in which genetic information can be stored. Gene libraries provide a means to conserve DNA sequences, even whole genomes, for research. The genetic material thus conserved can be introduced to other extant genotypes, but whole individuals cannot be regenerated independently. Although gene libraries are subject to the same problems of population genetics as other methods of ex situ conservation, at present they can more easily be stored indefinitely. This can be done by a variety of techniques available for reducing metabolic activity or, if desired, rendering them fully inert.

Some may consider these topics to be remote from the mission of a botanical garden or arboretum. Where, though, if not in a botanical garden, are these issues related to education and to research on and management of plant diversity to be addressed? In fact, with adequate funding botanical gardens and arboreta are well suited to take on this responsibility. After all, botanical gardens are museums without walls generally occupying, if not public space, then prime space from which the public can hardly be excluded. They are magnets for the curious as well as for seekers of tranquility. Most important, their mission is the cultivation and study of the diversity of plants. Furnished as a rule with a reference herbarium and a library as well as living plant material, they alone have the breadth of facilities required for the research needed to back up and refine the management of conserved plant populations in the wild.

THE GENETIC CONSEQUENCES OF EX SITU CONSERVATION

Although for practical reasons, including nomenclatural custom, we fall into the habit of giving priority to species conservation, the genetic nature of the biological species remains elusive for many kinds of organisms, including bacteria and other microorganisms and a large percentage of plant groups. The genetic mechanisms underlying the maintenance, increase, and loss of biological diversity proceed at the scale of the breeding population. For conservation planning, the unit of the breeding population is therefore clearly more desirable on biological grounds.

For very rare and local endemic species, this should not prove as serious a problem as among widespread and variable species, because, ipso facto, they are generally confined to one or a few populations and restricted habitat ranges in which out-breeding opportunities are already severely limited. Where widespread, variable species are to be conserved, a strong case can be made for preserving infraspecific variation (Antonovics, in press), particularly when eventual reintroduction into the wild is visualized. The restriction of useful heritable attributes to specific populations and races is well known among progenitors and landraces of crop plants and has been demonstrated in unique populations of wild taxa (Bradshaw, 1984). Wherever possible, it is important to distinguish genetic from environmental effects by observation of variation among plants grown in garden plots (Briggs and Walters, 1984).

It is generally agreed that most genes do not vary at the population, or even species, level. Most variable alleles are sufficiently abundant to be adequately sampled and conserved without danger of chance extinction through random drift in artificial populations as small as 50 (Marshall and Brown, 1975) to 100 (Frankel and Soulé, 1981) randomly selected individuals. However, the importance of rare, variable genes in the long-term survival of species is unknown. It can be expected that a few genes rare in one population, or ecotype, will be common in another. Selection varies quantitatively and qualitatively within the life of a population. Some of these changes will be unidirectional and are the stuff of evolution. Alleles will therefore vary in fitness over time scales exceeding the life cycle of the plant in conservation. Samples for conservation in gardens need to be taken from a range of populations, therefore, if genetic variability is to be adequately represented.

Yet more important, the characteristics by which we distinguish races and even some individual populations (Clausen and Hiesey, 1958) may often represent differences in gene complexes rather than in individual gene loci. Such variation between populations in the representation of gene complexes can occur over remarkably short distances (Antonovics and Bradshaw, 1970; Hamrick, 1983), although they can be remarkably constant within populations (Clay and Antonovics, 1985; Morishima et al., 1984). We know little about the heritability of such complexes and nothing concerning their reconstruction.

Thus, a serious problem is that natural selection cannot be simulated ex situ. If the artificial population is established from seed, the progeny have been released from natural selection from the start. Whether ex situ conservation takes the form of plants, seed, or vegetative propagules, including tissue culture, genetic rejuvenation through controlled cross-pollination of regenerated plants is periodically obligatory with present technology. New unselected gene combinations are inevitably introduced every time. However much care is taken in seed collection by controlled pollinations, natural selection cannot be simulated, particularly if it is mediated by interspecific competition. Some artificial selection is therefore unavoidable, being imposed through the methods adopted for pollination, germination, and continuing storage ex situ.

Ex situ conservation of small samples therefore can be expected to lead inevitably to unpredictable genetic change. Hybridization between different population samples grown for more than one generation in isolation, and more particularly between samples from different populations, will increase the rate of these changes. The proportion of fit genotypes in progeny intended for reintroduction into the wild can thereby be greatly reduced, especially if the original wild donor populations are already small and in decline, and hence already suffering an accumulation of deleterious gene combinations.

For these reasons, we still lack the knowledge to establish genetically representative population samples in cultivation. Nor have there been more than a handful of experimental reintroductions into the wild from cultivated cultures, and none that I know of plants derived from stock that has been in cultivation for a long time.

For all practical intents, therefore, ex situ conservation of species leads inexorably, and perhaps irrevocably, to domestication. But ex situ samples of breeding populations can effectively serve as means to conserve unique or important characteristics rather than species as such. This has been recognized by forest geneticists whose ex situ conservation programs for "provenances" represent just this (Zobel, 1978). Among crop plants, unique genes and gene combinations, which may be as important to the survival of endangered species as to the protection of domestic crops, can presently be recovered if lost from a culture only by resorting to wild populations. Ex situ conservation should also serve the reverse function: reintroduction of desirable alleles and gene combinations that become lost in wild populations. When extinction in the wild becomes a reality, the means to control artificial selection is greatly reduced, and the means to reintroduce alleles lost in culture is eliminated. Ex situ conservation must often be deemed preferable to loss nevertheless (Raven, 1976). But it is imperative for long-term conservation that managers of ex situ population samples define, as carefully as information allows, the characteristics they are intending to conserve: all alleles in the subsample or only some, and, if some, then which? Examples might be the diagnostic characteristics of a taxon or utilitarian attributes. Once a clear set of priorities has been defined, the role of what in effect will become a form of gene library will be considerable.

The role of interspecific competition in speciation has never been rigorously demonstrated in animal, let alone plant, populations. In plant populations it remains unclear whether species-specific natural selection is of common occurrence. Shugart and West (1981) indicated through mathematical modeling that the species composition of the southern U.S. hardwood forest could arise purely as a consequence of the hazards of island biogeography—the balance of immigration and extinction over time. If correct, their view would certainly make the task of ex situ conservation a whole lot easier. But many examples of mutual interdependence do exist. Oaks cannot be conserved without the rich symbiotic fungal flora on their roots on which they depend and which differs quantitatively, and sometimes qualitatively, between species. In some cases, such plants and their symbionts will prove exceptionally difficult to conserve in perpetuity ex situ. Furthermore, it is these keystone mutualists and mobile links whose minimum viable population areas are being used as a principal criterion for assessing the minimum area of in situ preserves (Wilcox and Murphy, 1985).

SOME PRACTICAL ISSUES

It takes an outbreeding species at least 200 generations to diversify to the point of speciation (Stebbins, 1950). Some outbreeding forest trees do not start flowering for 50 years, even when in cultivation (Ng, 1966). We need to think in evolutionary time scales if it is really our intention to conserve species as opposed to alleles or gene complexes ex situ (Frankel and Soulé, 1981). It is questionable whether seed dormancy can be induced for comparable periods in many species. What is the likelihood of a major power cut (perhaps a war?) during such a period? Could not most endangered species be conserved in situ more cheaply by careful planning and management? How is the cost of in situ conservation to be estimated?

The record to date for ex situ culture does not give grounds for optimism. As Raven (1981, p. 56) stated so eloquently:

> Unfortunately, such collections are often dismantled or simply deteriorate after the specialists who built them up are no longer active at the respective institutions. . . . Although they are often of very great value internationally, they may if they are not actively utilized come to be viewed as a drain upon the limited resources of the institution where they are housed. Even when financial considerations are not limiting, it is difficult to provide for such collections the meticulous and sustained care that is essential for their survival without the attention of a specialist who is deeply concerned with them.

Similarly, Peeters and Williams (1984) estimated that of the 2 million accessions of plant germplasm in seed banks worldwide, 65% lack even basic data on source; 80% lack data on useful characteristics, including methods of propagation; and 95% lack any evaluation data such as responses to germinability tests. There are extensive data only on 1%. It goes without saying that a substantial proportion of the accessions not tested for germinability may be dead.

WHAT THEN IS THE SPECIFIC ROLE FOR BOTANICAL GARDENS IN CONSERVATION?

Ex situ conservation is a refuge of last resort: a high-risk refuge, perhaps of no escape. The immediate role of botanical gardens in the ex situ culture of rare and endangered species lies in research and education rather than in conservation per se. This role is absolutely vital if we are to have knowledge about plant populations on the edge of extinction that provides a sufficient basis for their management.

This point is brought home most poignantly by consideration of the humid tropics, where species diversity is at its greatest. There, long-lived perennials are in the majority, making cultivation easier, but few plants possess seed dormancy and many are of formidable size. Knowledge of the cultural requirements of the minority is nonexistent. It is obvious that ex situ conservation cannot be an end in itself under such circumstances.

Yet the accessibility of the plants in cultivation presents all manner of research opportunities not possible with remote and dispersed wild populations as well as opportunities for education and for increasing public awareness that would not otherwise exist. The necessity to focus on strictest priorities leads to the identification of very specific roles in conservation for tropical botanical gardens. One of particular current importance is conservation of the myriad local landraces of tropical cultivated plants from cereals and vegetables, in many of which seed dormancy can be induced, to fruit trees, which must be grown in plantations. The National Botanical Garden of Indonesia is already a pioneer in many aspects of tropical conservation research based on such plantings. Landraces do not exist in nature: their habitat is therefore ephemeral over evolutionary time anyway and is therefore rapidly disappearing as world agriculture becomes increasingly market-oriented and industrialized.

The greatest use is made of gene banks by workers at the bank site itself. In 1981, for example, 4,376 samples were distributed to outside users worldwide from the 57,027 seed accessions at the International Rice Research Institute gene bank, whereas 29,056 samples were requested by Institute staff (IRRI, 1983). This further underlines the unique opportunities that botanical gardens have for promoting interdisciplinary research and instruction. There certainly will be a rapid increase in research on, and public understanding and support of, rare and endangered species once population samples are conserved ex situ in botanical gardens that have active programs of education and research.

The basic data on methods of seed storage can be best accumulated where there is ready access to captive populations. Many other aspects of reproductive biology will be advanced most rapidly where there are botanical gardens with captive populations as well as good laboratory and library facilities and research staff on site. All aspects of research that can be advanced through comparative study of closely related species, generally of different provenances, are aided when plants are grown side by side and at one accessible location. Even when the work must be carried out with populations in situ, as in demographic and population genetic research, much time and expense are saved if captive populations are easily accessible for use in the refinement of field and laboratory techniques prior to implementation (Ashton, 1984). The standard procedures for diagnosing the genetic component in phenotypic variation within and between populations, through culture of subsample populations under uniform garden conditions, would be an essential adjunct of ex situ cultivation methodology and one for which the botanical garden is best equipped. For all these purposes, the long-term maintenance of genetic integrity of the captive population, and indeed the long-term sustenance of the captive populations, are lesser priorities. The aim of the research is always to improve the techniques for managing wild populations in situ. Again, ex situ conservation is an ancillary to in situ conservation, not an alternative. Of course, the fact remains that plant species are becoming extinct in the wild, and some can and should be conserved in botanical gardens.

Purpose must then be strictly defined as a prerequisite to management planning. The result will likely be a program for conservation of chosen heritable attributes rather than of species or even demes.

Fundamental to the role of botanical gardens in conservation research is the development of meticulous, consistent, accessible, and hence computer-based record systems. Not only are details of source and information on the ecology of the wild population essential if this research is to be successfully applied, but plant material ex situ must be carefully inventoried together with comprehensive records of all treatments and the ensuing responses (Frankel and Soulé, 1981). Results of previous research must be retained with the records of ex situ living accessions, together with methods adopted and, preferably, copies of unprocessed results for evaluation and reanalysis by later workers intending to build on them.

REFERENCES

Allard, R. W. 1960. Principles of Plant Breeding. Wiley, New York. 485 pp.

Antonovics, J. In press. Genetically based measures of uniqueness. In G. Orians, W. Kunin, G. Brown, and J. Swierzbienski, eds. Proceedings of the Lake Wilderness Conference on Genetic Resources. University of Washington, Seattle, Wash.

Antonovics, J., and A. D. Bradshaw. 1970. Evolution in closely adjacent plant populations. VIII. Climal patterns at a mine boundary. Heredity 25:349–362.

Ashton, P. S. 1984. Botanic gardens and experimental grounds. Pp. 39–48 in V. H. Heywood and D. M. Moore, eds. Current Concepts in Plant Taxonomy. Academic Press, London.

Bradshaw, A. D. 1984. The ecological significance of genetic variation between populations. Pp. 213–228 in R. Dirzo and J. Sarukhan, eds. Perspectives in Plant Population Ecology. Sinauer Press, Sunderland, Mass.

Briggs, D., and S. M. Walters. 1984. Plant Variation and Evolution. Second edition. Cambridge University Press, New York. 412 pp.

Clausen, J., and W. M. Hiesey. 1958. Experimental Studies on the Nature of Species. IV. Genetic Structure of Ecological Races. Publication No. 615. Carnegie Institution of Washington, Washington, D.C.

Clay, K., and J. Antonovics. 1985. Quantitative variation of progeny from chasmogamous and cleistogamous flavors in the grass Danthonia specata. Evolution 39:335–368.

Diamond, J. M. 1975. The island dilemma: Lessons of modern biogeographic studies for the design of natural reserves. Bio. Conserv. 7:129–146.

Einset, J. W. 1985. Chemicals that regulate plants. Arnoldia 45:28–34.

Frankel, O. H., and M. E. Soulé. 1981. Conservation and Evolution. Cambridge University Press, New York. 327 pp.

Hamrick, J. L. 1983. The distribution of genetic variation within and among plant populations. Pp. 335–348 in C. M Schonewald-Cox, S. M. Chambers, B. MacBryde, and W. L. Thomas, eds. Genetics and Conservation. A Reference for Managing Wild Animal and Plant Populations. Benjamin/Cummings, Menlo Park, Calif.

Henshaw, G. G. 1975. Technical aspects of tissue culture storage for genetic conservation. Pp. 349–357 in O. H. Frankel and J. G. Hawkes, eds. Crop Genetic Resources for Today and Tomorrow. International Biological Programme 2. Cambridge University Press, New York.

IRRI (International Rice Research Institute). 1983. Annual Report for 1981. Los Banos, Philippines.

Marshall, D. R., and A. H. D. Brown. 1975. Optimum sampling strategies in genetic conservation. Pp. 53–80 in O. H. Frankel and J. G. Hawkes, eds. Crop Genetic Resources for Today and Tomorrow. International Biological Programme 2. Cambridge University Press, New York.

Morishima, H., Y. Sano, and H. J. Oka. 1984. Differentiation of perennial and annual types due to habitat conditions in the wild rice Oryza perennis. Plant Syst. Evol. 144:119–135.

Ng, F. S. P. 1966. Age at first flowing in dipterocarps. Malay. Forester 29:290–295.

Paabo, S. 1985. Molecular cloning of ancient Egyptian mummy DNA. Nature 314:644–645.

Peeters, J. P., and J. T. Williams. 1984. Towards better use of gene-banks with special reference to information. Plant Genet. Resour. News (FAO) 60:22–32.

Rackham, O. 1976. Trees and Woodland in the British Landscape. Dent, London. 204 pp.

Raven, P. H. 1976. Ethics and attitudes. Pp. 155–179 in J. B. Simmons, R. I. Beyer, P. E. Brandham, G. L. Lucas, and V. T. H. Parry, eds. Conservation of Threatened Plants. Plenum, New York.

Raven, P. H. 1981. Research in botanical gardens. Bot. Jahrb. 102:52–72.

Schonewald-Cox, C. M. 1983. Conclusions: Guidelines to management: A beginning attempt. Pp. 414–445 in C. M. Schonewald-Cox, S. M. Chambers, B. MacBryde, and W. L. Thomas, eds. Genetics and Conservation. A Reference for Managing Wild Animal and Plant Populations. Benjamin/Cummings, Menlo Park, Calif.

Senner, J. W. 1980. Inbreeding, depression and the survival of 300 populations. Pp. 209–224 in M. E. Soulé and B. A. Wilcox, eds. Conservation Biology: Evolutionary-Ecological Perspective. Sinauer Associates, Sunderland, Mass.

Shugart, H. H., Jr., and D. C. West. 1981. Long-term dynamics of forest ecosystems. Amer. Sci. 69:647–652.

Simberloff, G., and L. G. Abele. 1984. Conservation and obfuscation: Subdivision of reserves. Oikos 42:399–401.

Soulé, M. E., B. A. Wilcox, and C. Holtby. 1979. Benign neglect: A model of faunal collapse in the game reserves of East Africa. Biol. Conserv. 15:259–272.

Stebbins, G. L. 1950. Variation and Evolution in Plants. Columbia University Press, New York. 643 pp.

Wilcox, B. A., and D. D. Murphy. 1985. Conservation strategy: The effects of fragmentation on extinction. Amer. Nat. 125(6):879–887.

Wilkins, C. P., and J. H. Dodds. 1983. The application of tissue culture techniques to plant genetic conservation. Sci. Prog. 68:259–284.

Zobel, B. 1978. Gene conservation—as viewed by a forest tree breeder. Forest Ecol. Manage. 1:339–344.

USING SCIENCE AND TECHNOLOGY TO REESTABLISH SPECIES LOST IN NATURE

TOM J. CADE

Director of Raptor Research, Laboratory of Ornithology,
Cornell University, Ithaca, New York

Science and technology are hardly new to species conservation. Marco Polo related how the great Kublai Khan had an appreciation for their use to increase game-bird populations on his hawking preserves 700 years ago (Komroff, 1926), and the origin of wildlife management goes even further back in human culture to the early empirical practices of aboriginal peoples. The Kutchin Indians were still burning climax spruce forest to create willow browse for moose in the Yukon River Valley when I first visited the region in 1951, and primitive hunters practiced similar manipulations of habitat to favor big game species in many parts of the world.

In 1933, Aldo Leopold wrote the following lines in the preface to his classic textbook *Game Management*: "game can be restored by the *creative use* of the same tools which have heretofore destroyed it—axe, plow, cow, fire, and gun. A favorable alignment of these forces sometimes came about in pioneer days by accident. The result was a temporary wealth of game far greater than the red man ever saw. Management is their purposeful and continuing alignment" (Leopold, 1933, p. vii).

Some of the authors in this section of the book write about the development of highly advanced techniques involving cryogenic processes, embryo transplants, artificial insemination, and other sophisticated clinical procedures. Although these new techniques are certainly needed, we have not yet made the most of many of the older technologies that can be applied to species conservation, and I want to emphasize the creative use of some old methods as important tools for conservation.

Two that I have used are the captive propagation of wild species and the reintroduction of individuals produced in captivity into vacant habitats. I work with the diurnal birds of prey—some 290 species of condors, vultures, eagles, hawks, and falcons, but especially with falcons, principally the peregrine falcon (*Falco peregrinus*). Some of these raptors, such as the California condor (*Gymnogyps californianus*) and the Mauritius kestrel (*Falco punctatus*), are among the most endangered species in the world, and all seem to have attracted increasing human interest in recent years.

Given the long history of human involvement with birds of prey in the sport of falconry and as tribal and national totems, it is curious that the propagation of these species in captivity is a quite recent activity. The first peregrine falcon known to be raised from captive parents was produced as recently as 1942, and even as late as 1965 only about 23 species of diurnal raptors had been successfully bred in captivity, mostly on a casual or accidental basis (Cade, 1986a). In the early 1970s it was still widely believed that systematic large-scale propagation of raptors was an impossible feat, owing, I suppose, to the wild and fierce nature of such birds and to their apparent need for vast expanses of air space in which to perform their spectacular aerial courtship displays prior to mating and nesting.

The situation has changed markedly in the last 15 years. When it became evident in the late 1960s that populations of many species of raptors in north temperate regions had suffered major declines as a result of exposure to DDT and related pesticides or to other forms of environmental degradation, a nascent interest developed, especially among falconers, in perfecting techniques of captive breeding for some of these species. Of particular interest was the peregrine falcon, whose numbers had been severely reduced by pesticides over most of Europe and North America. This species has always been highly esteemed by naturalists and falconers alike for its near perfection as a flying machine.

In a summary of the worldwide effort to propagate raptors for the Fourth World Conference on Breeding Endangered Species in Captivity held in 1984, I came up with a list of 83 species that had reproduced in captivity—most of them since 1965 (Cade, 1986a). At least three species now need to be added to the list—the pondicherry vulture (*Sarcogyps calvus*), the white-tailed hawk (*Buteo albicaudatus*), and the crested eagle (*Morphnus guianensis*) of the Neotropics (ISIS, Captive Population Information, 31 December 1985). At present, more than 25% of all falconiform species have been bred in captivity. Some species have produced thousands of progeny; certainly the number of peregrine falcons raised in captivity exceeds 5,000 worldwide, and the American kestrel (*Falco sparverius*) is not far behind. At least 12 species have produced more than 100 progeny in captivity since 1975. It is safe to conclude that most if not all diurnal birds of prey can be bred in captivity given sufficient knowledge of their needs and sufficient resources to carry out the work.

What is the explanation for these rather sudden breakthroughs in the breeding of birds of prey in captivity? First, raptor breeders are true zealots with a single-mindedness of purpose to succeed against all odds and at any personal sacrifice. Although they come from all walks of life, most have in common the fact that they are falconers steeped in 3,000 years of accumulated knowledge about handling and training hawks and falcons and possess a special rapport with their birds that gives them an intuitive feel for what is needed in any particular situation.

A second part of the explanation has to do with the rapid and free exchange of information among raptor breeders throughout the world (Cade, 1986a). This exchange has followed the very best tradition of open scientific inquiry, unlike the secretive activities of some aviculturists.

Finally, successful raptor propagation owes much to the application of basic scientific information on avian reproductive physiology and breeding behavior and ecology. One quick example makes the point.

One of the chief bottlenecks hindering the mass production of raptors in captivity has been the frequent failure to achieve fertilization of eggs because of incompatibilities between mates. Thus, breeders must often resort to artificial insemination of egg-laying females. Obtaining usable quantities of semen from male raptors by the standard poultry method of physical massage and forced ejaculation is difficult and yields only small volumes of material over a short time (Weaver and Cade, 1983).

Since the work of the German ethologists Oskar Heinroth and Konrad Lorenz, it has been known that under certain conditions birds reared in isolation from members of their own species will become sexually imprinted on their human keepers and will actually attempt to mate with them. A fascinating body of scientific literature has developed on this subject (Brown, 1975; Immelmann, 1972; Lorenz, 1937). Raptor breeders have capitalized on this information and on their own experiences with hand-reared hawks and falcons raised in isolation to develop human-imprinted raptors into highly effective semen donors for artificial insemination. The birds are usually trained to copulate and ejaculate on a special hat worn by the human companion. Such birds produce copious volumes of high-quality semen for 2 to 3 months and have been great assets in increasing the percentage of fertile eggs laid in raptor breeding projects (Weaver and Cade, 1983).

Other examples of how basic scientific research has provided information useful to propagators involve incubation, photoperiod, hormone physiology, and psychobehavioral factors such as courtship displays, nest-site stimulation, and other factors, which Lack (1937) referred to as "psychological" in nature (Cade, 1980, 1986b).

REINTRODUCTIONS

There have been more than 1,670 attempts to establish avian species in outdoor environments, according to the excellent summary by Long (1981). At least 425 (25%) of these have been successful in establishing breeding populations. For example, in attempts to establish 119 species in North America up to 1980, success was achieved for 39 (37%) species. Comparable figures for Europe are 69 attempted and 27 (31%) established; for the Hawaiian Islands, 162 attempted and 45 (28%) established; for New Zealand, 133 and 38 (29%); and for Australia, 96 and 32 (33%).

Most of these have been *introductions* of exotic species into a new range. Some of these exotics have caused ecological or economic problems, but others appear to have filled vacant niches without seriously disrupting other species in the ecosystem (Long, 1981). There is much to be learned from careful study of these exotic introductions—both the successful and failed examples. We need to develop enough ecological foresight to predict and to effect benign exotic introduction as an additional technique of conservation. For example, the Mauritius kestrel badly needs a new home, and the neighboring Indian Ocean island of Runion could well be the place. This possibility needs study.

Reintroduction—the restoration of a species to its native habitat—has only recently been attempted, and there are fewer successful examples. In North America, one of the earliest successes was the restocking of some western ranges with American bison (*Bison bison*) derived from captive-bred stock held at the Bronx Zoo.

Much public attention has focused lately on the *possibility* of successful captive propagation and eventual reintroduction of the California condor—the most endangered bird in North America. It is perhaps less widely appreciated by the general public that successful reintroductions have already been achieved with several birds of prey. Since the 1950s efforts have been under way in the Federal Republic of Germany and in Sweden to reintroduce the eagle owl (*Bubo bubo*) into vacant range, and new breeding populations have been established in both countries through the release of captive-bred owls. Breeding populations of the white-tailed sea eagle (*Haliaeetus albicilla*) and goshawk (*Accipiter gentilis*) have been reestablished in Great Britain, and the griffon vulture (*Gyps fulvus*) has been restored to the Massif Central in France (Terrasse, 1985). Also as a result of reintroduction, several pairs of bald eagles (*Haliaeetus leucocephalus*) and, more recently, ospreys (*Pandion haliaetus*) are now nesting in formerly vacant haunts in the United States. Through the release of young birds produced mostly by falconers, a small nesting population of Harris' hawks (*Parabuteo unicinctus*) has been restored to the lower Colorado River valley of California and Arizona, where the species had disappeared around 1970, primarily because of habitat destruction. (See Cade, 1986b, for details and documentation.)

The best-known case is that of the peregrine falcon. The most widely distributed, naturally occurring avian species in the world, the peregrine suffered drastic reductions in numbers in both Europe and North America in the 1950s and 1960s, primarily because of the effects of organochlorine pesticides such as DDT and dieldrin. Since then there have been major international efforts to restore this species, in part by restricting the use of persistent organochlorine pesticides and by captive breeding and reintroduction. Successes have been achieved from both approaches.

In regions where some breeding pairs were able to persist, the falcon populations recovered dramatically on their own once restrictions were placed on use of these chemicals (Cade et al., in press). Great Britain is the best studied region where natural recovery has occurred (Ratcliffe, 1984). By 1963, the nesting population there had been reduced to 44% of its estimated pre-World War II size: there were some 350 surviving pairs. Once the use of dieldrin and DDT decreased in the latter part of the 1960s, the population began to increase, and today there are more than 1,000 occupied breeding territories in the British Isles. The species is, in fact, more numerous there now than at any time in this century.

In other regions where the species was more severely reduced or entirely extirpated as a breeding bird, restocking or reintroduction through the release of captive-reared falcons has been the main method for recovery. As a result of this work, small breeding populations have now been reestablished in the Federal Republic of Germany, Sweden, the eastern United States, the upper Mississippi River region in Wisconsin and Minnesota, the Rocky Mountain states, California, and several places in Canada (Cade, 1986b; Cade et al., in press). Three methods of release have been used: hacking, an old technique borrowed from falconry by which young are placed outdoors and cared for until they can fend for themselves; fostering, by placing captive-reared young in the nests of wild peregrines; and cross-fostering, by placing them in the nests of other species. The best results have been achieved with hacking and fostering (Cade et al., in press).

The eastern U.S. recovery program can serve as an example of the effort required for one of these regional reintroductions. The original nesting population of peregrines in the United States east of the Mississippi River in 1942 was estimated to be around 350 pairs on average each year (Hickey, 1942); some 250 eyries are actually known to have existed. By the early 1960s, this breeding population had completely disappeared, and no peregrines are known to have reproduced again in the eastern part of the country until 1980, when the first members of the reintroduced population fledged their young.

The Peregrine Fund began its captive breeding program at Cornell University in 1970. By 1974, there were sufficient numbers of young being raised to begin experimental releases. The following year, with substantial support from the U.S. Fish and Wildlife Service, we were able to release 16 young by hacking, and the number increased each year until we were putting out an average of 100 birds annually from 1981 onward (Barclay and Cade, 1983). At the end of the 1986 season we had released more than 850 captive-produced peregrines in 13 eastern states and the District of Columbia.

Falcons that survived their first winter immediately began establishing spring and summer territories, often at or near locales where they had been released. The first pairs formed in 1977 and 1978; the first eggs were laid at a nesting tower in New Jersey in 1979, but they were depredated by crows. In 1980, three pairs bred and raised young, and the nesting population has been increasing each year since. This past season, 43 pairs were known to be occupying territories from North Carolina to Montreal, Quebec; 30 pairs actually laid eggs, and 24 of them produced 52 young. There have been 83 nestings since 1979, and 71 of them produced 180 young.

The reintroduced breeding population has been doubling in size about every 2 years (Figure 32-1). If that rate of increase can be continued, by 1991 we will have achieved the official recovery goal of 175 nesting pairs established by the U.S. Fish and Wildlife Service in its recovery plan for the eastern United States. The complete restoration of the peregrine falcon in North America is now a predictable result of continuing to apply the technology developed for its recovery.

CAPTIVE BREEDING IN WILDLIFE MANAGEMENT

The rearing of captive animals and their use for restocking and reintroduction have not been favored techniques of modern wildlife management. Leopold (1933)

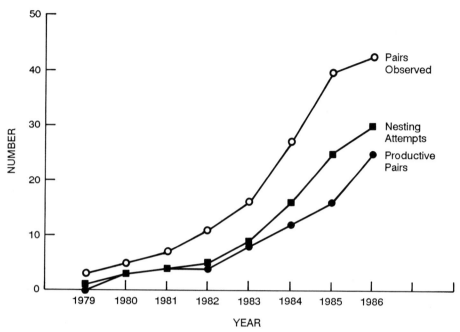

FIGURE 32-1   Growth of reestablished eastern peregrine population.

dismissed these subjects with 5 pages on "artificial propagation" and 7 pages on "transplantation" in a 481-page textbook. He stressed the cost of such artificial techniques and their aesthetic limitations. Leopold began what can be called the "ecological tradition" in wildlife management with its emphasis on habitat—the preservation and manipulation of all the environmental factors that are necessary to support wildlife populations as being more important than direct manipulation of the populations themselves. This approach has continued to the present day and certainly is the best policy wherever it can be pursued.

There have been good reasons for skepticism about propagation and release. Many captive stocks have not been reared in ways that are conducive to their survival after release in the wild (Cade, 1980). There have been many costly failures in attempts to restock or transplant animals into outdoor habitats and, particularly, to introduce exotic game birds by mass releases with little or no conditioning of the birds for their new environment (Bohl and Bump, 1970). One example is the repeated failure of many state game agencies to establish Japanese quail (*Coturnix coturnix*) in the United States. Hundreds of thousands of these birds were released—more than 350,000 from 1956 through 1958 alone—but no breeding populations ever became established (Long, 1981).

There are two main reasons for such failures. First is the use of genetic stocks or species that are poorly or not at all adapted to the environments into which they are introduced. Second is the lack of understanding of the behavioral, physiological, and ecological processes required for successful establishment (Cade,

1986b). Given the poor record of success with many species, especially when captive-produced individuals have been used, the recent reintroductions of raptors have been more successful than many biologists would have predicted.

These results with raptors and some other species in the last 10 years show that given enough biological knowledge about a species and enough time, effort, and money to apply that knowledge, reintroduction can be made to work. It is, however, an expensive and labor-intensive procedure, especially for captive-produced birds. It requires a tremendous amount of cooperation among numerous private individuals, government agencies, conservation organizations, corporations, and so on. Because so many different interests are usually involved, especially with an endangered species, it almost always becomes highly politicized.

The eastern peregrine recovery program is a case in point. Since 1970, the Peregrine Fund at Cornell (Figure 32-2) has spent about $2.8 million to propagate and release peregrine falcons in the eastern United States, but this figure does not include the expenses of cooperating agencies. The total cost has probably been between $3 and $3.5 million, perhaps more. More than 325 people—mostly temporary summertime helpers—have worked directly for The Peregrine Fund, not to mention dozens of federal and state agency personnel who have also helped. Although the U.S. Fish and Wildlife Service, the federal agency that administers the Endangered Species Act, is nominally in charge of the program, The Peregrine Fund has also worked closely with the U.S. Forest Service, National Park Service, and to some extent with other federal agencies, as well as with 13 state wildlife

FIGURE 32-2 Human-imprinted male peregrine feeding young at the Cornell University "Hawk Barn." Photo courtesy of The Peregrine Fund.

departments and, occasionally, with county and municipal authorities. Last year no fewer than 50 government agencies and offices, conservation organizations, universities, and businesses of one sort or another were directly involved in some aspect of the eastern program. Financial support has come from numerous conservation organizations, foundations, corporations, several thousand private individuals, and federal and state agencies.

Remember, this is only one of the *four* major regional recovery programs for the peregrine in the United States not to mention others in Canada and elsewhere. Similar accounts could also be given for the restoration of the Atlantic salmon (*Salmo salar*) in New England, Kemp's ridley sea turtle (*Lepidochelys kempi*) on the Texas coast, the whooping crane (*Grus americana*), the California condor, the Puerto Rican parrot (*Amazona vittata*), the golden lion tamarin (*Leontopithecus rosalia*) in Brazil, the Arabian oryx (*Oryx leucoryx*) in Oman, or the cheetah (*Acinoynx jubatus*) in South Africa. Each of these restoration programs is a huge enterprise, drawing on the skills and financial resources of many different individuals and institutions. For example, Snyder et al. (1987) give a detailed account of the work to restore the Puerto Rican parrot.

This new trend to rely more on technologies such as captive breeding and reintroduction for the restoration of species has become necessary because we human beings have not paid heed to the ecological imperative to preserve nature. We have not followed the philosophy of Leopold and others like him to keep the land and waters as fit habitats for wild animals and plants. And we are not likely to change our ways, as Norman Myers and many other authors emphasize in this book. Most human beings continue to be captivated more by what engineers do than by what ecologists say. As a consequence we are left with engineering solutions to environmental problems that have been largely created by engineers in the first place. Thus, I am led back to my opening quotation from Leopold: wildlife "can be restored by the *creative use* of the same tools which have heretofore destroyed it. . . ."

As natural environments become more fragmented and degraded in their capacity to support a diversity of living forms, many once common species will be reduced to isolated remnants, existing as small populations in marginally suitable habitats separated by vast expanses of uninhabitable area. Some species—especially large ones and predatory ones—will be unable to maintain populations indefinitely under these conditions, owing to genetic, demographic, or ecological problems. Such populations can be sustained in part through captive propagation and reintroduction or by translocating individuals from one island of habitat to another (Cade, 1986b). I believe we will see an increasing need for these sorts of techniques in the coming decades as the biosphere undergoes the final change from unmanaged wilderness ecosystems to managed nature preserves, megazoos, and zooparks, which will be the final refuges for some species in outdoor environments (Conway, 1983).

Although I have stressed the cost and labor-intensive nature of propagation and reintroduction programs, I hasten to add that they really are not very expensive relative to many of the other things we human beings are willing to spend our public and private wealth to obtain. Consider the hundreds of billions of dollars this nation is being urged to spend on star wars technology, or the billions spent annually on war-making machines. Consider the $10 million one person was willing to pay for an untrained race horse, and single purses of $15 million—win or lose—for a world championship heavyweight prize fight, or the billions of private dollars being spent to support cocaine and heroin addictions. When one takes account of the trillions of dollars human beings are willing to spend on foolish or trivial things, the cost of saving endangered species and endangered habitats pales to insignificance. All the work needed to restore the world's threatened birds of prey could be accomplished with $5 million a year, about what it costs to make one armored tank.

Conservationists spend entirely too much time talking to each other. We do not need to convince each other of the importance of what we are trying to do. We need to convince the vast majority of other folks! We need less talk in any case and more action.

We do not need more science and more technology nearly as much as we need more innovative application of the knowledge and techniques of wildlife management already available. We do not have more action to preserve and to restore threatened species and their environments, first of all, because conservationists have not been persuasive enough to obtain the money needed to do the work. But another and final reason is the excessive bureaucratic incumbrance associated with the work. The number of state and federal regulations impinging on conservation projects these days, the number of permits required for one action or another, the number of agencies that have to be mollified, the number of committees that have to meet—all sap strength from the effort and further dilute the already limited funds.

All endangered species programs suffer from these ills. The California condor must be the premier example, but the peregrine falcon is a close second. To ensure the survival of these and other endangered species, programs must be clearly focused and the responsibilities for their management carefully allocated to those who are best qualified to work with each species.

REFERENCES

Barclay, J. H., and T. J. Cade. 1983. Restoration of the peregrine falcon in the eastern United States. Bird Conserv. 1:8–40.

Bohl, W. H., and G. Bump. 1970. Summary of foreign game bird liberations 1960 to 1968 and propagation 1966 to 1968. Special Scientific Report—Wildlife, No. 130. U.S. Fish and Wildlife Service, Department of the Interior, Washington, D.C. 61 pp.

Brown, J. L. 1975. The evolution of behavior. W. W. Norton, New York. 761 pp.

Cade, T. J. 1980. The husbandry of falcons for return to the wild. Int. Zoo Yearb. 20:23–35.

Cade, T. J. 1986a. Propagating diurnal raptors in captivity; a review. Int. Zoo Yearb. 24/25:1–20.

Cade, T. J. 1986b. Reintroduction as a method of conservation. Pp. 72–84 in S. E. Senner, C. N. White, and J. R. Parrish, eds. Symposium on Raptor Conservation in the Next Fifty Years, October 1984. Hawk Mountain Sanctuary Association. Raptor Research Reports No. 5. Raptor Research Foundation, Inc., Provo, Utah.

Cade, T. J., C. M. White, J. H. Enderson, and C. G. Thelander, eds. In press. Peregrine Falcon Populations, Their Management and Recovery. The Peregrine Fund, Inc., Boise, Idaho.

Conway, W. 1983. Captive birds and conservation. Pp. 23–36 in A. H. Bush and G. A. Clark, Jr., eds. Perspectives in Ornithology. Cambridge University Press, Cambridge.

Hickey, J. J. 1942. Eastern population of the duck hawk. Auk 59(2):176–204.

Immelmann, K. 1972. The influence of early experience upon the development of social behavior in estrildine finches. Int. Ornithol. Congr. 15:291–313.

Komroff, M. ed. 1926. Travels of Marco Polo. Boni and Liveright, New York. 369 pp.

Lack, D. 1937. The psychological factor in bird distribution. Br. Birds 31:130–136.

Leopold, A. S. 1933. Game Management. Charles Scribner's Sons, New York. 481 pp.

Long, J. L. 1981. Introduced birds of the world. Universe Books, New York. 528 pp.

Lorenz, K. Z. 1937. The companion in the bird's world. Auk 54:245–273.

Ratcliffe, D. A. 1984. The Peregrine breeding population of the United Kingdom in 1981. Bird Study 31(1):1–18.

Snyder, N. F. R., J. W. Wiley, and C. B. Kepler. 1987. The Parrots of Luquillo: Natural History and Conservation of the Puerto Rican Parrot. The Western Foundation of Vertebrate Zoology, Los Angeles. 384 pp.

Terrasse, M. 1985. Le grande aventure de Vautors. Fonds D'Intervention Pour Les Rapaces 11:16–18.

Weaver, J. D., and T. J. Cade, eds. 1983. Falcon propagation, a manual on captive breeding. The Peregrine Fund, Inc., Ithaca, N.Y. 93 pp.

INTENSIVE TECHNOLOGY IN THE CARE OF EX SITU
POPULATIONS OF VANISHING SPECIES

ULYSSES S. SEAL

Veterans Administration Hospital, Minneapolis, Minnesota

I was tempted, for this brief survey, to think solely in terms of the use of object-oriented technology in captive breeding programs for endangered species. Instead, reflection upon recent activities suggested a focus on information, software, decision analysis, and social psychology.

The use of high technology in ex situ management of endangered species is part of a rescue operation undertaken in response to piecemeal habitat management and ecosystem collapses. With new techniques, we may be able to provide a temporary haven for some of the vertebrate components of these habitats. It is rather like a lifeboat operation. It is our wish, as professionals, to do the best possible job to assist in ensuring the survival of those species that happen to gain a seat.

The word *technology* encompasses applied science, means of achieving a practical purpose, and the totality of the means used to provide objects necessary for human sustenance and comfort. It's ancient Indo-European root is *tekh(s)*, means weaving or building (Shipley, 1984).

Over the past 20 years, major contributions to successful captive management programs include the following, which are presented in more or less temporal order:

- nutrition and prepared diets
- chemical immobilization and anesthesia
- vaccinations and antibiotics
- individual animal identification and records combined with a central data base, the International Species Inventory System (ISIS)
- reproductive control (contraception) and enhancement
- population biology and molecular genetics

- information technology and microcomputers
- decision analysis and crisis management

SCOPE OF EX SITU MANAGEMENT PROBLEM

The U.S. Congress passed the Endangered Species Act in 1973. Since then, more than 910 animal and plant species worldwide have been listed as threatened or endangered by the U.S. Fish and Wildlife Service. Of these, 410 occur in the United States or its territories. The famed Red Data Books, published by the International Union for the Conservation of Nature and Natural Resources (IUCN), list about 840 animal species and subspecies worldwide as endangered.

Soulé (1986) has suggested that about 1,500 nonfish vertebrate species are likely to become extinct by the year 2050. Inclusion of fish will likely double this number. This implies a growing extinction rate of 10 to 30 vertebrate species per year without consideration of subspecies and populations.

Current guidelines from population biologists indicate that it is desirable to maintain captive populations of 100 to 300 individuals of a species (depending upon the number of founders, i.e., individuals whose offspring form captive populations, and generation time), if it is to be viable and if it is to retain about 80 to 90% of its genetic diversity for 100 to 200 years (Frankel and Soulé, 1981; Ralls and Ballou, 1986). These numbers suggest that if all institutional spaces for mammals and birds were used for Species Survival Plan (SSP) programs, then only about 300 birds and mammal species could be managed in North American institutions. This is less than 10% of the number of species now exhibited in North America. Expansion of the programs to fill all the space in all the world's potentially cooperating exhibit institutions might increase the number to about 900 species. This is 20 to 25% of the number of species exhibited. Thus there is pressure to increase the efficiency of captive management programs and to find alternative means for captive preservation of genetic resources.

NUTRITION

Prepared diets, carefully tuned to the needs of particular groups of species, are available as commercial products, greatly reducing the variations inherent in local supply and introduced by individual preparers. Now attention is directed to special requirements of individual or particular collections, for example, to thwart reproductive failure or failure of neonates to thrive or survive and to meet the special needs of hand-reared animals.

ANIMAL RESTRAINT AND CAPTURE

Remotely administered chemical immobilization of captive and free-ranging animals is a major innovation that has revolutionized our ability to work with wild animals and has become acceptably safe for routine use by experienced personnel (Harthoorn, 1976). Animals can be captured in the wild for treatment, relocation, data collection, and fitting of radio collars (Mech et al., 1984). Modern delivery techniques range from dart guns to blow pipes. The most widely used drugs in

North America are ketamine, etorphine, and xylazine. We are very close to having ideal drugs and combinations of drugs for a wide range of vertebrate species (Seal and Kreeger, in press).

BIOMEDICAL RESOURCES

Individual animal treatment, preventive medicine, systematic vaccination regimens, antibiotics, and commercially prepared diets have resulted in a general improvement in the health, well-being, reproduction, and survival of most species maintained in captivity. The advantages gained by increased survival and longevity include reduced demand on captive habitat resources for maintaining a population, longer generations, and a slower loss of genetic diversity through time.

The wild population of 100 to 120 black-footed ferrets (*Mustela nigripes*) rediscovered in 1981 in Meeteetse, Wyoming, was driven to near extinction (approximately 12 animals) in 1 year by successive epidemics of plague striking the prairie dogs and of canine distemper decimating the ferrets. Now the ferrets survive only in captivity. Management of disease in field populations is being explored through the use of vaccines presented in baits as, for example, rabies vaccines administered to fox populations and, in the near future, parvovirus vaccines for wolves. Individual animal treatment programs have been applied to wild populations of bighorn sheep, mountain goats, elk, and, recently, mountain gorillas. They have also been attempted for black-footed ferrets.

IDENTIFICATION AND DATA COLLECTION

Individual identification remains a difficult problem in captive and free-ranging populations. Animal marking technology, now fairly sophisticated, requires the handling of individuals. This technique presents difficulties for some captive species. Transponders, although appealing in principle, have not been widely used because the animals must be approached closely for scanning.

Telemetry used to locate individual animals has proven to be an essential tool for studies of many species in free-ranging populations and in release programs for condors, Arabian oryx, tamarins, and others. Innovations to the simplest technology for free-ranging species have been remote detection of death, detection of parturition, detection under ground and under water, and continuous monitoring of position with automated recording.

Recapture of individual animals using recently developed radio-operated dart collars is now possible. Dart collars contain syringes loaded with anesthetic drugs that can be injected into the free-ranging animal on command by a radio signal (Mech et al., 1984). These devices are also planned for use in several carnivore release programs to allow elective recapture of dispersing animals.

REPRODUCTION

Discussions of the possible contribution of ex situ captive breeding programs to conservation refer hopefully to the near future of 100 to 200 years by which time methods of reproductive enhancement and long-term gamete or embryo storage techniques will allow large numbers of a species to be placed on hold in vials or straws. The cost of developing such technologies for cattle, turkeys, and humans has been in the millions of dollars over several decades, but only minimally effective transfer to other species has been accomplished to date. Cryopreservation techniques have been developed only for sperm in egg-laying species—primarily birds; invertebrates have not been given serious attention. Thus embryo and ovum preservation techniques are not yet available for all nonmammalian vertebrate classes and all invertebrates.

Reversible contraception for regulation of zoo mammal populations has been in use for 10 years (Seal et al., 1976). In females, oral preparations are used occasionally; more commonly, individuals are given hormonal implants lasting 2 to 5 years. Male contraception without substantial behavioral disruption or introduction of novel chemicals with uncertain long-term effects remains a major challenge.

Additional manipulations of reproduction and behavior in vertebrates are being achieved by altering photoperiod and by administering the pineal hormone melatonin. In selected cases, these techniques will double the number of annual reproductive periods in species that reproduce seasonally.

POPULATION BIOLOGY AND MOLECULAR GENETICS

The key problem is how to manage small, captive, and fragmented wild-living populations—both demographically and genetically—to ensure their viability and retention of genetic diversity through time (Soulé, 1986). The current flux in the concept of "species" has led to the suggestion that molecular, morphological, and geographic distribution data be examined to assess the genetic structure of the population and to determine whether separate population units should be considered for captive conservation management (Ralls and Ballou, 1986). Molecular techniques are also being used to measure heterozygosity in founders and captive-bred populations, to validate pedigree assignments, and to evaluate the origins of founder stock.

Molecular methods include karyotyping, many forms of electrophoresis, immunological comparison of proteins, sequencing of proteins and DNA, and comparison of DNA restriction fragment-length polymorphisms (O'Brien et al., 1985). It is possible to identify individuals and familial relationships with several of these techniques. The current large-scale automation of these techniques will allow the complete sequencing of any genome.

These methods have had a significant impact on the captive management of three species in particular: orangutans, with the establishment of separate subspecies from Borneo and Sumatra by karyotype and two serum proteins; Asiatic lions, with the electrophoretic demonstration of African lion founders in the lineage of nearly all the captive population; and tigers, with the molecular evidence failing to support the traditional separation into separate subspecies.

INFORMATION TECHNOLOGY

Digital technology is the foundation of our new age of information. Integrated circuits, computing software, and photonics are the key technologies for manipulating digits. Single-chip circuits can now hold about 1 million components; an upper limit of 10 to 100 million should be reached by the year 2000. Integrated circuits are the source of computing power.

This trend combined with access to unlimited, inexpensive random-access memory (RAM) and the availability of gigabyte storage devices with 20-millisecond access times permit full desktop access to all the world's printed information. We need to place high priority on facilitating access to the world's conservation data bases stored on magnetic, compact disk/read-only memory (CD-ROM), or laser disk media. Thought must now be given to management of the transfer, loan, or sale of these data bases.

The Conservation Monitoring Center of IUCN's Species Survival Commission (SSC) collects data worldwide on the status of species in the wild. This information is published in the Red Data Books (see Chapter 29) and is also available on selected printouts. These data are now also becoming available on line or on magnetic media—an urgent need.

The International Species Inventory System (ISIS) provides a continuously updated, centralized data base of census, demographic, pedigree, and laboratory data on wild animals kept in zoos. It has established agreed-upon minimum data standards for animal record keeping, and it reports to all participants at least annually.

International and regional studbooks have been compiled by zoo biologists for about 80 rare and endangered species. Currently, IUCN and the International Union of Directors of Zoological Gardens (IUDZG) require that the data in those books be placed in the ISIS data base for species of captive population management interest.

COMPUTER SOFTWARE AND ISIS

The pressure to develop suitable software is enormous and creates a major bottleneck, hampering progess in developing species survival plans. Programming productivity is very low and has not improved in pace with the development of technology. Serious efforts are under way to develop software application generators that will respond to English language requests and yield programming on a conversational level.

The development of rule-based systems for well-circumscribed problem solving (so-called expert systems) is well under way in the biomedical community, but no applications have yet appeared in the conservation or zoo communities. Training will be accomplished through interactive use of video disk and computer technology.

Contributions to the ISIS data base are now being made by 218 zoological institutions in 14 countries (Flesness et al., 1984). Recently, coverage has expanded substantially in Europe and India and is beginning in Australia. The U.S. Fish and Wildlife Service has agreed to sponsor participation by zoos in India.

Relevant ISIS summary census data are being sent annually to all SSC special groups. Software for in-house inventory records has been developed. This package, ARKS, has now been ordered by more than 60 zoological institutions in 7 countries.

A companion medical record system, MedARKS, is under construction. Phase 1, on anesthesia, will be released this year. ISIS is beginning to work on single species analysis and management software, i.e., studbook software.

New International Studbook keepers are now required to enter their data into the ISIS data base. The American Association of Zoological Parks and Aquariums (AAZPA) requires its studbook keepers to assist in maintaining their species data on ISIS. The European Species Survival Plan coordinators are also required to enter their species data into ISIS.

IN SITU AND EX SITU METHODS

Conservation of species is clearly best served by in situ methods based on biological community and habitat. It is equally clear that the pace of reduction in population numbers and complete loss of the larger vertebrate species—the charismatic megavertebrates—is frequently exceeding our ability to achieve their in situ survival. However, the habitat of many of these species is still present, e.g., for the Arabian oryx, Przewalski's horse, and black-footed ferret.

The application of ex situ methods and biotechnology may provide a buffer against final extinction. Ex situ methods also promote the use of the "rarity crisis mentality" to stimulate awareness and action on habitat protection schemes. Early and repeated reintroduction programs, as for the peregrine falcon (*Falco peregrinus*; see Chapter 32), then can encourage continued visibility and protection of the chosen habitat fragments. It will probably also be necessary to manage the species in these fragments for many years to come. Protection may thus be afforded to a wider range of biological diversity, including the plants and invertebrates in these refugia, which may eventually serve as sources for expansion.

ETHICS AND VALUE SYSTEMS

The clearest challenge is one of ethics. None of the technologies and strategies described above will ensure the continuing evolution of these species as part of a biological community. They are temporary expedients at best. Protection of biological diversity by resorting to aesthetic, economic, and other utilitarian values can always be discounted. Any single species can be discounted to near zero value by economic reasoning. The incremental ecological value of a single species, as a part of the interconnectedness of biological communities, is usually small except for keystone species. Thus many species or individual habitat units will be lost before their loss is felt. This characteristic of current human thinking requires catastrophe and crisis to capture awareness and motivate action. Thus much of our time as conservation biologists is spent in crisis management, negotiations, and hasty efforts to assemble sufficient data to retrieve yet another favored species. It is essential that humans be encouraged to expand their ethical awareness to encompass other species and the living land as a loan from our children which must be retained.

REFERENCES

Flesness, N. R., G. Garnatz, and U. S. Seal. 1984. ISIS—An international specimen information system. Pp. 103–112 in R. Allkin and F. A. Bisby, eds. Databases in Systematics. Academic Press, New York.

Frankel, O. H., and M. E. Soulé. 1981. Conservation and Evolution. Cambridge University Press, New York. 327 pp.

Harthoorn, A. M. 1976. The Chemical Capture of Animals. Bailliere Tindall, London. 416 pp.

Mech, L. D., R. C. Chapman, W. W. Cochran, L. Simmons, and U. S. Seal. 1984. Radio-triggered anesthetic-dart collar for recapturing large mammals. Wildl. Soc. Bull. 12:69–74.

O'Brien, S. J., M. E. Roelke, L. Marker, A. Newman, C. W. Winkler, D. Meltzer, L. Colly, J. Everman, M. Bush, and D. E. Wildt. 1985. Genetic basis for species vulnerability in the cheetah. Science 227:1428–1434.

Ralls, L., and J. D. Ballou, eds. 1986. Genetic management of captive populations. Zoo Biol. 5(2):84–238.

Seal, U. S., and T. J. Kreeger. In press. Chemical immobilization of furbearers. In M. Novak and J. A. Baker, eds. Wild Furbearer Management and Conservation in North America.

Seal, U. S., R. Barton, L. Mather, K. Olberding, E. D. Plotka, and C. W. Gray. 1976. Hormonal contraception in captive female lions (Panthera leo). Zoo Anim. Med. 7:12–20.

Shipley, J. T. 1984. The Origins of English Words. A Discursive Dictionary of Indo-European Roots. Johns Hopkins University Press, Baltimore. 636 pp.

Soulé, M. E. 1986. Conservation Biology: The Science of Scarcity and Diversity. Sinauer Associates, Sunderland, Mass. 584 pp.

CHAPTER

34

CRYOBIOLOGY, EMBRYO TRANSFER, AND ARTIFICIAL INSEMINATION IN EX SITU ANIMAL CONSERVATION PROGRAMS

BETSY L. DRESSER

Director of Research, Cincinnati Wildlife Research Federation, Cincinnati, Ohio

As a prelude to this discussion of the ex situ technology that is or could be available to maintain biological diversity in conservation programs, we must first explore why we should be interested in this technology at all. The greatest threat to the preservation of genetic diversity of wild animals comes from the destruction and degradation of their habitats (Croner, 1984). Some experts are estimating that if today's pace of habitat destruction continues, 1 million species of plants and animals may become extinct by the end of this century (AAZPA, 1983). Continued destruction of tropical rain forests, the most species-rich land environment on Earth, may result in mass extinctions that would permanently impoverish the planet (see Part 3 of this book: Diversity at Risk: Tropical Forests).

Other threats to species come from the destruction of animals and plants for food and trade. The Global 2000 Report to the President of the United States (Barney, 1982) states that the extinctions projected for the coming decades will be largely generated by humans and on a scale that renders the gradual process of natural extinction trivial by comparison. Extinction is a catastrophe, not only aesthetically or because of the effect it has on the ecological balance, but also because it deprives mankind of part of Nature's potential. Each species forms a genetic reservoir (genome) that may be of value in agriculture, medicine, or industry (Daniel, 1981). Forty percent of our present-day medicines are derived from wild plants and animals (Wolkomir, 1983; see also Farnsworth, Chapter 9 of this volume). It is impossible to know what other additional resources lay waiting to be tapped.

Animal and plant germplasm should not be preserved merely for altruistic reasons but also for reasons that are of direct benefit to mankind. It is not known which plants and animals may prove useful in the future. It is known, however, that certain animals have proven invaluable to our knowledge of diseases and human health. For example, armadillos (*Dasypus novemcinctus*) have taught us much about leprosy as they are the only animal model to acquire this disease when injected with *Mycobacterium leprae* experimentally (Storrs et al., 1980). The small cotton-top tamarins (*Saguinus oedipus*) from South America have the highest incidence of colon cancer of any animal and have proven to be an excellent species for the study of this fatal disease (Lushbaugh et al., 1984). Studies of spontaneous diabetes mellitus in the South African hamster (*Mystromys albicaudatus*) (Stuhlman, 1979) and AIDS (acquired immunodeficiency syndrome) in the macaque (*Macaca cyclopis*, *M. mulatta*, or both) may eventually lead to cures in humans (Letvin and King, 1984). Hepatitis B virus was discovered in a group of North American woodchucks (*Marmota monax*) by a zoo pathologist at the Penrose Research Laboratories in Philadelphia (Snyder et al., 1982). The study of the livers of these animals has given human medicine the greatest insight into the etiology of this often seen liver tumor. There are many more examples. Because of the diversity of species, there can be and, in fact, is a broader base of general medical knowledge that is benefiting mankind.

One bright spot in the otherwise gloomy picture of animal extinction is the new interest that has been generated in the field of wildlife reproduction. Fueled by the groundwork laid by the agricultural industry, zoo researchers have begun to study the reproductive processes of wild animals and to use technology such as embryo transfer and artificial insemination to help improve the reproductive potential of wild animals. Zoos worldwide have begun to pursue newfound roles as conservators rather than merely displayers of wildlife.

One of the biggest problems plaguing animals in today's zoos is the loss of genetic diversity due to inbreeding. This is particularly true with species whose numbers have been so depleted that they cause a genetic bottleneck. For example, the douc langur (*Pygathrix nemaeus*), a primate from Southeast Asia, is now virtually extinct in the wild, and there are not enough in captivity to ensure the genetic diversity necessary to keep the species alive (Gorman, 1980). With no new blood, so to speak, they can pass only a limited array of genes to their offspring, and as a result, they gradually become genetic carbon copies lacking the built-in adaptability to environmental change that would otherwise occur through natural selection in a genetically diverse group. Genetic diversity is a key to species survival. A species must have enough variation within its genome to enable it to adapt to environmental changes.

Inbreeding has taken its toll in the zoo world, resulting in problems such as decreased fertility, high juvenile mortality (known as inbreeding depression), and birth defects. In 1973, to help fight inbreeding, Ulysses Seal and Dale Mackey created ISIS (the International Species Inventory System), which is based at the Minnesota Zoo (see Chapter 33). ISIS computers catalog animal information from nearly 200 zoological institutions worldwide, including genealogy information for individual animals. One of the best functions of ISIS is that of a computerized matchmaker for participating zoos who wish to swap bloodlines. As zoo animal sperm and embryo banks become a reality, it is hoped that frozen embryo and semen samples can also be cataloged in the ISIS computer as an aid to zoos interested in transporting germplasm rather than animals themselves for the introduction of new bloodlines.

In consideration of the above concerns, this chapter discusses reproductive technology as it applies to the long-term preservation of animal germplasm and maintenance of biological diversity. Most ongoing wild animal research is directed toward the improvement of genetic and species diversity. Scientists have realized that the development of advanced reproductive technology, such as embryo transfer, gamete cryopreservation, and artificial insemination, may represent the real key to the future for many species who are currently threatened by extinction. They have also realized that this technology will do a great deal to improve and maintain the genetic diversity within captive populations. However, much of the application of this technology to wild species is still in its infancy.

EMBRYO TRANSFER

Embryo transfer is a technique by which fertilized ova and early embryos are recovered from the reproductive tract of a donor female, the genetic mother, and are transferred into the tract of a recipient female, the foster mother, in whom the embryos develop into full-term fetuses and live young. The first successful transfer of mammalian embryos was performed in the rabbit by Heape in 1891. His observations stimulated relatively little further research until about 1950. Since then, there has been an explosion of research in this area. Numerous published reviews and textbooks describe both the methods and the fundamental principles on which the technique of embryo transfer rests [see reviews by Betteridge (1977), Mapletoft (1984), Seidel (1981), and Sreenan (1983), and texts by Adams (1982), Cole and Cupps (1977), and Daniel (1978)]. Although the specifics of the methods depend upon the species used, the general principles are the same whether performed in laboratory animals such as mice and rabbits, in large domestic animals such as horses and cattle, or in wild species such as baboons and antelope.

Embryo transfer in two species, mice and cattle, has become absolutely routine. It is not an exaggeration to state that tens of thousands of living mice and cattle have been produced by embryo transfer. In general terms, the transfer of embryos of other species has been modeled on techniques devised for these two species. Although there are fewer live young, probably thousands of live rabbits, pigs, and horses have been produced by embryo transfer. The application of embryo transfer to wild species is a relatively recent event. Its history can be highlighted as follows:

1975   The first successful nonhuman primate surgical embryo transfer in a baboon (*Papio cynocephalus*)(Kraemer et al., 1976).

1976   The first successful wildlife surgical interspecies embryo transfer between mouflon (wild sheep; *Ovis musimon*) and domestic sheep (*Ovis aries*)(Bunch et al., 1977).

1981   Second successful surgical transfer of an embryo from a wild species into a domestic species—gaur (*Bos gaurus*) to Holstein (*Bos taurus*)(Stover et al., 1981).

1983   First successful nonsurgical embryo transfer performed with an eland antelope (*Tragelaphus oryx*)(Dresser et al., 1984a).

1983   First successful nonsurgical embryo transfer with the eland antelope (*Tragelaphus oryx*) involving a previously frozen embryo (Kramer et al., 1983).

1984   First successful primate interspecies embryo transfer—macaque (*Macaca fascicularis*) to rhesus (*M. mulatta*)—following in vitro fertilization (Balmaceda et al., 1986).

1984   First frozen embryo transfers in nonlaboratory species of primates accomplished in the common marmoset (*Callithrix jacchus*)(Hearn and Summers, 1986).

1984   First successful nonsurgical interspecies embryo transfer between two different species of wild animals—bongo (*Tragelaphus euryceros*) to eland (*Tragelaphus oryx*). These bongo antelope embryos were brought from Los Angeles to Cincinnati and transferred fresh, 12 hours after collection (Dresser et al., 1984b). Embryos from a wild species had never before been transported long distances.

1984   Nonsurgical interspecies embryo transfer from Grant's zebra (*Equus burchelli*) to horse (*E. caballus*)(Bennett and Foster, 1985; Foster and Bennett, 1984).

1984   First long-term frozen embryo transfer in a wild species: an eland (*Tragelaphus oryx*) embryo previously frozen for 1.5 years successfully transferred nonsurgically to an eland surrogate (Dresser, 1986).

1984   Interspecies embryo transfer from Przewalski's horse (*Equus przewalski*) to New Forest pony (*Equus caballus*)(Kydd et al., 1985).

1985   First successful embryo transfer in Dall sheep (*Ovis dalli*)(K. Mehren, Metro Toronto Zoo, personal communication, 1986).

1987   First nonsurgical embryo transfer between guar (*Bos gaurus*), an endangered species, and *Bos taurus*, a domestic Holstein, by Dresser and colleagues.

Synchronization of the donor and recipient animals in embryo transfer can be accomplished through precisely timed injections of prostaglandins, such as Lutalyse and Estrumate. These hormone analogs serve to stimulate the ovaries to begin a new cycle. Superovulation of the donor is accomplished through the injection of fertility hormones such as follicle-stimulating hormone (FSH). Superovulation has been fairly successful with the ungulates, but optimal drugs and dosages have yet to be refined for most other species. As many as 31 embryos have been collected from one FSH-stimulated eland cow (Dresser, 1983). On the other hand, fertility drugs seem to have little or no effect on the equids (such as zebras and Przewalski's horse) (Hearn and Summers, 1986). Clearly, the hormone regimen that produces the optimal superovulation response within a given species seems to be fairly individualized and much work is needed in this particular area.

In felines, the superovulation of donors and the synchronization of donors and recipients is complicated by the fact that most cats are induced or reflex ovulators, meaning that they ordinarily do not ovulate without the stimulation of copulation. Human chorionic gonadotropin (HCG) has been administered to domestic cats to cause ovulation to occur (sometimes in conjunction with stimulation by a vasectomized male) (Bowen, 1977). Researchers in several institutions have been working on embryo transfer in domestic cats with limited success (Dresser et al., 1987; Goodrowe et al., 1986; Kraemer et al., 1979), but little work has been done thus far on wild cats (Bowen et al., 1982; Reed et al., 1981). It is hoped that domestic cats may be able to serve as surrogates for incubating embryos from small endangered

Gaur calf born to Holstein surrogate on 5/25/87 at Kings Island Wild Animal Habitat. This was the first nonsurgical interspecies embryo transfer between an endangered species and a domestic species. Photo by Dr. Betsy L. Dresser.

wild cats such as the black-footed cat (*Felis nigripes*). A great deal more work needs to be done in this area to determine the best regimens for stimulating cat's ovaries to produce more than the usual number of follicles and to cause the ovulation of these follicles and, thus, to obtain embryos.

The development of embryo transfer techniques is essential if genetic diversity within captive populations is to be maximized. The ability to introduce new bloodlines into a captive population through the transfer of nonlocal embryos into surrogates would be far preferable to the transport of adult animals for breeding purposes, or to the depletion of wild herds to add new breeding stock to captive populations. In addition, it is a goal of many zoo researchers to develop interspecies embryo transfers to the point at which embryos can be collected from endangered species and transferred to surrogates of a more common species, thereby greatly increasing the reproductive potential of the donor species.

Other important benefits result from embryo transfer. For example, it has been found that disease transmission between different populations can be dramatically reduced by embryo transfer (see Hare, 1985, for review). This happens because the intact embryo collected from a diseased mother is almost always free of the microbial or viral disease agent and does not transmit the disease to the foster mother. Alternatively, the surrogate mother may confer passive immunity to her offspring that develop from transferred embryos. This may occur either through the placental blood supply or via the colostrum, the first milk. That is, it has been found that cattle of a given breed may quickly succumb to local diseases when imported into a new location. However, live calves produced by transfer of embryos from a foreign breed into another (native or endemic) domestic breed will usually be as disease-resistant as this domestic breed. This accident of biology should have important consequences for the transplant of both domestic and wild species from one location to another.

Interspecies embryo transfer has enjoyed limited success in wild animals, but much more research needs to be done in this area. Intergeneric, e.g., eland to cow (Dresser et al., 1982a) or water buffalo to cow (Drost, 1983), as opposed to interspecies, e.g., tiger to lion (Reed et al., 1981), embryo transfers have never been successful. It seems that most embryos can develop to the early blastocyst stage in the oviducts of unrelated species (Daniel, 1981), but further development requires a much closer relationship between donor and recipient and similarity in time and type of implantation, placenta formation, rate of ovum transport, length of gestation, birth weight, and both neonatal and maternal postpartum behavior. Several techniques have been tried to help overcome the surrogate mother's immune response and prevent the rejection of the foreign embryo in an interspecies embryo transfer. Recently, a domestic horse mare gave birth to a donkey foal after she had been injected with donkey white blood cells (Antczak, 1985). Prior to this treatment, all other donkey embryo transfers into domestic horses had failed. It is not yet clear why the procedure was successful.

Although various attempts have been made to collect embryos from nonhuman primates at various zoos and primate centers over the years, very few of these procedures have been reported in the scientific literature. For a review of the existing literature on superovulation and ova collection attempts in the more common nonhuman primates, see Bavister et al. (1985), Clayton and Kuehl (1984), Hodgen (1983), and Kraemer et al. (1979).

There has been dramatic progress in embryo micromanipulation in little over 6 years, especially as it applies to domestic animals. The research of Willadsen (1979, 1980) and Willadsen et al. (1981) is most notable in this regard.

In the most important application of micromanipulation to exotic animals, an embryo is microsurgically bisected into two or more pieces, thereby producing genetically identical twins or triplets. This technique, although not yet successful with exotic species, could help to quickly increase the numbers of endangered or rare animals.

Another result of micromanipulation has been the production of chimeras, which are embryos that are a product of combining two embryos at a relatively early stage of development. It is possible to prevent the surrogate uterus from recognizing the foreign embryo by combining different blastomeres. The younger cells tend to form the trophoblast that gives rise to the placenta, and the older cells tend to form the inner cell mass that will form the fetus. Chimeric embryos can be constructed so that the cells that constitute the trophoblast belong to the surrogate species and the cells of the inner cell mass belong to the donor species. A reupholstered embryo such as this can effectively trick the surrogate's uterus into thinking it is carrying a native embryo (Vietmeyer, 1984).

Interspecies embryo transfer and inner cell mass transfer should not be confused with hybridization of species. Offspring contain only genetic material of the original species. Many people now recognize that the method of chimera production might be utilized to rescue endangered species. It might be possible to construct chimeras consisting of an embryo from an endangered species plus an embryo of a common, but related species. The common species might then carry the fetus of the endangered species to term.

In summary, then, embryo transfer has become a widely used technique to produce live young animals, especially of domestic species. Although it is as yet still a novel procedure in wild species, continued research will inevitably make transfer of wild animal embryos as successful as the transfer of domestic animal embryos.

CRYOPRESERVATION OF EMBRYOS

The first successful freezing of mammalian embryos was reported in 1972 (Whittingham et al., 1972; Wilmut, 1972). Since those reports, more than 300 articles and 100 abstracts have been published on this one subject. There have also been three full meetings (Ciba Foundation, 1977; Muhlbock, 1976; Zeilmaker, 1981) and numerous symposia devoted exclusively to the freezing of mammalian embryos. Numerous reviews have also been published (Lehn-Jensen, 1981; Leibo, 1977, 1981; Maurer, 1978; Rall et al., 1982; Renard, 1982).

Freezing of mouse, rabbit, and bovine embryos has now become a routine procedure. Altogether, embryos of 11 mammalian species have now been successfully frozen. Again, success means that live young have been born from frozen embryos. To date, the species that have been successfully preserved include the mouse (Whittingham et al., 1972), cow (Wilmut and Rowson, 1973), rabbit (Bank and Maurer, 1974), sheep (Whittingham et al., 1972), rat (Whittingham, 1975), goat (Bilton and Moore, 1976), horse (Yamamoto et al., 1982), human (Trounson and Mohr, 1983), baboon (Pope et al., 1984), antelope (Dresser et al., 1984a), and cat (Dresser et al., 1987).

The freezing of mouse embryos has become so routine that banks of tens of hundreds of thousands of mouse embryos have been frozen to preserve valuable genetic stocks for extended times. For some species, most notably the mouse, rabbit, and cow, preservation by freezing has reached such a high level of sophistication and reliability that approximately 80 to 90% of frozen embryos will develop in vitro when thawed and cultured. Moreover, the procedures to freeze embryos have become increasingly simplified. Regardless of the methods used, it can be reasonably estimated that thousands, if not tens of thousands, of live young of domestic species have been produced from frozen-thawed embryos.

Again, because of extremely limited experimental material, only a few attempts have been made to transfer previously frozen wild animal embryos into recipients.

The procedure has yielded some limited success (Cherfas, 1984; Dresser et al., 1984a; Hearn and Summers, 1986; Kramer et al., 1983).

Cryopreservation will undoubtedly prove to be an important adjunct to reproductive research in nondomestic animals. Geneticists Thomas Foote and Ulysses Seal have determined that a population of 250 properly managed animals of a particular species can theoretically preserve 95% of the original genetic diversity of the group after 50 generations (400 years) (Myers, 1984). The world's zoos, however, have limited facilities and often cannot accommodate large numbers of animals of each species they maintain. Success in the area of cryopreservation will allow zoo professionals to overcome the limited space in zoos and wild animal preserves by maintaining the bulk of the desired genetic diversity in liquid nitrogen freezers. Embryos containing new bloodlines could be recovered in the wild and brought back to zoos in the frozen state to improve the bloodlines of captive populations without depleting the wild herds. U.S. government restrictions currently prohibit the importation of embryos from other countries, but scientists have been actively lobbying for a change in these restrictions. It appears that a change may be possible in the future.

ARTIFICIAL INSEMINATION

Artificial insemination is the introduction of semen into the vagina or cervix by artificial means. This procedure was supposedly used by the Arabs in ancient times, but the first documented success in the modern world occurred in 1784 with the artificial insemination of a dog (Betteridge, 1981). In the 1930s, artificial insemination of livestock was used extensively in Russia. Arthur Walton demonstrated its potential as an effective method to transport genes in the 1920s and 1930s by shipping fresh rabbit, sheep, and bull semen from England to other European countries (Betteridge, 1981). The ability to successfully freeze semen resulted from the discovery of the cryoprotective action of glycerol by Polge, Smith, and Parkes in 1949 (Betteridge, 1981). Artificial insemination is very common in the agricultural industry today. Foote (1981) estimated that close to 90 million head of cattle were produced worldwide in 1977 by artificial insemination with previously frozen semen samples. Artificial insemination has had limited success in wild animals thus far, especially with certain species of mammals and birds. Success has been attained for the following species:

Nondomestic Mammals

Addax[1]	Brown brocket deer	Ferret	Rhesus monkey
Guanaco	Reindeer	Fox[1]	Baboon
Llama	Red deer	Wolf[1]	Squirrel monkey
Blackbuck[1]	Speke's gazelle	Persian leopard	Chimpanzee[1]
Bighorn sheep	Giant panda[1]	Puma	Gorilla[1]

[1]Frozen semen.

Birds		Reptiles
Cranes (several species)	Waterfowl (ducks)	Tortoises[1]
Albino cockatiel	Pheasants	
Raptors		[1]Frozen semen.

Much more semen has been collected from wild species than has actually been evaluated and used. The first successful artificial insemination of a wild species with previously frozen semen occurred in 1973 with the wolf (Seager, 1981). This was followed by the successful insemination of a gorilla (Douglass and Gould, 1981). There has been a great deal of time and effort spent trying to artificially inseminate wild-caught felidae (cats); most attempts have resulted in failure (Dresser et al., 1982b). The London Zoo finally produced a puma in 1980 through surgical artificial insemination with a fresh semen sample (Moore et al., 1981). This was followed in 1981 by a successful nonsurgical artificial insemination of a Persian leopard with fresh semen at the Cincinnati Zoo (Dresser et al., 1982b).

There are three methods of semen collection: manual stimulation of the male reproductive tract, use of an artificial vagina, and electroejaculation (Cherfas, 1984). Electroejaculation was invented by two French workers, Jonet and Cassou, and is by far the most common collection mode for wild animals. Electroejaculation works by inserting a lubricated probe into the rectum of an anesthetized animal. This conveys mild pulsating electrical stimuli to the nerves of the reproductive tract, resulting in ejaculation. There is some question about the fertility of sperm collected through electroejaculation, but there is also a question about how viable a semen sample must be to be effective. For example, the successful gorilla insemination at the Memphis Zoo in 1981 was accomplished with a previously frozen sample whose motility was 10% and judged to be poor at the time of insemination (Douglass and Gould, 1981).

A great many problems are associated with artificial insemination as it applies to wild animals. First of all, it necessitates the use of anesthesia, which is always a risk, for both semen collection from the male and insemination of the female. In addition, as mentioned above for electroejaculation, the fertility of semen obtained from artificial collection techniques is sometimes questionable when compared to that produced in a natural ejaculation. Sperm usually begin to die as soon as they are collected. Even if a fresh sample is used for insemination, it is likely to have undergone a certain amount of sperm loss.

SEMEN CRYOPRESERVATION

The freezing process is somewhat detrimental to sperm, and it is very unlikely for sperm to come out of a thaw as motile as they were going into the freeze (Cochran et al., 1985). Much work needs to be done in the area of semen cryopreservation for exotic animals. To date, sperm from at least 200 different species has been frozen, but very little of it has actually been thawed and tested (Seager, 1981). The ultimate test is production of offspring. From semen that has been tested, cryobiologists have found that sperm from each species needs to be extended and frozen under slightly different conditions to produce the optimal results.

Extenders used to preserve the collected semen basically consist of a buffered solution that contains a cryoprotectant (e.g., glycerol), antibiotics, and either egg yolk or milk. Many variations have been tried in the basic recipe for semen extender. The agricultural industry has found that the optimal extender for a given species seems to be very species-specific and that is turning out to be true with wild animals also. Perhaps some clues for semen preservation can be found in the natural world from studying certain female reptiles who have the potential for keeping sperm viable within their bodies for up to 6 years after mating (Cherfas, 1984).

Other problems with artificial insemination include difficulty in predicting the optimal time for inseminating the female and the fact that artificial insemination cannot occur as frequently as a female would have been inseminated naturally for the duration of her estrous cycle.

On the positive side, artificial insemination can be a great boon toward improving the genetic diversity of a captive population of animals. As with embryo transfer, the risk and expense of transporting semen is far less than that of transporting a male animal for breeding purposes. Artificial insemination could also be used to overcome quarantine restrictions and the risks of disease. Often, the strict agricultural legislation has made transport of zoo animals more difficult and costly than is perhaps necessary.

FUTURE PROSPECTS

Since preservation of genetic material from a species is one of the keys to ensuring diversity, the development of reproductive technology for exotic species, such as cryopreservation of gametes, embryo transfer, and artificial insemination, should be emphasized and supported. Ex situ animal conservation programs that are dependent upon the long-term preservation of genetic variation should apply this technology as it becomes available because of the increasing realization that captive breeding programs are essential to prevent many species from becoming extinct. Loss of genetic diversity could also limit the potential of a population to adapt to new environments when reintroduced to the wild.

A large amount of basic research is urgently needed before application of new technology will be routine for maintaining captive populations. It is hoped that the urgency will be recognized by many more scientists worldwide than at present, and that ex situ conservation programs will become the nuclei of genetic material for dwindling populations of wild animals. Extinction for some may be softened by the frozen zoo concept, which may turn out to be the single most important reproductive technology developed for exotic animals during this decade. Its effects will reach centuries into the future for many species.

REFERENCES

AAZPA (American Association of Zoological Parks and Aquariums). 1983. Species Survival Plan Handbook Publication. Oglebay Park, Wheeling, W. Va. 24 pp.

Adams, C. E., ed. 1982. Mammalian Egg Transfer. C. R. C. Press Inc., Boca Raton, Fla.

Antczak, D. F. 1985. Cute cross-species birth. Sci. News 128:8.

Balmaceda, J., T. Heitman, A. Garcia, C. Pauerstein, and R. Pool. 1986. Embryo cryopreservation in cynamologus monkey. Fertil. Steril. 45(3):403–406.

Bank, H., and R. R. Maurer. 1974. Survival of frozen rabbit embryos. Exp. Cell Res. 89:188–196.

Barney, G. O. (Study Director). 1982. The Global 2000 Report to the President of the United States. Vol. 1: The Summary Report—Special Edition with the Environmental Projections and the Government's Global Model. Council on Environmental Quality, U.S. Government Printing Office, Washington, D.C. 360 pp.

Bavister, B. D., K. Collins, and S. Eisele. 1985. Non-surgical embryo transfer in the rhesus monkey. Theriogenology 23:177.

Bennett, S. D., and W. R. Foster. 1985. Successful transfer of a zebra embryo to a domestic horse. Equine Vet. J. Suppl. 3:53–62.

Betteridge, K. J. 1977. Embryo Transfer in Farm Animals. Monograph 16. Canada Department of Agriculture, Ottawa. 92 pp.

Betteridge, K. J. 1981. An historical look at embryo transfer. J. Reprod. Fertil. 62:1–13.

Bilton, R. J., and R. W. Moore. 1976. In vitro culture, storage, and transfer of goat embryos. Aust. J. Biol. Sci. 29:125–129.

Bowen, M. J., C. C. Platz, Jr., C. D. Brown, and D. C. Kraemer. 1982. Successful artificial insemination and embryo collection in the African lion (Panthera leo). Am. Assoc. Zoo Vet. 57–59.

Bowen, R. A. 1977. Fertilization in vitro of feline ova by spermatozoa from the ductus deferens. Biol. Reprod. 17:144–147.

Bunch, T. D., W. C. Foote, and B. Whitaker. 1977. Interspecies ovum transfer to propagate wild sheep. J. Wildl. Manage. 41(4):726–730.

Cherfas, J. 1984. Test-tube babies in the zoo. New Sci. 16–19.

Ciba Foundation. 1977. The Freezing of Mammalian Embryos. Symposium 52. Elsevier Excerpta Medica, Amsterdam. 330 pp.

Clayton, O., and T. J. Kuehl. 1984. The first successful in vitro fertilization and embryo transfer in a non-human primate. Theriogenology 21:228. (Abstract)

Cochran, R. C., J. K. Judy, C. F. Parker, and D. M. Hallford. 1985. Prefreezing and post-thaw semen characteristics of five ram breeds collected by electroejaculation. Theriogenology 23:431–440.

Cole, H. H., and P. T. Cupps, eds. 1977. Reproduction in Domestic Animals. Academic Press, New York. 665 pp.

Croner, S. 1984. An Introduction to the World Conservation Strategy. Prepared for the International Union for Conservation of Nature and Natural Resources (IUCN) by its Commission on Education. IUCN, Gland, Switzerland; United Nations Environmental Program, Nairobi, Kenya. 28 pp.

Daniel, J. C., Jr., ed. 1978. Methods in Mammalian Reproduction. Academic Press, New York. 566 pp.

Daniel, J. C. 1981. Preserving the genome of endangered species. SWARA—E. Afr. Wildl. Soc. 4:16–18.

Douglass, E. M., and K. G. Gould. 1981. Artificial insemination in lowland gorilla (Gorilla g. gorilla). Pp. 128–130 in Annual Proceedings—American Association of Zoo Veterinarians. Hills Division, Riviana Foods, Topeka, Kans.

Dresser, B. L. 1983. Embryos of the African eland (Taurotragus oryx). Pp. 9–11 in Proceedings of the Owners and Managers Workshop of the IXth Annual Meeting of the International Embryo Transfer Society. International Embryo Transfer Society, Fort Collins, Colo.

Dresser, B. L. 1986. Embryo transfer in exotic bovids. Int. Zoo Yearb. 24/25:138–142.

Dresser, B. L., L. Kramer, C. E. Pope, R. D. Dahlhausen, and C. Blauser. 1982a. Superovulation of African eland (Taurotragus oryx) and interspecies embryo transfer to holstein cattle. Theriogenology 17:86. (Abstract)

Dresser, B. L., L. Kramer, B. Reece, and P. T. Russell. 1982b. Induction of ovulation and successful artificial insemination in a Persian leopard (Panthera pardus saxicolor). J. Zoo Biol. 1(1):55–57.

Dresser, B. L., L. Kramer, R. D. Dahlhausen, C. E. Pope, and R. D. Baker. 1984a. Cryopreservation followed by successful transfer of African eland antelope (Tragelaphus oryx) embryos. Pp. 191–193 in Proceedings of the 10th International Congress on Animal Reproduction and Artificial Insemination. Standing Committee of the International Congress of Animal Reproduction and Artificial Insemination, University of Illinois at Urbana-Champaign, Ill.

Dresser, B. L., C. E. Pope, L. Kramer, G. Kuehn, R. D. Dahlhausen, E. J. Maruska, B. Reece, and W. D. Thomas. 1984b. Successful transcontinental and interspecies embryo transfer from bongo antelope (*Tragelaphus euryceros*) at the Los Angeles Zoo to eland (*Tragelaphus oryx*) and bongo at the Cincinnati Zoo. Pp. 166–168 in Annual Proceedings of the American Association of Zoological Parks and Aquariums (AAZPA). American Association of Zoological Parks and Aquariums, Wheeling, W. Va.

Dresser, B. L., C. S. Sehlhorst, G. Keller, L. W. Kramer, and R. W. Reece. 1987. Artificial insemination and embryo transfer in the Felidae. Pp. 287–293 in Tigers of the World: The Biology, Biopolitics, Management and Conservation of an Endangered Species. Noyes Publications, Park Ridge, N.J.

Drost, M. 1983. Reciprocal embryo transfer between water buffaloes and cattle. Pp. 63–71 in Proceedings of the Annual Meetings of the Society for Theriogenology. Society for Theriogenology, Hastings, Nebr.

Foote, R. H. 1981. The artificial insemination industry. Pp. 13–39 in B. G. Brackett, G. E. Seidel, and S. M. Seidel, eds. New Technologies in Animal Breeding. Academic Press, New York.

Foster, W. R., and S. D. Bennett. 1984. Non-surgical embryo transfer between a grant zebra and domestic quarterhorse: A tool for conservation. P. 181 in Annual Proceedings of the American Association of Zoo Veterinarians. Hills Division, Riviana Foods, Topeka, Kans.

Goodrowe, K. L., J. G. Howard, and D. E. Wildt. 1986. Embryo recovery and quality in the domestic cat: Natural versus induced estrus. Theriogenology 25:156. (Abstract)

Gorman, J. 1980. Brave new zoo. Discover 1:46–49.

Hare, W. C. D. 1985. Diseases Transmissable by Semen and Embryo Transfer. Office Internationale des Epizost, Paris. 36 pp.

Heape, W. 1891. Preliminary note on the transplantation and growth of mammalian ova within a uterine foster-mother. Proc. R. Soc. London 48:457–458.

Hearn, J. P., and P. M. Summers. 1986. Experimental manipulation of embryo implantation in the marmoset monkey and exotic equids. Theriogenology 25:3–11.

Hodgen, G. D. 1983. Surrogate embryo transfer combined with estrogen-progesterone therapy in monkeys. J. Am. Med. Assoc. 250:2167–2171.

Kraemer, D. C., G. T. Moore, and M. A. Kramen. 1976. Baboon infant produced by embryo transfer. Science 192:1246–1247.

Kraemer, D. C., B. L. Flow, M. D. Schriver, G. M. Kinney, and J. W. Pennycook. 1979. Embryo transfer in the nonhuman primate, feline and canine. Theriogenology 11:51–62.

Kramer, L., B. L. Dresser, C. E. Pope, R. D. Dahlhausen, and R. D. Baker. 1983. The nonsurgical transfer of frozen-thawed eland (*Tragelaphus oryx*) embryos. Pp. 104–105 in Annual Proceedings—American Association of Zoo Veterinarians. Hills Division, Riviana Foods, Topeka, Kans.

Kydd, J., M. S. Boyle, W. R. Allen, A. Shephard, and P. M. Summers. 1985. Transfer of exotic equine embryos to domestic horses and donkeys. Equine Vet. J. Suppl. 3:80–83.

Lehn-Jensen, H. 1981. Deep freezing of cow embryos, a review. Nord. Vet. Med. 32:523–532.

Leibo, S. P. 1977. Preservation of mammalian cells and embryos by freezing. Pp. 311–344 in D. Simatos, D. M. Strong, and J. M. Turc, eds. Cryoimmunologie: Colloque, Dijon, 17–18 Juin 1976. Organis Editions de l'Institut National de la Sant et de la Recherche Medicale, Paris.

Leibo, S. P. 1981. Physiological basis of the freezing of mammalian embryos. Pp. 353–379 in M. E. Gershwin and B. Merchant, eds. Immunologic Defects in Laboratory Animals, Vol. 2. Plenum, New York.

Letvin, N. L., and N. W. King. 1984. Human disease: Acquired immune deficiency syndrome. Animal model: Acquired immune deficiency syndrome-like disease in non-human primates. Comp. Pathol. Bull. 16:5–6.

Lushbaugh, C. C., G. L. Humason, and N. K. Clapp. 1984. Spontaneous colorectal adenocarcinoma in cotton-topped tamarins. Comp. Pathol. Bull. 15:2,4.

Mapletoft, R. F. 1984. Embryo transfer technology for the enhancement of animal reproduction. Biotechnology 2:149–160.

Mauer, R. R. 1978. Freezing mammalian embryos. A review of the techniques. Theriogenology 9:45–68.

Moore, H. D. M., R. C. Bonney, and D. M. Jones. 1981. Induction of oestrus and successful artificial insemination in the cougar, *Felis concolor*. Pp. 141–142 in Annual Proceedings—American Association of Zoo Veterinarians. Hills Division, Riviana Foods, Topeka, Kans.

Muhlbock, O., ed. 1976. Basic Aspects of Freeze Preservation of Mouse Strains. Gustav Fisher, Stuttgart, Federal Republic of Germany.

Myers, N. 1984. Cats in crisis! Int. Wildl. 14 (Nov-Dec):42–52.

Pope, C. E., V. Z. Pope, and L. R. Beck. 1984. Live birth following cryopreservation and transfer of a baboon embryo. Fertil. Steril. 42:143–145.

Rall, W. F., D. S. Reid, and C. Polge. 1982. Physical and temporal factors involved in the death of embryos that contain ice. Pp. 303–305 in F. Franks, ed. Biophysics of Water. Proceedings of a working conference held at Cirton College, Cambridge, June 29–July 3, 1981. John Wiley and Sons, New York.

Reed, G., B. Dresser, B. Reece, L. Kramer, P. Russell, K. Pindell, and P. Berringer. 1981. Superovulation and artificial insemination of Bengal tigers (*Panthera tigris*) and an interspecies embryo transfer to the African lion (*Panthera leo*). Pp. 136–138 in Annual Proceedings—American Association of Zoo Veterinarians. Hills Division, Riviana Foods, Topeka, Kans.

Renard, J. P., Y. Menezo, M. C. Pheron. 1982. Effect of lipid supplement on the viability of D7–D8 Ovine blastocysts after freezing and thawing (Abstract). In Embryo Transfer in Mammals, 2nd International Congress, Annecy. Foundation Marcel Merieux, Lyon, France.

Seager, S. W. J. 1981. A review of artificial methods of breeding in captive wild species. Dodo, Jersey Wildl. Preserv. Trust, 18:79–83.

Seidel, G. E., Jr. 1981. Superovulation and embryo transfer in cattle. Science 211:351–358.

Snyder, R. L., G. Tyler, and J. Summers. 1982. Chronic hepatitis and hepatocellular carcinoma associated with woodchuck hepatitis virus. Am. J. Pathol. 107:422–425.

Sreenan, J. M. 1983. Embryo transfer procedure and its use as a research technique. Vet. Rec. 112:494–500.

Storrs, E. E., C. H. Binford, and G. Migaki. 1980. Experimental lepromatous leprosy in nine-banded armadillos (*Dasypus novemcinctus* Linn). Am. J. Pathol. 92:813–816.

Stover, J., J. Evans, and E. P. Dolensek. 1981. Interspecies embryo transfer from the gaur to domestic holstein. Pp. 122–124 in Annual Proceedings—American Association of Zoo Veterinarians. Hills Division, Riviana Foods, Topeka, Kans.

Stuhlman, R. A. 1979. Diabetes mellitus. Animal model: Spontaneous diabetes mellitus in Mystromys albicaudatus. Am. J. Pathol. 94:685–688.

Trounson, A. O., and L. Mohr. 1983. Human pregnancy following cryopreservation, thawing and transfer of an eight-cell embryo. Nature 305:707–709.

Vietmeyer, N. 1984. From horses come zebras. Int. Wildl. 14(Nov-Dec):12–13.

Whittingham, D. G. 1975. Survival of rat embryos after freezing and thawing. J. Reprod. Fertil. 43:575–578.

Whittingham, D. G., S. P. Leibo, and P. Mazur. 1972. Survival of mouse embryos frozen to −196° and −269°C. Science 178:411–414.

Willadsen, S. M. 1979. A method for culture of micromanipulated sheep embryos and its use to produce monozygotic twins. Nature 277:298–300.

Willadsen, S. M. 1980. The viability of early cleavage stages containing half of the normal number of blastomeres in the sheep. J. Reprod. Fertil. 59:357–362.

Willadsen, S. M., C. Polge, L. E. A. Rowson, and R. M. Moor. 1974. Preservation of sheep embryos and liquid nitrogen. Cryobiology 11:560. (Abstract)

Willadsen, S. M., H. Lehn-Jensen, C. D. Fehilly, and R. Newcomb. 1981. The production of monozygotic twins on preselected parentage by micromanipulation of nonsurgically collected cow embyros. Theriogenology 15:23–30.

Wilmut, I. 1972. The effect of cooling rate, warming rate, cryoprotective agent and stage of development on survival of mouse embryos during freezing and thawing. Life Sci. 11:1071–1079.

Wilmut, I., and L. E. A. Rowson. 1973. Experiments on the low-temperature preservation of the cow embryos. Vet. Rec. 92:686–690.

Wolkomir, R. 1983. Draining the gene pool. Natl. Wildl. 21(Oct-Nov):24–28.

Yamamoto, Y., N. Oguri, Y. Tsutsumi, and Y. Hachinohe. 1982. Experiments in the freezing and storage of equine embryos. J. Reprod. Fertil. Suppl. 32:399–403.

Zeilmaker, G. H., ed. 1981. Frozen Storage of Laboratory Animals. Proceedings of a workshop at Harwell, U.K., May 6th–9th, 1980. Gustav Fisher, Stuttgart, Federal Republic of Germany. 193 pp.

RESTORATION ECOLOGY:

CAN WE RECOVER LOST GROUND?

Restored prairie at the University of Wisconsin Arboretum, Madison. Inset shows members of the Civilian Conservation Corps planting at this site in the late 1930s. *Photo courtesy of the University of Wisconsin Arboretum and Archives.*

ECOLOGICAL RESTORATION
Reflections on a Half-Century of Experience at the University of Wisconsin-Madison Arboretum

WILLIAM R. JORDAN III

Editor, *Restoration & Management Notes*, The University of Wisconsin-Madison
Arboretum, Madison, Wisconsin

So far, in this volume and in thinking and discussions about the conservation of biological diversity generally, the emphasis has been on preservation of what we already have. This makes sense. Preservation obviously has a critical role to play in the conservation of diversity. At the same time, however, it is clear that by itself preservation is not an adequate strategy for conserving diversity. At best, preservation can only hold on to what already exists. In a world of change, we need more than that. Ultimately, we need a way not only of saving what we have but also of putting the pieces back together when something has been altered, damaged, or even destroyed.

Consider, for example, that

- vast areas of both land and water have already been profoundly altered by human activities ranging from agriculture to mining and construction and to various forms of pollution;
- barring a catastrophe on the scale of nuclear war, human-caused alterations of natural and wilderness areas will continue indefinitely;
- certain kinds of change—notably changes in climate—are beyond human control, and they in turn will inevitably change even those areas we have succeeded in preserving;
- existing wilderness preserves are often inadequate in size or are suboptimal in shape or design; in many cases, their value as reservoirs of biodiversity could be dramatically increased by relatively modest increases in size, which could be achieved by active reconstruction of communities around their borders;

- numerous species are already on the brink of extinction and their habitats have been reduced to a remnant or perhaps eliminated completely, so that their only hope for long-term survival is the re-creation of their habitat by human beings; and

- the conservation of species ex situ will have little environmental value in the long run unless we find ways of providing habitat for them, often by creating it on disturbed sites.

All these considerations push us, unwillingly it seems at times, beyond a preoccupation with preservation, either in situ or ex situ, as the single strategy for the long-term conservation of diversity and toward a recognition of the importance of an active role for our species in reversing change or repairing damage. Unless, for example, we are prepared simply to write off disturbed lands as potential contributors to diversity, we are going to have to take seriously the problem of increasing diversity on these lands. Similarly, the inevitability of further change, including changes in climate, clearly implies that in order to preserve many communities over the long haul we are going to have to learn not only how to manage them but even how to move them around (Jordan et al., in press). And this brings us to the area of environmental healing, or ecological restoration, which is the subject of this section.

PIONEERING RESTORATION AT THE UNIVERSITY OF WISCONSIN-MADISON ARBORETUM

The starting point for this discussion will be the experience of the University of Wisconsin-Madison Arboretum, where research on restoration of ecological communities native to Wisconsin and the upper Midwest has been under way since 1934. Here, under the early leadership of Aldo Leopold and John Curtis, intensive restoration has been carried out on several hundred hectares of land, most of which had been seriously degraded by farming, logging, and sporadic development during the preceding century. Gradually, 40 hectares of tallgrass prairies have been restored on degraded pasture and plowland. A small xeric prairie has been created on an artificially constructed limestone outcropping. Red and white pine forests and boreal forests have been established on old pasture sites, and two types of maple forests are being developed by underplanting existing oak forests in which the understory had been depleted by grazing. The early stages of this effort were carried out by Civilian Conservation Corps crews working out of a camp on the site between 1935 and 1941. More recent work has been carried out by University of Wisconsin-Madison researchers and by the Arboretum staff. In general, the intensity of the restoration effort declined dramatically after 1941, though work continues, and indeed the need for ongoing restoration and management is one of the fundamental lessons that has emerged from the Arboretum's experiences.

Overall, this has been a pioneering effort, and the Arboretum's collection of restored and partially restored communities is now the oldest and most extensive of its kind anywhere in the world. Even more to the point, however, because of the Arboretum's experience, it is possible to make a number of observations about the nature of restoration, about its potential and its limitations as a strategy for conserving biological diversity, and about the environmental and social conditions under which it is likely to be feasible.

TECHNICAL, ECOLOGICAL FEASIBILITY

The first lesson that one might derive from this experience is that it is indeed possible, at least under certain circumstances, to re-create reasonably authentic replicas of some native ecological communities (Blewett, 1981). For example, the Arboretum's two restored tallgrass prairies (Curtis and Greene prairies) now include areas believed to resemble quite closely prairies native to the area—at least with respect to floristic composition. In other words, most of the appropriate vascular plants are present; they are present in more or less the right proportions and associations; and the number of inappropriate plants—that is, exotics or plants not native to the tallgrass prairies of this area—is small.

On the other hand, there are large areas on these prairies where ecological or historic authenticity is relatively low and where various exotic species are abundant. Certain of these species have proved to be extremely difficult to remove or control. Some have turned out to be capable of invading the more or less intact prairie community, often at the expense of the native plants. As a result, it is now abundantly clear that the problem of dealing with exotics is an ongoing one and that the struggle will in many instances be unending. Undisturbed natural communities are also vulnerable to invasions by exotic species but, in general, probably less so than communities in the process of being restored. Without doubt, this has turned out to be a major problem facing restorationists.

In addition, the restoration program at the Arboretum has strongly emphasized revegetation, far less attention being paid to the reintroduction of animal species. This is frequently the case in restoration and land reclamation projects, since the assumption is often made that the appropriate animals will find their way into the community once it has developed to a certain point. But this does not always happen for complex reasons that include the size of the communities, their uneven quality, and their isolation from existing animal populations. An instance of this now appears to have occurred in the Arboretum's restored southern maple forest, where ommission of an ant species that normally aids the dispersal of the seeds of certain herbaceous plants, such as bloodroot (*Sanguinaria canadensis*) and wild ginger (*Asarum canadense*), has resulted in the development of these species into peculiar, dense patches (Woods, 1984).

A related problem with restored communities generally is their small size, which can directly influence their ecological quality. Certain animals, for example, may not inhabit restored communities simply because these communities are often too small. This is a major reason why few if any restored prairies include buffalo, for example. At present, the prairie at Fermilab in suburban Chicago is probably the largest restored tallgrass prairie in existence (Nelson, 1987). Of course, this nearly 240-hectare prairie is still very small in comparison to the millions of hectares of prairie that existed in this area at the time of European settlement, and its ability to support populations of large native animals is at best problematic.

In addition to the more conspicuous defects in the composition of restored communities, there are numerous features, such as soil structure and chemistry, composition of soil flora, populations of less conspicuous animals (including insects), and various aspects of ecosystem function, that in many instances may not be authentic. Only rarely have these been studied in any detail.

On the positive side, however, the Arboretum's restored communities have brought back into the landscape numerous plants and animals that had become rare or had even been eliminated locally. The entire project certainly represents an enormous contribution to what might be called the native diversity of the Madison area. The Arboretum's restored tallgrass prairies, for example, are now among the largest prairies in Wisconsin, a state that had some 4.8 million hectares of prairie and savanna at the time of European settlement (Curtis, 1959). These prairies alone include more than 300 species of native plants. Some of them, including plants such as big bluestem grass (*Andropogon gerardi*), compass plant (*Silphium laciniatum*), and yellow coneflower (*Ratibida pinnata*), were extremely abundant in presettlement times, often dominating whole landscapes, but were virtually eliminated from the area by the time the restoration efforts at the Arboretum began. These now flourish in the restored communities, which also provide habitat for numerous rare species. Examples from the Arboretum's collection include such rarities as the white-fringed orchid (*Habenaria leucophaea*), prairie parsley (*Polytaenia nuttallii*), smooth phlox (*Phlox glaberrima*), and wild quinine (*Parthenium integrifolium*)—all considered threatened or endangered, at least for the state. In general, the Arboretum itself probably has more biological diversity than any other area of comparable size in the state. This is due largely to the presence of the various restored communities.

In short, the Arboretum's experience shows that restoration of some native communities may be technically feasible under certain conditions. The ecological quality of the resulting communities may vary, but under proper conditions, it may actually be quite high, and restored communities may often resemble the historic community chosen as a model quite closely, at least in floristic composition.

SOCIAL, ECONOMIC FACTORS

At the same time, the experience of the Arboretum raises a number of questions about the cost of such projects and the social, political, and economic feasibility of carrying them out. Thus, in considering the environmental significance of the Arboretum's restoration efforts, one should keep in mind that these efforts have been carried out under conditions that clearly limit their relevance to other situations. These conditions include first of all the fact that the Arboretum itself is part of a major university and that its work has been performed primarily for scientific and academic reasons. In other words, from the very beginning, this effort has benefited from its academic setting and has been justified as an experiment or as a way of creating communities for research, rather than as a way of coping with environmental, much less economic, problems.

The second set of conditions that have contributed to the success of the Arboretum project were those directly related to the economic and ecological conditions of the 1930s, notably the Great Depression and the Dustbowl. Together, these national calamities provided conditions (specifically, cheap land, free labor in the form of the Civilian Conservation Corps, and an incentive for ecological restoration) that proved crucial to the development of the Arboretum, but that have also reduced its value as a model for carrying out restoration projects in the real world outside academia. This point carries us outside the little world of the Arboretum to the larger world, where we have to ask a crucial question: What good is restoration? Is it likely to prove merely an academic pursuit or a pastime for environmentalists who happen to be interested in an unusual form of gardening? To just what extent and under what conditions can restoration be expected to contribute in a significant way to the conservation of diversity?

These questions have not yet been dealt with systematically, as far as I am aware. But it is important that we begin to take them seriously. In general, given the interrelatedness of everything on Earth and the inevitability of change, it would seem that an ineluctable logic argues for the importance of restoration as part of any comprehensive strategy for the conservation of biological diversity. Critical as it may be as part of such a strategy, preservation has serious defects. Basically, it is a one-way strategy that offers no way of responding to change or recouping losses. By itself, any such approach is clearly inadequate because in a changing world the quality of the environment is ultimately going to depend not simply upon the amount of land we manage to set aside and to preserve but upon the *equilibrium* we are able to maintain between the forces of destruction—or change—on the one hand and the forces of recovery on the other. All things considered, and despite its various limitations, it seems likely that restoration will ultimately play an important role in determining the position of this equilibrium.

This being the case, the questions raised above and a whole host of corollary questions and issues take on a great deal of urgency. Can we restore ecological systems? And if so, how authentic will the results be? Which communities lend themselves to restoration, and which are likely to prove more difficult—or even impossible—to restore? To what extent can we hope to re-create communities specifically designed to provide habitat for rare and endangered species? What needs to be known in order to restore a system effectively—and efficiently? What is the state of the art for the restoration of various communities, and what currently limits the effectiveness of restoration techniques for these systems? What sorts of research need to be undertaken in order to refine these techniques?

Beyond these questions about the technical feasibility of restoration, there are the various social, economic, and political questions: How much will it cost? Who will be expected to pay for it, and why? How will the costs compare with those of preservation or with the natural recovery of disturbed systems? What incentive will society have for restoring naturally diverse communities rather than for simply reclaiming land for some other purpose such as agriculture? In general, what incentives can be found for restoring communities—incentives that will ensure that restoration is actually accomplished and that its potential for contributing to biological diversity is effectively exploited?

In fact, there are a number of such incentives, including some traditional ones such as the creation of habitat for fish and game and the use of prairies as pasture and rangeland. There are also important aesthetic incentives in park and wilderness management and in landscape architecture.

But restored communities may well have other economic values that have not yet been fully identified or widely recognized. Examples include development of wetlands to control water distribution and quality (Holtz, 1986), of prairies to rehabilitate soils degraded by agriculture (Miller and Jastrow, 1986), and of forests as part of a program of sustained-yield timber production (Ashby, 1987). Applications such as these at least suggest ways in which restoration might eventually prove critical as a way of reintegrating native communities into the economies of developed nations, in the process returning them to the landscape on a large scale.

These questions are addressed in the four chapters that follow. The first two are devoted mainly to defining the state of the art of ecological restoration for two community types. In the first of these, Chapter 36, Joy Zedler discusses restoration of a temperate zone community, the tidal wetland. In Chapter 37, Chris Uhl addresses the much-neglected subject of tropical forest restoration. The following two chapters turn to the more socially oriented aspects of the business of restoration. In Chapter 38, John Cairns looks at disturbed lands as opportunities for increasing local and regional biodiversity through restoration. In Chapter 39, John Todd presents some ideas about creating a social, political, and economic context for restoration projects.

REFERENCES

Ashby, C. 1987. Forests. Pp. 89–108 in M. E. Gilpin, W. R. Jordan III, and J. D. Aber, eds. Restoration Ecology: A Synthetic Approach to Ecological Research. Cambridge University Press, New York.

Blewett, T. J. 1981. An Ordination Study of Plant Species Ecology in the Arboretum Prairies. Ph.D. Thesis, University of Wisconsin-Madison. 354 pp.

Curtis, J. T. 1959. The Vegetation of Wisconsin. University of Wisconsin-Madison Press. 657 pp.

Holtz, S. 1986. Bringing back a beautiful landscape—wetland restoration on the Des Plaines River, Illinois. Restoration & Management Notes 4:56–61.

Jordan, W. R. III, R. L. Peters, and E. B. Allen. In press. Ecological restoration as a strategy for conserving biological diversity. Environ. Manage.

Miller, R. M., and J. D. Jastrow. 1986. Soil studies at Fermilab support agricultural role for restored prairies. Restoration & Management Notes 4:62–63.

Nelson, H. L. 1987. Prairie restoration in the Chicago area. Restoration & Management Notes 5(2).

Woods, B. 1984. Ants disperse seed of herb species in a restored maple forest. Restoration & Management Notes 2:18.

CHAPTER

36

RESTORING DIVERSITY IN SALT MARSHES
Can We Do It?

JOY B. ZEDLER

Professor of Biology, San Diego State University, San Diego, California

Along the U.S. coastline, development has reduced the area of coastal wetlands and endangered certain wetland-dependent species. Despite the threats to biodiversity, development of wetland habitat is still permitted by regulatory agencies if project damages can be mitigated by improving degraded wetlands or creating new wetlands from uplands. For example, the California Coastal Act allows one-fourth of a degraded wetland to be destroyed if the remaining three-fourths is enhanced. The expectation is that increased habitat quality will compensate for decreased quantity.

The concept sounds reasonable, but biodiversity is continuing to decline. Why? First, the process allows a loss of habitat acreage. Second, there is no assurance that wetland ecosystems can be manipulated to fulfill restoration promises. The magnitude of the problem is well illustrated by examples from southern California, where more than 75% of the coastal wetland acreage has already been destroyed, where wetland-dependent species have become endangered with extinction, and where coastal development pressures rank highest in the nation. This chapter reviews several restoration plans and implementation projects and suggests measures needed to reverse the trend of declining diversity.

RESTORATION PLANS

Several large development projects in southern California wetlands have recently been approved by the California Coastal Commission (see Figure 36-1). Three federally endangered species are affected by such projects: the California least tern (*Sterna albifrons browni*), light-footed clapper rail (*Rallus longirostris levipes*), and salt marsh bird's beak (*Cordylanthus maritimus* spp. *maritimus*; see Figure 36-2).

317

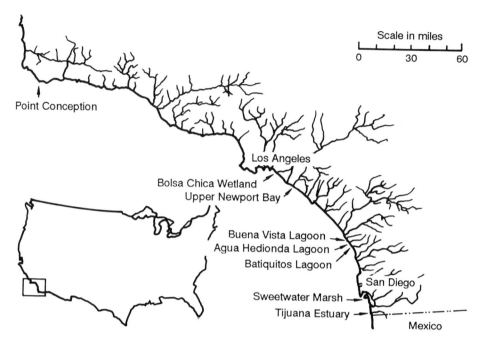

FIGURE 36-1   Sites of some coastal development projects in southern California. In all, there are 26 coastal wetlands between Point Conception and the Mexico-U.S. border.

FIGURE 36-2   The salt marsh bird's beak grows near the upper wetland edge. As an annual plant, its seeds germinate after winter rainfall to maintain the population; as a hemiparasite, its seedlings grow roots that can attach to those of other plants, thereby increasing its supplies of water and nutrients. Photo by J. Zedler.

There would also be an impact on the Belding's Savannah sparrow (*Passerculus sandwichensis beldingi*), which is listed as endangered by the state, and on several plant species of regional concern (Ferren, 1985).

Projects That Show Losses in Wetland Area

At Bolsa Chica Wetland, more than 1,200 acres (480 hectares) of lagoonal wetland will be reduced to 951 acres (366 hectares) of restored wetland (California State Coastal Conservancy, 1984). Mitigation plans are not final, but the draft concept includes cutting an ocean inlet to serve a new marina. Inland from the marina are sites for restored wetlands with controlled tidal flushing. Uplands designated as "environmentally sensitive habitat areas" that lie within the lowland area and that will be destroyed during development are to be relocated adjacent to the restored wetland in a bluff-edge (linear) park. The draft concept plan accommodates development, but does not ensure maintenance of biodiversity. The restoration activities are based on the assumption that habitat values can be created and moved about at will.

In Los Angeles Harbor, about 400 acres (160 hectares) of shallow water fisheries habitat will be filled to construct new port facilities. At this project site, there is no habitat available to be restored—all the wetlands have been filled or dredged. Thus, off-site mitigation has been approved. Batiquitos Lagoon, more than 80 miles (130 kilometers) south of Los Angeles, will be dredged to create deep-water habitat and increase tidal flushing. According to plans (California State Coastal Conservancy, 1986), the net loss of aquatic habitat in Los Angeles will be mitigated by altering (not increasing) habitat elsewhere. The dredging of Batiquitos Lagoon will remove sediments and, at least temporarily, solve the occasional problems of algal blooms (odors and fish kills after sewage spills). However, maximizing tidal flushing at Batiquitos Lagoon (to replace fisheries habitat in Los Angeles Harbor) will destroy existing salt marsh habitat (Figure 36-3) and reduce the area of shallow water and mudflat habitat. The mitigation plan contains two strikes against biodiversity—the loss of area and the loss of existing functional wetland types.

At Aqua Hedionda Lagoon, about 14 acres (5.6 hectares) of wetland were filled to build a four-lane road. The mitigation plan (U.S. Army Corps of Engineers, 1985) promised to enhance diversity and increase the functional capacity of the lagoon. Brackish-water ponds were planned for a wetland transitional area (itself a rare habitat type); a 2-acre (0.8-hectare) dredge spoil island was to be built for bird nesting; and a 7-acre (2.8-hectare) debris basin was proposed within a riparian area to reduce sedimentation into the lagoon. Flaws in the plan became clear when construction of the brackish ponds began. Pits were dug to a depth of 6 feet (1.8 meters) without encountering groundwater. Areas that were modified included transition habitat, pickleweed marsh (Figure 36-4), brackish marsh, and riparian habitat. The wetland lost both acreage and habitat quality.

All these projects show a net loss in wetland habitat area. Proponents argue that the lost areas are already degraded. However, they could be enhanced to maintain biodiversity. The fact that four wetland-dependent species have become endangered in Southern California while coastal wetlands have shrunk by 75% indicates a cause-effect relationship. There is some minimum area required to support regional biodiversity and a limit to the number of species that can be packed into individual wetlands. Populations are dynamic; some migrate and use several wetlands, whereas others experience local declines and must reinvade from another refuge. The need to maximize area available for wetland species is indicated by population declines that have followed human disturbance and environmental catastrophes.

Several species may be lost simultaneously if a wetland experiences multiple catastrophes. At Tijuana Estuary, the combination of destabilized dune sands (following long-term trampling), the winter storms of 1983, dune washovers, and channel sedimentation led to closure of the ocean inlet in April 1984. The drought of 1984 coincided with an 8-month nontidal period. The population of endangered light-footed clapper rails dropped from about 40 pairs to 0 and did not fully recover after tidal flushing was restored (16 pairs were present in 1987). In addition, there were major declines of three salt-marsh plant species—cordgrass (*Spartina foliosa*), annual pickleweed (*Salicornia bigelovii*), and sea-blite (*Suaeda esteroa*)—and none has recovered to pre-1983 levels. This salt marsh has shifted from the region's most-diverse to a species-poor wetland (Zedler and Nordby, 1986).

FIGURE 36-3 Salt flats may appear to have low habitat value, but many unusual insects, some of them threatened with extinction, are found only in these open areas. In the winter, runoff and high tides inundate the areas, and they become highly productive ecosystems. What appears to be barren in summer is heavily used by shorebirds and dabbling ducks in winter, as migrants visit the flats and feed on the abundant insects and algae. Photo by J. Zedler.

FIGURE 36-4 Pickleweed marsh may seem monotonous, but close inspection will reveal a variety of insects, invertebrates, and dozens of species of microscopic algae. Individuals and trails of the California horn snail (*Cerithidea californica*) are visible in the tidal pool. Photo by J. Zedler.

Maintenance of the region's resources through years of wet and dry periods, with and without closure to tidal flushing, requires that each habitat type be maintained at several different wetlands so there will be refuges during periods of environmental extremes. Further losses in habitat area cannot be justified.

Projects That Replace Functional Wetland Habitat with Modified Wetland Habitat

Some restoration projects retain acreage but exchange one type of habitat for another. In these cases, functional wetland habitats may be destroyed in order to create some other habitat type. Following are some examples.

The City of Chula Vista's Bayfront Development Plan calls for several developments near and in the last major salt marsh within San Diego Bay (90% of the Bay's wetland has already been developed). The plan includes a multistory hotel and a nature center to be built on an island that is surrounded by Sweetwater Marsh and San Diego Bay. Residential and commercial buildings would surround the marsh. To provide access, three roads are to be built over the wetland. The plan will also require modification of the wetland for the construction of debris basins. Wetland restoration is planned to mitigate impacts. The U.S. Fish and Wildlife Service has concluded that portions of the project jeopardize the following endangered species: California least tern, light-footed clapper rail, California brown pelican (*Pelicanus occidentalis*), and salt marsh bird's beak.

At Los Cerritos Wetland near Los Angeles, a complicated plan (California State Coastal Conservancy, 1982) proposes development of some wetland in exchange for an equal area of wetland creation. In all, 129 acres (51.6 hectares) of wetland will be retained. Some dikes that now prevent tidal flushing will be breached; some new areas will be graded to allow tidal flow. The restored wetlands will be divided into four segments and surrounded by high-density urban uses. Buffers between the wetland and development are as narrow as 25 feet (7.5 meters) for much of the project. A main concern is that existing wetland habitat will be lost and that the artificially created replacements cannot guarantee maintenance of biodiversity.

At Upper Newport Bay, a sediment control plan within a California State Ecological Reserve has received wide political support, in part because dredging in the upper bay reduces sedimentation in the lower bay's marina. Sedimentation in the upper bay is a long-term threat to the marsh habitat, but sudden changes in hydrology may have a negative impact on the habitat of endangered species. Upper Newport Bay has the highest density of light-footed clapper rails in the state of California and some of the region's most robust cordgrass vegetation. Shallow-water and transitional habitats are being traded for deeper channels, and the value of the new habitats to biodiversity is uncertain.

At Buena Vista Lagoon, sediment control measures were also taken. Shallow-water areas were deepened, and dredge spoils were placed alongside them in the wetland. The dredge spoil islands became hypersaline, bricklike substrates that have not developed the desired vegetation or significantly improved the status of the least tern population.

CONCLUSIONS CONCERNING RESTORATION PLANNING

Several observations on the status of restoration plans can be made:

• Many large projects result in a loss of wetland acreage.

• Mitigation measures for lost habitat often involve changing one type of wetland habitat into another, rather than creating wetland from upland habitat.

• Proposed projects are planned and reviewed individually rather than with a regional perspective. While cumulative impacts may be considered, there is no regional coordination to set priorities and guide decision making. There is no way to ensure that the wetlands with the greatest potential for maintaining clapper rails will be managed for clapper rails.

• There is no single source of information on restoration projects, no center that keeps records on changes in biodiversity to ensure that resource agencies are aware of changes in individual wetlands, no comprehensive monitoring programs to assess changes in biodiversity (although some endangered species are censused annually), and no mechanism to require suitable and comparable methods for the few monitoring programs that have been planned.

IMPLEMENTATION OF RESTORATION PROJECTS

To enhance, restore, or create wetland habitat requires manipulation of the physical environment (especially the topography and the degree and timing of fresh- and seawater influence) as well as the biota (e.g., by introducing target species and eliminating undesirable ones). Research in this area is just beginning; most of the work has been done on a trial-and-error basis, and the evaluation criteria are not yet standard.

Assessing Success

Restoration success must be measured in time scales that relate to the species being managed and to the periodicity of extreme environmental conditions characteristic of the region. Successful creation of clapper rail habitat cannot be measured by censusing mortality of cordgrass a few weeks after transplantation. Rather, such projects need to be followed at least until clapper rails establish breeding populations. Measures of restoration success must be done within spatial scales that relate to whole ecosystems. The degree to which breaching a dike and restoring tidal flushing can enhance a lagoon must be measured beyond channel biota and water quality, because there will also be substantial impacts on intertidal marshes. Likewise, dredging to improve fish diversity cannot be considered successful if endangered birds become extinct in the process. In short, restoration success must be measured at the ecosystem level and with long-term evaluation. To date, this has not been done.

Summary of Trials

All the projects described above incorporate some element of habitat creation or restoration, for which there are no guaranteed benefits to threatened species. Many projects have been designed to restore wetlands and mitigate losses, but in no case have ecosystem functions been duplicated, nor have endangered species been rescued from the threat of extinction. Projects to reduce sedimentation have as one goal the creation of fish and benthic invertebrate habitat (sometimes to provide food for the California least tern). Marsh restoration projects have focused on vegetation used by target bird species (cordgrass for clapper rails, pickleweed for Belding's Savannah sparrows).

While some wetland plant species can be transplanted successfully and others will invade voluntarily given suitable conditions (Zedler, 1984), there are only a few cases where the marsh ecosystem has been monitored for several years (Broome et al., 1986; Homziak et al., 1982) and no example of a threatened species that has been increased as desired. Attempts to restore wetlands in southern California have generally failed to attract target species. In a few cases, the California least tern has nested on dredge spoil islands—but not always where its use was planned. One briefly successful site was an 80-acre (132-hectare) island in south San Diego Bay, which was planned for salt marsh and fish habitat.

Conclusions Concerning Implementation

Four observations on the status of wetland habitat restoration in southern California can be made.

- Selected plant species (e.g., cordgrass) can be transplanted successfully.
- No plant or animal populations have been taken off the endangered list as a result of restoration projects.
- Wetland restoration assessments have not been made for entire ecosystems but have been limited to one or a few target species.
- No studies have been conducted to determine the minimum wetland area or configuration of multiple wetlands required to maintain regional biodiversity.

Therefore, it is premature to conclude that an artificial tidal wetland can develop and replace the functions of a natural one. Furthermore, there is no evidence that restoration of degraded wetland habitat can compensate for lost habitat area.

PROSPECTS FOR THE FUTURE

To restore biodiversity in the nation's coastal wetlands, we must understand the factors controlling these ecosystems and develop the ability to modify them to meet desired management goals. We must make substantial advances in ecotechnology—the scientifically sound manipulation of ecosystems to maintain natural diversity and achieve specific management objectives. The field is relatively new in ecology. Only one journal, *Restoration and Management Notes*, and a few books focus on the topic. Although most of the work in this area concerns disturbed ecosystems, all ecosystems need some management to maintain their natural hydrology as well as air and water quality.

Ecosystems of greatest concern tend to be those whose areas have been reduced and whose species are threatened with extinction. Rare species are difficult to study, because the conditions that allowed them to thrive no longer exist. Manipulative experimentation, required to establish cause-effect relationships, cannot always be done without threatening the endangered populations even further. Bringing animals or plants (even seeds) into the laboratory may reduce field populations to levels that jeopardize population recruitment. Thus, maintenance of biodiversity must be based on an understanding of the factors that control the ecosystems in which rare species persist—the type of long-term, ecosystem-level research now funded by the National Science Foundation, the National Oceanic and Atmospheric Administration through its Sea Grant Program, and other agencies. A new research emphasis could allow major advances to be made in wetland ecotechnology. I recommend manipulative experimentation, first in replicate mesocosms (medium-size artificial ecosystems), followed by experimental restoration at the ecosystem level. This approach was adopted by the U.S. Environmental Protection Agency in their research plan for the nation's wetlands (Zedler and Kentula, 1985).

Ecosystem-level experiments have not been incorporated into wetland restoration projects. The contention that artificial or restored wetlands can maintain biodiversity must be tested. Every restoration project can include experimentation in its design, e.g., to provide different tidal flows; to test different hydroperiods, salinities, and nutrient inputs; to use different transplantation regimes; or to vary the width of buffer zones, with treatments appropriately replicated. Detailed, long-term evaluation of the experiments will document success or failure to maintain natural diversity. In either event, we will learn whether it can be done and why it succeeds or fails. The present practice of poorly planned, unreplicated, undocumented *trials* leads mainly to *errors* whose causes cannot be identified. Only as our understanding of factors controlling wetland ecosystems improves can we ensure the restoration and maintenance of biodiversity.

ACKNOWLEDGMENTS

Research on wetland restoration was funded in part by NOAA, National Sea Grant College Program, Department of Commerce, under grant number NA80AA-D-00120, project number R/CZ-51, through the California Sea Grant College Program, and in part by the California State Resources Agency.

REFERENCES

Broome, S. W., E. D. Seneca, and W. W. Woodhouse, Jr. 1986. Long-term growth and development of transplants of the salt-marsh grass *Spartina alterniflora*. Estuaries 9:63–74.

California State Coastal Conservancy. 1982. Los Cerritos Wetlands: Alternative Wetland Restoration Plans Report. State of California—Resources Agency, State Coastal Conservancy, Oakland. 49 pp. + appendixes.

California State Coastal Conservancy. 1984. Staff Presentation for Public Hearing: Bolsa Chica Habitat Conservation Plan. State of California—Resources Agency, State Coastal Conservancy, Oakland. 31 pp. + exhibits.

California State Coastal Conservancy. 1986. Batiquitos Lagoon Enhancement Plan Draft. State of California—Resources Agency, State Coastal Conservancy, Oakland. 183 pp. + appendixes.

Ferren, W., Jr. 1985. Carpinteria Salt Marsh. Publication #4, Herbarium, University of California, Santa Barbara, Calif. 300 pp.

Homziak, J., M. S. Fonseca, and W. J. Kenworthy. 1982. Macrobenthic community structure in a transplanted eelgrass (*Zostera marina*) meadow. Mar. Ecol. Prog. Ser. 9:211–221.

U.S. Army Corps of Engineers. 1985. Public Notice of Application for Permit, Application No. 85–137–AA. U.S. Army Corps of Engineers, Los Angeles District. 13 pp.

Zedler, J. B. 1984. Salt Marsh Restoration: A Guidebook for Southern California. Report No. T-CSGCP-009. California Sea Grant, La Jolla, Calif. 46 pp.

Zedler, J. B. In press. Salt marsh restoration: Lessons from California. In J. Cairns, ed. Rehabilitating Damaged Ecosystems. CRC Press, Boca Raton, Fla.

Zedler, J. B., and M. Kentula. 1985. Wetland Research Plan. Corvallis Environmental Laboratory, U.S. Environmental Protection Agency, Corvallis, Oreg. 118 pp.

Zedler, J. B., and C. S. Nordby. 1986. The Ecology of Tijuana Estuary: An Estuarine Profile. U.S. Fish Wildl. Serv. Biol. Rep. 85(7.5). 104 pp.

CHAPTER

37

RESTORATION OF DEGRADED LANDS IN THE AMAZON BASIN

CHRISTOPHER UHL

Assistant Professor, Department of Biology, The Pennsylvania State University,
University Park, Pennsylvania

The deforestation and degradation of the Amazon ecosystem have important global implications. Conservation groups have responded by working to establish national parks and biological reserves in Amazonia. Meanwhile, scientists have been attempting to document the types and frequency of both natural and human-induced disturbances in Amazonia and the capacity of its ecosystems to recover from disturbances. Central to this ecological research is the question, How much can Amazon forests be abused and still recover? As discussed herein, forest communities in the Amazon reform naturally following natural disturbances; however, forest regeneration is slow and uncertain following some of the larger-scale, more-intensive human-induced disturbances that are becoming increasingly common in the area. In such cases, humans may have to change hats and become restorers rather than exploiters.

NATURAL DISTURBANCES IN AMAZONIA

There is a tendency to believe that the Amazon rain forest has existed in a pristine, cathedral-like state for tens of thousands or even millions of years and that this forest is just now being disturbed for the first time because of the development activities of modern human beings. There is ample reason to believe, however, that disturbance has always been a common feature of Amazon forest ecology.

Winds causing forest treefalls and forest fires have probably been the most important natural disturbances during Amazon forest history. Several studies have shown that treefall disturbances are common in Amazon forests. In fact, it would

326

not be out of the ordinary to see or hear a tree crashing to the ground on an afternoon walk through the forest. From 4 to 6% of any Amazon forest will be studded with canopy openings (light gaps) formed by treefalls. Recently formed light gaps have a ground layer of tree and vine seedlings. As a light gap patch ages, it enters the building phase during which it develops into a densely stocked patch of pole-sized trees. The patch reaches a mature phase when it contains a mix of large trees, poles, and seedlings. With a practiced eye, it is possible to walk through the rain forest and detect these light gaps and identify the building and mature phase patches—testimonies to past disturbances and to the dynamic nature of these ecosystems. Importantly, this type of small-scale disturbance is more of a subsidy than a stress to the plant community, because the resources critical for growth—light, water, and nutrients—are more readily available in treefall gaps than in the undisturbed forest understory.

Another natural disturbance that has no doubt been an important part of past Amazon disturbances is fire. For example, in the Upper Rio Negro region of Amazonia, charcoal is widespread and abundant in the soil. Radiocarbon dating of soil charcoal samples (Sanford et al., 1985) indicates that numerous fires have occurred during the past 6,000 years in this area. The radiocarbon dates correspond well with what are believed to have been dry episodes during recent Amazon history. The presence of abundant charcoal in Rio Negro soils is not an anomoly. Amazon researchers from EMBRAPA (Empresa Brasileira de Pesquisa Agropecuária) and INPA (Instituto Nacional de Pesquisas da Amazonia) in Brazil concur that charcoal is common in the soils of the central and eastern Amazon. Indeed, it appears to be much more difficult to find sites that do not have charcoal than to find sites that do.

Given the prevalence of fire throughout the history of the Amazon forest, how does Amazon vegetation respond to fire disturbance? Our studies of forest succession following forest cutting and burning disturbances at San Carlos de Rio Negro, Venezuela (Uhl and Jordan, 1984) provide an indication of regrowth potential following fire. We found that forest reforms quickly on burned sites. This is because many Amazonian-tree species have the ability to sprout after damage, and although fires do kill many stems outright, a pool of individuals survives burning and quickly sprouts. In addition, Amazon forests have a rich seed bank (from 500 to 1,000 seeds of successional woody species per square meter), and a portion of these seeds survive burning, germinate, and become established. The ability of some rain forest species to survive fire disturbances, e.g., when seeds are buried and thus protected and later sprout, may be the result of natural selection (i.e., fire may have been a selecting agent for these characteristics).

DISTURBANCE BY HUMANS IN AMAZONIA

Human-induced disturbances, such as slash-and-burn agriculture and the conversion of forest to pasture, are generally more intense than the natural disturbances just discussed. Nature's disturbances are over in an instant (a treefall, a wildfire), whereas humans prolong the disturbance period in their efforts to wrest some benefit from the land.

In slash-and-burn agriculture, the felled forest land is used to grow subsistence crops such as cassava, beans, and fruits. When these farm plots are abandoned and if they are not disturbed, they eventually return to forest, but the process takes a long time. The frequent weeding typical of slash-and-burn agriculture means that succession begins and is curtailed many times before sites are finally abandoned. These repeated weedings cause striking shifts in the composition of the regrowth. Most notably, the number of woody pioneer species declines after each weeding, whereas the density of forbs and grasses increases. Forbs and grasses are able to germinate, flower, and set seed in the interval between weedings and therefore can build up high plant densities and large seed banks. In contrast, the woody pioneer species that are established from seeds surviving the burn are weeded from the site before they have had time to produce more seeds. Because the agricultural practices of cutting, burning, and weeding largely eliminate mechanisms of on-site regeneration, the only way for forest species to establish on farms is through seed dispersal. This dependence on seed dispersal slows succession, because many of the animal species that routinely disperse seeds of forest species do not frequent large forest openings. Approximately 200 years are required for abandoned farm plots to reach mature forest proportions (Saldarriaga, 1985). That forest regeneration proceeds at all on abandoned farm sites is the result of two important factors:

- Slash-and-burn clearings are relatively small (usually =1 hectare). Hence, seed dispersal distances are short.
- The period of farm use is relatively short (± 3 years). Hence, safe germination sites, e.g., slash piles (the shaded, moist zones along the sides of decaying logs), and soil nutrients are still available to ensure the establishment of some forest trees.

Forest regeneration on intensively used cattle pasture in Amazonia is more problematic, because the disturbances there can have the following characteristics:

- They are frequently prolonged and can therefore result in a highly compacted soil and allow time for the decomposition of slash, which normally provides important microhabitats for the establishment of seedlings.
- They can involve repeated burning and weeding, which eventually destroy all means of on-site woody regeneration and further homogenize the site, eliminating establishment microhabitats.
- They can be very large and, hence, make seed dispersal from distant forests extremely unlikely.

A ROLE FOR RESTORATION ECOLOGY

Regeneration to forest occurs naturally in Amazonia following natural disturbances and small-scale human-induced disturbances, but where disturbances are severe, human intervention may be necessary to ensure reforestation. Restoration efforts are most needed on abandoned mine sites and highly degraded pastures.

The most extensive work on abandoned mines is being done by Oliver Henry Knowles, a resident ecologist for the Vale do Rio Dulce Bauxite Mining Company located near Porto Trombetas in Pará, Brazil. This company mines about 60 hectares per year. The topsoil (about 20 centimeters thick) is bulldozed into stockpiles during mining. After bauxite is removed, the mined trench is refilled with over-burden and the stockpiled topsoil is respread on the site. The soil is then sliced at 1-meter intervals to a 90-centimeter depth to increase aeration and facilitate root penetration. Finally, the area is planted with nursery-raised seedlings of native forest species. Although growth is slow (most species attain about 20 centimeters in height per year), survival is good. The principal impediments to restoration are insect pests (e.g., leaf-cutter ants, grasshoppers, and caterpillars) and soil nutrient deficiencies.

Restoration of highly degraded Amazon pasturelands is also receiving attention through a study I am conducting in collaboration with Robert Buschbacher, Daniel Nepstad, and Adilson Serrao near Paragominas in northern Pará, Brazil. The vegetation of our study area is lowland evergreen rain forest. Rainfall is approximately 1,700 millimeters per year, and there is a distinct dry season from June through November. As a first step in determining how to rehabilitate degraded pastures, we are studying how natural forest and degraded pasture environments differ with respect to microclimate, soil water and nutrient availability, and physical properties of soil. This research sets the stage for our central research goal: to determine how biological and physical forces act to retard the establishment of rain forest tree species in highly degraded Amazon pastures. Our data suggest that forest trees have difficulty establishing in degraded pastures because of three factors:

- few seeds of forest trees are being dispersed into pasture environments;
- most seeds that do arrive are killed by seed predators; and
- the few seeds that do manage to germinate eventually die because of harsh environmental conditions.

Seed Dispersal

The fact that few seeds of forest trees are being dispersed into pastureland is a fundamental impediment to regeneration of these areas. Our floristic survey has shown that less than 15% of the forest species have fruits with adaptations for long-distance autodispersal (e.g., for dispersal by wind). The majority of tree fruits are fleshy and appear to be dispersed by birds, bats, and both arboreal- and ground-dwelling mammals. Hence, if seeds of forest tree species are to arrive at pasture environments, they will usually have to be carried there by animals. With this in mind, we are censusing and trapping (i.e., mist netting) birds and bats in forest edge, forest second-growth, and open-degraded pasture environments. In the case of birds, we have identified more than 150 species in our area, but only a subset of them are frugivores, and of the frugivores, fewer than 10 will move out into large openings. We are now studying the movement patterns of the few species that we have identified as potential seed vectors. Our goal is to critically evaluate the role that these species play in seed movement and to begin to elucidate the determinants of movement for these species.

Postdispersal Seed Predation

Seed predators are also important roadblocks to the establishment of forest trees in degraded pastures. To study the fate of seeds that are artificially dispersed to degraded pastures, we placed seeds in groups of 8 that are 10 centimeters apart at 30 widely spaced stations. By repeatedly revisiting these stations, we are able to determine disappearance rates and, if we are lucky, the animal species that are removing the seeds. In studies on the tree *Inga* sp., leaf-cutter ants (*Atta sexdens*) were carrying off the seeds to their subterranean burrows within minutes of placement. Large seeds of other species were slowly consumed in situ, presumably by rodents, judging from teeth marks on partially consumed seeds.

In recently completed studies, we placed seeds of forest tree species simultaneously in pastures, in closed forest, and in canopy openings caused by treefalls. For all six species tested, seed removal was much more rapid in the degraded pasture than in the forest environments. All these findings suggest that seed predation is an important impediment to forest regeneration in degraded pastures.

Harsh Environmental Conditions in Pastures

Harsh environmental conditions present a final impediment to the establishment of forest trees in degraded pastures. The physical conditions of the pasture and forest contrast sharply and may significantly impede seedling survival and growth in the pasture. During the 6- to 7-month wet season usually ending in June or July, daily rains maintain pasture and forest soils near saturation. In the ensuing dry season, rain falls episodically, usually totalling less than 100 millimeters per month, and soil moisture levels decline between rain events. In the pastures, tensions of 2.0 megapascals are reached in soil 15 centimeters below the surface, whereas in forests, moisture is rarely held at tensions greater than 1.5 megapascals. This means that pasture plants rooted in the 15 centimeters of soil are subjected to intense drought (i.e., permanent wilting point conditions), whereas water availability in the surface soil of forests is adequate.

In short, the seasonal changes in moisture and energy conditions near the ground surface of degraded pastures are dramatic compared to conditions in the forest. During the wet season, the availability of soil moisture and radiation load in degraded pastures is very similar to conditions beneath closed-canopy forest. During drought episodes, soil and air moisture deficits and air temperatures in pastures greatly exceed those of the forest.

A summary of the factors that limit the establishment of forest trees in degraded pastures is provided in Figure 37-1.

RESTORATION ECOLOGY STRATEGIES

We have begun a series of field experiments to try to overcome these impediments to forest recovery on degraded lands. The *seed dispersal* limitation is straightforward and can be overcome, since humans can act as the seed dispersal agent. Alternatively, instead of individual seeds, humans could disperse onto degraded lands packets of forest surface soil with its component seed bank. This might allow whole

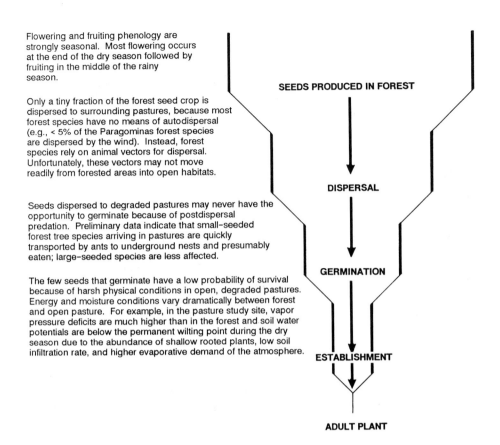

Flowering and fruiting phenology are strongly seasonal. Most flowering occurs at the end of the dry season followed by fruiting in the middle of the rainy season.

Only a tiny fraction of the forest seed crop is dispersed to surrounding pastures, because most forest species have no means of autodispersal (e.g., < 5% of the Paragominas forest species are dispersed by the wind). Instead, forest species rely on animal vectors for dispersal. Unfortunately, these vectors may not move readily from forested areas into open habitats.

Seeds dispersed to degraded pastures may never have the opportunity to germinate because of postdispersal predation. Preliminary data indicate that small–seeded forest tree species arriving in pastures are quickly transported by ants to underground nests and presumably eaten; large–seeded species are less affected.

The few seeds that germinate have a low probability of survival because of harsh physical conditions in open, degraded pastures. Energy and moisture conditions vary dramatically between forest and open pasture. For example, in the pasture study site, vapor pressure deficits are much higher than in the forest and soil water potentials are below the permanent wilting point during the dry season due to the abundance of shallow rooted plants, low soil infiltration rate, and higher evaporative demand of the atmosphere.

SEEDS PRODUCED IN FOREST

DISPERSAL

GERMINATION

ESTABLISHMENT

ADULT PLANT

FIGURE 37-1   A summary of the factors that limit the establishment of forest trees in degraded pastures in the Eastern Amazon, Paragominas, Pará.

community patches to develop in unison. Another approach is to consider new types of seed vectors. Daniel Janzen (Chapter 14 of this volume) is studying the use of cattle as forest seed dispersal agents in old pastures in Costa Rica. In a like manner, there may be certain bird or bat species that could provide valuable seed dispersal services if introduced into a landscape mosaic of forest and abandoned pastures.

Seed predation in highly degraded sites will be difficult to overcome. Conventional approaches are to bury introduced seeds, saturate the environment with seeds, soak introduced seeds in repellent chemicals, and introduce predators or parasites to reduce the population of the seed predator. Of course, this need to protect the seeds can be bypassed entirely by introducing seedlings, which, however, can be attacked by seed predators, particularly leaf-cutter ants. Given the cost involved in these approaches, it seems wiser to choose species for introduction on the basis of their resistance to predators. Hence, we need to learn enough about seeds and their predators to be able to predict what kinds of seeds are least vulnerable to predation. Our results to date show that tough-coated, heavy (e.g. ≥ 10g) seeds are the most predator resistant.

Harsh environmental conditions, particularly those related to drought and excessive radiation, are also not easily overcome. In field experiments, we have found that many forest species require shade (in the form of a shade cloth) and irrigation treatments (at least during their first dry season) to ensure establishment. Moreover, it appears that big-seeded forest species have a higher probability of establishment than small-seeded pioneer species in degraded pasture environments.

The critical first step in forest restoration is to foster the establishment of some predator-resistant, stress-tolerant tree species. Deep-rooted species that are able to extract water and nutrients from the lower soil horizons would be particularly appropriate. (In our forest studies we have found that root systems of some tree species extend beyond 10 meters in depth!) Once a few scattered trees are present in these pastures, the subsequent phases of forest development may occur naturally. This is because isolated trees attract bird and bat seed vectors by serving as perch and roosting sites and they provide favorable microsites for establishment (i.e., the shaded, moist conditions created by trees facilitate establishment of dispersed seeds).

Our studies should ultimately prove useful by providing guidelines for the reforestation of degraded pastures, should that become a land-use goal at some future time. For example, by coupling precise descriptions of pasture environments (in terms of barriers to the dispersal, germination, and establishment of tree seeds) with a knowledge of tree morphology and physiology, tree species can be selected for introduction into degraded pastures based on characteristics such as degree of tap rooting, ability to penetrate compacted soil horizons, drought tolerance, and competitive ability.

REFERENCES

Saldarriaga, J. G. 1985. Forest Succession in the Upper Rio Negro of Colombia and Venezuela. Ph.D. Thesis, University of Tennessee, Oak Ridge.

Sanford, R. L., Jr., J. Saldarriaga, K. Clark, C. Uhl, and R. Herrera. 1985. Amazon rain-forest fires. Science 227:53–55.

Uhl, C., and C. F. Jordan. 1984. Succession and nutrient dynamics following forest cutting and burning in Amazonia. Ecology 65:1476–1490.

CHAPTER

38

INCREASING DIVERSITY BY RESTORING DAMAGED ECOSYSTEMS

JOHN CAIRNS, JR.

University Center for Environmental Studies, and Department of Biology,
Virginia Polytechnic Institute and State University, Blacksburg, Virginia

The large and growing number of human beings on Earth together with rapid advances in technological development ensure extensive damage to ecosystems. For example, Vitousek et al. (1986) have shown that nearly 40% of potential net primary terrestrial productivity is used directly, co-opted, or foregone because of human activities. Mabbutt (1984) noted that 35 million square kilometers of land is at least moderately desertified. Surface and subsurface mining, hazardous substances, clear-cutting, and a variety of other human-induced stresses have taken their toll on ecosystems. Damaged ecosystems in all parts of the world, even where the human population is sparse, comprise abundant material for experimentation.

Ideally, all ecosystems should be restored to their original condition; however, this is often difficult or impossible for the following reasons: detailed ecological information about the original condition is not available, techniques for recolonizing the damaged ecosystem with original species are not adequate, and there is no satisfactory source of organisms for recolonizing to the original condition. Under these circumstances, one might reasonably consider alternative ecosystems that would be ecologically superior to the damaged condition but often markedly different ecologically from the original system (Cairns, 1980, 1985). In some cases, wetlands lost elsewhere to development can be established on disturbed sites where they did not formerly exist. In others, additional habitat can be provided for rare, threatened, or endangered species (see, for example, Bruns, in press). A plan for developing these capabilities has been described by Cairns (1987 and in press).

333

Devastation following logging near Pellston, Michigan, around the turn of the century. Photo courtesy of Gary R. Williams of Glen Ellyn, Illinois.

EVIDENCE FOR CAUTIOUS OPTIMISM

In view of the present modest capability to predict ecosystem development following remedial measures to repair anthropogenic stress, it may seem arrogant to propose rehabilitation to increase diversity. However, it is abundantly clear that we must both arrest and repair widespread ecological damage. By doing this, we should learn much about both the structure and function of ecosystems if only because there are, unfortunately, all too many experimental sites that have been seriously damaged. One might object that the existence of an ecosystem restoration capability would provide an excuse for further damage. This seems quite unlikely once the difficulties and costs of restoring damaged ecosystems are more thoroughly documented and widely understood. This will be immediately apparent if the restoration costs are the responsibility of the organization or persons causing the damage.

This 1978 photo clearly shows the restoration of a site where comparable damage occurred near that shown on the opposite page, now part of University of Michigan Biological Station at South Fishtail Bay. Photo courtesy of Gary R. Williams of Glen Ellyn, Illinois.

Many biologists are unaware that some of the world's leading biological field stations are located on sites that suffered severe ecological damage less than 100 years ago. For example, the University of Michigan Biological Station near Pellston occupies a site that was once extensively logged, followed by extensive burning of slash (e.g., unmarketable limbs). It is now a Biosphere Preserve under the Man and the Biosphere Program. Photographs of both the destruction and recovery of this site have been published (see, for example, Figures 38-1 and 38-2)(Cairns, 1980). Another station, the Rocky Mountain Biological Laboratory (RMBL), is located at Gothic, Colorado, a former mining town where a number of ecologically damaging mining activities, such as smelting, occurred (Vandenbusche, 1981). In fact, the Laboratory's 205 acres (82 hectares) of land at Gothic already function as a small nature preserve (Brussard, 1982). A number of highly regarded ecologists regularly return each summer to these field stations to conduct research that is published in respected professional journals.

Thus, the use of a formerly damaged ecosystem to preserve biological diversity is already an established practice. In both field stations just mentioned, restoration management practices appear to have been preventing further damage and letting natural processes occur. While I was writing this chapter, Boyce A. Drummond, Director of the Pikes Peak Research Station, and Mark G. Noble, Field Director of the Mountain Research Station, Institute of Arctic and Alpine Research, University of Colorado, visited RMBL. It occurred to me to learn whether the field stations where Drummond and Noble worked were situated on formerly damaged ecosystems.

The Mountain Research Station area was logged and burned around the turn of the century, and although there were no restoration management practices other than protecting the ecosystem from further damage, the lodgepole pine (*Pinus contorta*) once characteristic of the area has recolonized. It is an interesting western example of the same logging and burning practices that seriously affected the site of Michigan's Biological Station. The Pikes Peak Research Station is located on a former ranching site, and some ranching is still going on nearby. On the station site itself, grazing was stopped in spring 1985. Although neither station has become restored to its original condition, both have witnessed the reestablishment of species characteristic of the original site, such as ponderosa pine (*Pinus ponderosa*), mountain muhly grass (*Muhlenbergia*), Arizona fesque (*Festuca montana arizonica*), and purple milk vetch (*Astragalus dasyglottis*).

The Pikes Peak Research Station provides an excellent example of another opportunity, namely, the use of formerly damaged ecosystems to increase regional diversity of both habitat and species while affording research opportunities for ecologists. Surely, utilization of present ecological knowledge and theory would produce better results sooner. If not, we are in deep trouble!

On July 31, 1986, a letter was sent to 61 directors and other persons affiliated with field stations from a list of more than 200 names. To summarize a detailed letter, there were four major requests for information:

- Is the field station located on a formerly damaged ecosystem?
- If so, were any management practices used to enhance the recovery process?
- Is the facility formally designated as a nature preserve or a biosphere reserve?
- Is the field station an ecological oasis surrounded by ecosystems that are substantially altered or damaged?

As of September 18, 1986, there were nine responses to this request. One response covered 26 sites in the Natural Reserve System managed by the University of California (information from C. Ronald Carroll). Of these, 11 were nature reserves or preserves on formerly damaged ecosystems. The other eight respondents were the W. K. Kellogg Biological Station, operated by Michigan State University (information from George H. Lauff); the Tyson Research Center, operated by Washington University (information from Richard W. Coles); the Sagehen Creek Field Station, operated by the University of California at Berkeley (information from Don C. Erman); the Turtle Cove Biological Research Station, operated by

Southeastern Louisiana University (information from Robert W. Hastings); Mountain Lake Biological Station, operated by the University of Virginia (information from J. J. Murray); the Notre Dame Environmental Research Center (information from Stephen R. Carpenter); Alice L. Kibbe Life Science Station, operated by Western Illinois University (information from John E. Warnock); and Milwaukee Field Station, operated by the University of Wisconsin (information from Millicent S. Ficken). All these are located on formerly damaged ecosystems, ranging from worn-out farms to sites affected by lumbering and other disturbances. Of the eight, four were formally designated as nature reserves or some other similar category. Two had no official designation, although it is clear they were serving the purpose without the formal designation. The formality of the designation was in some doubt for two, although their uses were not. In short, all were effectively being used as nature preserves, although the formality of the designation varied.

For the damaged sites in the University of California Natural Reserve System, no particular management practices were in effect other than letting nature take its course and excluding the ecologically damaging influences. For the others, there were some management practices such as planting trees, but these were relatively low-key undertakings. Efforts did not, for example, involve large-scale interventions such as the colonization of species. Probably more than half the sites could be described as ecological oases, although only in a few cases was the contrast with neighboring land startling.

Although this is a relatively small sample, it provides substantive evidence that:

• Damaged ecosystems in temperate regions can recover rather rapidly in such a way that the biological diversity of the system is vastly improved. Much evidence suggests that tropical terrestrial ecosystems are far more fragile.

• Simple management practices, such as benign neglect, often result in a vastly improved condition.

• Good ecological research can be conducted on formerly damaged ecosystems, indicating that the processes in them are ecologically interesting and comparable in many respects to those in undisturbed ecosystems.

• The quality of these ecosystems relative to most of the surrounding countryside is sufficiently superior to justify such designations as nature reserves or preserves and experimental ecological preserves.

OBSTACLES TO RESTORATION RESEARCH

The few laws regarding the restoration of damaged ecosystems tend to be so prescriptive that they often impede implementation of obvious solutions to relatively simple problems (Cairns, 1985, 1986). It is even more regrettable that laws do not encourage the experimentation so necessary to the development of both the science and the art of ecological healing.

Bradshaw (in press) espouses a systematic approach to each wasteland problem. In the first step, the qualities of the site and the needs of the region and its people must be interrelated. Bradshaw believes that the process requires both imagination and science but that the science has to be a particularly practical, constructional form of ecology in many ways more allied to engineering than to much of modern analytical biology. The result is very rewarding. Noss (1983) cautions, however, that too much fragmentation may not produce the desired results and that a more comprehensive view is required for perpetuation of regional diversity. He recommends a regional *network* of preserves with sensitive habitats protected from human disturbance. This approach would enhance the *alpha, beta,* and *gamma*[1] diversity levels of Whittaker (1972). But even oases may serve in the restoration of damaged ecosystems (e.g., Marsh and Luey, 1982).

All this will require a substantial adjustment in the outlook of most ecologists; however, since they will have only damaged ecosystems to study if present trends continue, it is in their enlightened self-interest to prevent further environmental degradation and reverse the trend where degradation has occurred. Fortunately, there are now newsletters (e.g., the U.S. Forest Service's *Rehabilitation News*) and journals (e.g., *Restoration and Management Notes*, edited by William Jordan) to inform ecologists and other interested persons about this rapidly developing field.

Two zoologists and a medical doctor have explored an analogy between the practice of human medicine and the practice of stress ecology (Rapport et al., 1980, 1981). This approach may be quite useful in communicating ecosystem rehabilitation in societies now concerned with personal health.

PLANNING TO INCREASE DIVERSITY

Sites for restoration ecology may range from a few to hundreds of square kilometers. The Guanacaste National Park (GNP) in Costa Rica is 700 square kilometers. Janzen (in press) has a detailed plan to integrate this park into local and national Costa Rican society. The details of integration are discussed by Janzen in Chapter 14 of this volume. The park is large enough to maintain healthy populations of all animals, plants, and habitats known to have originally occupied the site and to contain an excess of needed habitat so that some can be intensively used by visitors and researchers. The GNP has three functions:

- to use existing dry forest fragments as seed sources to restore about 700 square kilometers of topographically diverse land to a dry forest sufficiently large and diverse to maintain into perpetuity all animal and plant species and their habitats known to have occupied the site originally;
- to restore and maintain a tropical wildland so that it can provide material goods such as plant and animal gene banks and stocking material; and
- to use a tropical wildland as the stimulus and factual base for a reawakening of the intellectual and cultural offerings of the natural world to audiences with a local, national, and international philosophy that is user-friendly.

---

[1] *alpha*: diversity measurements are those applied to samples from particular communities (Whittaker, 1972, p. 221); *beta*: the extent of species replacement or biotic change along environmental gradients or between habitat diversity (Whittaker, 1972, p. 230); *gamma*: richness in species of a range of habitats (a landscape, a geographic area, an island) is consequent on the *alpha* diversity of the individual communities and the range of differentiation on *beta* diversity among them (Whittaker, 1972, p. 231).

Janzen (in press) estimates a land purchase cost of $8.8 million and a start-up endowment of $3 million. As of May 12, 1987, Janzen (personal communication) states that 51% of the terrain had been purchased. These are relatively small costs for such a large project, and the world needs a large restoration pilot project with these goals. A few restoration efforts of this magnitude may be more cost-effective in increasing diversity than many smaller projects.

INTEGRATED RESOURCE MANAGEMENT

On July 8, 1986, I took my class on stressed ecosystems from RMBL to the Mexican Cut tract approximately 10 kilometers away. As of August 3, 1986, the total Mexican Cut tract owned by the Nature Conservancy and administered by RMBL was approximately 960 acres (384 hectares)—larger than the main site of the RMBL. The road through RMBL is no superhighway and was blocked east of Schofield Pass (elevation 10,707 feet, or approximately 3,200 meters) by 50 feet (15 meters) of snow from an avalanche. We hiked the remaining 4 miles (6.4 kilometers), fording an icy mountain stream and surmounting patches of snow in the process. Mexican Cut (elevation approximately 11,200 feet, or 3,360 meters) is not easily accessible by foot, yet John Harte, University of California at Berkeley, has been studying the effects of acid snow melt there for 6 years.

The students were astounded to find a stressed ecosystem in such a remote location. They carefully helped with a census of a declining population of salamanders (*Ambystoma tigrinum*). Who, the students asked, is supposed to prevent this from happening? A good question with an unsatisfactory answer.

Despite congressional intentions that one agency manage environmental problems, in reality that responsibility is fragmented and dispersed. This process of compartmentalization occurs within agencies as well. Turf battles occur within and between agencies. No quality control system can work effectively under these circumstances. At a meeting held in Charleston, South Carolina, in December 1985, I found that agency administrators were well aware of this problem and eager to do something constructive about it. Restructuring agency missions is not easy at the state level and is truly formidable at the federal level. Even then we are still in deep trouble, as demonstrated by discussions between Canada and the United States about acid rain.

We can make some progress in increasing diversity through restoration of damaged ecosystems without integrated resource management. However, as at Mexican Cut, deleterious materials (e.g., lead, aerosols; Elias et al., 1975) of known composition but uncertain and distant origin can affect and even destroy the best local and regional plans. Until we cope with the larger problems, the local successes will be temporary and therefore unsatisfactory.

ESTABLISHMENT OF ECOLOGICAL RESERVES

A committee of the National Research Council (NRC, 1981) recommended that *ecological reserves* be established throughout the country as sites for the field observations of baseline ecosystems. Its report further recommended, "Selected ecosystems representative of geographic regions and unique or fragile systems should be studied and the results used in identifying key species, properties of populations, and ecosystem processes" (NRC, 1981, p. 78). If these ecological reserves were sufficiently large, they might also be used as a source pool of species to colonize damaged ecosystems. The studies of properties and processes useful as baseline information would be equally useful in rehabilitating damaged ecosystems, which could then serve as source pools of species for additional rehabilitation efforts, leaving the ecological reserve to serve its original purpose.

GNOTOBIOTIC ECOSYSTEMS

Taub (1969) has established gnotobiotic (species-defined) microcosms for toxicity testing and other purposes. Although these assemblages of species might never occur together in nature, they appear to function reasonably well together and have many interactions typical of natural systems. There are a number of situations where such artificial assemblages might be used on a larger scale to increase diversity. Some illustrative examples follow.

Power Line Right of Ways

Enormous areas under electric power transmission lines afford opportunities for creative ecological management. If these corridors were put together, they would probably approach the land area of one of our smaller states. Even in strips, however, they represent a considerable challenge. Vegetation is kept low by spraying or cutting. Both measures are expensive, and both may affect adjacent ecosystems. Drift from sprays may be toxic, and the corridor may be a barrier to some species. Development of an artificial assemblage of plant species that would not exceed a particular height would obviate the use of sprays or cutting and, if carefully structured, would be less of an ecological barrier than the present system.

Hazardous Waste Site Closure

There are a large number of hazardous waste sites in the United States where the highest concentrations of chemicals can be removed and transformed but where, for a variety of reasons (e.g., Novak et al., 1984, state that "acid leachate will continue to be emitted from this area for some years in the future"), lower concentrations must be left in place. Selection of species for colonizing these areas should be based on their ability to decrease the intrusion of these chemicals into groundwater supplies, to immobilize the chemicals within the site, and to accelerate degradation of the compounds into less harmful materials. Successful colonization requires not only that the organisms contribute to these benefits but also that they be tolerant of residual concentrations. Selection on this basis might well produce a species assemblage that has never occurred together in nature. Nevertheless, ecological benefits would be substantial as would benefits to diversity. Furthermore, an appropriate selection of a few organisms with only marginal tolerance to the hazardous materials would provide biological sentinels warning of unfavorable trends on the site. These assemblages could be replaced with natural assemblages as hazardous conditions diminish.

Reintroduction of Societal Wastes into the Environment

All sewage and many industrial waste-treatment plants depend heavily on biological processes, although one might not reach this conclusion from examination of an operator's training manual. However, the ecological distance between the waste-treatment plant and natural ecosystems is too great. A gnotobiotic system designed to act as an ecological buffer between the two systems should help increase diversity in the natural system by more efficiently carrying out processes not completed by the waste-treatment system for which the natural system is not well equipped. Since most waste-treatment plants are not located in pristine ecological environments, these ecological buffers would be constructed on damaged or partly damaged ecosystems.

The selection of species for these assemblages should be based on their tolerance for the waste and on their ability to carry out the remaining transformation processes. They might or might not occur naturally together. The only contaminants one can easily process in this way are those from point sources. However, by appropriate design of run-off collection systems, nonpoint discharges, such as agricultural wastes and surface mine wastes, can be converted to point source discharges. Since most of these contaminants are waterborne, artificial wetlands are probably the best buffer ecosystems for coping with them. Such systems can tolerate heavy loadings of both organic wastes and some toxicants. Recent descriptions may be found in Brooks et al. (1985).

GROUNDWATER DETOXIFICATION

There is persuasive evidence of groundwater contamination in North America, but its extent has not been well defined. There seems to be little concern about the fate of the organisms that inhabit the aquifers, however, even though they occupy a unique habitat and are part of the global diversity. Some of the contaminated aquifers might be partly decontaminated by the introduction of engineered microorganisms to degrade the wastes. Although presumably less dangerous, the water would not yet be normal. In some cases, quality could be further improved by pumping the water to the surface and passing it through an artificial wetland before recharge through a pipe or by normal infiltration.

The many contaminated and depleted aquifers could be recharged by artificial wetlands designed for this purpose. These can be built on damaged ecosystems, as Brooks et al. (1985) have shown.

PROSPECTS FOR RESTORATION ECOLOGY

Damaged ecosystems can be changed from ecological liabilities to assets that are useful in both increasing diversity and protecting natural systems in a variety of ways. They can be developed as ecological buffers between waste-treatment systems and natural systems. Damaged ecosystems can be used to replace wetlands and other scarce habitats lost to development, even when such habitats did not originally exist on the damaged site. They can be developed as refuges for rare, endangered, or threatened species. Damaged ecosystems that may act as ecological barriers (e.g., power line right of ways) can be converted to "bridges" between the ecosystems on either side. Finally, they can be used to recharge and possibly purify our groundwater aquifers.

None of these desirable events will happen unless the present prescriptive regulations on rehabitation of damaged ecosystems are made more flexible to encourage experimentation. The sizable information base needed for effective ecosystem rehabilitation will not be generated until more ecologists are willing to work on damaged ecosystems.

It is probably not an exaggeration to say that much of the planet is occupied by partially or badly damaged ecosystems. Restoring them is probably the best means of increasing diversity. If we put as many resources and as much energy into restoring this planet as we have into the space program, Curry's (1977) vision of reinhabiting Earth might become a reality.

REFERENCES

Bradshaw, A. D. In press. Alternative end points for reclamation. In J. Cairns, Jr., ed. Rehabilitating Damaged Ecosystems. CRC Press, Boca Raton, Fla.

Brooks, R. P., D. E. Samuel, and J. B. Hill, eds. 1985. Wetlands and Water Management on Mined Lands. Proceedings of a conference, October 23–24, 1985, Pennsylvania State University. Pennsylvania State University, University Park. 393 pp.

Bruns, D. In press. Restoration and management of ecosystems for nature conservation in West Germany. In J. Cairns, Jr., ed. Rehabilitating Damaged Ecosystems. CRC Press, Boca Raton, Fla.

Brussard, P. F. 1982. The role of field stations in the preservation of biological diversity. BioScience 32(5):327–330.

Cairns, J., Jr., ed. 1980. The Recovery Process in Damaged Ecosystems. Ann Arbor Science Publishers, Ann Arbor, Mich. 167 pp.

Cairns, J., Jr. 1985. Keynote address: Facing some awkward questions concerning rehabilitation management practices on mined lands. Pp. 9–17 in R. P. Brooks, D. E. Samuel, and J. B. Hill, eds. Wetlands and Water Management on Mined Lands. Proceedings of a conference, October 23–24, 1985, Pennsylvania State University. Pennsylvania State University, University Park.

Cairns, J., Jr. 1986. Restoration, reclamation, and regeneration of degraded or destroyed habitats. Pp. 465–484 in M. Soulé, ed. Conservation Biology: The Science of Scarcity and Diversity. Sinauer Associates, Sunderland, Mass.

Cairns, J., Jr. 1987. Disturbed ecosystems as opportunities for research in restoration ecology. Pp. 307–320 in W. R. Jordan III, M. E. Gilpin, and J. D. Aber, eds. Restoration Ecology: A Synthetic Approach to Ecological Research. Cambridge University Press, New York.

Cairns, J., Jr. In press. Restoration ecology: The new research frontier. In J. Cairns, Jr., ed. Rehabilitating Damaged Ecosystems. CRC Press, Boca Raton, Fla.

Curry, R. R. 1977. Reinhabiting the earth: Life support and the future primitive. Pp. 1–23 in J. Cairns, Jr., K. L. Dickson, and E. E. Herricks, eds. Recovery and Restoration of Damaged Ecosystems. University Press of Virginia, Charlottesville.

Elias, R., Y. Hirao, and C. Patterson. 1975. Impact of present levels of aerosol Pb concentrations on both natural ecosystems and humans. Pp. 257–272 in International Conference on Heavy Metals in the Environment, October 27–31, 1975. Toronto.

Janzen, D. H. In press. Guanacaste National Park: Tropical ecological and cultural restoration. In J. Cairns, Jr., ed. Rehabilitating Damaged Ecosystems. CRC Press, Boca Raton, Fla.

Mabbutt, J. A. 1984. A new global assessment of the status and trends of desertification. Environ. Conserv. 11:103–115.

Marsh, P. C., and J. E. Luey. 1982. Oases for aquatic life within agricultural watersheds. Fisheries 7(6):16–19, 24.

NRC (National Research Council). 1981. Testing for Effects of Chemicals on Ecosystems. National Academy Press, Washington, D.C. 103 pp.

Noss, R. F. 1983. A regional landscape approach to maintain diversity. BioScience 33(11):700–706.

Novak, J. T., W. R. Knocke, M. S. Morris, G. L. Goodman, and T. Jett. 1985. Evaluation of an acidic waste site cleanup effort. Pp. 111–120 in Proceedings of the 40th Industrial Waste Conference. May 14–16, 1985, Purdue University, West Lafayette, Ind. Butterworth, Boston.

Rapport, D. J., C. Thorpe, and H. A. Regier. 1980. Ecosystem medicine. Pp. 179–182 in J. B. Calhoun, ed. Perspectives on Adaptation, Environment and Population. Praeger Scientific, New York.

Rapport, D. J., H. A. Regier, and C. Thorpe. 1981. Diagnosis, prognosis and treatment of ecosystems under stress. Pp. 269–280 in G. W. Barrett and R. Rosenberg, eds. Stress Effects on Natural Ecosystems. Wiley, Chichester, United Kingdom.

Taub, F. 1969. Gnotobiotic models of freshwater communities. Vehr. Internat. Verein. Limnol. 17:485–496.

Vandenbusche, D. 1981. The Gunnison Country. Vandenbusche, Gunnison, Colo. 472 pp.

Vitousek, P. M., P. R. Ehrlich, A. H. Ehrlich, and P. A. Mason. 1986. Human appropriation of the products of photosynthesis. BioScience 36(6):368–373.

Whittaker, R. H. 1972. Evolution and measurement of species diversity. Taxon 21:231–251.

RESTORING DIVERSITY
The Search for a Social and Economic Context

JOHN TODD
President, Ocean Arks International, Falmouth, Massachusetts

Restoration ecology is beginning to develop into two quite distinct disciplines. Although they have a lot in common, it is the differences between the two directions that are important. The first type of restoration ecology is primarily an academic field of activity in which there is an attempt to recreate authentic ecosystems of the past, particularly those that have been destroyed or modified by human alteration or abuse. There is an emphasis on selecting the correct species mix, and there is an effort to recreate the original ecological relationships at least as far as they are known. Exotic species or organisms intrinsic to other environments are shunned. Here, whether a forest or a prairie is the focus, restoration means recreating the original state both in terms of structure and species. These newly created ecosystems have the intrinsic value of maintaining important gene pools in regions where the organisms have previously flourished. They can also teach a great deal about the dynamic processes of ecosystems and succession. The pioneering work at the University of Wisconsin-Madison Arboretum (see Jordan, Chapter 35) falls into this category.

The second approach to restoration ecology operates from a different set of assumptions. The overall objective is different, although much of the knowledge and techniques are comparable. A hypothetical example might help to illustrate what I mean. If a mixed-forest hillside is logged and clear-cut, then turned into pasture for cattle, later grazed by sheep, and finally allowed to lose its topsoil and erode into gullies that eventually can support only coarse grasses and thistles, it is apparent that human activity has destroyed a complex and diverse habitat and replaced it with a degraded one. The hill's ability to support abundant life is reduced as well as its capability to underwrite economic activity. In this second category, the restoration ecologist is interested in the structure or architecture of the original forest because of its ability to do things that cannot be done by degraded and eroded soils. The forest can build soils, control the capture and release of moisture, and regulate nutrient cycles. It can withstand climatic fluxes and perturbations, and critically important to the naturalist, it can support an enormous spectrum of life forms. The forest's dimensions extend from the tree canopy to root depths often tens of meters below the soil surface.

In the hillside example, the structural, as distinct from the species-specific, restoration ecologist wants to build fertile soils, develop a water and nutrient regimen, and assemble an ecosystem that mimics the original structural integrity of the forest. The actual organisms selected to do the job may or may not be the original species. Often, the ecologist will seek out equivalent species of plants or animals that have secondary properties as well. For example, a tree species that is not adapted to the environment, but is highly valuable, provides an economic dimension to the process. Also the land restorer, like a farmer, will in most instances use a large array of biotechnologies and technological aids to orchestrate and even speed up successional and other biological processes. Sophisticated bioengineering is used to recreate the equivalent of hundreds of years of topsoil within a decade or two.

The mission of the two restoration ecologies can be quite different too. Structural ecologists are willing to tolerate wider margins of uncertainties and gaps in knowledge, thereby adopting a more applied viewpoint simply because they see the planetary environmental crisis as the backdrop against which they work. To reverse desertization and habitat destruction, it ultimately will be necessary to undertake ecological restoration on a vast, planetary scale. This means that the task cannot be guided by charity based on social conscience, since there isn't enough of either even to finance or underwrite the required backup ecological research. It is hard to avoid the conclusion that if there is to be any meaningful change, restoration ecology will have to become quite simply a major economic activity. Just as the activity seeks to recreate the forest on the hill, it will also be expecting the hill to become a sustainable and environmentally enhancing economy. A goal will be the provision of a wide variety of marketable products as a by-product of the restoration process. This is an extremely ambitious objective for a young field of endeavor, but one that is essential to its widespread application.

Fortunately there exists, albeit widely scattered, the biological knowledge and field experience to build a science and practice of habitat recovery. This knowledge needs to have an ecological framework whose elements will be found in geology, climatology, agriculture, forestry, horticulture, aquaculture, limnology, research ecology, landscape architecture, and natural history, to name some of the more relevant fields of endeavor. There will also be a need to develop and adapt tools and machines for use in agriculture, earth moving, forestry, waste management, and process engineering.

Returning to our hypothetical example of the eroded hillside, the first step in the restoration process would be to arrest rapid rainwater runoff and to reduce erosion. Among the many approaches that can be taken are various tillage techniques, microcatchments, contouring, and land sculpturing to regulate water movement. Most of the methods require sophisticated machines, careful engineering, and timed planting to be successful. Next, the degraded soils, if they are to be rebuilt quickly, will require the precise addition of minerals, fertilizers, organic matter, and vegetation to effect rapid stabilization and to increase its moisture retention. The subsequent step, intentionally recreating a forest, requires a knowledge not only of the original forest cover but also of a range of equivalent species that play an analogous role while serving as a key economic component. This ecosystem is in some respects a cross between a forest and an orchard. In our hypothetical example the wild deer of the original may be replaced by domestic animals—not necessarily cattle, however, but by the European fallow deer (*Cervinae dama*), which fit ecologically and are highly marketable because of the extremely low cholesterol levels in their flesh.

The goal of this ecological restoration is the production of food and fiber on a commercial scale. It is not agriculture but an ecology with agricultural elements within a broader biological framework. It does not have the environmental destructiveness of monocrop agriculture or simpler agricultural systems. Its function is to restore diversity and to be bountiful in terms directly useful to humans. Linking together nature's restoration requirements with the economic needs of people may be the only way the terrestrial fabric of the planet can be rebuilt.

Large amounts of human labor are essential to the restoration process. For example, labor is needed to plant new trees and to provide them with adequate moisture and protection from weed competition or predation by grazing animals. To support the required labor, restoration ecology will have to attract capital. The future hillside ecosystem will have to be seen as a prudent investment, possibly providing favorable returns within years rather than decades.

The goal of much of my research and planning over the last 15 years has been to find ways of economically underwriting the restoration and diversification process. This has involved the development of a family of biotechnologies that are in essence short-cycle ecosystems with economic by-products that also have the capacity to catalyze the longer-cycle restoration processes. These biotechnologies have been proven successful and in some contexts have been shown to be cost-effective and economically feasible (Todd and Todd, 1984). We have not yet had the opportunity to integrate all the subsystems into a full-scale restoration project, but an outline of a restoration project in the Mediterranean has been prepared (Todd, 1983, 1984). In addition, projects for the west coast of Costa Rica and the Atlantic coast of Morocco are now in the planning stage.

The Costa Rican project is intended to reclaim lands badly degraded because of earlier inappropriate agriculture. The proposed Moroccan project involves creating a diverse plant environment where the desert and the sea meet. In this instance, the newly created ecosystems will provide part of the underpinnings for a new human settlement.

In all the above examples, the land is not currently fit for intensive agriculture or easy restoration. A new biotechnology, which we have named the desert-farming module, provides the short-term ecological economy needed to initiate the restoration cycle. The development of this technology began at the New Alchemy Institute under my direction in 1974. In summary, a desert-farming module is a translucent solar-energy absorbing cylinder with a capacity up to 1,000 gallons (3,785 liters) that is filled with water and seeded with more than a dozen species of algae and a complement of microscopic organisms (Figures 39-1, 39-2, and 39-3). Within these cylinders, phytoplankton-feeding fishes and omnivorous fishes are cultured at very high densities. The species selected depend upon climate, region, and market opportunities: the range of species we have studied is broad, including African tilapia (*Tilapia* spp.), Chinese carps (*Cirrhinus molitorella, Ctenopharyngodan idellus, Hypophthalmichthys molitrix,* and *Aristichthys nobilis*), and North American catfish (*Ictalurus* spp.) and trout (*Saluelinus* spp.).

Dense populations, up to one actively growing fish per 2 gallons (7.6 liters), produce high levels of waste nutrients beyond the capability of the ecosystem to take up. The module recycles these nutrients in the following ways: nutrients released from the decomposed waste material are absorbed by the fish, the plankton, and the crop plants rafted on the cylinder surface. In addition, partially digested algae that floculate out and settle to the bottom are periodically discharged through a valve to fertilize and irrigate the surrounding ecosystem under restoration. The root systems of the vegetable crop plants take up the nutrients before they reach toxic levels and also capture detritus; they function as living filters by purifying the water.

These modules can yield more than 250 pounds (113.5 kilograms) of fish annually in a 25-square-foot (2.3-square-meter) area, depending on species and supplemental feeding rates. At the same time, each unit can produce 18 heads of lettuce weekly, i.e., more than 900 heads per year (Zweig, 1986). Tomatoes and cucumber crops

FIGURE 39-1   Solar aquaculture ponds. A dense population of fishes can be seen in the module in the foreground. Photo by J. Todd.

FIGURE 39-2   Solar aquaculture ponds in winter conditions on Cape Cod, Massachusetts. Photo by J. Todd.

can also be grown on the surface for even higher economic yields. An additional benefit of the modules is conservation of water. Since evaporation is almost eliminated from the surface, rates of water replenishment are based on plant evapotranspiration and the amount of water released from the module to irrigate and fertilize the adjacent area (Figures 39-3 and 39-4).

Desert-farming modules are an agroecology that require initial seed capital to construct and establish. But to a large extent, tillage, harvesting, fertilizing, and irrigation are a substitute for the heavy equipment that would otherwise have to be used for establishing and operating a farm on degraded soils. Not only are the

FIGURE 39-3 Desert-farming module in solar greenhouse with lettuce growing hydroponically on surface. Photo by R. Zweig.

modules less costly than the equipment, but it may well turn out that they are much more likely to support the restoration process technically and economically.

Within a given land restoration project, the modules could be established in rows in the most highly degraded areas. Young trees on the shaded side of these cylinders could be planted and subsequently nurtured by the periodic release of water and nutrients. On the sunny side of the modules, a variety of short-term economic crops could be established to add to the produce from the module. The labor needs for the module-based agriculture could also be used to tend the emerging ecosystems.

The approach based on the desert-farming module need not be static in the sense that the modules, having fed and watered the newly emergent vegetation including trees through their most vulnerable stages, could be shifted to new locations to repeat the process. In this way, the short-cycle biotechnology could spread its benefits to surrounding ecosystems over a larger geographic area.

For arid environments, such as the Moroccan coast, we have developed a bio-shelter system to assist with diversifying an area ecologically. The bioshelter is a transparent climatic envelope or greenhouse structure that houses the fish and vegetable modules. Our prototype is a circular geodesic structure that functions as a solar still and as an incubator for the early stages of the ecological diversification process (Figure 39-5). These bioshelters can even operate where there is no fresh water. In this extreme case, the aquaculture modules are placed inside the climatic envelope and water from the sea is pumped through them. During the day, the

350

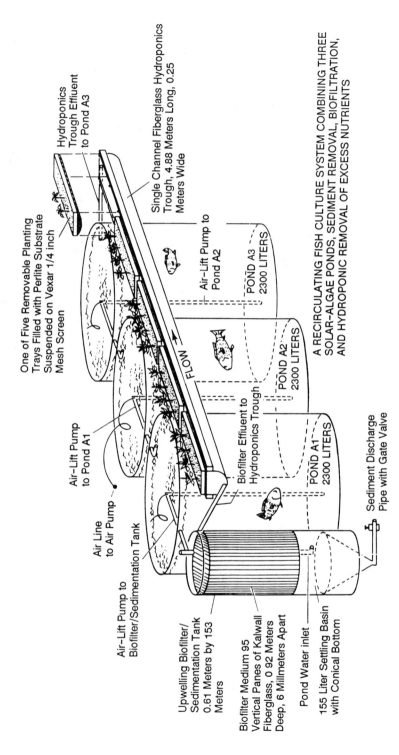

One of Five Removable Planting
Trays Filled with Perlite Substrate
Suspended on Vexar 1/4 inch
Mesh Screen

Hydroponics
Trough Effluent
to Pond A3

Single Channel Fiberglass Hydroponics
Trough, 4.88 Meters Long, 0.25
Meters Wide

Air-Lift Pump to
Pond A2

POND A3
2300 LITERS

Air-Lift Pump
to Pond A1

FLOW

POND A2
2300 LITERS

Air Line
to Air Pump

Biofilter Effluent to
Hydroponics Trough

POND A1
2300 LITERS

Air-Lift Pump to
Biofilter/Sedimentation Tank

Sediment Discharge
Pipe with Gate Valve

Upwelling Biofilter/
Sedimentation Tank
0.61 Meters by 153
Meters

Biofilter Medium 95
Vertical Panes of Kalwall
Fiberglass, 0.92 Meters
Deep, 6 Millimeters Apart

Pond Water inlet

155 Liter Settling Basin
with Conical Bottom

A RECIRCULATING FISH CULTURE SYSTEM COMBINING THREE
SOLAR-ALGAE PONDS, SEDIMENT REMOVAL, BIOFILTRATION,
AND HYDROPONIC REMOVAL OF EXCESS NUTRIENTS

FIGURE 39-4   A recirculating fish culture system combined with hydroponic cultivation of vegetables. From Baum, 1981.

FIGURE 39-5   Prototype bioshelter. A geodesic dome housing fish and vegetable modules. Photo by J. Todd.

structure heats up until the temperature differential between the seawater in the tanks and the air is great enough to cause the tanks to sweat fresh water, which irrigates the ground around them. Tree seedlings are then planted in this moist zone. Drought-tolerant trees can also be established around the outer periphery of the structure. At night, the moisture-laden air cools to the desert sky, causing water droplets to form on the interior skin of the climatic envelope. With the prototype shelter, we found that drumming on the structure's membrane in the early morning caused the droplets to fall like rain inside, thereby making it possible to plant the entire interior of the module. Inside the tanks, marine fish and crustacea such as mullet and shrimp can be cultured to form the basis of an economy. After a few years, the original cluster of climatic envelopes can be moved to a new location to repeat the cycle, leaving an established, semiarid agroecosystem behind.

These are two examples drawn from a range of biotechnological options that could help reverse environmental degradation and restore diversity and bounty to a region. These advanced technologies may well prove to be essential tools in creating sustainable environments.

In all of this, there is of course the fundamental question of land tenureship and social constraints that will ultimately drive any ecological changes. One option for countries with private land holdings and a willingness to tackle serious long-range environmental issues might be the creation of restoration corporations. In this particular model, a corporation would have the financial capability of buying large blocks of ruined land and to hire and train local farmworkers to operate the desert-farming modules, process the foods and other by-products, and tend the emerging ecosystems. When the ecosystems are ready for agriculture, natural resource management, and conservation on a given section of land, the corporation could divest itself of the land, selling it to the formerly landless workers who would be trained land stewards and farmers. The original corporation would become a mortgage banking company for the new landholders and at the same time continue to own and operate part of the infrastructure, including the training component, and to share in processing and marketing. Its earnings would be used in turn to acquire new blocks of land and repeat the cycle. In this particular model, there is the potential theoretically to link capital, institutional structures, information, and family-held productive lands to effect land restoration. There is also an opportunity to set aside lands as wilderness, since the intensive agriculture will reduce pressure on the overall environment. Even in poorer parts of the world, there is the chance to generate enough wealth to underwrite continuing ecological research out of which new models of Earth stewardship will arise.

REFERENCES

Baum, C. M. 1981. Gardening in fertile waters. New Alchemy Q. Summer (5):3–8.

Todd, J. 1983. Planetary healing. Annals of Earth Stewardship 1(1):7–9.

Todd, J. 1984. The practise of stewardship. Pp. 152–159 in W. Jackson, W. Berry, and B. Coleman, eds. Meeting the Expectations of the Land. North Point Press, San Francisco.

Todd, N., and J. Todd. 1984. Bioshelters, Ocean Arks, City Farming: Ecology as the Basis for Design. Sierra Club Books, San Francisco. 210 pp.

Zweig, R. 1986. An integrated fish culture hydroponic vegetable production system. Aquaculture 12(3):34–40.

P A R T

9

ALTERNATIVES TO DESTRUCTION

Traditional Indonesian rice paddies, which
contain numerous indigenous varieties of rice.
Photo courtesy of Miguel A. Altieri.

ARE THERE ALTERNATIVES TO DESTRUCTION?

MICHAEL H. ROBINSON
Director, National Zoological Park, Washington, D.C.

Any consideration of the forces tending to produce drastic reductions in biodiversity must concentrate on habitat destruction. To slow down, ameliorate, or prevent habitat destruction we need to understand the forces that cause it. There is a widespread assumption that habitat destruction, particularly in the Third World, is caused by ignorance or stupidity or both. This assumption is reflected in the philanthropic funding of environmental education programs and in attempts to seed environmental defense organizations in Africa, Latin America, and Asia. It is often assumed that these efforts are constructive. The contrary viewpoint is that environmental destruction results from economic pressures that have nothing to do with stupidity or ignorance. It is certainly arguable that given the present international distribution of wealth, the mainly tropical less-developed countries will be forced to exploit their natural resources on a massive scale in order to try to raise living standards quickly. A discussion of whether they can do this is beyond the scope of this chapter and beyond my competence; the majority of the countries involved are certainly trying.

In this volume are a number of analyses of the causes of habitat destruction and the economic realities facing the Third World. Most tropical deforestation, and with it the major threat to biological diversity, comes from efforts to increase the level of subsistence and to generate foreign exchange for the purchase of goods manufactured in the developed world. To halt destruction we must find alternative means of providing subsistence goods (food, fuel, and construction materials) and alternative commodities to replace those resulting from environmental rape. There is a third logical possibility that is almost certainly not a real one. That is for the entire world to adopt voluntary restraints on resource exploitation and living standards. This is the pathway of so-called green politics and the small-is-beautiful concept. On this there seems to be unanimity of opposition from all conventional political movements from the capitalist right to the Marxist left.

I would contend that the discovery of alternatives to destruction can come from the results of purely basic (i.e., nonapplied) studies. I have developed this theme in detail elsewhere (Robinson, 1985, 1986). In Chapter 42 of this volume, Rubinoff deals in detail with the scheme that we developed in Panama—a scheme based on the principle that native species of plants and animals should be considered for domestication and cultivation, as opposed to the use of exotics introduced at the behest of the developed world. In Panama, the animal studies include the green iguana (*Iguana iguana*), a large folivorous lizard, and the paca (*Cuniculus paca*), a caviomorph rodent widely esteemed for its succulence. Both these animals had been the subject of numerous academic studies before being considered as potential alternatives to destruction. The National Research Council's Board on Science and Technology for International Development, on the other hand, has deliberately set out to search for animals that have a high potential for domestication or husbandry (NRC, 1983). This approach was thus project-oriented from the start. Both approaches are valid and potentially productive. The purist approach has some exciting implications. Tropical regions are often the scene of highly coadapted/coevolved communities of plants and animals. It would seem that these communities should, a priori, be the place to search for exploitable bioresources that can be used by humans. They can be regarded as having a high potential for containing species that are preadapted for human use.

As an example of this we can briefly consider the paca. This rodent lives on the floor of rain forests in Central and South America. Its activity is nocturnal. These facts set the scene for its potential utility. Because it lives on the forest floor, does not climb trees, and is a rodent, it has a proscribed food source. It feeds on fruits that fall to the ground from the trees and to a much lesser extent on roots and seedlings. It utilizes the secondary products of the forest that are unavailable to humans and most grazing animals (exotic or indigenous). Fruitfall in the humid tropics is sporadic, but the paca subsists on an intermittent food supply by scatter hoarding during times of plenty (intermittent fruitfall may be an interspecific adaptation to ensure dispersal; Smythe, 1970). Thus pacas can prosper without destroying forest. They are almost tailor-made for domestication. Being nocturnal they do not need the capacity to run long distances to escape predators. They can be fat, unlike their diurnal complementary species the agouti (*Dasyprocta* spp.). Their disadvantages are a low reproductive rate, small families, and solitary social life.

In other parts of the world, there have been similar adaptations in response to similar environmental and biological pressures. The bearded pig (*Sus barbatus*), found in Borneo, Java, and the Philippines, is a forest floor fruit eater, grows to a substantial size, and copes with intermittent or seasonal fruit supplies by migration from area to area. It is clearly a candidate for exploitation within the forest (NRC, 1983; Robinson, 1986). There are similar adapted animals in Australasia, despite the absence of large placental animals there. The tree kangaroo (*Dendrolagus* spp.) is a leaf-eating arboreal animal that might prove to be as suitable for domestication as the iguana. In Malaysia and throughout the Indian subcontinent and Southeast

Asia there are giant squirrels, and so on. As we contemplate the mass extinction of tropical animals, we can speculate that if it were not for a dramatic historical extinction we might be considering the dodo (*Raphus cuculatus*) as a candidate for domestication and transtropical exploitation. This giant (25-kilogram) pigeon was flightless and probably subsisted on fruits, just like the paca! Just as the dispersal strategies of some Neotropical trees seem to have coevolved with the pacas, so, it seems, have some of the Mauritian fruits coevolved with the dodo. Temple (1977) suggested that the hard, nutlike seed of *Calvaria major* depended for its germination on its abrasion in the gizzard of the dodos.

Janzen and Martin (1982) suggested that at least 30 species of trees in Costa Rica depended for their dispersal on the digestive processes of some now extinct, large (sometimes giant), herbivores of the Neotropics. These, like the dodo, were probably eminently edible. Similarly, the giant ground sloths of South America and the giant birds of New Zealand, known from historic times, could have been candidates for domestication. This is not fruitless speculation but a notice that our regrets about these extinctions are certainly likely to be matched soon by more massive regrets about the ongoing and imminent extinctions.

Our knowledge of plants with a high potential for future exploitation is very sketchy indeed. The Board on Science and Technology for International Development within the National Research Council has highlighted some examples (NRC, 1975, 1979, 1981, 1984, 1985). In Chapter 10 of this volume, Iltis has drawn attention to some other dramatic examples. Again on a priori grounds we can predict that there must be many plants that have or are in themselves valuable, undiscovered products.

The tropical forests of the world are the most intense battlegrounds for species competition on the face of the Earth. This intensity is a direct result of extraordinary biodiversity and the continuingly ferocious evolutionary arms race. Huge assemblages of animal species eat leaves, and it is no exaggeration to say that tropical plants must have evolved a formidable array of insecticidal compounds and insect deterrents. We already use some of these. The picture is similar in the case of fungicides. The moist tropical forest is a superb environment for fungi of all kinds. Fungi relish warmth, wetness, and an abundance of organic substrates. The leaves of tropical trees are not only assaulted by insects and other animals but also by fungi and epiphyllic plants. What a place to look for fungicides and herbicides!

An illustration of how fundamental studies lead to discoveries in this field comes from a report of the food preferences of leaf-cutter ants by Hubbell et al. (1983). They found that leaf-cutter ants (*Atta* sp.) rejected the leaves of the leguminous tree *Hymenaea courbaril*. Such leaves kill the fungus that the ants grow in their nests as a food resource. It is entirely improbable that the tree evolved the fungicide to deter leaf-cutter ants. Rather, it probably had to cope with a broad spectrum of epiphyllic fungi and evolved a broad-spectrum fungicide to (literally) keep its leaves clean. Laboratory tests on the fungicide, a terpene-caryophyllene epoxide, show antifungal activity against a wide range of pathogenic fungi. This precisely illustrates the points I make above. It surely means that we can confidently predict, because of the nature of tropical forest ecology, that its plants should be a source of a wide range of useful products. These await discovery. There is little doubt that among them are further alternatives to destruction. In terms of food sources we already exploit tropical plants as sources of oils, fats, carbohydrates (sugars as well as starches), and proteins, but the current number of species that we use for food is very small. Discovering potential forest crops is going to require a massive research effort, but we can expect some shortcuts. These are outlined below.

First, we have a great potential source of insights in ethnobotany and ethno-zoology. Populations of humans throughout the world have existed in essentially harmonious relationships with tropical rain forests for centuries. Many of these relationships involved sustained yield subsistence use of the forests, and the Mayans produced a surplus of foodstuffs without destroying vast areas of forest. There are thus a wide range of peoples with invaluable knowledge about plants and animals. This knowledge may not be expressed in the terminology of modern science; it may be interlaced with magic, myth, and superstition, but it is certainly extractable. Just as certainly, its extraction is a matter of extreme urgency, since the folkways of forest peoples are disappearing more rapidly than the plants and animals that they have learned to exploit. In Part 2 of this volume, Nations, Farnsworth, Iltis, and Plotkin address various aspects of our need to utilize traditional knowledge—knowledge that is threatened and fragile. It is not merely knowledge of species but also knowledge of practices of husbandry, gardening, and agriculture. In Chapter 41, Altieri argues persuasively for the modern applications of ancient systems of mixed cropping as means of ex situ preservation of diversity. In addition to tra-ditional knowledge, there is another shortcut to research into alternatives. This second method is implied in much of what I have already argued above. We have, in my opinion, a virtually untapped resource in the insights of tropical biologists. Experienced tropical biologists are potentially a source of major advances; they should be able to identify systems that are sources of evolutionary strategies that are preadapted to nondestructive parasitization by humans.

To apply such intuitions is going to require a change of attitude on the part of many academic biologists. They will have to recognize that elegant theoretical generalizations, intellectually exciting and satisfying as they may be, are not their only responsibility to science and society. We must become concerned about the future of tropical mankind, even if only because this is the only way we can preserve tropical nature. Janzen's application of his extensive theoretical insights to the practical problems involved in regrowing tropical forests is a shining example to us all (see Chapter 14). The involvement of several Smithsonian biologists in research on alternatives to destruction is another excellent sign. We should also realize that sciences other than biology may be able to make substantial contri-butions to halting environmental destruction. An incalculable pressure is exerted on forests worldwide by fuelwood gathering. Eckholm (1975) has estimated that 1.5 billion people derive more than 90% of their fuel needs from wood, while another billion derive at least 50% from wood. Most of these people use extremely inefficient stoves, many with efficiencies less than 10%. The invention of an efficient and inexpensive wood-burning stove could greatly reduce the subsistence pressures on tropical forests.

While all the initiatives and possibilities mentioned above give cause for hope, although certainly not cause for optimism, there is still another area from which alternatives to destruction might be derived. This is the area of radical innovation. Instead of trying to solve problems by applying existing techniques or improving them, it is possible that we can discover new approaches that are so disjoined from present approaches that they qualify as breakthroughs or new scientific revolutions. The woodgrass system proposed by Shen in Chapter 43 has all the earmarks of a revolutionary approach. It is very simple in concept: in trees, there is a comparatively large increase of biomass in the first years of growth. The growth curve is most efficiently cropped at its maximum angle. At this stage, trees are thin and useless as sources of board timber. But they can be harvested by techniques more appropriate to hay production and the product then subjected to treatment as cattle fodder, composition boards and beams, fuelwood, and so on. The agricultural management of trees as if they were essentially grasses is surely revolutionary. This kind of approach needs to be applied to our whole approach to tropical problems.

Finally, there must be a realization that present levels of research into all the matters relating to tropical ecosystems, natural and man-made, terrestrial and marine, is totally and fundamentally inadequate. It is scientifically invidious to compare research expenditures on astronomy and tropical biology, but this does serve to point to the neglect of studies of life on Earth. Astronomical studies are important; they are contributing fundamentally to knowledge, but the stars are not about to disappear. A world expenditure on fundamental studies of tropical biology that is less than half the cost of a Boeing 747 airliner is a sad reflection on both our priorities and our values.

REFERENCES

Eckholm, E. 1975. The Other Energy Crisis: Firewood. Worldwatch Paper 1. Worldwatch Institute, Washington, D.C. 20 pp.

Hubbell, S. P., D. F. Wiemer, and A. Adejare. 1983. An antifungal terpenoid defends a neotropical tree (Hymenaea) against attack by fungus-growing ants (Atta). Oecologia 60:321–327.

Janzen, D. H., and P. S. Martin. 1982. Neotropical anachronisms: The fruits the gomphotheres ate. Science 215:19–27.

NRC (National Research Council). 1975. Underexploited Tropical Plants with Promising Economic Value. Board on Science and Technology for International Development Report 16. National Academy Press, Washington, D.C. 187 pp.

NRC (National Research Council). 1979. Tropical Legumes: Resources for the Future. Board on Science and Technology for International Development Report 25. National Academy Press, Washington, D.C. 331 pp.

NRC (National Research Council). 1981. The Winged Bean: A High Protein Crop for the Tropics. Board on Science and Technology for International Development Report 37. National Academy Press, Washington, D.C. 49 pp.

NRC (National Research Council). 1983. Little-Known Asian Animals with a Promising Economic Future. Board on Science and Technology for International Development Report 46. National Academy Press, Washington, D.C. 133 pp.

NRC (National Research Council). 1984. Amaranth: Modern Prospects for an Ancient Crop. Board on Science and Technology for International Development Report 47. National Academy Press, Washington, D.C. 76 pp.

NRC (National Research Council). 1985. Jojoba: New Crop for Arid Lands. Board on Science and Technology for International Development Report 53. National Academy Press, Washington, D.C. 102 pp.

Robinson, M. H. 1985. Alternatives to destruction: Investigations into the use of tropical forest resources with comments on repairing the effects of destruction. Environ. Prof. 7:232–239.

Robinson, M. H. 1986. The Biological Resources of Southeast Asia and Future Development. Paper presented at ASEAN Science and Technology Conference, Kuala Lumpur, Malaysia, 29 April 1986.

Smythe, N. 1970. Relationships between fruiting seasons and seed dispersal methods in a neotropical forest. Am. Nat. 104(935):25–35.

Temple, S. A. 1977. Plant-animal mutualism: Coevolution with Dodo leads to near extinction of plant. Science 197:885–886.

CHAPTER

41

AGROECOLOGY AND IN SITU CONSERVATION OF NATIVE CROP DIVERSITY IN THE THIRD WORLD

MIGUEL A. ALTIERI

Associate Professor, Division of Biological Control, University of California, Berkeley

LAURA C. MERRICK

Graduate Research Assistant, Department of Vegetable Crops, University of California, Davis, and L. H. Bailey Hortorium, Cornell University, Ithaca, N.Y.

Today, the foundation and health of agriculture in industrial countries largely depend on their access to the rich crop genetic diversity found in Third-World countries. Yet the very same germplasm resources most sought after for their potential applications in biotechnology are constantly threatened by the spread of modern agriculture. On the one hand, the adoption of high-yielding, uniform cultivars over broad areas has resulted in the abandonment of genetically variable, indigenous varieties by subsistence farmers (Frankel and Hawkes, 1975; Harlan, 1975). The new varieties are often less dependable than the varieties they have replaced when grown under traditional agricultural management (Barlett, 1980). On the other hand, the planting of vast areas with monocultures of genetically uniform cultivars makes agricultural productivity extremely vulnerable to yield-limiting factors, as illustrated by the southern corn leaf blight epidemic in the United States in 1969–1970 (Adams et al., 1971). Agroecosystems established far from centers of origin tend to have simpler genetic defenses against pathogens and insect pests, rendering crops more vulnerable to epidemic attack—a situation that rarely occurs in an unmodified traditional agroecosystem (Segal et al., 1980).

Concern for this rapid loss of genetic resources and crop vulnerability consolidated at the international level about 13 years ago with the establishment of the International Board for Plant Genetic Resources (IBPGR), which coordinates a global network of gene banks to provide plant breeders with the genetic resources necessary to develop better crops. International efforts have so far placed more emphasis on increasing yield than on maintaining stable harvests (Plucknett et al., 1983)—an emphasis that has provided the justification for technological innovation and transfer in a manner not reflecting indigenous social, ecological, and ethnobotanical considerations. Landraces[1] and wild relatives of major crops are collected from their native habitats, and the seed or vegetative material is placed in gene banks for storage or breeding collections for evaluation and potential use (Frankel and Bennett, 1970). Although ex situ conservation methods have contributed to the improvement of certain crops and the storage of the germplasm of a variety of major crops (Frankel and Bennett, 1970), they do not provide a panacea for conserving natural sources of crop genetic resources (Oldfield, 1984). A major problem is that seed storage freezes the evolutionary processes by preventing new types or levels of adaptations or resistance to evolve, because plants are not allowed to respond to the selective pressures of the environment (Simmonds, 1962). In addition, ex situ methods remove crops from their original cultural-ecological context (Nabhan, 1985)—the human-modified systems in which they evolved.

As Wilkes (1983, p. 136) stated, "The centers of genetic variability are moving from natural systems and primitive agriculture to gene banks and breeders' working collections with the liabilities that a concentration of resource (power) implies." Controversy has already erupted around the control of gene banks, since countries such as Colombia, Cuba, Libya, and Mexico question the free access to genetic resources by industrial countries. In the industrial countries, breeders develop new commercial varieties, often using valuable genes derived from landraces or wild species originally collected in the Third World. Then, the new commercial varieties are sold back to the Third World at considerable profit (Wolf, 1985).

A number of scientists have emphasized the need for in situ conservation of crop genetic resources and the environments in which they occur, since in situ conservation allows for continued, dynamic adaptation of plants to the environment (Nabhan, 1985; Prescott-Allen and Prescott-Allen, 1982; Wilkes, 1983). For agriculture, this phenomenon is particularly important in areas under traditional farming, where crops are often enriched by gene exchange with wild or weedy relatives (Harlan, 1965). However, most researchers consider that in situ preservation of landraces would require a return to or the preservation of microcosms of primitive agricultural systems—to many, an unacceptable and impracticable proposition (Ingram and Williams, 1984). Nevertheless, we contend that maintenance of traditional agroecosystems is the only sensible strategy to preserve in situ repositories of crop germplasm. Although most traditional agroecosystems are undergoing some process of modernization or drastic modification, conservation of crop genetic resources can still be integrated with agricultural development, especially in regions where rural development projects preserve the vegetational diversity of traditional agroecosystems and depend upon the peasants' rationale to utilize local resources and their intimate knowledge of the environment (Alcorn, 1984; Nabhan, 1985; Sarukhan, 1985).

---

[1]Landrace populations consist of mixtures of genetic lines, all of which are reasonably adapted to the region in which they evolved but which differ in reaction to diseases and insect pests.

In alternative strategies, the conservation of plant genetic resources and agricultural development by peasants can be considered simultaneously. In a recent article (Altieri and Merrick, 1987), we suggested the best ways in which traditional varieties, agroecological patterns, and management systems can be integrated into rural development programs to salvage crop genetic resources. These are reviewed below.

PEASANT AGRICULTURE AND CROP GERMPLASM RESOURCES

The stability and sustainability of traditional agriculture are based on crop diversity (Altieri and Merrick, 1987; Chang, 1977; Clawson, 1985; Egger, 1981; Harwood, 1979). The peasant's strategy of hedging against risk by planting several species and varieties of crops in different spatial and temporal cropping systems designs is the most effective long-lasting means of stabilizing yields. Although improved varieties are distributed throughout Third-World countries, they have made serious inroads in areas strongly linked to commercial agriculture and the national market, where they have hastened the disappearance of wild relatives and traditional varieties of crops (Brush, 1980). Thus today, the rural landscapes consist of mosaics of modern and traditional varieties and technologies (Figure 41-1). As areas become more marginal in natural resources and in infrastructural support, however, the use of improved varieties declines; farmers abandon them because of

FIGURE 41-1 A traditional small farm system in Tlaxcala, Mexico, exhibiting a corn-alfalfa strip-cropping pattern, borders of Maguey and Capulin trees, and a number of wild plants both within and around the crop area. Photo by M. A. Altieri.

the risk and expense and rely on their century-tested, regionally adapted stocks (Figure 41-2). In Peru, for example, as altitude increases, the percentage of native potatoes in the field increases steadily (Brush, 1980). In Thailand, rice farmers plant the modern semidwarf varieties in part of their land during the dry season and sow traditional varieties during the monsoon season. They have thus established a system that allows them to take advantage of the productivity of irrigated modern varieties during dry months and the stability of the traditional varieties in the wet season when pest outbreaks are common (Grigg, 1974).

As the economic crisis deepens in most developing countries, and rural populations become increasingly impoverished, a sizeable portion of the peasantry is renewing use of the traditional varieties and low-input management practices needed for subsistence agriculture (Altieri and Anderson, 1986). Opting for less crop uniformity may mean lower yields for farmers, but it gives them the extra margin of resistance to pests, diseases, and other environmental hazards—an important

FIGURE 41-2 Bean seeds of different colors expressing high genetic diversity. Harvested from a single field in a rain-fed traditional cropping system in Tlaxcala, Mexico. Photo by M. A. Altieri.

consideration when working under conditions of economic uncertainty. In many areas, unfortunately, the return of some peasant communities to native varieties has been difficult because of genetic erosion.

Several factors have contributed to this loss of crop genetic resources (Nabhan, 1986): decrease in the number of growers, decrease in crop size per field, decrease in planting frequency, loss of seed-saving and seed-collection skills, and changes in the crop's vulnerability to pests and weeds. In such regions, the implementation of in situ crop genetic conservation will be more complex, since both the folk science and the genetic heritage necessary to nurture such programs can only be retrieved very slowly (Altieri and Merrick, 1987).

TOWARD A STRATEGY FOR IN SITU CROP GENETIC CONSERVATION

Recommendations for in situ conservation of crop germplasm have emphasized the development of a large system of village-level landrace custodians (a farmer-curator system) whose purpose would be to continue to grow a limited sample of endangered landraces native to the region (Nabhan, 1985). To preserve crop-plant diversity, Wilkes (1983) has suggested that the governments set aside carefully chosen 5-by-20-kilometer strips at as few as 100 sites around the world where native agriculture is still practiced, areas where both indigenous crops and their close wild relatives may interbreed periodically. The idea of setting aside parks for crop relatives and landraces is obviously a luxury in countries where farmland is already at a premium, but to some this may be less costly than allowing native crop varieties to disappear (Brush, 1980; Wolf, 1985). In many areas, the urgent short-term issue is survival, and it would therefore be totally inappropriate to divert the limited land available to peasants for conservation purposes per se so that the germplasm could be used by industrialized nations. Such a position could be viewed as an undesirable form of neocolonialism.

Supporters of in situ strategies argue that farmers should be incorporated into conservation programs by creating biosphere reserves where peasants can continue their traditional agriculture under subsidy (Wilkes, 1983). Once identified, these areas would be designated as germplasm centers and would qualify for special agricultural assistance aimed at promoting the cultivation of native varieties. Industrialized countries using this germplasm would subsidize farmers cultivating native varieties and would help them in marketing the produce. Brush (1980) believes that these support programs could also include the machinery and financial aid needed to compensate for monetary losses incurred by farmers who maintain germplasm and do not reap the rewards of growing improved varieties. Some means of computing the opportunity cost of maintaining these cultivars must be established.

Although Nabhan (1985) agrees with the creation of centers of traditional agriculture, he believes that conservation measures will be most effective when native farmers are cognizant of, and involved in, their planning and implementation. Such efforts are likely to succeed, he argues, as long as members of a culture identify their own reasons for maintaining their crop heritage and persevere in conducting the practices for nurturing these plants. An obvious incentive for resource-poor peasants is to enhance harvest security and to make them independent of the market for seeds and inputs. It is for this type of farmer that preservation efforts should be linked to the overall rural development agenda. Design of sustainable farming systems and appropriate technologies aimed at upgrading peasant food production for self-sufficiency should incorporate locally adapted, native crops, and wild and weedy relatives, to complement the various production processes (Altieri and Merrick, 1987).

At present, there are a number of assistance programs temporarily directed at meeting the subsistence needs of peasants (Altieri and Anderson, 1986; Altieri and Merrick, 1987). These efforts are intended to minimize dependency on purchased inputs and industrialized technology; improve the efficiency with which local resources, including local vegetation, is used; achieve the production goals needed to satisfy home consumption; and favor peasant organization to enhance their capacity for economic reproduction (de Janvry, 1981). The approaches consist generally of taking existing peasant production systems and technologies as starting points and then using modern agricultural science to improve, progressively and carefully, on the productivity of these systems (Altieri, 1983).

Thus, proposed agricultural models are based on the peasants' skills in utilizing the environment and their ability to cope with change, as well as their knowledge of the plant resources and the general biology of the area. The programs have a definite ecological bent and rely on resource-conserving and yield-sustaining production technologies. Through the design of crop associations and regionally adapted patterns, functions of nutrient recycling, natural pest control, and soil conservation can be optimized (Altieri, 1983; Gliessman et al., 1981). As subsistence needs are met, most programs emphasize channeling of excess production to local markets. Income generation is also achieved by promoting nonagricultural activities (e.g., basketry) within the villages.

When valuable crop genetic resources are incorporated into farming systems designed to encourage self-sufficiency of the rural poor, important conservation gains can be achieved. This is illustrated by the efforts by Nabhan (1984) and associates to improve arid land agriculture for native Americans in the U.S.-Mexico borderlands and by a number of groups in Latin America (Altieri and Anderson, 1986; Altieri and Merrick, 1987).

In Mexico, for example, researchers designed for peasants production modules based on the pre-Hispanic traditional *chinampas* (a meadow or garden developed from a reclaimed lake or pond) and multilayered, species-rich kitchen gardens that once characterized the original agroecosystems of Tabasco, Mexico. Diverse arrays of crop and noncrop species were utilized in the various modular systems. In a parallel project, integrated farms were established in Veracruz to help farmers make better use of their local resources (Morales, 1984). In unique designs based on the *chinampas* and on Asiatic systems, vegetable production and animal husbandry, including aquaculture, were integrated through the management and recycling of organic matter. The intensive cultivation of corn, beans, and squash for local consumption and of vegetables with high commercial value, e.g., Swiss chard (*Beta vulgaris cicla*), cilantro (*Cordiandrum sativum*), chilies (*Capsicum* spp.), and cabbage (*Brassica oleracea capitata*), provided abundant plant wastes and cuttings used as cattle and horse feed; all animal wastes were reintegrated as fertilizer for the fields (Altieri and Merrick, 1987).

In the highlands of Bolivia, the project AGRUCO is attempting to maintain the ecological diversity of the Andean agropastoral economy by helping peasants recover their production autonomy. To replace the use of fertilizers and meet the nitrogen requirements of potatoes and cereals, intercropping and rotational systems utilizing a native species, wild lupin (*Lupinus mutabilis*), have been designed. Wild lupin has been cultivated in the high Andes for several thousand years (Smith, 1976). It can fix 200 kilograms of nitrogen per hectare, part of which becomes available to the associated or subsequent potato crop, thus significantly minimizing the need for fertilizers (Augstburger, 1983).

In Chile, where lately the peasantry has been subjected to a process of systematic impoverishment, the Centro de Educacion y Tecnologa (CET) is helping peasants become self-sufficient, thus reducing their dependence on credit demands and fluctuating market prices. CET's approach has been to establish several 0.5-hectare model farms where most of the food requirements for a family with scarce capital and land can be met (Altieri, 1983). Peasant community leaders live in CET farms for variable periods, thus learning through direct participation farm design, management technologies, and resource allocation recommendations. CET farms are composed of a diversified combination of crops, trees, and animals. The main components are vegetables, staple crops (corn, beans, potatoes, fava beans), cereals, forage crops, fruit trees, forest trees (e.g., *Robinia*, *Gleditsia*, *Salix*), and domestic animals all assembled in a 7-year rotational system designed to produce the maximum variety of basic crops in six plots, taking advantage of the soil-restoring properties of the legumes included in the rotation (Altieri, 1983). Most species are locally adapted varieties traditionally grown and consumed by rural populations (Altieri and Merrick, 1987).

THE FUTURE

A number of people have stressed the importance of in situ preservation of crop genetic resources but have failed to suggest practical avenues to achieve this goal in Third-World countries (Prescott-Allen and Prescott-Allen, 1982). If the conservation of crop genetic resources is indeed to succeed among small farmers, the process must be linked to rural development efforts that give equal importance to local resource conservation and to food self-sufficiency and market participation. Any attempt at in situ crop genetic conservation must struggle to preserve the agroecosystem in which these resources occur (Nabhan, 1985, 1986). In the same vein, preservation of traditional agroecosystems cannot succeed if not tied to the maintenance of the sociocultural organization of the local people (Altieri, 1983). The few examples of grassroots rural development programs currently functioning in the Third World suggest that the process of agricultural betterment must utilize and promote autochthonous knowledge and resource-efficient technologies; emphasize the use of local and indigenous resources, including valuable crop germplasm and essentials such as firewood resources and medicinal plants; and be based on self-contained villages and the active participation of the peasants (Altieri and

Anderson, 1986; Altieri and Merrick, 1987). The subsidizing of a peasant agricultural system with external resources (e.g., pesticides, fertilizers, improved seeds, irrigation water) can bring high levels of productivity through dominance of the production system, but these systems are sustainable only at high external cost and depend on the uninterrupted availability of commercial inputs. An agricultural strategy based on native crop diversity can bring moderate to high levels of productivity through manipulation and exploitation of the resources internal to the farm and can be sustainable at a much lower cost and for a longer period.

Ecologists, agronomists, anthropologists, and ethnobotanists have an important, as yet unrealized role in agricultural development and genetic resource conservation (Alcorn, 1984). Through interdisciplinary efforts they can assess traditional know-how to guide the use of modern agricultural science in the improvement of small farm productivity. Ethnobotanists and ecologists can provide critical information for policy makers about resources needing protection and about the ecological and management factors that determine the persistence of natural vegetation in the traditional agroecosystems (Alcorn, 1984).

It is time to recognize the active role of peasants in genetic resource conservation (Alcorn, 1984), and it behooves the industrial nations interested in the germplasm to provide fair subsidies to the peasants for their ecological service of maintaining native cultivars. Farmers should be made aware that reciprocal exchanges of seeds with gene banks are possible and that they have unconditional access to seed held in gene banks, which they can tap if they lose their remaining seeds. Farmers must also know the reasons why others are interested in their seeds. Agricultural education programs including information from elders on traditional planting techniques, seed saving, and seed selection should also be established (Nabhan, 1985).

Incorporation of indigenous crops and other native plant germplasm in the design of self-sustained agroecosystems should ensure the maintenance of local genetic diversity available to farmers. This approach sharply contrasts with current efforts by international centers that tend to concentrate on fewer varieties, potentially eroding genetic diversity and making farmers increasingly dependent on seed companies for their seasonal seed supply. A major concern is that when impoverished peasants become dependent on distant institutions for inputs, rural communities tend to lose control over their production systems (Altieri and Merrick, 1987).

REFERENCES

Adams, M. W., A. H. Ellingbae, and E. C. Rossineau. 1971. Biological uniformity and disease epidemics. BioScience 21:1067–1070.

Alcorn, J. B. 1984. Development policy, forests and peasant farms: Reflections on Haustec-managed forests' contributions to commercial production and resource conservation. Econ. Bot. 38:389–406.

Altieri, M. A. 1983. Agroecology: The Scientific Basis of Alternative Agriculture. Division of Biological Control, University of California, Berkeley. 162 pp.

Altieri, M. A., and M. K. Anderson. 1986. An ecological basis for the development of alternative agricultural systems for small farmers in the Third World. Am. J. Alt. Agric. 1:30–38.

Altieri, M. A., and L. C. Merrick. 1987. In situ conservation of crop genetic resources through maintenance of traditional farming systems. Econ. Bot. 41(1):86–96.

Augstburger, F. 1983. Agronomic and economic potential of manure in Bolivian valleys and highlands. Agric. Ecosys. Environ. 10:335–346.

Barlett, P. F. 1980. Adaptation strategies in peasant agricultural production. Annu. Rev. Anthropol. 9:545–573.

Brush, S. B. 1980. The environment and native Andean agriculture. Am. Indigena 40:161–172.

Chang, J. H. 1977. Tropical agriculture: Crop diversity and crop yields. Econ. Geogr. 53:241–254.

Clawson, D. L. 1985. Harvest security and intraspecific diversity in traditional tropical agriculture. Econ. Bot. 39:56–67.

de Janvry, A. 1981. The Agrarian Question and Reformism in Latin America. Johns Hopkins University Press, Baltimore. 311 pp.

Egger, K. 1981. Ecofarming in the tropics—characteristics and potentialities. Plant Res. Devel. 13:96–106.

Frankel, O. H., and E. Bennett, eds. 1970. Genetic Resources in Plants—Their Exploration and Conservation. International Biological Programme Handbook #11. Blackwell, Oxford, United Kingdom. 554 pp.

Frankel, O. H., and J. G. Hawkes, eds. 1975. Crop Genetic Resources for Today and Tomorrow. Cambridge University Press, Cambridge, United Kingdom. 422 pp.

Gliessman, S. R., R. E. Garcia, and M. A. Amador. 1981. The ecological basis for the application of traditional agricultural technology in the management of tropical agro-ecosystems. Agro-Ecosystems 7:173–185.

Grigg, D. B. 1974. The agricultural systems of the world: An evolutionary approach. Cambridge University Press, Cambridge. 358 pp.

Harlan, J. R. 1965. The possible role of weed races in the evolution of cultivated plants. Euphytica 14:173–176.

Harlan, J. R. 1975. Our vanishing genetic resources. Science 188:618–622.

Harwood, R. R. 1979. Small Farm Development—Understanding and Improving Farming Systems in the Humid Tropics. Westview Press, Boulder, Colo. 160 pp.

Ingram, F. B., and J. T. Williams. 1984. In situ conservation of wild relatives of crops. Pp. 163–178 in J. H. W. Holden and S. T. Williams, eds. Crop Genetic Resources. Conservation & Evaluation. G. Allen and Unwin, London.

King, F. H. 1927. Farmers of Forty Centuries; or, Permanent Agriculture in China, Korea and Japan. Cape, London. 379 pp.

Morales, H. L. 1984. Chinampas and integrated farms: Learning from the rural traditional experience. Pp. 188–195 in F. De Castri, F. W. G. Baker, and M. Hadley, eds. Ecology in Practice. Vol. 1. Ecosystem Management. Tycooly, Dublin.

Nabhan, G. P. 1984. Replenishing desert agriculture with native plants and their symbionts. Pp. 172–182 in W. Jackson, W. Berry, and B. Colman, eds. Meeting the Expectations of the Land. North Point Press, San Francisco.

Nabhan, G. P. 1985. Native crop diversity in Aridoamerica: Conservation of regional gene pools. Econ. Bot. 39:387–399.

Nabhan, G. P. 1986. Native American crop diversity, genetic resource conservation and the policy of neglect. Agric. Human Val. 2:14–17.

Oldfield, M. L. 1984. The Value of Conserving Genetic Resources. U.S. Department of the Interior, National Park Service, Washington, D.C. 360 pp.

Plucknett, D. L., N. J. H. Smith, J. T. Williams, and N. M. Anishetty. 1983. Crop germplasm conservation and developing countries. Science 220:163–169.

Prescott-Allen, R., and C. Prescott-Allen. 1982. The case for in situ conservation of crop genetic resources. Nat. Resour. 23:15–20.

Sarukhan, J. 1985. Ecological and social overviews of ethnobotanical research. Econ. Bot. 39:431–435.

Segal, A., J. Manisterski, G. Fischbeck, and I. Wahl. 1980. How plant populations defend themselves in natural ecosystems. Pp. 75–102 in J. G. Horsfall and E. B. Cowling, eds. Plant Disease: An Advanced Treatise. Academic Press, New York.

Simmonds, N. W. 1962. Variability in crop plants, its use and conservation. Biol. Rev. 37:422–465.

Smith, P. M. 1976. Minor crops. Pp. 312–313 in N. W. Simmonds, ed. Evolution of Crop Plants. Longman, London.

Wilkes, H. G. 1983. Current status of crop plant germplasm. CRC Crit. Rev. Plant Sci. 1:133–181.

Wolf, E. C. 1985. Conserving biological diversity. Pp. 124–126 in The State of the World. A Worldwatch Institute Report on Progress Toward a Sustainable Society. W. W. Norton, New York.

CHAPTER

42

ALTERNATIVES TO DESTRUCTION
Research in Panama

GILBERTO OCANA
Conservation Resources Manager

IRA RUBINOFF
Director

NICHOLAS SMYTHE
Staff Scientist

DAGMAR WERNER
Research Supervisor, Iguana Management Project
Smithsonian Tropical Research Institute, Balboa, Republic of Panama

In 1983, the Smithsonian Tropical Research Institute (STRI) in the Republic of Panama initiated a program called Alternatives to Destruction to develop practices that could help to feed human populations and to provide some alternatives to cattle rearing and other extractive practices that are responsible for much of the destruction of tropical forests in Central and South America. STRI has approached the problem in ways involving both plants and animals with grants from the W. Alton Jones Foundation, the Exxon Corporation, the James Smithson Society, and the World Wildlife Fund. Its goals are either to enhance production in association with native vegetation or to enhance vegetation development and soil stabilization in deforested areas. Animal species selected for management projects feed directly on forest vegetation, so there is an incentive to native populations to preserve some of the forest. STRI is using knowledge derived from basic research to explore approaches that will not only yield harvestable products but will also protect natural forest.

Tropical deforestation and the concomitant loss of biological diversity in tropical rain forests have received much recent attention. The causes of tropical deforestation are complex, and no single response will be sufficient to counteract its effects. The problem has been identified, but solutions will have to incorporate sound biological principles and must be reasonable within a social and economic context. A research organization devoted to tropical biology cannot directly resolve social and economic problems, but an understanding of the biological functioning of the system is a prerequisite to amelioration of such problems.

The slash-and-burn farmer is perhaps the principal agent of rain forest destruction, and it is commonplace to hear that he must be stopped. The key question is: "How to stop him?" Simply enacting legislation that would prevent him from further cutting cannot work, for the need for food grows with the blossoming population, and slash-and-burn is the only proven way of exploiting the land. It will continue until suitable alternatives are provided.

In the search for alternatives to destruction, four major projects were selected by STRI researchers, who believe that their knowledge of the species involved combined with the possibilities of finding less destructive crops could be productive. These projects are management of the green iguana, captive breeding of the paca, forest gardening or experimentation with crops that cause minimum perturbation in the existing forest, and management of game mammals. It was, of course, realized that quick fixes can not be expected in the face of so many complex problems, but success in even one or two projects could contribute greatly toward decreasing the seriousness of those problems.

MANAGEMENT OF THE GREEN IGUANA (*IGUANA IGUANA*)

For over two decades, basic research at STRI has led to great increases in knowledge of the biology of the green iguana (Burghardt and Rand, 1982). Formerly a wide-ranging reptile, and prized as a protein source for more than 7,000 years, this animal is now drastically decreasing in numbers due to habitat destruction from slash-and-burn farming, conversion of forested lands to pasture, increased use of biocides, and uncontrolled hunting. Through much of the natural range of the green iguana, there is a widespread belief that the about-to-be-laid eggs have aphrodisiac properties, and the gravid females, which are easily discovered as they converge on traditional, communal nesting areas, are especially vulnerable.

An early goal in this project was to reduce egg and hatchling mortality. Eggs laid in the wild have about 50% hatching success, and only 5% of the hatchlings survive their first year. Techniques developed at STRI have yielded captive hatching success in excess of 95%. Moreover, survival to the yearling stage is near 100%. Captive-raised hatchlings are fed a low-cost, high-protein diet and gain weight twice as fast as wild iguanas. An artificial nest developed by STRI to facilitate egg collection is preferred by both wild and captive females over nests of their own construction. Techniques developed at STRI ensure production of a predictable number of young iguanas and their regulated multiplication in captivity (Werner, 1986).

Several new directions, including studies of nutrition and disease control, as well as experimental releases of yearlings into the wild, are now being explored. In December 1985, reintroduction experiments using 1,200 7- to 10-month-old, captive-raised iguanas were begun in appropriate habitats in rural communities. Their survival from the age at which they were released appears to be comparable to that of wild-born animals. The repopulation experiment is accompanied by an education project designed to encourage rational resource management and to introduce the techniques to the rural communities. Results to date indicate that wild iguana populations can be reestablished in 3 years, after which harvesting can begin.

Our calculations show that meat production from iguanas matches that of cattle. Beef production may be higher in the first years after the forest is cleared to create pasture. However, the annual cattle yield drops to 15 kilograms per hectare after 10 to 15 years because the quality of the pasture is lower. By comparison, the results of the Iguana Management Project indicate that iguanas could provide a sustainable yield of more than 230 kilograms per hectare annually. Moreover, iguana management has the potential to maintain or to improve soils on land degraded by intensive cultivation or cattle ranching. Production costs are estimated at $0.66 per kilogram, and the meat is presently sold (albeit illegally) for between $1 and $6 per kilogram. A secondary benefit of iguana management is that simultaneous reforestation with high-quality lumber and fruit trees could also become attractive. Development of a reforestation scheme including plant species that also support iguana populations could yield a harvest of iguanas after 5 to 6 years and thus provide an early return that compensates for the reforestation investment.

CAPTIVE BREEDING OF THE PACA (*CUNICULUS PACA*)

The paca is a large nocturnal rodent related to the guinea pig and is native to the broad-leaved forests of Central and South America. Its meat is very highly prized by people of both rural and urban populations. An adult paca weighs about 10 kilograms, 60 to 70% of which is edible meat. Domestication of pacas could provide an inexpensive supply of high-quality meat for local consumption or for use as a cash crop. The purposes of this project are to evaluate the possibilities of breeding pacas in captivity and to develop techniques that rural people can use for captive management. Once regular captive breeding is established, the process of domestication should follow as a matter of course. Some characteristics of the paca favor domestication; some do not (Smythe, 1987). Precocial young undergo a period of early learning that makes individuals very easy to tame. In addition, the animals subsist on food that comes from the forest and thus would be inexpensive to acquire.

Among the characteristics that do not favor domestication are the fact that pacas naturally live in pairs that defend territories and are thus highly intolerant of other pacas. They are also strictly nocturnal, retiring to burrows or retreats, which are aggressively defended against other pacas or other potential intruders in the daytime. Furthermore, pacas generally produce only one or two young per year. (The fact that their natural reproductive *rate* is low does not necessarily indicate a low reproductive *potential*. Artificial selection to achieve the reproductive potential would be expected to occur early in the domestication process. The reproductive potential of a female paca should be between 12 and 16 per year.)

All young mammals are more socially tolerant than adults of the same species. By taking young pacas away from their mothers when they are only a few days old and raising them in contact with other pacas, animals have been taught to be nonaggressive toward other members of their own group. The goal is to raise them in groups of five females per male. At present, we have two groups of two females with one male, and two of three females with one male. None of these animals has shown any aggression toward other group members, and all are easier to handle than those raised by their mothers.

It might be expected that denial of a retreat to a naturally burrowing animal would give rise to pathological trauma, but researchers in comparative psychology have shown with other burrowing mammals (Price, 1984) that raising them in cages without any form of retreat causes the animals to mature faster and show less stress than those raised with retreats. Pacas raised without burrows are definitely more tranquil and are also more diurnal, which facilitates observation of their health and behavior. Psychologists have also demonstrated that handling makes an animal's subsequent behavior toward its handlers less aggressive. The different groups of pacas have been treated according to different regimens, and all are easier to handle than animals left alone. The ease of manipulation is inversely proportional to the age at which they were removed from their mothers. It is hoped that the young born to human-acclimated females will learn their attitudes from their parents.

Pacas born early in the project gained weight at an average rate of 13 grams per kilogram of body weight per day (g/kg bw/day) and reached a weight of 6 kilograms (approximately the weight at maturity) at 18 months. As a result of improved nutrition and a rigorous antiparasite program, those most recently born gained 20 g/kg bw/day, and some have reached 6 kilograms at 4 months. By comparison, cattle in temperate regions gain about 7 to 10 g/kg bw/day during the same age span (Seigmund, 1979).

Pacas are naturally frugivorous. The gathering of wild fruits is not only highly laborious but is also subject to the seasonality of production. Many forest leaves are as nutritious as fruit and may be higher in proteins. One of the benefits of hand-rearing pacas is that they can be taught to eat diets that they would not encounter in the natural state and that adults only learn to eat reluctantly. Experiments are presently under way to develop a diet that consists of a high proportion of easily obtained leaves supplemented with a readily available, aseasonal, high-carbohydrate food such as manioc (*Manihot esculentia*), which is locally called yucca.

Only a few wild species have been truly domesticated, and most of these have been social species such as dogs and ungulates. Thus, the attempt to domesticate (or even semidomesticate) the paca is faced with long odds. But none of the traditional domestic meat animals thrive in the lowland tropical areas; thus there is a need for a domestic species that will do well in the area. In the 3.5 years since the inception of this project, significant progress has been made.

FOREST GARDENING

This 3-year project has several objectives. Using species that are not traditional in Panama, STRI hopes to restore and maintain soil fertility by selecting and establishing highly adaptable, hardy, and competitive leguminous plant species with emphasis on multipurpose shrubs and trees that can be used for green manure, forage, firewood, and timber production; to evaluate the adaptability of potential food crops and of native fruit and leguminous trees to poor soil and other conditions found in small forest clearings; and to transfer field plots and plant material to Panamanian government agencies responsible for the development of agricultural technology to enable them to continue with the most promising results obtained by this project.

Soils in the project area are very poor, like much soil under tropical forests. They are characterized by very low to medium pH values (4.3–5.4); a low to high aluminum content (traces–7.4 ppm), and low to extremely low phosphorus concentrations (traces–9.5 ppm).

The quality of these soils did not seem to vary with successional stage of vegetation. Seeds of leguminous shrubs, trees, and vines with the required adaptation and performance potential (NRC, 1979) were obtained from several sources. Promising results in terms of vigor, competitiveness, and stress tolerance against drought or impeded drainage were obtained with fast-growing trees, e.g., Sabah salwood (Acacia mangium) in Malaysia, wattle (A. auriculiformis) in Papua New Guinea, and mata raton, madre del cacao (Gliricidia sepium) in Latin America; shrubs, e.g., Desmodium gyroides, Tephrosia candida, and Townsville clover (Stylosanthes guianensis) in Trinidad; and creeping vines, e.g., tropical kudzu (Pueraria phaseoloides) and Desmodium ovalifolium. Of these, only G. sepium and S. guianensis are native to the Neotropics.

The response of Acacia mangium was impressive under the local conditions; no species hitherto proposed for lumber production in similar soil in Panama even approaches the growth figures obtained with this species. The trees showed a striking response to the addition of phosphorus, and rhizobium-induced root nodules were large and prolific. A symbiotic relationship with mycorrhizal fungi was clearly indicated in growth trials using forest and pasture soils. The performance of A. mangium and of other leguminous species under severely adverse soil conditions opens up a wide range of options for the restoration and maintenance of soil fertility and for the development of agroforestry systems.

Results of trials in small forest clearings showed that the following species could constitute the core of a highly productive forest garden in tropical soils of low fertility: arrow root (Maranta arundinacea), tuber-bearing yam (Dioscorea bulbifera), Mexican yam bean (Pachyrhizus erosus), American peach palm (Bactris gasipaes), hardy banana clones (Musa spp.), a shade-tolerant, soil-covering vine (Desmodium ovalifolium), and leguminous shade trees (Erythrina spp. and Gliricidia sepium) (cf. Huxley, 1983).

MANAGEMENT OF GAME MAMMALS

Native populations in the Neotropical humid forests have traditionally relied on native wildlife for their high-quality protein. Since the advent of firearms and headlamps, more efficient hunting methods combined with habitat destruction have resulted in the extermination of game species in many areas. It is thus urgent to determine the sustainable yield of protein that can be obtained from these forests and to devise the means to manage the game populations appropriately.

In many areas, forest remains only in small patches such as on hilltops, along rivers, or in areas where the topography makes exploitation difficult. Populations of game animals are usually reduced in such areas, and since they are isolated, it is difficult or impossible for the game species to return. Wildlife populations could be maintained (or established) in these isolated patches, and in the absence of their natural predators (which the patches are too small to support), they could be cropped as a sustainable yield protein.

Most forest game animals are frugivorous, and as a result, they experience high juvenile mortality during seasonal scarcity of fruit (Smythe, 1986). Pre-Hispanic forest natives may have left waste crops in the fields, inadvertently or intentionally. Game animals that came to feed were exploited as a supplementary source of protein. The natives thus practiced provisioning of the population and may, in so doing, have reduced seasonal juvenile mortality of the game animals. A principal aim in this project is to develop methods to enhance juvenile survivorship in the field by provisioning in times of fruit scarcity.

In any wild animal management scheme, it is necessary to census the populations as accurately as possible. Since no single method of population censusing appears to be adequate in tropical forests, two different methods are used where possible. Populations of potential meat animals are being estimated with strip-census techniques in a 62-hectare peninsula adjacent to Barro Colorado Island, which is isolated from the contiguous forests by a 1-kilometer-long, five-strand electric fence. The animals being censused in this way are the agoutis (Dasyprocta punctata), pacas, spiny rats (Proechimys semispinosus), collared peccaries (Tayassu tajacu), and white-tailed deer (Odocoileus virginianus). The three species of rodents are also being censused in a trap-mark-release program. This program will be used to make more accurate estimates of the rodent populations but is confined to the period when food is scarce in the forest (between September and March) and animals are willing to enter traps.

The study area is divided into quadrants, one of which is artificially provisioned during the season of scarcity. The effect of the provisioning on the survival of juvenile animals is being determined by the censuses.

Progress in this project has been slower than in those discussed earlier, chiefly due to the unpredictable elements inherent in working with wild animals and the necessary restrictions on manipulation of predators or competing species imposed by working within a protected area. Furthermore, the electric fence is less effective than expected: predators have moved in and taken advantage of trapped animals, and unwanted animals have settled in the provisioned areas and used provisions intended for the targeted species. Nonetheless, much valuable knowledge is being gained.

STRI personnel undertook these projects several years ago with the full understanding that instant solutions to the problems that they were trying to solve were unlikely. But significant progress has been made in these high-risk ventures. Further progress, no matter how slow it is in coming, and the dedication necessary to achieve that progress are essential if conservation goals are to be achieved at the same time as the livelihood of the rural poor is improved.

REFERENCES

Burghardt, G. M., and A. S. Rand, eds. 1982. Iguanas of the World: Their Behavior, Ecology, and Conservation. Noyes, Park Ridge, N.J. 472 pp.

Huxley, P. A., ed. 1983. Plant Research and Agroforestry: Proceedings of a Consultative Meeting Held in Nairobi, 8 to 15 April, 1981. International Council for Research in Agroforestry, Nairobi, Kenya. 617 pp.

NRC (National Research Council). 1979. Tropical Legumes: Resources for the Future. National Academy of Sciences, Washington, D.C. 331 pp.

Price, E. O. 1984. Behavioral aspects of animal domestication. Q. Rev. Biol. 59(1):1–32.

Siegmund, O. H., ed. 1979. Nutrition. Pp. 1268–1269 in The Merck Veterinary Manual, Part V. Merck, Rahway, N.J.

Smythe, N. 1986. Competition and resource partitioning in the guild of Neotropical, terrestrial, frugivorous mammals. Annu. Rev. Ecol. Syst. 17:169–188.

Smythe, N. 1987. The paca (*Cuniculus paca*) as a domestic source of protein for the Neotropical, humid lowlands. Appl. Anim. Behav. Sci. 17(1–2):155–170.

Werner, D. I. 1986. Iguana management in Central America. BOSTID Dev. 6(1):1, 4–6.

BIOLOGICAL ENGINEERING FOR SUSTAINABLE BIOMASS PRODUCTION

SINYAN SHEN

Program Manager, Energy and Environmental Systems Division,
Argonne National Laboratory, Argonne, Illinois

Since the 1950s, the new science of molecular biology has produced a remarkable outpouring of new ideas and powerful techniques. From this revolution has sprung a new discipline called genetic engineering, which gives us the power to alter living organisms for important purposes in agriculture, energy, industry, and medicine. Applications of resulting technologies span the range from the ancient arts of fermentation to the most esoteric use of gene splicing and monoclonal antibodies. During the last decade, progress has continued with the rapid development of techniques to produce chemical feedstock using renewable sources of material. In particular, modern and traditional genetic engineering techniques have been combined with modern agricultural methods and chemical engineering know-how to produce high-volume biomass feedstocks at low cost for use as energy, chemicals, and building materials.

A recently evolved engineering discipline requires the design of biological systems that accumulate organic material photosynthetically more efficiently than those used in food production and with positive environmental impacts. In a fairly broad context, biological engineering can be defined as the engineering and use of biological systems in the production of goods and services for industry, trade, and commerce. The underlying fundamental scientific principles, engineering considerations, and government regulations dealing with the development and application of biological engineering all need to be addressed. Since this is a relatively new discipline, the best way to describe its integral components is through a number of examples. The examples discussed in this chapter include the engineering of biological production systems for biomass feedstocks with environmental benefits in mind.

Every region of the world has its own ways of managing its energy and material resources. The criteria for selecting the appropriate mixture of energy and material technologies are usually based on resource constraints imposed on the region and the structure of its industry, e.g., whether it is capital-intensive or labor-intensive. Fossil energy resources and mineral resources are accumulated over a very long time and can only be managed. Biomass resources, on the other hand, are composed of organic material accumulated by photosynthesis and can be enhanced and managed.

Very few things in this world have a global impact on us as great as the forest/grassland ecosystem. A balanced tree and grass ecosystem controls runoff, supplies water, and supports irrigation, soil fertility, and oxygen production. For every given ecological zone, there is a minimum ratio of tree acreage to food-crop acreage that must be maintained locally. Otherwise, the long-term productivity of land and the supply of water and nutrients cannot be maintained.

Since the introduction of large-scale mechanization, plant breeding, irrigation, and fertilization, there has been a tendency to move away from the ecological balance needed to sustain productivity. We often speak of land productivity, but we are really only talking about land fertility. The overall productivity can only be measured by considering the entire ecosystem, which supports the life of all zoological species, trees, legumes, and grasses. The study of genetics and plant breeding has resulted in the production of superior varieties of trees, legumes, and grasses in terms of product yields. To increase the economic output of agriculture and forestry, however, we have often selected ways to design production systems that are not compatible with long-term productivity.

The pursuit of short-term profitability at any price, justified by the technocratic concept of consumer needs, is leading to the replacement of genetically diversified natural forest, for example, by plantations consisting of row upon row of trees belonging to a very small number of different species. These species are chosen for their rate of growth and consist mainly of conifers, poplars, or eucalyptus. The complete lack of diversity in monoculture plantations makes them particularly vulnerable to attack by pests, that is, by pathogenic insects and fungi. Proliferation of bark beetles (family Scolytidae), for example, which is generally haphazard and localized among naturally diversified conifers, assumes disastrous proportions in spruce plantations that have spread so extensively over France during the last decades.

One of the results of the systematic introduction of monoculture plantations is a significant alteration of the soil structure, including a reduction in porosity of upper soil horizons, which makes them less and less permeable. Thus, there is a twofold or threefold increase in runoff. Loss by evaporation also increases, because pine needles retain a large amount of rainwater. Coniferous forests are therefore drier than broadleaved woodlands, other conditions being equal. Soil fauna is also affected. The number of earthworms under conifers is smaller by a factor of 100 to 500 than under broadleaved trees. Decomposers are also affected, and a marked reduction in bacteria occurs, especially in the microorganisms involved in the nitrogen cycle. The current practice of reforestation with conifers in temperate, and even in tropical, regions would seem to constitute an astonishing perseverance with mistaken methods on the part of the authorities responsible for forests in the countries involved.

There is nothing wrong with using genetically superior material, and there is also nothing wrong with increasing our economic output, as long as we fully understand what it takes ecologically to sustain long-term balance and productivity.

The horticultural agriculture of Asia provides some food for thought. The intensive agriculture of Eastern China, for example, has been sustained for thousands of years. The key to its success is genetic diversity in any one locality. With sufficient genetic diversity in the agriculture of any one locality and a balance of trees and grasses (most food crops are tall grasses), Eastern China maintains its ecological balance and long-term productivity. The indiscriminate exploitation of forests in parts of Northern China, however, has resulted in the rapid expansion of deserts there.

WOOD GRASS

High-yield, fast-growing multipurpose trees have been considered by many countries to increase both economic and natural resource development. A biologically engineered system developed in the United States produces a large volume of woody feedstock at low cost. Modern agricultural techniques are combined with our understanding of the tree growth to produce feedstocks that are conveniently handled in conversion processes. One biologically engineered system produces two varieties of wood grass products: the dry Dushen and the wet Dushen (Shen, 1982; Shen et al., 1984).

Dry Dushen is made by chopping the wood grass after it is crushed. It is a good combustion feedstock for use in thermochemical conversion processes such as combustion, liquefaction, gasification, and pyrolysis. Wet Dushen is made by chopping wood grass into 2.5-centimeter segments as the wood grass is harvested. This product makes an excellent feedstock for such biochemical conversion processes as acid hydrolysis, enzymatic hydrolysis, and anaerobic digestion. Wood grass was given that name because its stumpage is a thin woody material (the wood grass) and it is managed by techniques used for managing grasses. The production system is biologically engineered for maximum productivity and efficiency in land utilization. Optimal plant spacing is used for the product desired, and woody biomass is harvested on an annual, biannual, or seasonal basis.

In traditional forestry, tree seedlings are spaced relatively far apart. Thus, the trees are not biologically affected by each other until they are near the end of the rotation, which could be many years after planting. The efficiency of land use is therefore low. The wood grass system uses specially selected fast-growing and coppicing species. The coppicing capability removes the need for frequent replanting, and the fast-growing characteristics are crucial since only species with a sufficiently narrow coppice growth curve can be used in the wood grass system.

Agricultural management techniques are used, optimal fertilization schedules are followed, and the soil is enriched to compensate for nutrient removal. Irrigation may be needed in certain parts of the world during root establishment. The methods of planting include vegetative propagation with unrooted cuttings, direct seeding, and other methods. The completely engineered wood grass system includes planting and harvesting. For some countries that do not have mechanized agriculture, these activities will most likely be handled manually. In mechanized agriculture, manure spreaders have been effectively used to spread cuttings and tractors have been fitted with specially made sickle bars for use in harvesting.

Good-quality and low-cost structural building material can be made from wood grass by extrusion. In this process, the fresh lignin embodied in the wood grass is used as a primary binder. The term *wood grass 2 × 4* is applied to the product and is used generically to refer to the extruded structural material. The shape of the die can be designed so that a variety of irregular cross-section shapes can be made in different sizes. Wood grass 2 × 4 utilizes the original molecular structure of wood and fuses the boundaries of adjacent wood grass. This structural material resists bending and twisting, since all joints between adjacent pieces of wood grass need to be broken before such movements are possible. Furthermore, since the extrusion process works better with thin trees and fresh lignin, 1-year-old wood grass is ideal as its feedstock. Figure 43-1 shows the feedstock applications of wood grass. In addition to wood grass 2 × 4, products include liquid hydrocarbons, alcohols, other chemical feedstocks, and protein. The energy products include heat, electricity, low- and medium-Btu gas, pipeline gas, and liquid fuels. The wood grass production system also stabilizes soil—a tremendous implication for erosion control and watershed management in many countries.

The biological foundation of the wood grass production system is shown in Figure 43-2. The right-hand curve indicates growth on a typical plantation planted with a certain number of trees (N) per hectare. The tree is a fast-growing species and has a narrow growth curve, which is characterized by three distinct phases:

- an initial establishment period in which the trees are developing their root structures;
- a growth period in which the established trees undergo steady growth—a period recognized by an almost constant rate of growth; and
- a period in which the growth rate declines as a result of competition for sunlight and nutrients. This third phase is known as closure. A point near inflection of the growth curve indicates the onset of closure.

During closure, competition for sunlight and nutrients begins to result in a decreased mean annual increment of biomass. After the onset of closure, not only is the growth rate of each individual tree affected but the average mortality also increases. In this chapter, the term *closure* is used to encompass all the phenomena that affect the total biomass yield per hectare. The onset of closure is defined quantitatively as the point before which the total quantity of biomass in a given area is proportional to the number of trees in that area.

The left-hand curve in Figure 43-2 shows the growth curve if twice the number of trees (2N) are planted per hectare. Initially, the quantity of biomass from 2N trees is almost twice that from N trees, but closure begins earlier. Because of the limits on the ultimate productivity of the land, the maximum quantity of biomass produced from 2N trees appears to be similar to that achieved from N trees after total closure.

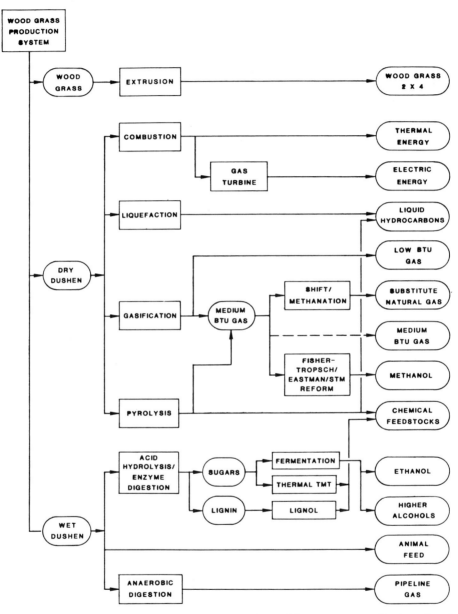

FIGURE 43-1   Feedstock applications of dry and wet Dushen and direct extrusion applications of wood grass.

Two observations can be made from Figure 43-2. First, closure begins sooner if tree density is increased. Second, the limit on the quantity of biomass that can be obtained can be ascertained from a growth curve for any density, provided N is not so low that closure does not take place. The implications of these observations are that the mean annual increment of biomass at harvest can be enhanced by increasing the planting density and that there is an optimal planting density for each rotation (age of trees at harvest), which can be determined.

For a short-rotation forest, the optimal harvesting time is some time before the onset of closure. The exact economically optimal harvesting time is a function of the discount rate, which determines the cost of money. Higher discount rates tend to favor an earlier harvesting age, since the value of future harvests is markedly reduced.

The decision to grow a particular tree as a crop can be hindered by many considerations. For example, tree growers usually cannot realize their revenue until several (or many) years after planting, and harvested forest land is expensive to clear. Thus, the use of land for wood production involves a long-term commitment. With the current analysis, however, appropriate planting density and production methods can be engineered and trees need not have these disadvantages.

Figure 43-3 shows the results of an experiment involving determination of the closure age of trees as a function of the number of trees planted (see Shen et al., 1984). As shown in the figure, if you want to harvest trees after 4 years to obtain a sufficiently large diameter, the planting density is about 1,700 trees per hectare. If you are harvesting every 3 years, you probably want to plant 6,000 to 7,400 trees. And if you are hoping to harvest every 2 years, you need a minimum of 25,000 trees. For an annual harvest, the planting density could be as high as 120,000 trees. And that's approaching the density of corn or rice.

The actual planting density should be based on two considerations. If one is not concerned about committing the land to trees, one could start with about one tree per thousand square centimeters. After the first cutting, the coppice ratio (the number of new stems per old stem) in the second spring could be around 5 to 1. And after the second cutting, the coppice ratio could be around 3 to 1. And the ratio approaches 1 to 1 after that. The desired steady-state wood grass density may

FIGURE 43-2 The onset of closure as a function of planting density for a typical plantation.

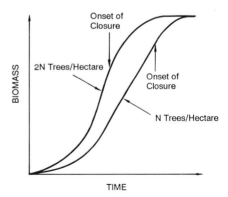

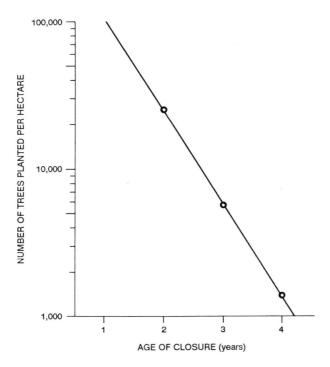

FIGURE 43-3 Closure age as a function of planting density in a forest located in the Great Lakes region of the United States.

thus be achieved after two or three cuttings. But by then the root system may be too large to be conveniently plowed under. If one is concerned about crop flexibility, the initial planting density should be chosen to control the size of the root system as well.

Wood grass production involves a degree of management as high as that in land consumption activities such as those found in urban areas or in association with roads, mining, and water impoundments. Wood grass is part of modern agriculture, except that in addition to its food and feed products, it produces energy and materials as well. Figure 43-4 shows wood grass grown at a density of one per thousand square centimeters for feed, electricity, and steam.

To determine the economics of wood grass production, we analyzed a plantation of wood grass (*Populus*) in the Pacific Northwest of the United States, allowing a range of uncertainties in the wood grass yield. The cost to produce the Dushen as a function of the discount rate (the cost of money) was estimated. At a 7% real interest rate, for instance, the range of uncertainties is from $12.00 to $25.00 per dry ton. At that rate, the average wood cost is $18.00 per ton, which is $1.00 per million Btu measured in energy value (Jones and Shen, 1982; Shen 1983; Shen et al., 1982; Vyas and Shen, 1982).

Which areas are best for growing wood grass? In the United States, the states bordering the five Great Lakes, the Eastern United States, and the Pacific North-

FIGURE 43-4 Wood grass in Livingston, California, grown as a low-cost source of protein for poultry feed and as a biomass fuel for cogeneration. Photo courtesy of Foster Farms.

west all have excellent soil and climate to support the wood grass production system. At the lower bound of the yield, wood grass competes economically in the Corn Belt at current prices. Its most important role, however, is its potential use in energy, materials, and feed markets of the world.

Due to differences in agricultural systems and practices and in topographical and climatic conditions, there are different schemes for producing Dushen feedstock using the wood grass production system. There are also considerations concerning capital versus labor intensity and availability. These can be illustrated by comparing Dushen production in two countries: the United States and China. Field preparation is mostly mechanical in both the United States and China. Planting is mechanical in the United States and manual in China. Figure 43-5 shows farmers planting wood grass at a density of four per thousand square centimeters in Southern China. In both countries, the cuttings are planted vertically or are placed horizontally on the ground and covered with a moisture-preserving mulch such as wheat straw. A manure spreader has proved effective in random horizontal planting. Vertical planting is used for rice seedlings or tree cuttings. Each cutting may produce more than one tree, depending on how many buds are present on the cutting and on the survival rate. The herbicide and fertilizer applications are mostly mechanical in the United States and manual in China. Both countries use chemicals and manure as fertilizer, but the Chinese use more manure, whereas chemicals predominate in the United States. Harvesting is mechanical in the United States and is still manual in China. The Dushen harvester chops wood grass directly into 2.5-centimeter pieces, the wet Dushen. The dry Dushen is produced if the wood grass is crushed with rollers before it is chopped in the harvesting process. A single harvester is needed to produce the Dushen.

The near-term uses of wood grass in the United States are heating, animal feed, industrial fuel, the coproduction of steam and electricity, and, most importantly, structural material—material that does not require 50 to 100 years of growing, such as pine trees. The mid-term applications include pipeline-quality gas production, which the Gas Research Institute (GRI) is developing, and liquid fuels. In his work on pipeline-quality gas performed for GRI, David Chynoweth of the University of Florida has found that wet Dushen approaches the performance of the most digestible cellulosic samples used as reference in his experiment. The long-term applications include its use as chemical feedstock, which has the highest market value.

In China, the regions of most intensive applications are located in the Northeast, where there is need for industrial fuel and heating, in Eastern China, and in Northwestern China. The near-term applications are cooking, heating, industrial

FIGURE 43-5 Establishment of a wood grass multipurpose plot in Fujian Province of Southern China. Chicken manure diluted with rice hull is used as fertilizer. The wood grass will be used as cattle feed. Photo courtesy of Yangting Farm.

fuel, cogeneration, animal feed, and construction materials. As in the United States, its long-term use is as chemical feedstock. There are tremendous possibilities here for the production of various materials, including building material, animal feed, and chemical feedstock.

At present, the Dushen products from the wood grass production system look attractive because of the system's rapid return on investment. In the long run, the soil conservation characteristics of the production system may be the most important benefit of wood grass. We recently analyzed the impact of large-scale production in the United States and found wood grass to be a significant soil-conserving crop (Shen and Turhollow, 1983). In contrast to conventional food crops, the wood grass system provides surface coverage throughout the year. In addition, the root system provides a soil-stabilizing matrix. Furthermore, selected wood grass configurations reduce groundwater runoff, increase groundwater infiltration, and recharge groundwater reservoirs.

In designing the wood grass production system, we first applied traditional genetic selection to hybrid trees, which were cloned for a number of attributes. We focused on a narrow growth curve and a rapid closure phase. The average growth curves of course are those under selected fertilization and other soil amendment schedules for a specific species-site match. For all clones selected for these two attributes, the coppice growth curves are examined, since sustained steady-state coppice growth is our objective function. In the next step, the selection criteria include disease resistance, drought resistance, and ease of establishment. Then, depending on the desired market, additional attributes directly related to product properties are identified. In practice, this last step is performed at a very early stage for general screening, because the objective of the biological engineering is really the product. On the other hand, without the growth attributes established in the genetic selection, the biological system has no basis.

Through the wood grass production system, we have shown that biological engineering with components based on genetic selection, species propagation techniques, modern agriculture, and chemical or mechanical processing methods could produce energy and novel materials that fit within the framework of the existing market infrastructure. This applied engineering approach will allow multidisciplinary teams to produce products and technologies that can be marketed more rapidly than most new technology, while taking environmental benefits into full consideration as a primary long-term objective.

CONTOUR HEDGEROWS

Contour hedgerows proposed for the highland regions of Nepal, Pakistan, India, and other countries (Benge, 1984) constitute a biologically engineered system for the production of reforestation planting stock. The Department of Soil and Water Conservation of Nepal estimates that between 30 and 75 tons of soil are washed away annually from each hectare of deforested land. This means that Nepal altogether loses as much as 249 million cubic meters of soil per year to India (Cool, 1980). The contour hedgerows alleviate the problem of the short supply of appropriate planting stock, and at the same time effectively check erosion, reduce

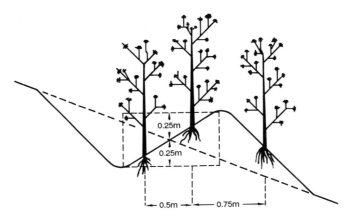

FIGURE 43-6   Cross-sectional view of hedgerows.

groundwater runoff, increase groundwater infiltration, recharge groundwater reservoirs, and produce the by-products of fuelwood and fodder.

A typical design of a clonal hedgerow system entails the construction of hillside ditches and the planting of two rows of trees or shrubs on the rise (bund) of the ditches, which follow the contour of the hillsides, and one row planted in the depression (Figure 43-6). The hillside ditches increase the system's effectiveness in controlling erosion while increasing water percolation into the soil. Soil erosion control is more effective when the spacing of the plants is alternated between hedgerows (Figure 43-7). Soil could still be eroded from around the stems. It is desirable to place some type of biomass uphill and behind the live stems to make a more effective barrier to hold the soil and accelerate the formation of natural terraces. Rows of plants on the rise and in the ditch would provide greater access to the forage by browsing animals, if the functions of plant material production and the cut-and-carry system of forage production were phased into a browse-pasture system.

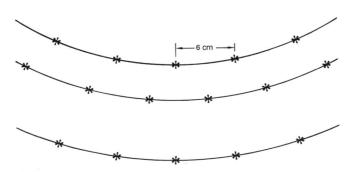

FIGURE 43-7   The spacing of plants is alternated between hedgerows.

According to Benge (1984), 1 hectare of contour hedgerows could provide enough plant material to reforest approximately 600 hectares (at a spacing of 4 meters between trees) the first year and 3,000 hectares the second year. To arrive at this production figure, a 6-centimeter in-row spacing of cuttings, a distance between rows of 0.5 meter and 0.75 meter, respectively (center to center—3 rows), and a spacing between hedgerows of 5 meters (edge to edge) was used (Figure 43-8). The production capacity of this contour hedgerow system used for clonal propagation of planting stock could be as high as 450,000 cuttings per hectare per year. The actual per-hectare production rate for cuttings from contour hedgerows depends upon the genetic capacity of the clones as well as the environmental conditions in which they are grown. The system benefits from high in-row density and edge effect, i.e., plants near the edge of each hedgerow benefit from additional sunlight and nutrients outside the hedgerow.

It is important to incorporate genetic diversity in contour hedgerow systems that cover large areas by the use of a wide variety of species or provenances in order to decrease the possibility of widespread disease and insect infestations and the production losses that would result. Many tree species that have been propagated from cuttings either will not develop a tap root or will develop a modified tap root with extensive lateral roots but without deep penetration into the subsoil. Therefore, these trees will draw nutrients only from the upper soil strata, which may already be nutrient poor from the mining effect of many food crops and other surface feeding plants. Thus, it may be desirable to plant between the contour hedgerows trees that have been propagated from seeds and that develop a deeper tap root system (Figure 43-8). This would increase the per-hectare productivity of the

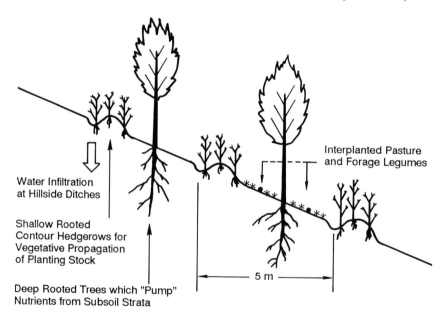

FIGURE 43-8   Complete biologically engineered system of contour hedgerows.

system, since these trees would pump from a deeper soil strata nutrients that may not be available in the shallower soil.

REFERENCES

Benge, M. D. 1984. Contour hedgerows for soil erosion control and planting stock, forage and fuelwood production in highland regions. Pp. 117–123 in Energy from Biomass: Building on a Generic Technology Base. Proceedings of the Second Technical Review Meeting. Report No. ANL/CNSV-TM–146. Argonne National Laboratory, Argonne, Ill.

Cool, J. C. 1980. Stability and Survival—The Himalayan Challenge. Ford Foundation, New York. 22 pp.

Jones, P. C., and S. Y. Shen. 1982. A Framework for Evaluating the Economics of Short-Rotation Forestry Research and Development. Report No. ANL/CNSV–35. Argonne National Laboratory, Argonne, Ill. 56 pp.

Shen, S. Y. 1982. Wood Grass Production Systems for Biomass. Paper presented at the 1982 Midwest Forest Economist Meeting, Duluth, Minn., August 17–19, 1982.

Shen, S. Y. 1983. Regional Economic Impacts of Woody and Herbaceous Biomass Production. Paper presented at the 1983 Joint National Meeting of the Institute of Management Science and the Operations Research Society of America, Chicago, April 24–27, 1983.

Shen, S. Y., and A. F. Turhollow. 1983. Regional impacts of herbaceous and woody biomass production on U.S. agriculture. Pp. 207–234 in Symposium Papers: Energy from Biomass and Wastes VII. Presented January 24–28, 1983, Lake Buena Vista, Fla. Institute of Gas Technology, Chicago.

Shen, S. Y., P. C. Jones, and A. D. Vyas. 1982. Economic Analysis of Short-Rotation Forestry. Paper presented at the 1982 Joint National Meeting of the Operations Research Society of America and the Institute of Management Sciences, San Diego, Calif. October 1982.

Shen, S. Y., A. D. Vyas, and P. C. Jones. 1984. Economic analysis of short and ultra-short rotation forestry. Resour. Conserv. 10:255–270.

Vyas, A.D., and S. Y. Shen. 1982. Analysis of Short-Rotation Forests Using the Argonne Model for Selecting Economic Strategy (MOSES). Report No. ANL/CNSV–36. Argonne National Laboratory, Argonne, Ill. 50 pp.

POLICIES TO PROTECT DIVERSITY

**Tropical Forests:
A Call for Action**

United Nations Environment Programme   UNEP

**"PARTNERSHIP AMONG PEOPLE
TO REPLENISH THE EARTH"**
STATEMENT
BY
DR. MOSTAFA KAMAL TOLBA
R SECRETARY GENERAL OF THE UNITED NATIONS
AND
EXECUTIVE DIRECTOR
INITED NATIONS ENVIRONMENT PROGRAMME
FOR
JAL SOCIETY OF AMERICAN FORESTERS MEETING
BIRMINGHAM, ALABAMA

**WORLD
CONSERVATION
STRATEGY**
Living Resource Conservation
for Sustainable Development

Prepared by the International Union for
Conservation of Nature and Natural Resources (IUCN)

with the advice, cooperation and financial assistance of
the United Nations Environment Programme (UNEP)
and the World Wildlife Fund (WWF)

and in collaboration with
the Food and Agriculture Organization of the
United Nations (FAO)
and the United Nations Educational, Scientific
and Cultural Organization (Unesco)

**GLOBAL RESOURCES, ENVIRONMENT,
AND
POPULATION ACT OF 1983**

HEARING
BEFORE THE
SUBCOMMITTEE ON CENSUS AND POPULATION
OF THE
**COMMITTEE ON
POST OFFICE AND CIVIL SERVICE
HOUSE OF REPRESENTATIVES**
NINETY-EIGHTH CONGRESS
SECOND SESSION
ON
H.R. 2491
A BILL TO ESTABLISH IN THE FEDERAL GOVERNMENT A GLOBAL
FORESIGHT CAPABILITY WITH RESPECT TO NATURAL RESOURCES,
THE ENVIRONMENT, AND POPULATION; TO ESTABLISH A NATIONAL
POPULATION POLICY; TO ESTABLISH AN INTERAGENCY COUNCIL ON
GLOBAL RESOURCES, ENVIRONMENT, AND POPULATION; AND FOR
OTHER PURPOSES

JULY 26, 1984

Serial No. 98–49

Printed for the use of the Committee on Post Office and Civil Service

U.S. GOVERNMENT PRINTING OFFICE
38-566 O   WASHINGTON : 1984

Examples of documents that address global
environmental problems, including the
preservation of biodiversity.

PRESERVING BIOLOGICAL DIVERSITY IN THE TROPICAL FORESTS OF THE ASIAN REGION

JOHN SPEARS
Forestry Adviser, World Bank, Washington, D.C.

A range of agriculture, forestry, fiscal, and other policy initiatives will be needed if the Asian Region is to preserve a significant proportion of its remaining 300 million hectares of biologically diverse tropical forests and to sustain a timber industry beyond the year 2000. These forests have been reduced by one-half since the turn of the century. To contain this forest encroachment effectively, a multidisciplinary approach and top-level political commitment will be needed.

STATUS OF THE REGION'S CLOSED FORESTS

Closed forests, i.e., tropical forests with a closed canopy as opposed to open grasslands savannah, cover 300 million hectares and account for about one-third of the land area of Asia. About 20% of the forest area has been logged over. Approximately 500 million cubic meters of fuelwood and 100 million cubic meters of industrial wood are produced each year. The forest industries of the Region annually generate more than US$5 billion in foreign-exchange earnings.

Wood removals in most countries exceed the sustainable yield of the forests, and investment in forest management and reforestation is annually running less than one-third that needed to replace what is being cut. According to the Food and Agriculture Organization (FAO/UNEP, 1981), only 13% of the forest area is being managed for sustained yield. Annually, some 1.8 million hectares are being lost to agriculture. In some countries of the Region, such as Nepal and Bangladesh, virtually all the natural forests will disappear before year 2000 unless swift remedial action is taken.

The principal causes of deforestation include the following:

- increasing population pressure and the need for additional land for cultivation;
- land ownership patterns that force peasant families and landless people into forests and marginal lands;
- commercial agriculture operations (particularly plantation agriculture for crops such as palm oil, rubber, and coconut); and
- commercial logging, which opens up previously inaccessible forests to cultivation and fuelwood harvesting at a rate far exceeding the regenerative capacity of the forests.

SOLUTIONS TO DEFORESTATION

Some of the solutions to deforestation will have to come from outside the forestry sector. The three key areas for intervention are:

- measures aimed at increasing agricultural productivity and providing the Region's 100-million forest dwellers and people living adjacent to the forests with an alternative to further forest encroachment;
- intensification of forest management and creation of compensatory plantations of fast-growing species that can provide an alternative to continued exploitation of natural forests; and
- a vigorous forest-conservation policy that will set aside substantial areas of the remaining tropical forests as ecological reserves to be protected from all forms of encroachment.

Intensification of Agricultural Productivity

Support for land reform and land-titling programs that address the issues of inequitable land distribution and encourage a more permanent and sustainable agriculture could do much to relieve pressure on forest land. A government assistance program is under way in northern Thailand to achieve this. Intensification of perennial tree crop yields (mainly oil palm, rubber, and coconut) on plantations that cover some 20 million hectares in the Asian Region is receiving high priority in the agricultural development plans of Indonesia, Malaysia, and the Philippines. While from one point of view it can be argued that the establishment of such plantations has contributed to deforestation, there are some positive aspects to this development. Malaysian experience, such as that in the Jengka Triangle, has demonstrated that well-managed agricultural tree crops can provide attractive income for settled families and help to reduce dependence on shifting agriculture. From an ecological viewpoint, such tree crops do a good job in protecting soil and water resources. A well-thought-out government land-use plan for that region carried out in the 1960s demarcated more favorable bottom lands for agriculture and set aside some 60% of the area as forest reserves. After some 20 years, the village population of the Jengka region remains relatively stable. Furthermore, the forests originally excluded from settlement have been protected and are still there today.

Market constraints for commodities such as palm oil and rubber will curtail future expansion, and it seems unlikely that by the year 2000 such agricultural plantations could exceed 30 million hectares (10% of the remaining closed-forest area of the region). By contrast, the planting of perennial tree crops in settlement projects such as the Indonesia Transmigration program, in which there is a much higher degree of dependence on annual food cropping, are experiencing considerable problems. A high-priority area for the future is more intensive research and field-scale trials of new technologies for sustaining annual food cropping in the acid latosols (bleached red and yellow tropical soils) that underlie rain-forest lands. Particularly important are agroforestry techniques such as alley-cropping (a system of growing crops interspersed with lines of fast growing leguminous woody species), which can reduce dependence on artificial fertilizers, and zero tillage technology, which leads to the retention of organic matter in soil.

Another high-priority area for research is the food-cropping potential of the thousands of unresearched plants that grow within the tropical rain forests. Such research could help to widen the range of food crops available for indigenous consumption. A good example is the winged bean (*Psophocarpus tetragonolobus*). Although this plant has been known for centuries to the forest tribes of New Guinea, it was hardly recognized elsewhere. Research has demonstrated that the plant has a nutritional value equivalent to soybean (40% protein and 17% edible oil). It is now being cultivated for food production in some 50 developing countries.

Intensification of Forest Productivity and Reforestation

Of the more than 200 potential timber species occurring in the natural forests of Asia, more than two-thirds are currently regarded as weed species of little commercial value. Intensive research and market promotion have done much to introduce some of the lesser known species to the market and could do much more. Between 1977 and 1981, for example, the utilization of lesser known species more than doubled in Peninsular Malaysia. They now account for 27% of the log intake of plywood/veneer mills.

Investment in more intensive natural forest management and, in particular, the establishment of compensatory plantations of fast-growing tree species can also help to take the pressure off natural forests. Species such as *Gmelina arborea*, *Albizzia falcataria*, *Leucaena* spp., and, in appropriate locations, *Eucalyptus* and *Pinus caribaea* grow at 4 to 5 times the rate of slower-growing indigenous forest trees. Fuelwood crops mature in 10 years, and timber, in 20 to 25 years, compared with 60- to 80-year rotations more typical of the natural forest species. All the needs for industrial wood in the region by the year 2000 could in theory be supplied from fast-growing plantations covering about 25 million hectares, i.e., less than 10% of the remaining forest area. In the region as a whole, about 2 million hectares of fast-growing industrial plantations have been established. Burma, Indonesia, and India have substantial ongoing programs. Malaysia, the Philippines, and Thailand are preparing for an increased rate of planting.

Support for social and agroforestry programs outside the forests can help to provide farmers and village communities with the fuelwood, poles, fodder, and other forest products they need for their survival, thereby reducing the pressure on natural forests. Countries such as China, Korea, and India have been in the forefront of such programs. In the State of Gujarat in India, for example, schools and local farmers have played a lead role in the production of more than 600 million tree seedlings, which have been distributed in the last 5 years.

A most promising area for increased support is research into technologies that have the potential for increasing productivity of the principal tree species being planted in the Region. Tree improvement and breeding programs can more than double the yield of natural forest trees. Pakistan, China, and India have achieved spectacular results with species of poplar. In the Philippines, a giant species of *Leucaena* produced more than 3 times the yield of indigenous stocks. Participation at a recent International Union of Forest Research Organization workshop identified 10 multipurpose species that will be given high priority in future tree improvement work.

Forest Conservation Policies

About 1.8 million hectares (some 6% of the Region's forest) have already been set aside as national parks or nature reserves. This is nowhere near sufficient to ensure preservation of the unique germplasm and wildlife resources of the Region or the survival of the many endangered species. Nor is it sufficient to ensure that the natural forests of the Region continue to play a vital role in protecting soil and water resources on which downstream farmers depend for both irrigation and drinking water. According to a recent World Resources Institute study (WRI et al., 1985), some 50 million hectares of deforested watersheds in the Region are in urgent need of rehabilitation.

Creation of interministerial land-use boards with enough political clout to direct future agricultural settlement away from threatened forest areas is a key policy issue for the future. Peninsular Malaysia has led the way. Achieving such effective coordination is frequently hampered by localized political and vested financial interests. It requires top-level political commitment to follow through on what are often unpopular decisions that restrain peoples' access to forest lands and reduce the level of potential profit to timber enterprises.

The introduction of improved mechanisms for consulting with forest-dwelling communities on land-use issues and forest reservation policies are a feature of government policy in northern Thailand. More specific measures need to be taken to involve local communities throughout the Region, especially to involve non-government organizations and environmental agencies in national land-use policy planning and in the implementation of development programs. Recent research by the Government of Indonesia with assistance from the International Institute for Environment and Development is moving in that direction.

Considerable further strengthening of the agriculture, forestry, and national park administrations in the Region is required to enable them to intensify scientific inventory research in national parks and nature reserves, in order to develop parks as a tangible source of revenue and recreation outlet both for local and overseas tourists.

POLICY REFORMS

Raising political awareness and mobilizing sources of support for forest conservation and development will achieve little unless at the same time high priority is given to the fundamental policy reforms that are needed to ensure natural resource conservation. Among the more important needs for the immediate future are the following:

• Strong political commitment to land reform and redistribution policies by national governments of the Region, particularly in areas adjacent to threatened forest lands. Land consolidation and titling programs encourage a shift to more stable agriculture.

• Strong commitment by agricultural ministries to the diversion of a specific part of agricultural development plan resources to the development of intensified agriculture buffer zones around the most imminently threatened forest lands and to the rehabilitation of degraded watersheds. In India under the late Mrs. Ghandi's leadership, steps were taken to increase the proportion of the agricultural budget devoted to rural afforestation and to upland watershed rehabilitation in the Himalayan range.

• The commitment of log-exporting countries to the raising of timber taxes to

TABLE 44-1   Strategies for Relieving Pressure on Tropical Forests of the Asian Region

Strategy	Examples of World Bank Forestry Projects Applying This Strategy
Reservation of discrete areas of tropical rain forests as ecological reserves or national parks using World Bank lending policy as a lever to encourage more systematic reservation policies.	Indonesia, Dumoga Bone National Park
Agriculture/buffer zone development that can provide shifting cultivators or small farmers with an alternative to forest encroachment.	Malaysia, Jengka oil palm
Allocation of land title or leasehold rights plus credits, submitted seedling, or other incentives to encourage people living in or adjacent to forest lands to take up cash crop tree farming.	Philippines, smallholder tree farming
Planting of fast-growing industrial plantation trees that can provide an alternative source of wood supply to the natural forest and thereby reduce pressure for opening up new logging concessions.	Burma, Forestry Extraction and Management
Involvement of local communities in natural forest management.	Nepal, Terai Forest Development
Social/agroforestry schemes to ensure farmers and local communities an adequate supply of fuelwood, fodder, poles, etc., and to reduce dependence on natural forests.	India, Gujarat Forestry
Strengthening of forest administrations, particularly in the area of resource conservation and protection.	Sri Lanka Forest Development

something more in line with the timber's real economic value. The introduction of fiscal and incentive policies that would encourage greater private sector participation in forest management and reforestation and the introduction of measures to ensure that a higher proportion of forest taxes and revenues are returned to forest management.

In many of the countries of the Region, overly generous past timber concessions and fiscal incentives have encouraged wasteful timber exploitation and accelerated destruction of the forests. They encourage high grading of valuable species and discourage investment in reforestation. Adjustment of timber taxation and fiscal policies in most countries of the Region is an essential complementary measure in addition to direct investment in forest-development programs if currently extensive and wasteful harvesting are to be brought under control.

ROLE OF THE DEVELOPMENT AGENCIES

The development agencies, particularly within the Region of the Asian Development Bank, as well as United Nations Development Programme, the Food and Agriculture Organization, the World Bank, and bilateral assistance agencies, have been supporting a range of initiatives covering some of the above topics. Examples of several forest conservation and development projects financed by the World Bank are given in Tables 44-1 and 44-2.

Greater support is needed from the development agencies to set aside specific tracts of natural forests as protected areas or national parks as an integral part of their agriculture and forest lending policies. To date, the majority of such aid-agency schemes have been project-oriented, affecting one small part of the overall problem and often confined to a small geographic region. To play a more effective role in this area, the development agencies need to shift toward better-coordinated lending programs incorporating policy measures such as those cited above as a condition of their development aid support.

REFERENCES

FAO/UNEP (Food and Agriculture Organization/United Nations Environment Program). 1981. Tropical Forests Resources Assessment Project: Tropical Africa, Tropical Asia, Tropical America. Food and Agriculture Organization of the United Nations, Rome. 3 volumes.

WRI (World Resources Institute), The World Bank, and United Nations Development Programme. 1985. Tropical Forests: A Call for Action. World Resources Institute, Washington, D.C. 3 vols.

TABLE 44-2 Examples of World Bank-Financed Forestry/Agricultural Projects That Could Help to Relieve Pressure on Tropical Moist Forests

Country	Summary Project Description	Year of Loan/Credit and Amount (US$)	How the Project Has Relieved (or Was Intended to Relieve) Pressure on Tropical Forests and Reasons for Success	Ranking of Effectiveness to Date[a]	Comment
Indonesia	Setting aside and protecting 300,000 hectares of Dumoga Bone National Park as an integral component of an irrigation project.	1984; US$1 million (national park component only). This represents less than 2% of total cost.	Reservation of an area of forest that will be protected and managed as national park. Government of Indonesia strongly supportive of reservation policies and committed to increasing area of protected reserves.	1	None
Malaysia	Involvement of about 9,000 families in an oil palm and rubber development in the Jengka Triangle.	1968; US$20 million	Careful land-use planning prior to the project demarcated 60% of the project area as permanent forest reserves and succeeded in channeling agricultural development to flatter terrain and higher potential soils. Use of a perennial agricultural tree crop (palm oil) provides effective soil and water protection. Oil palm has proved a profitable cash crop for settlers, and in first-phase development involving	1	Project design could have been improved by including provisions for scientific inventory of the resource area and evaluation of critical areas of unique wildlife or plant species that were deserving of special protection.

cont'd

TABLE 44-2 Continued

Country	Summary Project Description	Year of Loan/Credit and Amount (US$)	How the Project Has Relieved (or Was Intended to Relieve) Pressure on Tropical Forests and Reasons for Success	Ranking of Effectiveness to Date[a]	Comment
			9,000 settlers, there has been less than a 2% turnover in settlement villages.		
Philippines	Smallholder tree farming. Involvement of some 8,000 smallholders in growing pulpwood and other forest products as a cash crop.	1973; US$10 million	Government was prepared to consider giving security of land tenure to people formally regarded as illegal squatters on forest land. The local pulp mill guaranteed an attractive price for wood and provided seedlings and extension advice. The highly productive, intensive pulpwood tree farms have avoided the necessity of cutting an area of natural tropical forest 20 to 25 times larger to obtain the same yield.	2	A typhoon wiped out some of the plantations. (Replanting is in progress.) Replicability of this project is highly sensitive to transport distance from markets.
Burma	As an integral component of the teak harvesting project, financing of improved forest management and	1980; US$35 million	Government is committed to a policy of forest protection and management. Technical package for teak	1	None

cont'd

			replanting of some 15,000 hectares of teak plantations.	plantation establishment was well proven prior to project development. Planting rate has exceeded targets. Program is being expanded to a second phase.		Scale of operation so far completed is very small in relation to the magnitude of the problem. The entire program envisages at least 1.5 million hectares being handed over to local panchayats.
Nepal	As part of a community forestry project, handing over of some 8,000 hectares of government-controlled forest to local panchayats (village councils) for protection and management.	1980; US$17 million		Local communities have responded to this initiative and are beginning to organize protection and management of forest lands. Some 7,000 hectares have been reforested.	2	None
India, Gujarat State	As part of a social forestry program, involvement of local families and communities in restoration of forest cover on degraded forest land. The scheme was started in 1980. By 1982, some 18,000 hectares of degraded forest had been completed under this scheme.	1980; US$37 million		Strong government commitment to the scheme. A flexible forest service, which has shifted more than 60% of its resources to promoting social forestry through a variety of farm forestry, village woodlots, and forest protection programs. Conditions of wood scarcity and rising prices for fuelwood and other products have made	1	

TABLE 44-2 Continued

Country	Summary Project Description	Year of Loan/Credit and Amount (US$)	How the Project Has Relieved (or Was Intended to Relieve) Pressure on Tropical Forests and Reasons for Success	Ranking of Effectiveness to Date[a]	Comment
			forestry a profitable cash crop for small farmers and local communities.		
Sri Lanka	Strengthening of the Forest Department and preparation of a master plan for long-term forest conservation and development.	1983; US$9 million	Government strongly supported this initiative. Forest Department training now includes more emphasis on conservation and natural forest protection.	2	Bank support to date has been heavily oriented toward plantation establishment. No systematic funding has been provided for protection and management of the remaining natural forests. Phase II project currently under preparation will begin to address some of these deficiencies.

[a]1, effective; 2, partially effective; 3, ineffective.

THE TROPICAL FORESTRY ACTION PLAN
Recent Progress and New Initiatives

F. WILLIAM BURLEY

Senior Associate, World Resources Institute, Washington, D.C.

Approximately 50 acres (20 hectares) of tropical forest are being converted or destroyed every minute (FAO/UNEP, 1981; Myers, 1980; Sommer, 1976). This is roughly equivalent to 400 square miles (1,040 square kilometers) of forest lost in less time than it took to present all the papers at the National Forum on BioDiversity.

For every 10 acres (4 hectares) of forest lost in the tropics, less than 1 acre (0.4 hectare) is developed as plantation forest or reforested in any way (OTA, 1984). In all regions, the rate of replacement falls far short of the rate of exploitation.

Deteriorating economies and international debt force many tropical countries to literally mine their forests and other biological capital rather than to treat them as renewable resources. The need to improve social conditions and to go after short-term gain simply overwhelms them.

The quality of the daily lives of more and more people in the tropics is getting worse, especially in rural areas (Asian Development Bank, 1980; World Bank, 1978). For example, demand for fuelwood is rising dramatically. Already, 70% of the people in developing countries, most of whom live in rural areas, depend on fuelwood to meet their household needs (WRI and ITED, 1986). Yet population growth cancels many of the gains in production. How can we expect to solve fuelwood needs in Kenya with projects that double fuelwood supplies in 20 years when population there is doubling every 17 years?

REASONS TO BE OPTIMISTIC

Despite all this, there are reasons to be optimistic. I believe we are reaching a critical mass to solve this global problem, and there are many encouraging prospects.

Compared to only 5 years ago, the public is much more aware of the value of conserving biological diversity and the need to stop the deterioration of major ecosystems. This growing awareness reaches into all levels of government and the private sector. Recent campaigns in Europe, North America, Brazil, and Indonesia have been very effective in bringing these issues to the public's attention. But we need similar movements in *all* countries.

TROPICAL FORESTRY ACTION PLAN

There have been several proposals to stop destruction of tropical forests. A few years ago, for example, Rubinoff (1983) proposed a scheme in which major development assistance would flow from developed countries into the tropics to conserve and protect tropical forests. Similarly, Guppy (1984) recently proposed an Organization of Timber Exporting Countries—a plan to force timber prices higher and to account for the real costs of managing the global timber resource sustainably.

In 1985, a global Tropical Forestry Action Plan (FAO, 1985; WRI et al., 1985) was developed jointly by the UN Food and Agriculture Organization, the development agencies, nongovernment organizations (NGOs), and representatives from more than 60 countries in which most of the tropical forests are found. This plan looks at the solutions needed in five areas: fuelwood and social forestry, conservation of forest ecosystems, and institution building, which includes research, education, and training. For 56 tropical countries, specific recommendations were made for the types of projects needed to slow or stop deforestation. The Action Plan has gained considerable international attention, and although it is far too early to judge its success, there are several very encouraging signs that at last we may have a concerted international effort to meet the problem.

The Plan recommends that $8 billion be spent in the areas just mentioned over the next 5 years to substantially reduce or eliminate forest loss in the tropics. This figure may appear enormous, but it would be only a doubling of current levels of developing assistance in forestry and related fields of agriculture. There are signs the agencies already are moving. For the first time, the development agencies are coordinating their grants and loans in forestry. This reduces waste and duplication in development assistance, but more important, it also is resulting in more funding for forestry and conservation projects. As part of this effort, Forestry Sector Reviews are now planned for 30 tropical nations in the next 23 years. These reviews already are under way in 11 countries, and reviews for the Sudan, Kenya, and Ghana have already been completed.

National Forestry Plans are being written or revised in more than a dozen countries with technical and financial help from the international development agencies. By July 1986, forestry and agriculture projects designed to meet the demand for forest products while also conserving the remaining tropical forest were being planned and developed in over 24 countries.

In July 1987, a high-level forestry conference of more than 15 nations was held in Bellagio, Italy. The goal is to get commitments to solve the deforestation crisis and to put tropical forest management and exploitation on a sustainable basis in those countries. Since the release of the Action Plan in October 1985, the governments of France, the Federal Republic of Germany, and the Netherlands already have committed to doubling their bilateral development assistance in forestry in the next 3 years.

The development agencies often are criticized for being as much a part of the problem as a part of the solution. Yet they are responding in a very tangible way to the implementation of the global Action Plan. For example, the World Bank has revised and expanded its entire African loan portfolio in forestry in line with the Action Plan recommendations.

In the Ivory Coast, the Bank is helping forestry officials to identify and map the remaining tropical moist forests. In this case, more than a hundred individual forest management plans are being developed, and they include major conservation zones and agricultural development to relieve exploitation pressures on the remaining forests.

The World Bank recently developed an innovative set of policy guidelines for wildlands. If these are carefully implemented, they will go far in improving the environmental aspects of development projects funded by the Bank. The other development banks and bilateral agencies such as the U.S. Agency for International Development and the Canadian International Development Agency (CIDA) are moving in similar ways to implement the Tropical Forestry Action Plan. Not only are fuelwood and industrial forestry projects being developed, but several of these agencies are now financing conservation projects, including such traditional activities as establishing parks and developing management plans for critical wildlife species. The European development agencies have taken the lead in this.

International and local nongovernmental organizations are also becoming much more involved. More than 5,000 forestry and conservation NGOs exist worldwide. Many are working not only to conserve biological diversity but also to develop forestry and agriculture so that critical forest ecosystems are protected while at the same time local demands for wood products are met.

SUCCESSFUL DEVELOPMENT PROJECTS

One underlying assumption in the Action Plan is that there are some good examples of successful projects in forestry and agriculture. But taken together, they represent far too small a response to the deforestation crisis. Four of these projects are listed below:

• The Dumoga Bone National Park in Indonesia is a major watershed being protected and managed as a park primarily to ensure the success of a large hydroelectric project downstream. In this case, a good deal of forest and its biological diversity is protected because it contributes to development nearby.

• Brazil recently developed an impressive system of parks and other conservation areas that cover nearly 15 million hectares. These areas include good examples of many of Brazil's major ecosystem types, and the government now has plans to establish other areas to fill the gaps in ecosystem coverage. International development assistance is helping to build this system.

• Brazilian rubber tappers are rapidly mobilizing to protect large areas in the Amazon Basin as extractive reserves. In this way they can work with the government to conserve major areas of tropical forest while continuing to maintain their livelihood.

• China has replanted 30 million hectares with trees—an area more than double that established for fuelwood and tree plantations in all other developing countries. Although it is too early to judge the real success of this example, it shows what can be done in a short period given sufficient national resolve.

We need hundreds more of these examples. With only slight changes, many of them could be repeated elsewhere. But the information on how to conduct these projects needs to get out—it needs to be available to governments and NGOs everywhere. And of course the financing needed to get them started must begin to flow.

WHAT ARE SOME OF THE SOLUTIONS?

First, agriculture must be improved and intensified in many ways all over the world. Most losses of tropical forest can be attributed to land clearance by millions of families who are simply trying to eke out a living as you and I would do in their places (Furtado, 1986). Until they can get enough forest products to meet their daily needs, absolutely nothing will stop them from moving further into the forests. There are examples all over the world where conservation areas are threatened by people who need food, fuel, or simply a little more space to put in a crop.

Some of the more innovative agricultural technologies should be tried on a large scale. Efficient agriculture and agroforestry techniques, some new and others ancient, must become easily available everywhere in the tropics if the landscape is not to be turned into unproductive wasteland.

Second, forestry and forest management need to change in many ways (Whitmore, 1984). These changes are detailed in the Tropical Forestry Action Plan, and many forestry projects are incorporating them already. Management of natural forest for timber *and* for nonwood products (e.g., rattan, medicinal plants, wildlife) must be developed and used more widely. Plantation forestry should not be rejected, as some naively suggest, but it must be improved and expanded. Small, well-managed plantations can take much of the pressure off natural forests, and there are good examples of this in Kenya, Chile, and Thailand. Deforested areas must be replanted, and degraded lands must be rehabilitated. What we really need is a universal tree-planting ethic, a second but different type of Green Revolution, and this has already begun in India, Indonesia, and Colombia.

Third, conservation in the traditional sense must receive more attention and funding. The global system of conservation units must be much larger and better designed. Those of us involved in developing this system must be more flexible and innovative in how it is designed to conserve biological diversity. Larger, multiple-use conservation areas in which the local people can participate by managing and exploiting the resource should become a standard part of the conservation arsenal and land-use planning.

Fourth, and last, economic and fiscal policies and incentives must be changed to reduce or eliminate the forces we all know exacerbate loss of forests and biological diversity. For example, most developing countries with large mature forests have failed to adopt revenue systems that come close to capturing real timber values for the public treasury. Fees charged the timber harvester typically represent a very low percentage of the real cost of replacement and forest management. In the Philippines, for example, this is an incredibly low 10%.

Other policies relating to transmigration and cattle ranching often result in major habitat disruption. If developing nations continue to encourage these projects, then the least we can do is to make known the real costs of environmental degradation and lost future harvests, in addition to whatever values we can assign to the loss of biological diversity.

But in the last analysis, will all these things be done? I worry about the scale of our response. Money certainly is not the solution to all this. Education and public commitment are fundamental. But without much more money, much of it in the form of development assistance, we are simply not going to conserve and manage the biological diversity we are all talking about.

Consider these figures, for example:

- The World Wildlife Fund has spent only *$110 million dollars* on conservation projects over its 25-year history.
- The biodiversity legislation passed by the U.S. Congress in 1986 designates less than *$10 million dollars* per year for conservation projects outside the United States.
- By comparison, the Tropical Forestry Action Plan estimates that 10% of the global effort, or *$800 million dollars*, will be needed over the next 5 years in tropical forest conservation. And this is only for tropical forests—it does not include coral reefs, wetlands, prairies, and all the other ecosystems (WRI et al., 1985).
- Total development assistance now amounts to about *$32 billion dollars* per year worldwide, only a tiny fraction of which, less than 2%, goes into forest conservation and development. The United States leads the way with about $9 billion dollars per year, mostly to Egypt and Israel. But we are only eighteenth in foreign assistance in terms of percent of our GNP (Tropical Forestry Task Force, unpublished data, 1985).
- Contrast all this with the *$80 billion dollars per year* contributed by Americans to their favorite charities. Nearly half this amount is given to support their churches. Far less than 1% is spent on conservation in the broadest definition of the word. Even most of the conservation money is spent here in the United States, rather than in the tropics where it is badly needed to save biological diversity (Thomas and Garred, 1986).
- Finally, at the end of the scale, about *$800 billion dollars per year* is spent worldwide on armaments and defense. This *annual* amount is 100 times the *5-year* amount needed to carry out the Tropical Forestry Action Plan.

The real irony is that often only a few thousand dollars is all that is needed to help a local community in Peru or Zambia get to the point where it can meet its own food, wood, and fuel needs without continuing to destroy the nearby forest.

The figures and the trends for biological diversity and tropical forest *are* discouraging. Yet change can come faster than we think. We will lose much more tropical forest and many more species because of the inexorable—yet reasonable—march of development (Prance and Elias, 1977; Raven, 1980). What we end up with at the beginning of the twenty-first century and beyond depends on the changes we make today. And fortunately, many of those changes already are under way.

REFERENCES

Asian Development Bank. 1980. Sector Paper on Forestry and Forest Industry. Asian Development Bank, Manila, Philippines. 112 pp.

FAO (Food and Agriculture Organization). 1985. Tropical Forestry Action Plan. Food and Agriculture Organization of the United Nations, Committee on Forest Development in the Tropics, Rome. 159 pp.

FAO/UNEP (Food and Agriculture Organization/United Nations Environment Program). 1981. Tropical Forests Resources Assessment Project: Tropical Africa, Tropical Asia, Tropical America. Food and Agriculture Organization of the United Nations, Rome. 3 volumes.

Furtado, J. I. 1986. The future of tropical forests. Pp. 145–171 in N. Polunin, ed. Ecosystem Theory and Application. John Wiley & Sons, London.

Guppy, N. 1984. Tropical deforestation: A global view. Foreign Affairs 62(4):929–965.

Myers, N. 1980. Conversion of Tropical Moist Forests. National Academy Press, Washington, D.C. 205 pp.

OTA (Office of Technology Assessment). 1984. Technologies to Sustain Tropical Forest Resources. Congress of the United States, Office of Technology Assessment, Washington, D.C. 344 pp.

Prance, G. T., and T. S. Elias, eds. 1977. Extinction Is Forever: Threatened and Endangered Species of Plants in the Americas and Their Significance in Ecosystems Today and in the Future. Proceedings of a symposium held at the New York Botanical Garden, May 11–13, 1976, in commemoration of the Bicentennial of the United States of America. New York Botanical Garden, Bronx, N.Y. 437 pp.

Raven, P. H. 1980. Research Priorities in Tropical Biology. National Academy Press, Washington, D.C. 116 pp.

Rubinoff, I. 1983. A strategy for preserving tropical forests. Pp. 465–476 in S. L. Sutton, T. C. Whitmore, and A. C. Shadwick, eds. Tropical Rain Forest: Ecology and Management. Blackwell Scientific Publications, Boston.

Sommer, A. 1976. An attempt at an assessment of the world's tropical moist forests. Unasylva 28(112–113):5–24.

Thomas, E., and E. Garred. 1986. Deep pockets for doing good. Time (June 16) 127(24):51.

Whitmore, T. C. 1984. Tropical Rain Forests of the Far East. Second edition, revised. Clarendon Press, Oxford. 353 pp.

World Bank. 1978. Forestry, A Sector Policy Paper. The World Bank, Washington, D.C. 65 pp.

WRI (World Resources Institute), The World Bank, and United Nations Development Programme. 1985. Tropical Forests: A Call for Action. World Resources Institute, Washington, D.C. 3 vols.

WRI (World Resources Institute) and IIED (International Institute for Environment and Development). 1986. World Resources 1986. Basic Books, New York.

INTERNATIONAL DEVELOPMENT AND THE PROTECTION OF BIOLOGICAL DIVERSITY

NYLE C. BRADY

Senior Assistant Administrator for Science and Technology, U.S. Agency for
International Development, Washington, D.C.

It has become clear that a significant proportion of the diversity of life on Earth could well be lost in the next half century. It is also clear that this loss could have serious negative impacts on society. The number of species currently available for use could be reduced by the loss of both wild germplasm and gene pools, and potential new genetic resources could be lost before their utility is discovered. Essential ecological services such as regulation of water quality and quantity, regeneration of plants and animals, cycling of nutrients, and buffering climate extremes could be impaired or lost altogether.

Society should face this issue squarely and should make a concerted effort to minimize the projected loss of biological diversity. But how can this be done? More importantly, how can it be done in the developing nations of the world where there is competition between meeting the basic human needs of burgeoning populations and maintaining biological diversity?

THE DEVELOPING COUNTRY

Developing countries face severe challenges in dealing with biological diversity. It is difficult for them to focus on long-term needs when they are faced with pressing, immediate needs for food and fuelwood and some means for earning foreign exchange to buy essential products and pay existing, mounting debts. This is a particularly urgent problem for those developing nations located in the tropics where the level of biological diversity is the highest and the threats to its maintenance are the greatest. Possibly up to 50% of all species on Earth may be native to the 6 to 7% of the Earth's land area that is covered by tropical moist forests

(Office of Technology Assessment, 1987). Yet, at current rates of tropical deforestation and conversion, virtually all accessible primary tropical moist forest areas will be gone within 50 to 70 years. At the same time, many of the tropical developing countries are among the poorest on Earth, often with large and rapidly growing populations. These countries have become increasingly dependent on external assistance to address their food and economic development needs as well as to help them conserve their biological resources. Without increased attention to both, the global community stands to lose living resources of truly inestimable value. It is my conviction that biological diversity concerns cut across a wide range of sectors. Furthermore, sustained economic development requires the conservation of biological resources, and conversely, conservation of these resources in the developing world is dependent upon their ability to achieve sustained economic growth.

The perspective of the U.S. Agency for International Development's (USAID) Mission in Bangkok is illustrative of this point. In responding to our request for guidance from the field as to how the agency should address this concern, the mission replied:

> In Thailand, the principal threat to long-term maintenance of biological diversity and tropical forest resources is agricultural encroachment on already-designated conservation areas by nomadic hill tribe groups and landless lowland Thais. Overcoming this problem is fundamental to the long-term viability of much of Thailand's biological resources, and will require concerted efforts in agricultural, rural, and economic development, as well as in reforestation and protected area management. For those situations where basic human needs must be met for conservation efforts to succeed, AID might be able to play a role both in conserving biological diversity and tropical forests and in fostering sustainable economic and social progress of the poor (U.S. Department of State, 1986).

This, to me, epitomizes the dilemma that this group of international scientists must come to grips with if conservation efforts are to succeed on a global scale. In meeting this challenge, we are going to have to find ways to conserve more natural habitats, better manage those that already exist, ensure that development projects are ecologically sound, improve our methods of economic analysis of the costs and benefits of natural resource deteriorations and investments in natural resources conservation, and increase food and fuelwood production on land already cleared in order to reduce pressure on the remaining wild areas.

HABITAT CONSERVATION—KEY TO THE PROBLEM AND THE SOLUTION

Although pollution and overexploitation are serious threats to many wild plant and animal species, the continuing loss of habitats, especially tropical forests, is the major cause of current and projected rates of species extinction. Consequently, habitat conservation is the key to the effective conservation of the world's biological diversity. The utility or necessity of a species from the standpoint of humans is not necessarily a corollary of a species' adaptability. Therefore, conserving biological diversity for human benefit means conserving sufficient natural habitat for those species incapable of surviving elsewhere.

Habitat conservation can be addressed in two major ways. First, we can critically analyze why the habitat is changing and can identify steps to be taken to arrest these changes. For example, we must determine the human needs currently being met by those who slash and burn tropical forests, and we must find alternatives to this devastating practice. These alternatives must meet those human needs without destroying the natural habitat and, in turn, the biological resources that depend upon that habitat. Perhaps no other scientific achievement would make a greater contribution to the maintenance of biological diversity in the tropics than would the development of practicable alternatives to slash-and-burn agriculture.

The second way of enhancing biological diversity is to set aside specific areas in which the current habitat is to be maintained. The establishment and maintenance of conservation reserves, parks, and wildlife refuges are examples of this means of maintaining biological diversity. Actions necessary to ensure the success of these protected areas must be taken. But we and others in the industrialized nations must not underestimate the difficulty of doing so. Questions such as the following must be addressed:

- How much habitat and what kinds of habitats must be maintained?
- Who will establish and maintain such habitat areas and at what cost in both economic and human terms?
- Who will pay?

THE NEED FOR A SYSTEMATIC APPROACH TO CONSERVATION

Systematic conservation can be defined as the conscious maintenance of the full range of natural diversity, e.g., species, communities, habitats, and ecosystems on a representative basis. Because we lack full knowledge of the identity and number of all species, let alone their distribution and habitat requirements, efforts to achieve systematic conservation must necessarily focus on higher levels of organization such as the habitat or ecosystem.

Although this is a much more tractable procedure than a species-based approach, it will nevertheless require both new research and a comprehensive synthesis of available information. Especially necessary are the review and integration of habitat classification systems and existing conservation area systems. It will also require the development of habitat size criteria to ensure the long-term security of those species each habitat type contains.

Worldwide, there are currently some 3,500 major conservation areas totaling approximately 4.25 million square kilometers (Office of Technology Assessment, 1987). These conservation areas can be grouped into 10 broad classifications, ranging from strict nature reserves to multiple-use natural resource lands. Broadly speaking, these areas represent some 178 of the 193 biogeographical provinces recognized by the International Union for the Conservation of Nature and Natural Resources as a first approximation of the diversity of Earth's major habitat types. Globally, therefore, the nations of the world already have made a significant investment of land for the maintenance of natural communities and the perpetuation of their biological diversity.

But this current investment is not yet adequate. First, not all habitat types, even very broad ones such as the biogeographic provinces mentioned above, are represented in the current system, and these biogeographic provinces themselves are so coarse as to miss a great deal of diversity.

Second, the size of the areas set aside for some habitat types may be inadequate. For example, of the 178 biogeographic provinces represented in the current system of conservation areas, 28 are represented by five or fewer individual units, for an aggregate area of 1,000 square kilometers or less (Office of Technology Assessment, 1987). There is growing concern that such a limited number of small conservation areas may not forestall the extinction of some species, particularly those that are large, wide-ranging, or especially susceptible to the chance variations in climatic and environmental factors.

Third, conservation areas in many countries lack adequate management for many of the units. Without professional, trained staff, adequately equipped and operating under an explicit management plan, many of these areas represent an idea rather than reality.

Nevertheless, there are hopeful signs. Since 1950, there has been a rapid increase in the number and extent of conservation areas worldwide. Eight of the nine countries reported to have set aside more than 10% of their land in protected areas are located in the developing world (Harrison et al., 1984). The U.S. Fish and Wildlife Service and the National Park Service are receiving a growing number of requests for assistance in conservation-area establishment and management from developing countries. Also, the U.S. Peace Corps now has about 170 volunteers afield working on parks and protected-area projects at the request of developing country governments.

All signs point to an increased awareness of the need for conserving natural systems for the potential resources they may contain, for the environmental services they provide, and for the educational and aesthetic needs they meet. Yet much remains to be done, and the development community can and should play a role.

THE OBJECTIVES AND ROLE OF INTERNATIONAL DEVELOPMENT ASSISTANCE

The goal of U.S. development assistance programs is to help people of the developing countries enhance their human, social, and economic conditions. The conservation ethic, which implies the rational and sustained use of resources, seems to prosper in both traditional cultures and in the highly developed and successful industrialized nations. But most of the world's people live in the developing nations, which are in transition between older traditional cultures and a more economically developed state. If the North American and Western European experiences are a guide for the future, the conservation ethic, especially as it relates to land use, is more likely to prosper in the developing nations as this transition progresses. In a very real sense, therefore, successful economic and social development can enhance conservation efforts in the developing world.

However, conservation efforts need not wait until development is complete. But make no mistake, such efforts must be part of the framework of overall economic development to fully succeed. Protected areas will not survive in a human environment of futility and dire need. Poor peasants do not burn the forests just because they like to see fires. They need fuelwood, or they need cleared land and the wood ashes to help them produce food for their families. Only if they are provided with feasible alternatives to the wasteful slash-and-burn system will the world be able to protect the natural habitats and, in turn, to maintain biological diversity.

Although the provision of such alternatives and better livelihoods for the world's poorest people is a major contribution of development assistance to conservation efforts, more can and must be done. First, it is encumbent upon the development-assistance community to ensure that our policies, programs, and projects are environmentally and ecologically sound as well as economically feasible. USAID's policy on natural resources and the environment and our environmental review procedures are designed to fulfill these requirements. Our policies and procedures are also designed to ensure that our program directions and our project review system adequately address the impacts of our assistance activities on endangered species and their critical habitats. The World Bank has recently adopted an operating policy on wildlands, which are discussed by Goodland in Chapter 49.

Second, USAID is playing a lead role within the donor community in establishing a dialog with the governments of developing countries on the importance of wise natural resource management, including the conservation of biological diversity. As an initial step in this dialog, we have assisted some 23 developing countries in preparing country environmental or natural resource profiles. These profiles outline the current status of renewable natural resources and identify major environmental issues, information on the biological diversity resources of countries, existing conservation programs, and future needs. We plan to accelerate the preparation of environmental profiles in the remaining countries in which USAID has programs and to ensure that biological diversity is better addressed.

USAID is also assisting several countries in the preparation of national conservation strategies that establish explicit conservation objectives, which can be integrated into overall economic development goals. This is a logical followup to the profiling effort and can lead to the identification of gaps in a national protected area system and of land use options (parks, multiple-use areas) that most realistically fit the country's socioeconomic status.

Without explicit plans and budgets, conservation efforts can be ineffective and development schemes can founder for lack of a thorough understanding of the environment and the resource base. For example, one government ministry may be planning a major hydroelectric project that happens to coincide with another ministry's plans for the project area as a wildlife refuge. A third ministry may also be planning the sale of timber concessions in the same area. With proper interagency consultation, project design, and implementation, the dam could provide power, well-managed logging could produce timber, and the overall reservoir catchment area could be set aside to protect the reservoir's watershed and help maintain the regional biota.

USAID is supporting the development of a national conservation strategy for Nepal and has assisted in the process of planning and initiating conservation strategies in the Philippines, Sri Lanka, and Zimbabwe. Again, this process needs to be accelerated, and we are working to do this.

Coupled with the integration of conservation and development is the need to develop or strengthen indigenous institutions within the developing countries that can assemble, analyze, and monitor data on plant and animal species, on land use, and on protected area systems. Both conservation and development agencies need the information and technical support that such institutions can provide on a continuing basis.

The Nature Conservancy's International Program has been working with several Latin American countries to develop Conservation Data Centers to meet this need. We believe their efforts are a good prototype for such institutions, and we are pleased to be assisting in supporting this program.

Development assistance agencies can also assist conservation efforts directly through grant or loan support and through the regular and systematic incorporation of conservation components into development assistance projects. In Peru, for example, the Central Selva Resource Management Project was started in 1982 in the Palcazu valley of the high jungle. It is being funded by USAID and the Government of Peru at a total of $30 million over 6 years. Its purpose is to test and institutionalize a method to promote sustained productivity of the Palcazu watershed and build an institutional capability within the country to plan and implement integrated regional development. Covenants included in the loan agreement require that the Government of Peru designate a national park and a protected forest in the watershed and assign technical staff to the area. The project design was derived from an environmental and social assessment that classified land-use capability. The USAID mission concluded on the basis of these studies that production forestry held the greatest potential for development, that previous plans for extensive resettlement of people to increase food production were not feasible, and that major attention should be given to managing the area for the existing inhabitants, many of whom were native peoples (USAID, 1982).

The project contains 10 components, two of which are of primary interest here: natural forest management for sustained yield and the establishment of protected areas. The forest management plan involves the testing of rotational, narrow clear-cut areas based on new knowledge about gap-phase dynamics and the regeneration requirements of tropical forest canopy tree species. It is hoped that this will permit natural regeneration over 30-year cycles. Current estimates are that logging 1 to 2 hectares a year will generate an adequate annual family income from 80-hectare holdings (Hartshorn, 1985).

The park and forest reserve areas were defined on the basis not only of land capability but also of their economic returns to downstream production forestry and agriculture. The project has also supported the work of an economic botanist to identify and store in local gene banks plant species of potential market value to the Palcazu people.

The example demonstrates progress, but it also raises the issue of the duration of projects that deal with renewable natural resources. Three- to six-year time frames appear inappropriate, given the severity of the problems, the time required before returns on investments may be realized, and the need to test new and sometimes risky technologies. Twenty-year projects or programs may be necessary, requiring that governments and lending agencies rethink existing policies and approaches.

USAID is now designing 10-year projects—a significant breakthrough. An example is a new natural resources project in Panama, which has four components: watershed management, natural forest management, private industrial plantations, and farm woodlots. The project rationale is based on the need to protect the economic values of existing agriculture and commercial investments, including the Panama Canal, to maintain electricity and a water supply to major urban areas, to reduce dependence on wood imports, and to enhance employment.

Because a principal rationale for conserving biological diversity is the need to conserve wild germplasm for improving agriculture, it is appropriate that development assistance also help in land acquisitions. For example, in the past decade or so, Costa Rica has undertaken an aggressive and comprehensive conservation area program. Last year, the Government of Costa Rica requested and received permission from the USAID Mission to use U.S.-controlled local currency generated by the sale of U.S. commodities (P.L. 480 Program) for the establishment of a new conservation area (the Zona Protectora) and the enhanced management of an existing area (Cano Negro). Also, USAID has and will continue to support the National Research Council in the National Academy of Sciences in its exploration of little-known or underutilized tropical plant and animal resources. Numerous publications have been produced on topics ranging from multipurpose tree species to butterfly and crocodile farming.

USAID is assisting conservation efforts in a variety of other ways as well. These range from supporting U.S. Peace Corps Volunteer wildlife biologists who are conducting endangered species surveys in Burundi to the development of a management plan for Tarutao National Park in Thailand and the provision of technical assistance to and training of personnel of various host countries in parks and wildlife management.

These types of assistance will continue and are likely to increase as more nations realize the importance of conserving the diversity of their living resources and seek our assistance in their own efforts.

FUTURE DIRECTIONS

We are witnessing a convergence of interests that could be a powerful force in the coming years—a growing consensus between the conservation community and the development institutions that maintenance of biological diversity and sound economic development are not only compatible but mutually interdependent. In the long run, economic growth is heavily dependent on the conservation of these resources. In turn, the conservation of these resources is not likely to take place, especially in the tropics, without quantum leaps in economic development. This means that we must work more closely together to promote both our goals. The support of U.S. environmental groups for foreign development assistance last fall is strong evidence of this trend. This support will help stimulate our development efforts.

USAID has been defining an action plan that has grown out of the Strategy on Biological Diversity submitted to Congress in 1985. Our draft action plan focuses on seven areas and is based on the critical assumption that in the final analysis, conservation of biological diversity in the developing countries is the responsibility of the governments and people in those countries (USAID, 1986).

In summary, the major themes for action are:

- policy dialog and strengthening national policies;
- public awareness and education, building the capacity of indigenous environmental organizations;
- strengthening natural resource management institutions and training host-country people;
- research on ecosystem dynamics and inventories of plant and animal species; and
- natural resource management programs such as those in Peru and Panama.

Our draft action plan builds on the USAID's institutional strengths, which lie mainly in the area of technical assistance as opposed to large capital investment projects, and seeks innovative ways to increase or leverage our investments in the limited number of countries in which we work.

I will be recommending the following actions to improve our approach:

- Priority countries should be identified in each of the three geographic regions (Africa, Asia and the Near East, and Latin America) in which USAID works. The guidance of the scientific and nongovernment communities represented here would be helpful in this process.
- Within the high-priority countries, the most responsive interventions should be identified and supported, interventions that could be any of the theme areas mentioned above.
- We should improve present methods of economic analysis to better address the real costs of natural resource depletion and the economic benefits of investments in maintaining ecosystem processes and conserving wildlands. (One of the most pressing practical needs is to back up investments in these areas by national governments and donor organizations. The economic costs of watershed deterioration and the loss of tropical forests and wildlife are subjects of wide speculation but have rarely been quantified in relation to national economic and development budgets. For example, rough estimates by some economists indicate that unsustainable forest depletions by major tropical hardwood exporting countries could be costing the countries more than they gain by the sale of the wood.)
- We should expand our research efforts to help us understand biological diversity and the means for maintaining it. Biological and physical science studies will be complemented by social science research, since after all, human activity is largely responsible for the loss of natural habitats.

- We should initiate an assessment of how best the International Agricultural Research Centers can play a role in ecosystems research in different biogeographic zones.
- We should intensify our efforts to develop alternatives to unsustainable agricultural practices, such as slash-and-burn agriculture, and to incorporate the use of multipurpose tree species in all agricultural projects to reduce pressure on natural habitats. The work under way by the International Institute for Tropical Agriculture in Nigeria on alley cropping (i.e., mixing trees with annual food crops in different ecological zones in Africa) is a prototype that could be adapted in other regions of the world.
- We should encourage and support expansion of the U.S. Peace Corps program on environmental education and protected area management and the greater use of P.L. 480 funding for conservation activities.
- We should use the experience we have gained to encourage and gain the cooperation of other development agencies with a view toward increased investment in conservation.

Many of you are aware that the U.S. Congress is now considering earmarking funds in the USAID budget for biological diversity. While USAID, as a matter of principle, is opposed to all earmarking of funds, we understand and appreciate that this possible allocation of funds is a clear indication of the priority that the conservation community in the United States places on this problem. In response to these interests, we expect to initiate a matching grant program for on-the-ground conservation activities in priority countries. We will also strengthen USAID's technical capabilities to provide overseas assistance for the design and implementation of these programs. The matching grant programs will attempt to use our funds to leverage investments by national, international, and nongovernment organizations and by developing country governments in programs related to wild plant and animal management and the inventory and assessment of biological diversity resources. The United States has a wealth of talent from which we will draw in our new endeavors to help improve conservation of biological diversity in our client countries.

The Forum on BioDiversity was an important event. It was an important opportunity to refine our understanding and approaches to solving the problems and to heighten the awareness of the public and our political leaders concerning the importance of and threats to the great variety of life on this planet. This awareness is crucial to build support, both here and abroad, for safeguarding this diversity at a time when government expenditures for all activities are declining in most every country in the world.

REFERENCES

Harrison, J., K. Miller, and J. McNeely. 1984. The world coverage of protected areas: Development goals and environmental needs. Pp. 24–33 in J. McNeely and K. Miller, eds. National Parks, Conservation and Development: The Role of Protected Areas in Sustaining Society. Smithsonian Institution, Washington, D.C.

Hartshorn, G. S. 1985. Sustained Yield Management of Natural Forests: A Synopsis of the Palcazu Development Project in the Peruvian Amazon. Tropical Science Center, San Jose, Costa Rica.

Office of Technology Assessment. 1987. Technologies to Maintain Biological Diversity OTA-F-330. U.S. Government Printing Office, Washington, D.C. 334 pp.

USAID (U.S. Agency for International Development). 1982. Central Selva Resource Management (Peru) Project Paper No. 527–0240, Washington, D.C. 145 pp.

USAID (U.S. Agency for International Development). 1986. Draft Action Plan on Conserving Biological Diversity in Developing Countries. Prepared by the Bureau for Science and Technology, Office of Forestry, Environment and Natural Resources, Washington, D.C. 27 pp.

U.S. Department of State. 1986. Comments on Biological Diversity Action Plan. Bangkok cable No. 23811. May 16, 1986.

PRESENT PROBLEMS AND

FUTURE PROSPECTS

Nest and eggs of the endangered light-footed clapper rail in intertidal cord grass marsh in Tijuana Estuary, California. The nest floats but is anchored to the grass so that it can rise and fall with the tide but not be flooded or carried away. *Photo courtesy of Christopher S. Nordby.*

DIVERSE CONSIDERATIONS

THOMAS E. LOVEJOY*

Executive Vice President, World Wildlife Fund, Washington, D.C.

Before dwelling on the economic, social, and political problems that are fundamental to present problems and future prospects, there are two aspects of natural science that require attention but have not yet been mentioned in this volume: the abundance of relatively few of the many species on Earth and the limitations deriving from our shallow knowledge of diversity.

SPECIES ABUNDANCE

One of the great questions of biological science arises when biological diversity is viewed through ecological glasses: Why are ecosystems generally made up of a large number of species of which only a few are abundant? While the roster of rarer species in an ecosystem is much longer in tropical regions than at higher latitudes, there is a general tendency to accumulate large numbers of species in all but the most simple ecosystems.

This pattern can be generally portrayed by graphing the relative abundance (for example, the percentage of total individuals or of total biomass) of species against the order of species from most to least abundant (Figures 47-1 and 47-2). In early successional communities, there is a smaller number of species and the most abundant ones constitute a larger fraction of the community, i.e., are more dominant.

---

*After the forum, Dr. Lovejoy joined the Smithsonian Institution as Assistant Secretary for External Affairs.

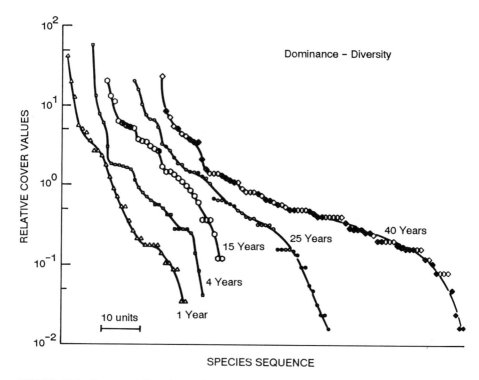

FIGURE 47-1   Patterns of the relative abundance of species at five different stages of abandonment in old fields in southern Illinois. The patterns are expressed as the percentage that a given species contributes to the total area covered by all species in a community, plotted against the species' rank and ordered from most to least abundant. The symbols are open for herbs, half open for shrubs, and closed for trees. From May, 1985, with permission.

Figure 47-1 shows such an accumulation of species after an old field was abandoned in southern Illinois. Conversely, if a community is subjected to stress, e.g., from temperature, toxic substances, or the arrival of an alien species, this process is reversed: diversity declines and a small number of species become dominant and are often called nuisance growths. Figure 47-2 shows such a response to heavy fertilizer applications in experimental grass plots in England. There are excellent aquatic analogs to these patterns, many of which have emanated from the work of Ruth Patrick and colleagues at the Philadelphia Academy of Natural Sciences (Patrick, 1949, 1975; Patrick et al., 1969).

The meaning of *pattern* in relative abundance and diversity in communities has only begun to be initially understood and remains largely tantalizing (Fisher et al., 1943; Hutchinson, 1958; MacArthur, 1957, 1965; May, 1975; Patrick 1961, 1984; Preston, 1948). Furthermore, it is very difficult to understand the function of any of the large number of rare species in an ecosystem precisely because they are so rare. Indeed, there has been a tendency when considering endangered vertebrate species to think of them as ecologically nonfunctional or even ecologically extinct. I believe this is a highly dangerous and inappropriate attitude not only because *rarity in an ecosystem is in fact the common condition* but also because the true meaning of rare species in the system cannot easily be assessed for a moment but only when viewed over periods of environmental change.

The role of an obscure species of yeast in the genus *Cryptococcus* (Brunner and Bott, 1974) is very telling in this regard. This species is rare in the aquatic communities of eastern Pennsylvania, presumably because of competition from other species. Its role and value are not immediately apparent. When either natural or human-generated mercury contamination occur, this yeast suddenly is at great advantage, because it is able to short-circuit a particular metabolic pathway along which mercury has toxic effects for organisms generally. The yeast is actually able to reduce methyl mercury to the elemental state and store the quicksilver in a vacuole that it subsequently deposits on a convenient surface such as a rock. Under conditions of elevated mercury concentration, the yeast temporarily becomes very abundant, while many previously abundant species in the community are depressed and diversity declines. The yeast literally cleans up the mercury contamination, thereby making itself rare again.

The number of species that respond in this fashion to environmental change in a way that keeps them rare is probably not great. Nonetheless, all rare species in ecosystems are likely to be able to respond with population increases, given the

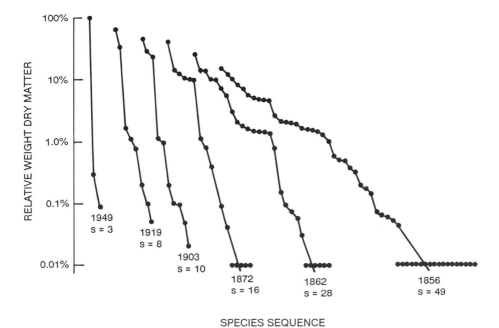

FIGURE 47-2 Changes in the patterns of relative abundance of species in an experimental plot of permanent pasture at Parkgrass, Rothamsted, England, following continuous application of nitrogen fertilizer since 1856. (Species with abundance less than 0.01% were recorded as 0.01%.) Note that here time runs from right to left, so that the patterns look like the successional patterns of Figure 47-1 running backward in time. From May, 1985, with permission.

right changes in environmental conditions. Many of these rare species are likely to be part of local habitats at any given moment, because the ecological conditions under which they flourish have occurred sufficiently frequently or recently for them to be represented. This means that the rare species in a sense reflect a variety of past conditions and confer some measure of ability for the ecosystem to respond to changing conditions.

Survival strategies of this kind have as yet only been superficially explored. Nevertheless, the role rare species may play under changing conditions suggests that the accumulation of species in an ecosystem, while perhaps only an accident of the history of any particular system, can turn out to be of value. We thus have the beginning of an argument in favor of rare species in general—species attuned to conditions that are not prevalent at the moment but that may well return.

At the present time, we are faced with a situation in which not just local ecosystems respond to stress by losing their diversity and by simplification; rather, the entire biosphere is subject to impoverishment. This inevitably means that individual ecosystems in a number of instances will be less diverse, because *some* of the species that would otherwise form part of the long list of rare species are doomed to extinction.

LIMITATIONS OF OUR KNOWLEDGE

To approach the global problem from a scientific perspective, we are immediately confronted by the second problem commanding attention, namely, the limitations deriving from our relatively shallow knowledge of flora and fauna. Recent discoveries of insect diversity in the canopy of South American tropical forests (Erwin, 1982, 1983, and Chapter 13 of this volume; May, 1986) warn us that we do not even know the extent of biological diversity on our planet to the nearest order of magnitude (Wilson, 1985, and Chapter 1 of this volume). Given such a poor inventory of life on Earth, biologists can say relatively little about which species occur where, which are in danger of extinction, where protected areas should be established, and where heavy environmental modification for development is permissible.

What is desperately required is a revitalization of the science of biological systematics, with all the ancillary strength modern technology and molecular biology can provide, combined with a crash program of biological exploration. A decade or two of intensive biological mapping is needed while development is halted, or at least severely curtailed, in areas that are evidently the richest but least explored.

SOCIOLOGICAL ISSUES

I would now like to switch very abruptly to matters of the social sciences, which drive our societies in a multiplicity of little-appreciated ways. Consider the price of shoes, which on the face of it, seems but a matter of domestic detail far removed from the threat of massive extinctions in the tropics. Yet the U.S. shoe industry, far from strong, has been seeking tariff protection from the Brazilian shoe industry. If the U.S. government were to respond in a protectionist manner, what might be the consequences? Brazil's struggles with its international debt would be the more difficult. Furthermore, its drive to increase exports would be even stronger, and as already apparent, the emphasis on export crops would lead to the intrusion of small farmers into the vital remaining natural areas.

Shoe prices and similar isolated indices may not be a large factor. They might even prove to have no measurable impact on conservation, or they could even be a positive increment, if decreased Brazilian shoe exports reduced incentives for cattle ranchers. The point, however, is that a myriad of such unrelated issues swirl about our everyday lives and together with decisions made on Wall Street and in Washington, have a great deal to do with fueling the biological diversity crisis. Each and every one of us is more tightly connected to the global conservation problem than generally realized. This Gordian Knot of economic and political linkages deserves far greater attention than it has received up to the present time.

Of course, there remain the great overarching issues such as population growth. I feel this acutely as I write from the great Indian subcontinent. When Indira Gandhi came to power in 1966, there were 480 million Indians, and when she left in violence in 1984, there were 250 million more. Since this trend continues, my environmental colleagues in New Delhi worry deeply about the long-term security of parks and reserves.

The problem of population growth sometimes seems so enormous as to be intractable. Even with stringent birth control, future population growth is inevitable because of the very youthful age structure of populations in developing countries. Short of coercion, the only permanently ameliorative approach is to improve living standards while providing birth control devices and aggressive public education programs. That is not an instant solution, because human conditions cannot be improved overnight and a response in the form of reduced fertility does not happen right away either. There is no question that the association between development, population control, and conservation needs our continuing preoccupation.

ECONOMIC PRESSURES

Looming large in the picture is the massive international debt, now expressed in trillions of dollars. This economic goliath binds the industrialized and developing nations as surely as their shared interest in the protection of biological diversity. In many instances, all earned foreign exchange must be devoted to interest payments. This leaves little capacity to use foreign exchange to invest in the economic growth so necessary for improved living standards and all that goes with them, including a greater awareness of, and willingness to do something about, conservation problems.

For some time, I have advocated the notion that some portion of the debt be allocated to these pressing problems of biological diversity. The money is already in the developing nations where the greatest conservation problems are, and much of it is unlikely to be returned to the lenders. Why not then arrange for some form of conservation credits where tropical debtor nations can use their own local currencies for national conservation projects? Already there is interest in buying debt at market-dollar values considerably less than face value and redeeming the paper in local currencies. Legislation is developing to encourage donation of such debt by commercial banks for international purposes. Whatever form these activities take, it must of course be fully sensitive to national sovereignty, but this should not be difficult to achieve since conservationists in the developing nations are as ardent as conservationists anywhere.

Perhaps the most important point about the possibility of exploiting the international debt is not so much the example itself but rather the need to constantly seek new ways to elevate and expand the conservation effort. The scale of the problem is so great compared to the conservation efforts now under way that we will not succeed in safeguarding the majority of biological diversity without major innovation and major infusion of resources. It is all too easy to feel content with some hard-won conservation victory when problems crop up all around us like warriors sown from dragon's teeth.

From time to time in a flight of apparent silliness, I advance the notion of buying a major piece of west African real estate with substantial populations of gorillas and chimpanzees and setting up a new political entity, the "Kingdom of the Apes." I do so not because such a possibility really exists. Rather, I do so because however fanciful the idea of a great silverback male gorilla being present at the UN General Assembly, it does suggest how apart we have set ourselves from other living things, and how awry our system of governing the world really is.

We and our fellow vertebrates are largely along for the ride on this planet. If we want to perpetuate the dream that we are in charge of our destiny and that of our planet, it can only be by maintaining biological diversity—not by destroying it. In the end, we impoverish ourselves if we impoverish the biota.

REFERENCES

Brunner, R. L., and T. L. Bott. 1974. Reduction of mercury to the elemental state by a yeast. Appl. Microbiol. 27(5):870–873.

Erwin, T. L. 1982. Tropical forests: Their richness in Coleoptera and other arthropod species. Coleop. Bull. 36(1):74–75.

Erwin, T. L. 1983. Beetles and other insects of tropical forest canopies at Manaus, Brazil, sampled by insecticidal fogging. Pp. 59–75 in S. L. Sutton, T. C. Whitmore, and H. C. Chadwick, eds. Tropical Rain Forest: Ecology and Management. Blackwell Science Publishers, Palo Alto, Calif.

Fisher, R. A., A. S. Corbett, and C. B. Williams. 1943. The relation between the number of species and the number of individuals in a random sample of an animal population. J. Anim. Ecol. 12:42–58.

Hutchinson, G. E. 1958. Concluding remarks. Cold Spring Harbor Symp. Quant. Biol. 22:415–427.

MacArthur, R. H. 1957. On the relative abundance of bird species. Proc. Natl. Acad. Sci. USA 43:293–295.

MacArthur, R. H. 1965. Patterns of species diversity. Cambridge Philos. Soc. Biol. Rev. 40:510–533.

May, R. M. 1975. Patterns of species abundance and diversity. Pp. 81–120 in M. L. Cody and J. M. Diamond, eds. Ecology and Evolution of Communities. Belknap Press, Harvard University, Cambridge, Mass.

May, R. M. 1985. Population dynamics: Communities. Pp. 31–44 in H. Messell, ed. The Study of Populations. Pergamon Press, Rushcutters Bay, Australia.

May, R. M. 1986. How many species are there? Nature 324:514–515.

Patrick, R. 1949. A proposed biological measure of stream conditions based on a survey of the Conestoga Basin, Lancaster County, Pennsylvania. Proc. Acad. Nat. Sci. Phila. 101:277–341.

Patrick, R. 1961. A study of the number and kinds of species found in rivers in eastern United States. Proc. Acad. Nat. Sci. Phila. 113(10):215–258.

Patrick, R. 1975. Structure of stream communities. Pp. 445–459 in M. L. Cody and J. M. Diamond, eds. Ecology and Evolution of Communities. Belknap Press, Harvard University, Cambridge, Mass.

Patrick, R. 1984. Some thoughts concerning the importance of pattern in diverse systems. Proc. Am. Philos. Soc. 128:48–78.

Patrick, R., B. Crum, and J. Coles. 1969. Temperature and manganese as determining factors in the presence of diatom or blue-green algal floras in streams. Proc. Natl. Acad. Sci. U.S.A. 64:472–478.

Preston, F. W. 1948. The commonness, and rarity, of species. Ecology 29:254–283.

Wilson, E. O. 1985. The biological diversity crisis: A challenge to science. Issues Sci. Technol. 2(1):20–29.

THE CONSERVATION OF BIODIVERSITY IN LATIN AMERICA
A Perspective

MARIO A. RAMOS

Director, Wildlife Program, Instituto Nacional de Investigaciones,
Sobre Recursos Bioticos, San Cristobal Las Casas, Chiapas, Mexico

T he increasing awareness of the need to preserve biological diversity in the world is demonstrated by the meetings that have been convened in developed countries. In the United States, for example, there were the Smithsonian Institution meeting in December 1985, the World Wildlife Fund meeting in September 1986, the National Academy of Sciences and Smithsonian Institution joint meeting in September 1986, and the meeting of the New York Zoological Society in October 1986. All are clear examples of this general trend. Although this awareness has come slowly to countries in the Third World, some important steps have been taken. In Latin America, for example, there has been a dramatic increase in the number of established protected areas (Harrison et al., 1984).

What has been slow to come, in my opinion, is recognition that the preservation of the biological diversity in the world is a shared commitment between rich and poor countries and that major responsibilities fall into the hands of the countries where this diversity is found. Since the greatest diversity exists in the tropical areas of the world, these responsibilities generally lie within the developing countries. In these countries, however, social, economic, and political problems often make conservation of their diversity very difficult. The riddle of balancing development; stability in economic, social, and political terms; and conservation of their natural resources is difficult for any of the countries to solve by themselves. Because Mexico provides an instructive example of these problems in the Latin American region, I will summarize the general conditions that determine the context in which this conservation is now and will be taking place in the near future. I will also consider means to achieve the sharing of responsibilities among countries in the world. Here I pull together the knowledge and experience of many other people working in the region who have expressed their views in the recent meeting during the World Wildlife Fund's Twenty-fifth Anniversary.

CONSERVATION AWARENESS

In Mexico, there has been a growing awareness of conservation issues, and new leadership has emerged within the scientific community and nongovernment organizations (NGOs) such as Monarca A. C. and Pronatura A. C. The new leaders have been responsible for the inclusion of environmental considerations in part of the government planning process and in political agendas. NGOs have flourished in the last 5 to 10 years. At present, there are probably 200 of these organizations located primarily in major cities throughout the country. In 1985, 22 NGOs established a federation in Mexico City. These organizations, supported by the scientific community, have pushed for improving environmental legislation and the establishment and protection of the conservation units within different habitats. In addition, they have discussed wildlife trade and have helped to organize and spread the conservation message among citizens in the country. A striking example is the controversial establishment and management of the Montes Azules Biosphere Reserve in Chiapas. International organizations such as the World Wildlife Fund/ Conservation Foundation and the Nature Conservancy International Program are supporting this grass-roots movement through direct economic support, official recognition, training of personnel, and infrastructure to help NGOs carry out their functions.

THE SOCIOECONOMIC ENVIRONMENT

Mexico is a clear example of a country in the process of development. It has slowed down its population growth from 3.3 to 2.3% per year, and it has two basic faces: an industrial one and a rural one. In addition to the problems stemming from industrial development, it has large rural and urban populations living in very marginal conditions. A comparison between 1940 and 1980 figures is revealing. There were about 19 million Mexicans in 1940, and in 1980, there were 68 million. By 1985, the population exceeded 79 million. Four million people lived in urban areas in 1940 and 41 million by 1980—a 10-fold increase, whereas rural populations changed from 15 to 27 million during the same period. Now, 65% of the population is concentrated in the central and southern part of the Mexican plateau (altiplano), where 80% of the total industry is also located.

In addition, Mexico probably has the largest city in the world, Mexico City, which is home to about 18 million people. The city faces many environmental problems related to its size. Among the more pressing ones are air pollution control, solid waste and sewage disposal, water supply, encroachment on green areas, and the movement of rural populations into urban areas, where they live under extremely marginal conditions in huge slum areas within and around the city.

The most important and fastest-growing industry has been the petrochemical industry, which generates not only oil, plastics, pesticides, and fertilizers but also air, water, and soil pollution. Modern industry generates large quantities of toxic materials, commonly storing them on site, dumping them down sewers or landfills, or burning them without regard for environmental concerns.

The country also faces enormous economic problems. Most of the capital is concentrated in a few hands, leaving the majority of people living under substandard conditions. Inflation has increased tremendously; consumer prices rose no less than 1,822% between late 1979 and February 1986 (Story, 1986), and real salaries dropped 27% in 1985 alone (Tello, 1986). Unemployment and underemployment have both soared, making it difficult for an important segment of the population to cope with the economic crisis. Devaluation of the Mexican currency has been rapid during the last 10 years, becoming critical in the last 4. A comparison between 1976 and 1986 figures is revealing. Our currency was worth 19.95 pesos to 1 U.S. dollar at the end of 1976, 473.60 to 1 in March 1986, approximately 700 to 1 in August 1986, and still continues to drop.

The international economic crisis has brought additional problems to Mexico. It was because of the impacts in Mexico that the international community first recognized the major problems confronting Third-World countries in relation to the international debt. A great part of the borrowing from the international community was wasteful or unjustified; it was not used for financing investment but for consumption and government budget deficits. Capital flight increased. Estimates from the 1979–1982 period suggest extraordinary amounts of capital flight in Mexico—about US$55 billion (Dornbusch and Fischer, 1986). The issue of the foreign debt has become politicized, partly because of the stagnation of the economy, which has a growth rate of less than 1%, and because the inflation appears to have become structural. Inflation in 1986 reached 105%, and hyperinflation seems to be knocking at the door. The situation has been exacerbated by trade deficits, showing the need for deep and significant structural changes.

In general terms, the foreign debt is being serviced but at increasing social costs (Lustig, 1986). The international commercial banks played a significant part in causing the economic crisis by imprudent lending to imprudent government and private-sector borrowers in Mexico, but they have paid a relatively low price for doing so. The country now suffers from a drop in economic activity, a worsening of the standards of living, and such a large increase in the foreign debt that it can not even pay the interest. Estimates for the foreign debt servicing bill reached US$11.5 billion in 1986. There have been some structural adjustments, but they have not gone as far as the international community requests. However, the Mexican government should get credit for taking some initiatives, such as the privatization of decentralized national companies. Changes of this magnitude, as requested by the International Monetary Fund, the World Bank, and other banking organizations, do not happen overnight.

A major concern of environmentalists has been the possibility that payment of the foreign debt may generate additional environmental problems and that the environmental costs may be too high. The reaction of the government so far has been similar to the strategy for payments of the foreign debt: negotiating reschedulings for future payments and hoping that someone will then solve the problem. Present administrators do worry about the environmental problems, but not at the high level or with the priority required to solve them. They plan and carry out economic and social development activities without major environmental constraints, letting others worry about them in the future.

Of special environmental concern is the production of oil—the main Mexican export product. For years, the Mexican government has depended on oil export revenues to strengthen its economy. Major loans were given and are still coming in now from the international banking community to support exploration, infrastructure building, and exploitation of oil wells, without apparent concern for environmental issues. However, oil prices keep dropping—from more than US$30 per barrel in 1984 to US$20 in early January 1986, approximately US$12 in late 1986 and US$15 in mid-1987. Story (1986, p. 30) reported, "Revenues from oil exports declined by $2 billion (in 1985) compared with a 1984 total of 16.5 billion. . . ." The total loss to Mexico due to the decline in oil prices and lower sales volume (in 1985) was US$3.3 billion. The situation became even worse in 1986 and 1987.

There has also been increased pressure from the developed world to obtain as many natural resources from Third-World countries as cheaply as possible. This has generated conflicts that are building and will continue to do so if present conditions prevail. The trade of wild flora and fauna at the international level is a clear example. Although Mexicans try to control the supply of natural resource goods, the international community must help to control the demand from their countries. Prices soar into the thousands of dollars for rare and endangered species of wild animals and plants, and while the demand exists, someone in the developing world will find, trap, and sell them.

THE POLITICAL ENVIRONMENT

The politics of the environment and its conservation is poorly defined in Mexico. There are no clear policies. Legislation is frequently a response to conflicting priorities and emergency situations. In some areas where there are laws (e.g., pollution control), there are no regulations to implement them (Albert, 1985). In other areas, the law is rarely applied, and when it is, the sanctions and fines are ridiculously low (Ramos, 1985). Government officers are politically appointed, and communication among them or their agencies is limited, because of strong territoriality at that level.

Traditionally, the management of environmental issues in Mexico was dispersed among different ministries. For example, pollution and its control were under the Health Ministry and fisheries under the Fisheries Ministry, while some protected areas (e.g., parks) were under the Agriculture Ministry and others under the Urban Development Ministry. Since late 1982, most of the government's environmental considerations were brought under a single ministry—the Urban Development and Ecology Ministry (SEDUE). Although the authorities of SEDUE have tried to cope with Mexico's environmental problems, they have had little success. Unfortunately, authorities at the cabinet level and below have changed three to five times since

SEDUE's establishment, producing inevitable instability in the ministry. All ministries are under the control of the Budget and Planning Ministry (SSP), which does not have staff well trained in environmental issues and thus faces serious conflict when trying to establish priorities. At the budgetary level, ecological issues are at the bottom of the agenda, if they are included at all. Changes in priorities need to start at this level.

During the past 65 years, the Mexican government has essentially been run by the official party—Partido Revolucionario Institucional (PRI). The environmental platform of the different political parties, including the official one, are not well known and in some cases, probably do not exist. Recently, before the 1982 presidential elections, Mexico's current president, Miguel de la Madrid Hurtado, did make references to environmental policy during his political campaign; he included them in his National Development Plan for 1982–1988. An evaluation of his accomplishments on the subject is still to be made. The democratization process now occurring in the official political party may facilitate the inclusion of more environmental considerations in the official party platform in the years to come. Environmental policy statements were also issued by the left-wing political parties in late 1982. It seems likely, however, that environmental policy and its implications will play a more powerful role in the elections to come in 1988 and beyond. The environment is becoming an economic issue politicians cannot afford to ignore. The health impact of the thermal inversions during the winters of 1986 and 1987, for example, are just now being appreciated. Environmentalists should work to improve their performance and relations with political groups now in order to be effective in the near future.

In terms of international affairs, our foreign policy has not been very active. In fact, it was more rhetorical than real until the 1970s when it began changing. Unfortunately, with the current economic crisis, the Mexican international profile has in general been diminishing. Furthermore, Mexico's definition of foreign policy depends on the president in office, and it changes with every new president elected. The situation is improving, however. The new leadership emerging from the official political party seems to be more technology oriented and apparently better able to cope with international agreements and relations.

Mexican international environmental policy is not well defined. Mexico has signed some international conventions, e.g., the Convention on Nature Protection and Wildlife Preservation in the Western Hemisphere (the Washington Convention) and the Convention Concerning the Protection of the World Cultural and Natural Heritage. However, it has not signed other, very important ones, e.g., the Convention on Wetlands of International Importance especially as Waterfowl Habitat (Ramsar Convention), the Convention on International Trade in Endangered Species of Wild Fauna and Flora (Cites Convention), and the Convention on the Conservation of Migratory Species of Wild Animals (Bonn Convention)(UNEP, 1984). The acceptance of many of these conventions generally depends on favorable opinions from different agencies within the government (e.g., SEDUE, the Foreign Relations Ministry, and the Mexican Senate), which have trouble reaching agreement. Analysis of the conventions has not been completed, and their value is not well understood.

THE TECHNICAL ENVIRONMENT

From the scientific and technical standpoint, Mexico has limited capabilities for environmental protection. Of primary importance, its scientific community is slowly eroding because of attrition. Salaries in this sector have dropped in real value to 30% of what they were 7 years ago. Operating budgets for research institutions have dropped dramatically too. Although these budgets for 1986 were the same as those for 1985, inflation has increased to about 65 to 70%, and because of dropping international oil prices, they suffered additional cutbacks of 23 to 27% for the whole year. This left most research institutions stranded. The personnel and infrastructure budgets have not been funded for the last 3 years, except in very selective cases approved by the Planning and Budget Ministry (SSP). This means that there are no possibilities for hiring personnel or replacing old infrastructure. Job openings are frozen as soon as they become vacated so that funds for those personnel costs can be used for other government expenditures. All these cutbacks are especially critical in research centers outside the main cities. The government has tried to decentralize some areas, science included, but the present economic and structural crisis has not helped this process.

Traditionally, most of the scientific endeavor in Mexico has been oriented toward basic sciences. Applied science does not carry the prestige or the curricular weight when academic evaluations are made. As in other countries, basic research tends to follow international academic fashions, and at present, molecular triumphalism overrides systematics and taxonomy, biogeography, and paleontology—some of the sciences necessary for mapping biological diversity on Earth.

In addition, Mexican scientists have commonly remained outside the political sphere and to some extent have neglected their roles as citizens. Scientists have not taken serious and clear stands on political issues or on policy related to the environment. Most of the technical work in this area has been designed or implemented by professionals in other areas such as economics, sociology, agriculture, or policy, without a clear understanding of the biologist's role and the importance of adequate scientific expertise in the decision-making process.

Moreover, only a few institutions are qualified to address issues of conservation and development, topics that commonly require interdisciplinary work and the infrastructure to carry basic and applied research to develop new technologies and apply the new knowledge to human communities. Among the most prominent institutions with these capabilities are the Instituto Nacional de Investigaciones sobre Recursos Bioticos (INIREB), the Centro de Investigaciones de Quintana Roo (CIQRO), the Instituto de Ecologia (IE), and to some extent, the Laboratorio Natural las Joyas de la Sierra de Manantlan in the University of Guadalajara (LNLJ). These organizations have been conducting research concerning conservation and development issues in well-defined natural areas. They are all very young organizations but have promising, capable people on their staffs.

The role of environmental education has been underestimated at both professional and nonprofessional levels. University curricula lag behind technical knowledge, and academic standards have been dropping for the last two decades. Conservation as a subject has not been commonly included in university curricula either in biology or in the social sciences. Important changes are necessary in this area too.

Amidst all this difficulty, we should recognize that as correctly pointed out by the International Union for the Conservation of Nature and Natural Resources in its World Conservation Strategy (IUCN, 1980), the way out is the achievement of a proper balance between conservation of natural resources and socioeconomic development. Economic development needs to offer alternative options for the human populations living under marginal conditions without damaging the natural resource base. Examples of this kind come from ancient Indian cultures, such as the Mayans in southern Mexico and Central America and many of the Indian groups in Brazil. Recent examples are difficult to find, but important efforts are being made in the Biosphere Reserve in Sian Ka'an in Quintana Roo, Mexico, and in the Planada project in Colombia. We have to recognize that there must be a sustainable balance between human populations and natural resources.

CONSIDERATIONS FOR SHARED RESPONSIBILITIES

Conservation of biological diversity on a global scale cannot take place only through the efforts of the developed world. Moreover, the general conditions of countries like Mexico and the real possibilities of preserving biological diversity under those conditions appear very grim. Major changes would have to be made within Mexico and in the international community to produce well-defined strategies designed to avoid or diminish the possibility of failure and the duplication of efforts. Both need to concentrate on specific plans of action for successful conservation and development efforts. Thus, although there is a tremendous need to address the issues of conservation more globally, it is essential to include the Third-World countries as equal partners in the process.

Mexico will have to respond effectively to the great challenge that includes the need for structural changes, constituency building, legislative reforms, and bureaucratic implementation. We will also require strong support from foreign and international groups. There are many ways countries like Mexico could be helped. Economic support is critical but is not the only important factor. Following are some of the basic ways developed nations and international organizations can provide the greatest assistance.

Developed Nations

• Developed nations should be aware of the issues involved in conserving biological diversity and their magnitude at the local, national, regional, and international levels. There is an extreme need to understand more globally the problems faced by Third-World countries. Money alone will not lead to the successful conservation of biological diversity in those regions.

• They must also be aware of our partnership and shared responsibilities as well as the need for mutual respect. For example, the conservation of migratory species of wildlife requires collaboration, understanding, and mutual respect.

• They must be cognizant of the need to spread the conservation message and educate citizens of developed nations on environmental issues. The international trade of wildlife demonstrates the urgent need to educate people where the demand is greatest.

• They must be sensitive to the need for a more balanced exchange in all interactions with the Third World, be it in policy, economy, or scientific expertise. Examples of such exchanges can be found in multilateral and transnational organizations.

International Organizations

International organizations can help in the following ways:

• They can facilitate the flow of information at all levels, as is done now by the Food and Agriculture Organization's Latin-American network of protected areas, other reserves, and wildlife.

• They can cooperate among themselves when working within certain countries. For example, close collaboration between the World Wildlife Fund and the Conservation Foundation with the Nature Conservancy International Program in Latin America would benefit them as well as the region.

• They can support leadership development in all fields related to the environment, considering local needs within a particular country or geographical region. Local needs should be specifically addressed with local vision. Train (1986, p. 1) wrote, "We often are so busy developing our own solutions to other peoples' problems as we see them that we sometimes become a bit deaf and a bit insensitive to the ideas of those very people whose problems we are trying to solve."

• They can cooperate in the definition of plans to coordinate conservation and development programs. The large international banks such as the World Bank or the Inter-American Development Bank would benefit a great deal from a closer association with local, national, and international conservation organizations.

• They can cooperate in the identification of alternative ways natural resources can be used that contribute to economic growth without compromising their survival. Projects such as the Green Iguana Project of the Smithsonian Tropical Research Center in Panama (see Chapter 42) and INIREB's captive-breeding programs for crocodiles and white-collared peccaries in Chiapas, Mexico, should be encouraged.

• They can collaborate in the application of ecological concepts, such as sustainability and carrying capacity, to development programs.

• They can help to build local capabilities with well-defined strategies. It is cheaper and ethically more correct in the long run. Good examples are the Conservation Data Centers of the Nature Conservancy International Program in Latin America and the support given to Restauracion Ambiental, Inc., in Mexico by the Conservation Foundation and the World Wildlife Fund.

• They can support the establishment of functional conservation units in every country within the region. Many protected areas in those countries are protected only on paper. Biological diversity will not survive under these conditions.

• They can promote research on basic and applied sciences related to environmental issues in the Third World. Support should be given preferentially to interdisciplinary groups that address topics on conservation and development. These groups should include qualified nationals from the country in which the work is performed.

• They can help to increase the prestige of particular research areas such as taxonomy, systematics, biogeography, and paleontology. This could be done by awarding prizes, fellowships, and scholarships and establishing exchange programs between research centers and universities.

We face enormous tasks that in the long run can be performed only by reaching a basic understanding of our shared responsibilities. Time is running out. The challenges to biodiversity can only be met with mutual understanding, help, and respect.

ACKNOWLEDGMENTS

I would like to acknowledge the assistance of Curtis Freese, Robert Healy, Martha Hays-Cooper, Oscar Flores, Patricia Gerez, and Alexander and Barbara Stevenson for reviewing early drafts of the manuscript.

REFERENCES

Albert, L. A. 1985. Legislacion y Plaguicidas. Jan. 4, 1985. Subcomite de Plaguicidas, Comite de Ecologia, Consejo General de Salubridad. Mexico, D.F.

Dornbusch, R., and S. Fischer. 1986. Third World debt. Science 234:836–841.

Harrison, J., K. Miller, and J. McNeeley. 1984. The world coverage of protected areas: Development goals and environmental needs. Pp. 24–33 in J. McNeeley and K. Miller, eds. National Parks: Conservation and Development. Proceedings of the 2nd World Conference on National Parks. Smithsonian Institution Press, Washington, D.C.

IUCN (International Union for Conservation of Nature and Natural Resources). 1980. World Conservation Strategy: Living Resource Conservation for Sustainable Development. International Union for Conservation of Nature and Natural Resources, Gland, Switzerland.

Lustig, N. 1986. El precio social del ajuste Mexicano: Balance de sombras. Nexos 106(Oct.):27–31.

Ramos, M. A. 1985. Problems hindering the conservation of tropical forest birds in Mexico and Central America and steps toward a conservation strategy. Pp. 305–318 in Technical Publication No. 4. International Council on Bird Preservation, Cambridge, England.

Story, C., ed. 1986. News and commentaries on Latin America. The Latin American Times (London) 73:29–31.

Tello, C. 1986. La deuda externa. Nexos 106(Oct.):19–26.

Train, R. 1986. Closing remarks, Partners in Conservation. Twenty-fifth Anniversary Conference of the World Wildlife Fund, September 17, 1986, Mayflower Hotel, Washington, D.C.

UNEP (United Nations Environment Programme). 1984. Register of International Treaties and Other Agreements in the Field of the Environment. UNEP/GC/Information/11. United Nations Environment Programme, Nairobi, Kenya. 205 pp.

CHAPTER

49

A MAJOR NEW OPPORTUNITY TO FINANCE THE PRESERVATION OF BIODIVERSITY

ROBERT J. A. GOODLAND
Ecologist, Office of Environmental Affairs, The World Bank, Washington, D.C.

In June 1985, the World Bank[1] promulgated a major new policy (World Bank, 1986) regarding wildlands, which are defined as natural land and water areas in a state virtually unmodified by human activity. This policy focuses mainly on the preservation of wild plants and animals and their habitats. It is very significant to the financing of the preservation of biological diversity.[2] There is no need to justify this policy to those committed to biological diversity, although such justification has been provided by the World Bank (1986, 1987), by Ehrlich and Ehrlich (1981), by Goodland (1985), and by Norton (1986).

The World Bank's general policy is to avoid eliminating wildlands—following the first injunction of Hippocrates 2,000 years ago: "Non noli nocere" (First do no harm). Rather, the Bank will assist in the preservation of wildlands. This translates into six specific policy elements that are addressed at various stages of the Bank's project cycle (see Appendix A at the end of this chapter).

The first policy element states that the Bank normally declines to finance projects involving the conversion of wildlands of special concern, which are areas recognized as exceptionally important in conserving biological diversity or perpetuating en-

---

[1] Includes the International Bank for Reconstruction and Development (IBRD), the International Development Association (IDA), and the International Finance Corporation (IFC).

[2] Preservation of genetic diversity, particularly varieties of economic plants and animals, is acknowledged to be important but is not discussed separately from biotic diversity in this chapter. The necessary complement, reducing poverty, population growth, and other pressures forcing people to destroy wildlands, is being addressed by, for example, the World Bank and World Resources Institute. See Chapter 44 by Spears and Chapter 45 by Burley.

437

vironmental services. They can be classified into two types. First are wildlands officially designated as protected areas by governments, sometimes in collaboration with the United Nations or the international scientific community. These are national parks, biosphere reserves, world heritage sites, wetlands of international importance, areas designated for protected status in national conservation strategies or master plans, and similar wildland management areas (WMAs), i.e., areas where wildlands are protected and managed to retain a relatively unmodified state. Second are wildlands not yet protected by legislation but recognized by the national or international scientific and conservation communities, often in collaboration with the United Nations, as exceptionally endangered ecosystems, known sites of rare or endangered species, or important wildlife breeding, feeding, or staging areas. These include certain types of wildlands that are threatened throughout much of the world, yet are biologically unique, ecologically fragile, or especially important for local people or for environmental services. Tribal people almost always manage their areas sustainably and in a manner entirely compatible with environmental conservation; this important link is discussed by the World Bank (1982).

Wildlands of special concern often occur in tropical forests, Mediterranean-type brushlands, mangrove swamps, coastal marshes, estuaries, sea grass beds, coral reefs, small oceanic islands, and certain tropical freshwater lakes and riverine areas. Within the spectrum of tropical forests, lowland moist or wet forests are the richest in species and are often the most vulnerable. Wildlands of special concern also occur in certain geographical regions where they have been reduced to comparatively small patches and continue to undergo rapid attrition. As a result, these regions harbor some of the most threatened species in the world. Following is a list of some tropical wildlands of special concern:

EAST AFRICA

Madagascar: significant proportions of the northern and eastern moist forests.
Ethiopia: much of the remaining highland forest.
Tanzania: Usambara, Pare, and Uluguru Mountains.
Rwanda: mountain forests along the Zaire and Uganda borders.
Kenya: Kakamega, Nandi, and Arabuko-Sokoke forests.

WEST AFRICA

Cameroon: particularly Cameroon Mountain and the moist forested area extending into Gabon and to the Cross River in southeastern Nigeria, including the Oban Hills.
Ivory Coast: southwestern forests (including the Tai forest) and adjacent parts of Liberia and Sierra Leone.

EAST ASIA AND PACIFIC

The Malay Peninsula (including parts of Thailand): lowland forests, especially along the northwestern and eastern coasts.
Indonesia: much of the remaining lowland forests of Kalimantan, Sumatra, Sulawesi (especially the two southern peninsulas), and many smaller islands (e.g., Siberut).
Philippines: much lowland forest on all larger islands.

SOUTH ASIA
Sri Lanka: the coastal hills of the southwest and the Sinharaja forest of the wet zone.

India: most of the forests remaining on the Western Ghats.

Burma: the untouched teak in the northern regions.

LATIN AMERICA AND CARIBBEAN
Ecuador: lowland coastal forests.

Mexico: Lacandon forest in Chiapas.

Honduras-Nicaragua border: Mosquitia forest.

Panama: Darien province.

Colombia: the Choco region adjacent to Darien province.

Brazil: coastal forests of the Cocoa Region in the southeastern extension of Bahia, between the coast and an area located 41° 30′ W and somewhere between 13° 00′ and 18° 15′ S, and an outlier near Linhares, Espiritu Santo.

Brazil: parts of the eastern and southern Amazon region.

TROPICAL AQUATIC AREAS
Brazil, Peru, Colombia, Ecuador, and Bolivia: Amazon River and associated wetlands (including varzea forests); Mato Grosso, Pantanal Swamp.

Venezuela and Colombia: Orinoco River and Delta.

Papua New Guinea: Purari River.

Sumatra, Indonesia: Musi River and Lake Toba.

Malawi: Lake Malawi and other Rift Valley lakes.

Sudan: Sudd Swamp.

Guatemala: Lake Atitlan.

Before a project is begun, a brief wildland survey is needed in order to determine whether the proposed project area fits one of these categories (see sample form on p. 445). This means that more tropical biologists in applied ecology and more biodiversity data will be needed. The World Bank normally declines to finance projects within wildlands of special concern even if they were partly converted prior to the Bank being invited to consider financing. The phrase "normally declines" may appear to be a glaring deficiency, but it provides for mutually useful trade-offs in certain cases. For example, for a request that a typical 1% of a desert national park be inundated by a reservoir, the quid pro quo or trade-off would normally be financing for the whole park plus the addition of an equivalent 1% area. This should be equivalent both biologically and in size. A needed, underrepresented habitat could be the compensatory tract if that is the national priority. Furthermore, the desert national park would get a permanent aquatic habitat and a source of perennial water. Financing the elevation of unprotected or unmanaged conservation units into well-managed parks can often be part of a mutually acceptable trade-off.

The second policy element pertains to wildlands other than those of special concern. The Bank prefers projects on lands already converted (e.g., logged over, abandoned, degraded, or already cultivated) some time in the past rather than in anticipation of a Bank project. This means that biotic surveys are necessary in order to select logged-over sites rather than intact sites. Rehabilitation of degraded sites therefore must take preference over conversion of wildlands.

The third element concerns deviations from these policies. All deviations must be explicitly justified. This need not necessarily include public hearings or town meetings, which are required in applications of the U.S. National Environmental Policy Act (CEQ, 1978). However, explicit justification implies a more open decision-making process than has occurred in the past. The options, particularly site selection, should be discussed outside the main ministry responsible for implementation and with the ministry of the environment (or equivalent), academia, and environmental groups.

The fourth element applies where development of wildlands is justified. In this case, the conversion of less-valuable wildlands is preferable to the conversion of more-valuable ones. This also means that detailed biotic inventory and resource surveys are essential for two reasons: to avoid more-valuable wildlands such as centers of endemism and to locate less-valuable wildlands. At the very least, such essential surveys will strengthen the environmental data base on wildlife and the nation's biological diversity. The need for experienced tropical biologists in applied ecology will therefore increase.

The fifth element applies when significant conversion of wildlands is explicitly justified, as described in the third element. Somewhat arbitrarily, "significant" conversion is defined as the conversion of 100 square kilometers (which is smaller than many development projects). "Significant" conversion also includes tracts smaller than 100 square kilometers when they are a large proportion of the remaining wildland area of a specific ecosystem. The loss of such wildlands must be compensated by including a wildlands management component in the project concerned—usually direct financing for the preservation of an area similar to that converted for the project, both ecologically and in size (it must not be smaller than the area converted by the project). Thus, biotic surveys are essential to define the area slated for conversion and to find an ecologically similar area. Occasionally, the compensatory wildland may be an underprotected habitat rather than another sample of an already adequately protected habitat. These policies pertain to any project in which the Bank is involved, irrespective of whether the Bank is financing the project component that affects wildlands.

The sixth element concerns projects in which conversion of wildlands is not contemplated. The policy is to preserve wildlands for the value of their environmental service alone. The success of projects that do not eliminate any wildlands often depends on the environmental services that those wildlands provide. In such cases, the Bank's policy is to include a project component to conserve the relevant land in a wildland management area (WMA) rather than to leave its preservation to chance. The best example of an environmental services component is in Indonesia, where the water supply for a 197-square kilometer irrigated rice-growing scheme in Sulawesi is protected by designating an entire watershed outside the rice fields as a national park. This 3,200-square-kilometer area, now called Dumoga Bone National Park, received $1.1 million (1.5% of project costs) from the Bank-assisted loan. Since then, the park has become the site of the world's biggest-ever scientific expedition, supported by the government of Indonesia and the Royal

Entomological Society (UK). This expedition, called Operation Wallace, lasted 3 years and involved over 150 scientists of many disciplines, mainly taxonomists and ecologists.

Four types of economic development projects need compensatory wildland management components, as required under the fifth policy element:

- *Agriculture and livestock projects* involving wildland clearing, wetland elimination, wildland inundation for irrigation storage reservoirs, watershed protection for irrigation, displacement of wildlife by fences or domestic livestock; *fishery projects* involving elimination of important fish nursery, breeding, or feeding sites; overfishing or introduction of ecologically risky exotic species within aquatic wildlands; *forestry projects* involving access roads, clear-cutting or other intensive logging of wildlands, wildland elimination. Plantations of fast-growing tree species are often an important complement to more direct wildland management activities by reducing the economic pressures for cutting the remaining forest wildland. They should be sited preferentially on already deforested land.

- *Transportation projects* involving construction of highways, rural roads, railways, or canals that penetrate wildlands or that facilitate access and spontaneous settlement in or near wildlands; channelization of rivers for fluvial navigation; dredging and filling of coastal wetlands for port projects.

- *Hydroelectric projects* involving large-scale water development, including reservoirs and power and water diversion schemes; inundation or other major transformation of aquatic or terrestrial wildlands; watershed protection for enhanced power output; construction of power transmission corridors. Wetlands (such as ponds, marshes, swamps, flood plain forests, estuaries, and mangroves) can be eliminated inadvertently through water diversions upstream or deliberately through drainage, diking, or filling.

- *Industry projects* involving chemical and thermal pollution, which may damage wildlands; wildland loss from large-scale mining; wildland conversion for industrial fuels or feedstocks.

Essentially, any project affecting wildlands, even indirectly, is a candidate (Ledec and Goodland, in press). The policy also states that projects with wildland management as the sole objective should be encouraged. This does not mean the Bank has changed from being an economic development lender to an environmental granting agency. The preservation of wildlands can often be justified with economic criteria alone, such as the income generated by tourism or scientific and educational activities. Moreover, it means that conversion of wildlands will be increasingly unlikely in Bank-financed projects. More importantly, it means that where some conversion will occur, then the trade-off translates into major financing for the preservation of biological diversity.

TYPES OF WILDLAND MANAGEMENT COMPONENTS

The most effective component of wildland management is financial support for the conservation of ecologically similar wildlands in one or more WMA. Where a WMA has already been established in the same type of ecosystem that is to be converted by a Bank-supported project, then it may be preferable, for administrative or biological conservation reasons, to enlarge the existing WMA rather than to establish a new one. Biological conservation is usually more effective in one large WMA than in several small ones that combined have the same total size and the same types of natural habitats. The government's wildland agencies, local university wildlife departments, and various international organizations can often assist in making such judgments.

A wildland management component could also involve the creation of a new wildland habitat rather than the preservation of existing habitat. For example, marginal land on the fringes of irrigation projects could be converted to wildlife reserves by taking advantage of the water supply created by the projects. Natural depressions or seasonal swamps could be exploited by diverting water from the canal systems (probably a very small part of the total supply). Such reserves could attract significant numbers of migratory and resident waterfowl with minimal additional project costs and land. The Wildfowl Trust in Slimbridge, England, has set up such reserves, ranging in size from 5 to 8 square kilometers.

A useful option is to improve the quality of the management of existing WMAs. Many WMAs in Bank-member countries receive insufficient on-the-ground management due to a lack of adequately paid staff; deficiencies in training, staff housing, equipment, spare parts, and fuel; or the absence of a well-developed management plan through which efficient resource-allocation decisions can be made. Small components can often help correct these deficiencies. In countries where effective management is clearly lacking, it is generally preferable to improve the management of existing WMAs than to create new units on paper, thereby further overextending the limited capabilities of the responsible agencies. Whenever a new WMA is established as a project component, provisions are needed to ensure effective management. Since many wildland agencies (e.g., departments of national parks or wildlife) are not as operationally effective as necessary, institutional strengthening (particularly support for training) should be an important element of Bank-supported wildland management components.

The establishment or strengthening of WMAs is particularly effective when the government includes these wildland areas in a national conservation or land-use plan. A growing number of Bank-member governments have undertaken some type of systematic land use planning for wildland management. Such planning can take various forms, ranging from master plans for a system of national parks and other WMAs to National Conservation Strategies, which address wildland management as only one component of a broad range of natural-resource planning concerns, and in which policy interventions such as economic incentives are used to influence resource utilization. Bank assistance with such planning efforts greatly strengthens wildland management at the national level. When member governments agree to develop appropriate land-use plans, the Bank refrains from supporting projects that involve the elimination of wildlands and run counter to these plans.

SIGNIFICANCE FOR BIOLOGICAL DIVERSITY

The significance of these new policies for biological diversity is that the World Bank will finance the preservation of wildlands. The World Bank is owned by 152 developed and developing member nations, so it is likely that the species or habitats with which most readers are concerned occur in a country that is a member of the Bank. This demonstrates a clear need for nongovernment organizations (NGOs). Communications between governmental agencies, NGOs, and academia are not yet efficient and systematic in the environmental arena. The Bank seeks to foster such communications.

The Bank lends money mainly for specific economic development projects—more than US$16 billion in fiscal 1986. This means that the country must be willing to borrow at near-prevailing global interest rates. If World Bank lending conditions become too onerous, a borrower may seek less-conditional financing elsewhere (and there are indications that this has already started to occur). Also, if the project does not have an impact on a wildland or is not influenced by a wildland, financing for wildlands becomes more difficult. If a project needs only 1% of a wildland, for example, a mutually beneficial trade-off can be negotiated. If you hear of a project needing such wildland financing, you may want to contact the government concerned or the Bank's project officer. Improved cooperation between government, academia, NGOs, and financing agencies is a key element in preserving biological diversity. Although this is not easy, the alternative is accelerating extinctions.

REFERENCES

CEQ (Council on Environmental Quality). 1978. Regulations (for Implementing the Procedura Provision of the) National Environmental Policy Act. U.S. Government Printing Office, Washington, D.C. 44 pp.

Ehrlich, P. R., and A. Ehrlich. 1981. Extinction: The Causes and Consequences of the Disappearance of Species. Random House, New York. 305 pp.

Goodland, R. 1985. Wildlands Management in Economic Development. First International Wildlife Symposium. IX World Forestry Congress and the Wildlife Society of Mexico, Maria Isabel Sheraton Hotel, Mexico, 14 May 1985. The World Bank, Office of Environmental and Scientific Affairs, Washington, D.C. 34 pp.

Ledec, G., and R. Goodland. In press. An environmental perspective on tropical land settlement. In D. A. Schumann and W. L. Partridge, eds. Human Ecology of Tropical Land Settlement in Latin America. Westview Press, Boulder, Colo.

Norton, B. G., ed. 1986. The Preservation of Species. Princeton University Press, Princeton, N.J. 305 pp.

World Bank. 1982. Tribal Peoples and Economic Development, Human Ecologic Considerations. The World Bank, Washington, D.C. 111 pp.

World Bank. 1984. Environmental Policies and Procedures of the World Bank. The World Bank, Office of Environmental and Scientific Affairs, Washington, D.C. 8 pp.

World Bank. 1986. The World Bank's Operational Policy on Wildlands: Their Protection and Management in Economic Development. The World Bank, Washington, D.C. 21 pp.

World Bank. 1987. Wildlands: Their Protection and Management in Economic Development. The World Bank, Washington, D.C. 300+ pp.

APPENDIX A:
THE WORLD BANK'S PROJECT CYCLE

Responsibility for implementing wildland management projects or components within the World Bank rests primarily with the Bank's regional operations staff, which receives advice and operational support from the Office of Environmental and Scientific Affairs, Projects Policy Division (PPDES). The key contact person is the project officer.

Projects under consideration are reviewed by the Bank's regional staff in conjunction with PPDES to identify, as early as possible, the need to avoid converting a wildland tract or to preserve such a tract as part of the project. To determine whether a proposed project will have an impact on or be in close proximity to environmentally important wildlands, Bank staff consult those government agencies with jurisdiction over wildland management authority. PPDES maintains contacts with such agencies and will assist upon request. Additional sources of information on ecologically important wildlands include computerized data bases maintained by some nongovernment organizations (NGOs) and several published directories, which are available from PPDES. In this manner, it will often be possible to learn quickly whether a proposed project site contains existing or proposed WMAs; known endangered species; major wildlife or fish breeding, feeding, or staging areas; important watershed catchments; or living resources of major importance to local people. If none of these mechanisms reveal the existence of ecologically important wildlands in the project area, a brief preproject field survey is necessary since many important wildlands have not yet been identified. This field survey should be undertaken by relevant specialists from the government's environmental ministry, wildlife agency, national university, or similar institution. This brief survey indicates the nature and extent of impacts on critical wildlands that would result from the implementation of the project and puts the information in a national context. The results should be recorded on the sample form provided below.[3]

During project preparation (or feasibility study stage), the Bank's project staff (or their consultants) may assist the borrower (e.g., the Ministry of Highways) or the project sponsor in carrying out the necessary environmental studies, including those pertaining to wildlands. PPDES can recommend consultants or other experts who can identify important wildland areas, carry out necessary field surveys, or help design appropriate wildland management project components. At the completion of any necessary studies, information on whether the project involves the conversion or disintegration of a relatively unmodified ecosystem and alternative suggestions for achieving the goals of the government should be added to the project brief. If conversion is justified, the brief should explain why and should identify the wildland management components needed.

---

[3]This type of information is expected as part of project identification and can be used for the project brief. This form can be completed by the government's environmental ministry or wildlife agency, or by the project prefeasibility team's wildlands specialist.

As part of the appraisal process, the Bank's project staff assesses the planned wildland management and other environmental measures. The Staff Appraisal Report specifically describes any planned wildland management measures, including budgets and agency responsibilities. Once wildland measures are identified as necessary, timely action should be ensured by loan conditionality such as loan effectiveness or disbursement. Since wildlands must be managed in perpetuity to be effective, the loan agreement should specify long-term measures that the borrowing nation has agreed to implement.

Supervision should routinely review implementation of the wildland component with the borrower. As a general principle, the wildland component of the project should be well under way before major land clearing or construction are allowed to proceed.

WILDLAND SURVEY AND MANAGEMENT FORM
(Sample Only)

Name of project: ————————————————————

Expected appraisal (or other) date: ——————————————

Date of this survey: ———— Surveyor: ———— Affiliation: ————

Methodology(ies) (circle one): site inspection / library research / both / other (specify)

1. Specific subcategory(ies) of ecosystem that proposed project will affect (e.g., tropical semievergreen moist forest, salt marsh, wet savanna):

2. Important environmental and biological features of ecosystem(s) (e.g., water-catchment area for large agricultural valley and habitat for the endangered mountain gorilla):

3. Projected general type of impact on ecosystem(s) of proposed project (e.g., deforestation, flooding, draining):

4. Proportion (%) of the region's remaining ecosystem(s) (as in No. 1 above) to be converted (and/or impacted, if different) (e.g., this project will flood about 10% of this country's remaining lowland riparian swamp forest):

5. Estimated annual rates of attrition of affected ecosystem(s) in this country and historical trend of this rate (e.g., the current annual rate of attrition of semimontane forest is 3% a year. This rate was 0.5% in 1975 and 1% in 1980):

Maps and more complete reports used or available should be appended or cited.

AND TODAY WE'RE GOING TO TALK ABOUT BIODIVERSITY . . . THAT'S RIGHT, BIODIVERSITY

LESTER R. BROWN

President, Worldwatch Institute, Washington, D.C.

In Chapter 1, E. O. Wilson says that we are locked into a race concerning biodiversity. He implies that we are in a race against time and that the "we" is humanity. Unfortunately, the only people who are actively engaged in the race to preserve our rich evolutionary inheritance of plant and animal life are a handful of concerned scientists and environmentalists. To make the sort of headway that the situation calls for will require a lot more people expressing concern and working on the issue.

The problem is difficult in that the loss of biodiversity is largely the indirect result of other activities, such as producing food and using energy. It is similar to carbon dioxide-induced climate change, the depletion of the ozone layer, and the acidification of lakes and forests. Often there are no obvious bad guys. And the issue is not new: *The Sinking Ark* was published in 1979 (Myers, 1979); *Extinction* came out in 1981 (Ehrlich and Ehrlich, 1981). Looking at the problem and comparing it with other issues that Worldwatch has looked at over the years, I can identify three things that might help in generating response on the scale needed: better information, internationalization of the issue, and heightened public awareness.

It has been impressive and confusing to see the number of estimates on the rate of future extinction—just looking to the end of the century. The estimates, from reputable scientists who are concerned about the issue, vary not only widely, but wildly. This uncertainty provides an excuse for inaction by those who prefer not to be doing anything. It might be useful to bring scientists together to try to narrow the range of estimates, to clarify the assumptions, and to provide some numbers that can be used. In addition to the rate of species loss, we need a better understanding than what we now have of the economic implications.

The question concerning internationalization of the issue is how. One way would be to look to the United Nations (UN) and its special conferences, a procedure that has been followed for the last 15 years with other emerging global problems. The 1972 UN conference on the environment in Stockholm gave the environmental issue international legitimacy. Then came the UN conference on population in Bucharest and one on food in Rome, both in 1974. Following this came conferences on habitat in Vancouver, on water in Rio de Plata, on deserts in Nairobi, and on technology in Vienna.

Maybe the UN should have a conference on biodiversity. Not because the conferences themselves achieve great things in their own right but because they do serve important educational functions. Their first function is education of the bureaucracy. This occurs when governments have to analyze their own national situation and prepare their position papers for the conference. Their second important educational function is that they provide an excuse for *Time* magazine to carry a story on biodiversity. And that's important.

Another option related to the UN is a possible special session at the UN General Assembly. Special sessions in the last few years have addressed disarmament and Africa. Perhaps biodiversity should also be an integral part of the new, major, international initiative that the International Council for Scientific Unions (ICSU) will be orchestrating on global habitability.

These activities can raise public awareness. There is a long-standing principle in Washington, D.C., that you can convince everyone within the beltway to move in a particular direction on a particular issue, but if no one outside the beltway knows the issue exists, nothing will happen. To illustrate, the United States over the last half century has had two soil erosion crises. The first was the Dust Bowl in the thirties—a highly visible crisis, which is now an era in U.S. history. The clincher in getting a response came when Congress left a session late one afternoon in 1934 and discovered their cars covered with dust. Someone pointed out that the soil came from Oklahoma and Kansas. Only then did the urban Congressmen from the East, who previously had been unmoved by the issue, take action by creating the Soil Conservation Service (SCS), which launched an effective response to that threat.

The more recent erosion problem unfolded in the late 1970s. This resulted from the doubling of grain prices in 1972 after the Soviet wheat purchase and the fencerow-to-fencerow planting, which increased the amount of cultivated land by about 10%. Much of that land should not have been brought under the plow. Coupled with this was a second issue: the need for information. Congress passed the legislation requiring a natural resources inventory. The SCS conducted two inventories of soil erosion, the first in 1977 and a second in 1982. With the two points, we could begin to get a fix on trends. Over a million readings were involved in the 1982 inventory, literally almost down to the individual farm. The inventory told us precisely where the soil erosion was occurring, under what conditions, and how much soil was being lost. We also knew from a number of research projects what that loss translated into economically. On both corn land and wheat land, a loss of one inch of topsoil reduces yields by 6%. For corn, the loss was about 6 bushels per acre; for wheat, about 2 bushels per acre. Thus erosion and economics were indisputably linked.

Many other studies followed. The American Farmland Trust (1984) did an excellent study using the data from the National Resources Inventory. The National Audubon Society made soil erosion one of their projects. At the Worldwatch Institute, our very first book over a decade ago was entitled *Losing Ground* (Eckholm, 1976). We've since done two Worldwatch Papers, and each *State of the World* (Brown et al., 1984, 1985, 1986) report has had either a chapter or a section of a chapter dealing with the issue. We have in our files something close to a thousand articles generated or bylined by staff members in periodicals, newspapers, magazines, and journals around the world. It has been an enormous educational process involving many groups, of which we are one.

I remember being invited to appear on the *Today Show* to talk about this issue. The lead that Jane Pauley used beginning at 7:00 a.m. was, "And today we're going to have an author who's written a book about soil erosion. THAT'S RIGHT, soil erosion." They repeated this in incredulous tones as though it was not an issue that would be found on the *Today Show*. For weeks after that show, my two children entertained themselves around the house saying, "And our dad's written a book on soil erosion. THAT'S RIGHT, soil erosion."

Once public concern was aroused and there was a public awareness of the issue, it became possible for politicians to act. In addition to the traditional organizations (e.g., the Soil Conservation Service of America), environmental groups, such as Audubon and Sierra—large membership groups with skilled lobbyists who write position papers and work with members of Congress—became active. Suddenly it became politically the thing to do, and both Houses voted overwhelmingly for a new farm bill that called for taking 45 million acres (18 million hectares) of our most highly erodible cropland out of production over the next 5 years and putting it into either grass or trees under 10-year contracts. The second thing they did would have been anathema only a few years ago. The farmers who were cropping highly erosive land or managing it badly were told they would not be eligible for farm support programs! That was new.

We are in a race. Maybe we should call it a contest. A lot of the things that are happening, like the threats to biodiversity, are the products of rather innocent actions. Population growth is the result of reproductive patterns that were an integral part of our survival as a species. It has now become a real threat. In July our number reached 5 billion. There was no dancing in the streets: rather, there was a profound sense of unease, not only because we had reached 5 billion but also because there is so much momentum inherent in population growth now. With the young age structure, there will be 3 billion people entering the reproductive age group over the next generation—an enormous push. Nonetheless, the U.S. government recently announced that it was withdrawing all support for the UN Fund for Population Activities—the major agency responsible for working with over 199 countries to try to get the brakes on population growth.

One of the lessons I draw from this sort of exercise is that scientists are going to have to become activists. The recently formed Club of Earth has pointed out that some of the leading scientific minds working on biodiversity are deeply concerned about the issue and believe that we need to take action on a number of fronts. Maybe the Club of Earth,[1] the purpose of which is to bring scientific attention more quickly to important but neglected environmental problems, should write the White House and explain why it is important to restore full support for the UN Fund for Population Activities in order to protect biodiversity.

We've got to move the issue from the scientific journals into the magazines and the popular press, so that maybe someday Jane Pauley will say, "And today we have a scientist who's going to discuss biodiversity. THAT'S RIGHT, biodiversity."

REFERENCES

American Farmland Trust. 1984. Soil Conservation in America: What Do We Have to Lose? American Farmland Trust, Washington, D.C. 133 pp.

Brown, L. R., W. U. Chandler, C. Flavin, J. Jacobson, C. Pollock, S. Postel, L. Starke, E. C. Wolf. 1984. State of the World 1984. W. W. Norton, New York. 250 pp.

Brown, L. R., W. U. Chandler, C. Flavin, J. Jacobson, C. Pollock, S. Postel, L. Starke, E. C. Wolf. 1985. State of the World 1985. W. W. Norton, New York. 301 pp.

Brown, L. R., W. U. Chandler, C. Flavin, J. Jacobson, C. Pollock, S. Postel, L. Starke, E. C. Wolf. 1986. State of the World 1986. W. W. Norton, New York. 263 pp.

Eckholm, E. P. 1976. Losing Ground. Environmental Stress and World Food Prospects. W. W. Norton, New York. 223 pp.

Ehrlich, P. R., and A. Ehrlich. 1981. Extinction. The Causes and Consequences of the Disappearance of Species. Random House, New York. 305 pp.

Myers, N. 1979. The Sinking Ark. Pergamon, Oxford. 307 pp.

---

[1] Its members are Jared Diamond, Paul R. Ehrlich, Thomas Eisner, G. Evelyn Hutchinson, Ernst Mayr, Charles D. Michener, Harold A. Mooney, Peter H. Raven, and Edward O. Wilson.

CHAPTER

51

THE EFFECT OF GLOBAL CLIMATIC CHANGE ON NATURAL COMMUNITIES

ROBERT L. PETERS II

Research Associate, World Wildlife Fund-Conservation Foundation,
Washington, D.C.

Current human population and development pressures are breaking wild biological communities into fragments surrounded by human-dominated urban or agricultural lands. The result is that many wild species, perhaps hundreds of thousands by the end of this century, will be lost because of habitat disturbance (Lovejoy, 1980; Myers, 1979). Recent advances in conservation biology have demonstrated that even some species we thought would be protected within reserves may still be lost because the reserves are too small to maintain viable populations of all the species within them (Frankel and Soulé, 1981; Schonewald-Cox et al., 1983; Soulé, 1986; Soulé and Wilcox, 1980). To this daunting picture must be added a newly recognized threat, one with potentially disastrous consequences for biological diversity. This threat is global warming, commonly called the greenhouse effect.

It now seems very likely that ecologically significant climate change will occur within the next century and that many natural populations of wild organisms will be unable to exist within their present ranges. They will be lost, unless they are able to colonize new habitat where the climate is suitable, either on their own or with human help. Simply because many species survived past natural climate changes does not mean that they will survive this one without aid. The coming change promises to be very big and very fast, and because human activities will increasingly fragment and isolate populations, it will be more difficult for many species to successfully colonize new habitat when the old one becomes unsuitable.

FUTURE CLIMATE

What do we know about the climatic future? In the last several years a virtual consensus has been reached among atmospheric scientists that the planet will warm significantly during the next hundred years as the result of the production of carbon dioxide and other so-called greenhouse gases by humans (NRC, 1983; Schneider and Londer, 1984). Because molecules of these gases absorb infrared radiation, preventing it from radiating into space, increases in their concentration will cause increases in average global temperature. Exactly how large the warming will be and how fast it will come are still uncertain, but best estimates are of a magnitude sufficient to have profound effects on natural biological systems. In 1983 the National Research Council concluded that $3 \pm 1.5°C$ of warming by the end of the next century was most likely (NRC, 1983), based on effects due to carbon dioxide concentration alone (see Figure 51-1A for one model's predictions). More recent analyses of the contributions of other greenhouse gases, including methane and the chlorofluorocarbons, suggest that the total greenhouse effect may be double that of carbon dioxide alone—cutting in half the amount of time necessary to reach a particular level of warming (Machta, 1983; Ramanathan et al., 1985). In short, warming of several degrees is likely within the next 100 years, perhaps the next 50 years. While a warming of this amount may seem small, it is not. Even a 2°C change is very large compared to normal fluctuations and would leave us with a planet warmer than at any time in the past 100,000 years (Schneider and Londer, 1984).

In thinking about the effects of climate change on natural communities, it is important to realize that the effects do not suddenly begin at some arbitrary threshold, such as the commonly used benchmark of doubled carbon dioxide concentration. Rather, ecological responses will begin with small amounts of warming and will increase as the warming does. Thus, a species like the dwarf birch (*Betula nana*), which exists in Britain only at sites where the temperature never exceeds 22°C (Ford, 1982), might begin retracting the southernmost portion of its range as soon as the local temperature climbs over 22°C.

In the long term, temperatures may rise above the several degrees predicted for a doubled carbon dioxide scenario, for there is no reason that concentrations of carbon dioxide and other greenhouse gases, and hence warming, should stabilize when the benchmark of doubled carbon dioxide is reached. If people continue to put more gases into the atmosphere, temperatures will continue to climb.

At least as important as temperature rise itself in affecting the distributions of species and the stability of biological communities will be the widespread changes in precipitation it causes (Hansen et al., 1981; Manabe et al., 1981; Wigley et al., 1980). Thus, the southern limit of the European beech tree (*Fagus sylvatica*) is determined by the point at which rainfall is less than 600 millimeters annually (Seddon, 1971), and a change in rainfall would be expected to cause a change in range. Although models of future rainfall distribution based on projected temperature increases are still rough, their implications are cause for concern. One model predicts that global warming will cause rainfall decreases of up to 40% for the American Great Plains by the year 2040 (Figure 51-1B; Kellogg and Schware,

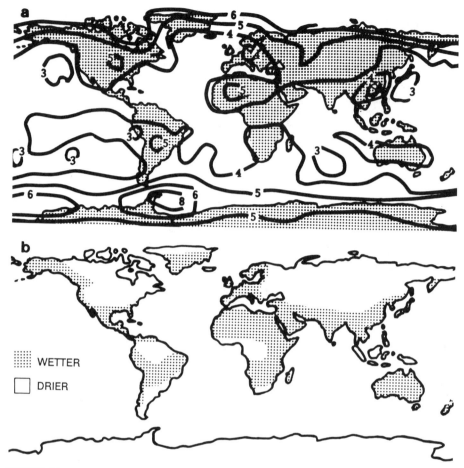

FIGURE 51-1 (a) Global patterns of surface temperature increase, as projected by the Goddard Institute for Space Studies (GISS) model (Hansen et al., in press). Numbers are in degrees Celsius. (b) Global changes in moisture patterns. After Kellogg and Schware (1981).

1981). Other factors associated with rising temperatures that have biological implications include the direct physiological effects of rising atmospheric carbon dioxide concentration itself on plants (in Lemon, 1983) and a moderate sea level rise, variously estimated to be between 144 and 217 centimeters by 2100, according to the U.S. Environmental Protection Agency (EPA) (Hoffman et al., 1983). Plants will vary according to the way carbon dioxide concentrations affect their photosynthetic efficiencies and water requirements, thus altering interspecific relationships. In addition, changes in both precipitation and elevated carbon dioxide levels would alter soil chemistry (Emanuel et al., 1985; Kellison and Weir, in press).

SPECIES RANGES SHIFT IN RESPONSE TO CLIMATE CHANGE

By using the fossil record to study past responses of communities to similar climate changes, we can get some idea of how species ranges might respond to the physiological and competitive stresses imposed by future change. The most important observation is that, not surprisingly, species tend to track their climatic optima, retracting their ranges where conditions become unsuitable while expanding them where conditions improve (Ford, 1982; Peters and Darling, 1985). A general observation is that during past warming trends, species have shifted both toward higher latitudes and higher elevations (Baker, 1983; Bernabo and Webb, 1977; Flohn, 1979; Van Devender and Spaulding, 1979). During several Pleistocene interglacial periods when the temperature in North America was only 2° to 3°C higher than at present, osage oranges (*Maclura* sp.) and pawpaws (*Asimina* sp.) grew near Toronto, several hundred miles north of their present distribution; manatees (*Trichechus* sp.) swam off the New Jersey shore; tapirs (*Tapirus* sp.) and peccaries (*Tayassu* sp.) foraged in Pennsylvania; and Cape Cod had a forest like that of present-day North Carolina (Dorf, 1976). As to altitudinal shifting, during the middle Holocene when temperatures in eastern North America were 2°C warmer than at present, hemlock (*Tsuga canadensis*) and white pine (*Pinus strobus*), for example, were found 350 meters higher on mountains than they are today (Davis, 1983). In general, a short climb in altitude corresponds to a major shift in latitude, so that 3°C of cooling may be found by traveling either 500 meters up a mountain or 250 kilometers toward a pole (MacArthur, 1972).

Evidence of such range shifts during periods of warming in the past, together with projections of range shifts based on physiological tolerances and computer-modeled future climatic conditions, suggest that in the United States, the oncoming warming trend may shift the area within which a particular species may flourish by as much as several hundred kilometers to the north. A projection for loblolly pine (*Pinus taeda*), for example, suggests that the southern limit of this species in the United States may shift more than 300 kilometers to the north by the year 2080 because of moisture stress (Miller et al., in press). Another simulation indicates that the doubling of atmospheric carbon dioxide concentrations expected by the early part of the next century would result in elimination of Douglas fir (*Pseudotsuga taxifolia*) from the lowlands of California and Oregon, because rising temperatures would preclude the seasonal chilling this species requires for seed germination and shoot growth (Leverenz and Lev, in press). On a larger scale, other simulations indicate that projected temperature changes (exclusive of changes in precipitation and soil characteristics) caused by a doubling of carbon dioxide concentration would result in the shifting of entire ecosystem complexes, including the loss of as much as 37% of boreal forest (Emanuel et al., 1985).

Because each species disperses at a different rate, major climatic changes typically result in a resorting of the species constituting natural communities and the creation of new plant and animal associations (e.g., Van Devender and Spaulding, 1979; see also Figure 51-2), thereby causing new, sometimes stressful interactions among species.

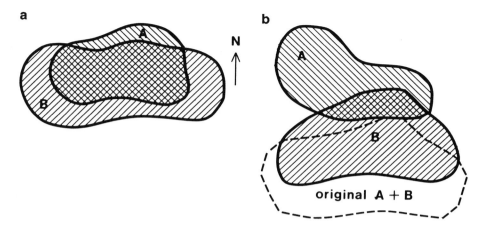

FIGURE 51-2 (a) Initial distribution of two species, A and B, whose ranges largely overlap. (b) In response to climatic change, latitudinal shifting occurs at species-specific rates and the ranges disassociate.

LOCALIZED SPECIES MAY NOT BE ABLE TO COLONIZE NEW HABITAT

If the entire range occupied by a species becomes unsuitable because of climate change, the species must either colonize a new, more suitable habitat or become extinct. The smaller the present range, the more likely it will be that the species will find the entire habitat unsuitable and therefore that extinction will occur. As discussed below, the vulnerability of many species will be increased by human encroachment that restricts them to small areas. Species restricted to reserves, like the one illustrated in Figure 51-3, are good examples.

Imagine a restricted population like that represented in Figure 51-3. What is the chance that colonists, such as seeds or migrating animals, from the original population will find new habitat before the parent population becomes extinct? It will, of course, depend upon a number of factors: how much suitable area there is (i.e., the size of the target the colonists must reach), how far away the suitable area is, how many potential colonists are sent out (which will be a function of how large the original population is and the reproductive strategy of the species), how efficient these colonists are at dispersing themselves, how many physical barriers to dispersal exist, and how long some individuals within the original population can survive to reproduce.

Although the number of colonists produced per parent and their intrinsic dispersal ability are likely to be essentially the same as during past times when species had to respond to climate change, this is not so for the other variables. For many species, the target areas to be reached will be reduced by development, the number of potential colonists will be reduced through reduction of the parent population, the length of time the parent population is allowed to exist may be reduced both through the rapidity of the climate change and development pressures, and, importantly, many more barriers to dispersal in the form of agriculture, urbanization, and other types of habitat degradation will be added to the natural physical barriers of mountains, oceans, and deserts. The predicament faced by a species in this situation is illustrated in Figure 51-4 for the Engelmann spruce (*Picea engelmanni*).

For a plant, the Engelmann spruce is probably a moderate disperser. It has small, wind-dispersed seeds, and its natural dispersal rate, in the absence of barriers, has been estimated to be between 1 and 20 kilometers per century (Seddon, 1971). If we assume that climate change will cause a several-hundred kilometer shift in the potential range of many species in the United States during the next century, say 30 kilometers per year, a plant with the 1- to 20-kilometer per century rate of the Engelmann spruce would be in trouble. Although some species, such as plants propagated by spores, may be able to match the 30 kilometers per year needed, many other species could not disperse fast enough to compensate for the expected climatic change without human assistance. Even some large animals that are physically capable of rapid dispersal do not travel far for behavioral reasons. Rates for several species of deer, for example, have been observed to be less than 2 kilometers per year (Rapoport, 1982).

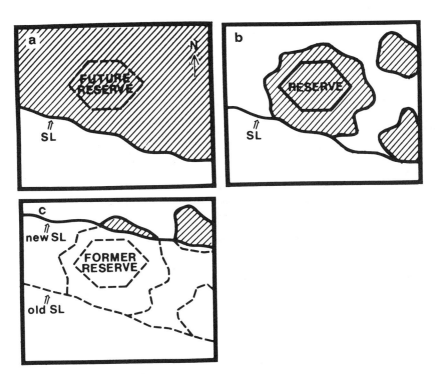

FIGURE 51-3 How climatic warming may turn biological reserves into former reserves. Hatching indicates: (a) species distribution before human habitation; (b) fragmented species distribution after human habitation; (c) species distribution after warming. SL indicates the southern limit of species range.

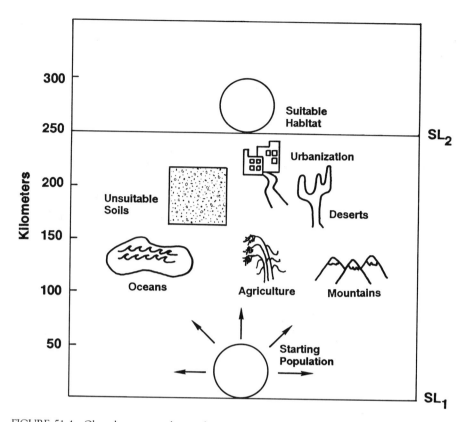

FIGURE 51-4 Obstacle course to be run by species facing climatic change in a human-altered environment. To win, a population must track its shifting climatic optimum and reach suitable habitat north of the new southern limit of the species range. SL_1 is the species southern range limit under initial conditions. SL_2 is the southern limit after climate change. The model assumes a plant population consisting of a single species, whose distribution is determined solely by temperature. After a 3°C rise in temperature, the population must have shifted 250 kilometers to the north to survive, based on Hopkins bioclimatic law (MacArthur, 1972). Shifting will occur by simultaneous range contraction from the south and expansion by dispersion and colonization to the north. Progressive shifting depends upon propagules that can find suitable habitat in which to mature and in turn produce propagules that can colonize more habitat to the north. Propagules must pass around natural and artificial obstacles like mountains, lakes, cities, and farm fields. The Engelmann spruce has an estimated, unimpeded dispersal rate of 20 kilometers/100 years (Seddon, 1971). Therefore, for this species to win by colonizing habitat to the north of the shifted hypothetical limit would require a minimum of 1,250 years.

We know these threats are more than speculation, because the fossil record provides evidence that not only have ranges shifted in response to climate change, but in some cases their total extent was drastically reduced. For example, a large and diverse group of plant genera, including watershield (*Brasenia*), sweetgum (*Liquidambar*), yellow poplar (*Liriodendron*), magnolia (*Magnolia*), moonseed (*Menispermum*), hemlock (*Tsuga*), cedar (*Thuja*), and cypress (*Chamaecyparis*), were found in both Europe and North America during the Tertiary period. But during the Pleistocene ice ages, these all became extinct in Europe, presumably because the east-west orientation of such barriers as the Pyrenees, the Alps, and the Mediterranean blocked southward migration, while they persisted in North America, which has longitudinally oriented mountain ranges (Tralau, 1973).

MANAGEMENT IMPLICATIONS

How might the threats posed by climatic change to natural communities be mitigated? One basic truth is that the less populations are reduced by development now, the more resilient they will be to climate change. Thus, an excellent way to start planning for climate change would be sound conservation now, in which we try to conserve more than just the minimum number of individuals of a species necessary for present survival.

In terms of responses specifically directed at the effects of climate change, the most environmentally conservative action would be to halt or slow global warming. Granted, this would be difficult, not only because fossil fuel use will probably increase as the world's population grows but also because effective action would demand a high degree of international cooperation. If efforts to prevent global warming fail, however, and if global temperatures continue to rise, then ameliorating the negative effects of climatic change on biological resources will require substantially increased investment in the purchase and management of reserves.

To make intelligent plans for siting and managing reserves, we need more knowledge. We must refine our ability to predict future conditions in reserves. We also need to know more about how temperature, precipitation, carbon dioxide concentrations, and interspecific interactions determine range limits (see, for example, Picton, 1984, and Randall, 1982) and, most important, how they can cause local extinctions.

Reserves that suffer from the stresses of altered climate will require carefully planned and increasingly intensive management to minimize species loss. To preserve some species, for example, it may be necessary to modify conditions within reserves, such as irrigation or drainage in response to new moisture patterns. Because of changes in interspecific interactions, competitors and predators may need to be controlled and invading species weeded out. The goal would be to stabilize the composition of existing communities, much as the habitat of Kirtland's warbler (*Dendroica kirtlandii*) is periodically burned to maintain pine woods (Leopold, 1978).

In attempting to understand how climatically stressed communities may respond and how they might be managed to prevent the gradual depauperization of their constituents, restoration studies, or more properly, community creation experiments can help. Communities may be created outside their normal climatic ranges to mimic the effects of climate change. One such relocation community is the Leopold Pines experimental area at the University of Wisconsin Arboretum in Madison, where there is periodically less rainfall than in the normal pine range several hundred kilometers to the north (W. R. Jordan III, University of Wisconsin, Madison, personal communication, 1985). Researchers have found that although the pines themselves do fairly well once established at the Madison site, many of the other species that would normally occur in a pine forest, especially the various herbs and small shrubs, have not flourished, despite several attempts to introduce them.

If management measures are unsuccessful, and old reserves do not retain necessary thermal or moisture characteristics, individuals of disappearing species might be transferred to new reserves. For example, warmth-intolerant ecotypes or subspecies might be transplanted to reserves nearer the poles. Other species may have to be periodically reintroduced in reserves that experience occasional climate extremes severe enough to cause extinction, but where the climate would ordinarily allow the species to survive with minimal management. Such transplantations and reintroductions, particularly involving complexes of species, will often be difficult, but some applicable technologies are being developed (Botkin, 1977; Lovejoy, 1985).

To the extent that we can still establish reserves, pertinent information about changing climate and subsequent ecological response should be used in deciding how to design and locate them to minimize the effects of changing temperature and moisture. Considerations include:

• The existence of multiple reserves for a given species or community type increases the probability that if one reserve becomes unsuitable for climatic reasons, the organisms may still be represented in another reserve.

• Reserves should be heterogeneous with respect to topography and soil types, so that even given climatic change, remnant populations may be able to survive in suitable microclimatic areas. Species may survive better in reserves with wide variations in altitude, since from a climatic point of view, a small altitudinal shift corresponds to a large latitudinal one. Thus, to compensate for a 2°C rise in temperature, a Northern Hemisphere species can achieve almost the same result by increasing its altitude only some 500 meters as it would by moving 300 kilometers to the north (MacArthur, 1972).

• As models of climate become more refined, pertinent information should be considered in making decisions about where to site reserves in order to minimize the effects of temperature and moisture changes. In the Northern Hemisphere, for example, where a northward shift in climate zones is likely, it makes sense to locate reserves as near the northern limit of a species' or community's range as possible, rather than farther south, where conditions are likely to become unsuitable more rapidly.

• Maximizing the size of reserves will increase long-term persistence of species by increasing the probability that suitable microclimates exist, by increasing the probability of altitudinal variation, and by increasing the latitudinal distance available to shifting populations.

• In the future, flexible zoning around reserves could allow us to move reserve boundaries in response to changing climatic conditions. Also, as habitat inside a reserve becomes unsuitable for the species or communities within, reserve land might be traded for nonreserve land that either remains suitable or becomes so as the climate changes. The success of these strategies, however, would depend on a highly developed restoration technology that is capable of guaranteeing, in effect, the portability of species and whole communities.

ACTIONS THAT CAN BE TAKEN

The best solutions to the ecological upheaval resulting from climatic change are not yet clear. In fact, little attention has been paid to the problem. What is clear, however, is that these changes in climate would have tremendous impact on communities and populations isolated by development and that by the middle of the next century, they may dwarf any other consideration in planning for reserve management. The problem may seem overwhelming. One thing is worth keeping in mind, however: the more fragmented and smaller populations of species will be less resilient to the new stresses brought about by climate change. Thus, one of the best things that can be done in the short term is to minimize further encroachment of development upon existing natural ecosystems. Furthermore, we must refine our climatological predictions and increase our understanding of how climate affects species, both individually and in their interactions with each other. Such studies may allow us to identify those areas where communities will be most stressed as well as alternative areas where they might best be saved. Meanwhile, efforts to improve techniques for managing communities and ecosystems under stress and for restoring them when necessary must be carried forward energetically.

ACKNOWLEDGMENTS

Ideas presented in this paper are based in part on an article, "The Greenhouse Effect and Nature Reserves," by R. L. Peters II and J. D. S. Darling, published in *BioScience* 35(11):707–717, 1985. Research was supported by the Conservation Foundation. I wish to thank Joan Darling for her thoughtful collaboration, and I am grateful to Kathy Freas, James Hansen, Bill Jordan, William Kellogg, Thomas Lovejoy, Norman Myers, Elliott Norse, Pamela Parker, Christine Schonewald-Cox, James Titus, and Bruce Wilcox for encouragement, helpful comments, and review of manuscript drafts during the development of these ideas.

REFERENCES

Baker, R. G. 1983. Holocene vegetational history of the western United States. Pp. 109–125 in H. E. Wright, Jr., ed. Late-Quaternary Environments of the United States. Volume 2. The Holocene. University of Minnesota Press, Minneapolis.

Bernabo, J. C., and T. Webb III. 1977. Changing patterns in the Holocene pollen record of northeastern North America: A mapped summary. Quat. Res. 8:64–96.

Botkin, D. B. 1977. Strategies for the reintroduction of species into damaged ecosystems. Pp. 241–260 in J. Cairns, Jr., K. L. Dickson, and E. E. Herricks, eds. Recovery and Restoration of Damaged Ecosystems. University Press of Virginia, Charlottesville, Va.

Davis, M. B. 1983. Holocene vegetational history of the eastern United States. Pp. 166–181 in H. E. Wright, Jr., ed. Late-Quaternary Environments of the United States. Volume 2. The Holocene. University of Minnesota Press, Minneapolis.

Dorf, E. 1976. Climatic changes of the past and present. Pp. 384–412 in C. A. Ross, ed. Paleobiogeography: Benchmark Papers in Geology 31. Dowden, Hutchinson, and Ross, Stroudsburg, Pa.

Emanuel, W. R., H. H. Shugart, and M. P. Stevenson. 1985. Response to comment: Climatic change and the broadscale distribution of terrestrial ecosystem complexes. Clim. Change 7:457–460.

Flohn, H. 1979. Can climate history repeat itself? Possible climatic warming and the case of paleoclimatic warm phases. Pp. 15–28 in W. Bach, J. Pankrath, and W. W. Kellogg, eds. Man's Impact on Climate. Elsevier Scientific Publishing, Amsterdam.

Ford, M. J. 1982. The Changing Climate: Responses of the Natural Fauna and Flora. George Allen and Unwin, London. 190 pp.

Frankel, O. H., and M. E. Soulé. 1981. Conservation and Evolution. Cambridge University Press, Cambridge. 327 pp.

Hansen, J., D. Johnson, A. Lacis, S. Lebedeff, P. Lee, D. Rind, and G. Russell. 1981. Climate impact of increasing atmospheric carbon dioxide. Science 213:957–966.

Hansen, J., A. Lacis., D. Rind, G. Russell, I. Fung, and S. Lebedeff. In press. Evidence for future warming: How large and when. In W. E. Shands and J. S. Hoffman, eds. The Greenhouse Effect, Climate Change, and U.S. Forests. Conservation Foundation, Washington, D.C.

Hoffman, J. S., D. Keyes, and J. G. Titus. 1983. Projecting Future Sea Level Rise: Methodology, Estimates to the Year 2100, and Research Needs. Strategic Studies Staff, Office of Policy Analysis, Office of Policy and Resource Management, U.S. Environmental Protection Agency, Washington, D.C. 121 pp.

Kellison, R. C., and R. J. Weir. In press. Selection and breeding strategies in tree improvement programs for elevated atmospheric carbon dioxide levels. In W. E. Shands and J. S. Hoffman, eds. The Greenhouse Effect, Climate Change, and U.S. Forests. Conservation Foundation, Washington, D.C.

Kellogg, W. W., and R. Schware. 1981. Climate Change and Society: Consequences of Increasing Atmospheric Carbon Dioxide. Westview, Boulder, Colo. 178 pp.

Lemon, E. R. 1983. CO_2 and Plants: The Response of Plants to Rising Levels of Atmospheric Carbon Dioxide. Westview, Boulder, Colo. 280 pp.

Leopold, A. S. 1978. Wildlife and forest practice. Pp. 108–120 in H. P. Brokaw, ed. Wildlife and America. Council on Environmental Quality, U.S. Government Printing Office, Washington, D.C.

Leverenz, J. W., and D. J. Lev. In press. Effects of CO_2-induced climate changes on the natural ranges of six major commercial tree species in the western U.S. In W. E. Shands and J. S. Hoffman, eds. The Greenhouse Effect, Climate Change, and U.S. Forests. Conservation Foundation, Washington, D.C.

Lovejoy, T. E. 1980. A projection of species extinctions. Pp. 328–331 in The Global 2000 Report to the President: Entering the Twenty-First Century. Council on Environmental Quality and the U.S. Department of State. U.S. Government Printing Office, Washington, D.C.

Lovejoy, T. E. 1985. Rehabilitation of Degraded Tropical Rainforest Lands. Commission on Ecology Occasional Paper No. 5. International Union for the Conservation of Nature and Natural Resources, Gland, Switzerland.

MacArthur, R. H. 1972. Geographical Ecology. Harper & Row, New York. 269 pp.

Machta, L. 1983. Effects of non-CO_2 greenhouse gases. Pp. 285–291 in Changing Climate: Report of the Carbon Dioxide Assessment Committee. National Academy Press, Washington, D.C.

Manabe, S., R. T. Wetherald, and R. J. Stouffer. 1981. Summer dryness due to an increase of atmospheric CO_2 concentration. Clim. Change 3:347–386.

Miller, W. F., P. M. Dougherty, and G. L. Switzer. In press. Effect of rising CO_2 and potential climate change on loblolly pine distribution, growth, survival, and productivity. In W. E. Shands and J. S. Hoffman, eds. The Greenhouse Effect, Climate Change, and U.S. Forests. Conservation Foundation, Washington, D.C.

Myers, N. 1979. The Sinking Ark. Pergamon Press, New York. 307 pp.

NRC (National Research Council). 1983. Changing Climate: Report of the Carbon Dioxide Assessment Committee. National Academy Press, Washington, D.C. 496 pp.

Peters, R. L., II, and J. D. S. Darling. 1985. The greenhouse effect and nature reserves: Global warming would diminish biological diversity by causing extinctions among reserve species. BioScience 35(11):707–717.

Picton, H. D. 1984. Climate and the prediction of reproduction of three ungulate species. J. Appl. Ecol. 21:869–879.

Ramanathan, V., R. J. Cicerone, H. B. Singh, and J. T. Kiehl. 1985. Trace gas trends and their potential role in climate change. J. Geophys. Res. 90:5547–5566.

Randall, M. G. M. 1982. The dynamics of an insect population throughout its altitudinal distribution: Coleophora alticolella (Lepidoptera) in northern England. J. Anim. Ecol. 51:993–1016.

Rapoport, E. H. 1982. Areography. Geographical Strategies of Species. Pergamon Press, New York. 269 pp.

Schneider, S. H., and R. Londer. 1984. The Coevolution of Climate and Life. Sierra Club Books, San Francisco. 563 pp.

Schonewald-Cox, C. M., S. M. Chambers, B. MacBryde, and W. L. Thomas. 1983. Genetics and Conservation. A Reference for Managing Wild Animal and Plant Populations. Benjamin-Cummings Publishing, Menlo Park, Calif. 722 pp.

Seddon, B. 1971. Introduction to Biogeography. Barnes and Noble, New York. 220 pp.

Soulé, M. E. 1986. Conservation Biology: The Science of Scarcity and Diversity. Sinauer Associates, Sunderland, Mass. 584 pp.

Soulé, M. E., and B. A. Wilcox. 1980. Conservation Biology: An Evolutionary-Ecological Perspective. Sinauer Associates, Sunderland, Mass. 395 pp.

Tralau, H. 1973. Some Quaternary plants. Pp. 499–503 in A. Hallam, ed. Atlas of Palaeobiogeography. Elsevier Scientific Publishing, Amsterdam.

Van Devender, T. R., and W. G. Spaulding. 1979. Development of vegetation and climate in the southwestern United States. Science 204:701–710.

Wigley, T. M. L., P. D. Jones, and P. M. Kelly. 1980. Scenario for a warm, high CO_2 world. Nature 283:17.

WAYS OF SEEING THE BIOSPHERE

A view of Puget Sound from Waldron Island, Washington, part of which is owned by The Nature Conservancy. *Photo courtesy of Susan Bournique.*

MIND IN THE BIOSPHERE; MIND OF THE BIOSPHERE

MICHAEL E. SOULÉ

Adjunct Professor, School of Natural Resources, University of Michigan,
Ann Arbor, Michigan

Say you want to convince your father-in-law to get involved in conservation—in rescuing biodiversity. How would you start? Would you tell him about the genes for disease resistance in wild relatives of crops plants? Would you mention the probable existence of undiscovered, valuable pharmaceuticals, talk of tropical rain forests and their rates of conversion, or describe a personal experience of nature that still brings tears to your eyes or goose bumps to your skin? That is, would you appeal to his intelligence or to his emotions? The chapters in this section may help to inform us about this choice.

There are many ways of seeing the biosphere. Each of us is a unique lens, a lens ground and coated by nature and nurture. And our responses to nature—to the world—are as diverse as our personalities, though each of us, at different times, may be awed, horrified, dazzled, or just amused by nature.

Most such experiences are quite ordinary, everyday encounters with suburban birds, street trees, garden pests, or domesticated plants and animals. But some of these experiences leave vivid memories and can change our behavior. These so-called peak experiences can fuse our separate selves to nature, establishing a lifetime bond.

Ordinary or sublime, such encounters constitute just one of several dimensions of our total involvement with the natural world. It is the fundamental dimension, though, because experience provides the raw material out of which the more conceptual dimensions are formulated.

What are these ordinary dimensions? Previous sections in this volume deal with some of them, including the value dimension, which is dominated by the polarity

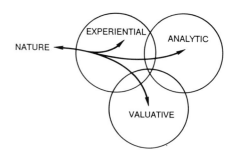

FIGURE 52-1   Three dimensions of mind involved in our perception of nature. The basis for the overlap of these dimensions is both neurophysiological and experiential.

between utilitarian values on the one hand and intrinsic (spiritual-ethical) values on the other. Another is the scientific-analytical one, in which the mind perceives biodiversity as a phenomenon to be organized and explained.

The relationships among these three dimensions are shown in Figure 52-1. First, there is one's immediate, sensory experience of nature; it is mediated by the sensory-neural apparatus of the nervous system. Next, this input is categorized, interpreted, and analyzed by the mind (mostly the limbic and neocortical organs of the brain). If the input is particularly arousing, the limbic-hypothalamic centers may trigger emotional responses such as fear, disgust, or sublime joy. In addition, there may be physiological changes such as sweating, goose bumps, and tears, or attack, flight, and exclamations.

Mental activity of another sort may be launched. One sort of activity is normative or judgmental; this is the value dimension mentioned above. The judgments and classification are partly learned. At some stage in our life we may make a generic judgment about nature, deciding whether it is, on the whole, good or bad, or whether it is a part of *me*[1] or, at the other extreme, is a hostile but useful *other*. Many neural structures, including the highest cortical centers, play a role in the normative process.

Finally, the scientific-analytical dimension of mental activity mentioned above occurs in the greatly expanded human cortex. This structure, called the neocortex in humans because of its evolutionary newness, occupies about 70% of the cranial vault, but it is almost nonexistent in the reptile brain. It is in this structure that complex associations are made, theories are conceived, and conceptual systems are born.

When biologists function in this dimension, their desire is to generalize and predict and ultimately to control. Their self-appointed task is to narrow and channel the Amazon of input from nature—to somehow place it into a few, manageable categories. Experience teaches that this process of pigeon-holing can lead to interesting or useful ideas.

The intellectual's standard operating procedure, therefore, is to discriminate, dissect, and simplify, reducing the infinite variety of things and processes to a

---

[1]Such identification with nature is probably the emotional root of the cognitive experience of intrinsic value. See Naess, 1985.

manageable number of categories and to the simplest atomistic parts and processes. This reductionistic approach has worked well in physics, chemistry, and in much of biology. Quite often, when we are finally able to reassemble the whole, it makes more sense (and is more beautiful) than before. For the scientist, in other words, understanding, not ignorance, is bliss.

MOTIVATION

Clearly, then, doing science, a characteristically neocortical-analytical activity, is not the same as loving nature, a limbic-emotional process. But this distinction between scientific activity and our appreciation of nature is a rock that often trips up many of us. Biologists wish to convince others of the importance of protecting biodiversity, including ecological and evolutionary processes. The problem is that very little thought and research has gone into the best ways to accomplish this vital goal.

Scientists, like everyone, usually revert to habitual ways of communicating. Their favorite format is the lecture. Facts, mixed with inductive or deductive reasoning, are presented with the idea of convincing the listener by the power of evidence and logic that nature is important and deserving of support. To the biologist, it may appear to be perfectly obvious that knowledge will lead to action—that once another human being (including a father-in-law or politician) understands the dimensions of the current spasm of species extinction and understands the agricultural, economic, and climatic implications of deforestation and desertization, that human being will *have* to do something about it and will simply be *forced* to join conservation organizations, change his or her life-style, contribute lots of money to the right causes, and vote the right way.

There are two lines of evidence, however, suggesting that such a didactic approach—the lecture-hall model—is inefficient and insufficient. The first is motivational science, the most frequent application of which is commercial advertising and promotion. I won't say much about this, because it isn't my field. But I hope that I won't be too far off the mark if I point out that the content of advertising is rarely informative or logical. Instead, commercials are designed to arouse and to evoke pleasurable emotions and desires. More precisely, they bypass the cognitive centers, communicating through our basic physical desires (oral, sexual) and emotional needs (security, status, control, potential profits).

For example, I recently came across a promotional brochure for a Caribbean cruise. The cover was a frontal view of a well-endowed blond in a minimalist bathing suit. She is kneeling on the beach, looking straight into my eyes with lips apart and an expression of intense desire. The accompanying text reads, "Come Aboard." Inside the brochure, one is told that passengers are called *ShipMates*. The crewmembers are described as *SeaMates*, "free-spirits . . . the most affectionate and entertaining people you'll ever meet . . . whose challenge is to bring you total pleasure, . . . and they don't accept tips." The brochure adds that "Never before has a cruise ship been so free," that there will be "feasts with unlimited beer and wine" and a "Monte Carlo-styled casino." Lavish color photographs display around 40 "mates," none over 25, all nearly naked, nubile, or muscular (if male). There isn't one word about the opportunity to meet local people or to learn about racial, social, or economic problems in the Lesser Antilles.

This is not to argue that one should try to equate extinction of species with sexual gratification;[2] rather, the point is that bad news by itself is not motivating— just the reverse. Physiologically, bad news is depressing, and depression inhibits arousal in the limbic-emotional system. Advertisers and politicians know this tacitly: consumers don't buy coffins, even when on sale, and voters don't elect prophets of doom. Perhaps this is one reason why there are no biologists in the U.S. Congress.

On the other hand, if our objective is to motivate people, the best way to do this is probably with pleasurable experiences and memories. If neurobiology has told us anything about the mammalian brain, especially the human brain, it is that the mind and the body are not separate. Furthermore, the most direct and powerful pathways to pleasurable emotions are *not* via the thought centers of the neocortex but through the sensory-motor centers of the brain stem and cerebellum, and from there into the emotional centers of the limbic system. This is also the region that houses the playful, nurturing, and social behaviors that we find so pleasurable and that must be evoked in the people we wish to involve in the cause of biodiversity.

Perhaps it would be more effective politically to stress that the members of the movement to save nature can have special, positive experiences—peak experiences that flow from participating with others in doing something of great importance and value. Furthermore, the new motivators for nature might take a page from the advertiser's book, promoting a wider love of nature with a sensory, physical experience of nature in the convivial company of like-minded friends.

One reason for the apparent frustration of the conservation educator may be inattention to the distinction between mentation and motivation, between the neocortex and the cerebellar-limbic axis. Students and others may be convinced cognitively, neocortically, of the value of life and diversity, but somehow our audiences don't follow through. The urgency isn't there. It is as if the organ of learning were not hooked up to the organ of doing.

The hypothesis is that if our pedagogy is purely cognitive, our chances of motivating a change in values and behavior are nil. We can't succeed in teaching people biophilia (Wilson, 1984) (i.e., the love of life), with economic arguments and ecological reasoning alone. We must see to it that they have limbic experiences, not just neocortical ones. We must learn from the experts—politicians and advertising consultants who have mastered the art of motivation. They will tell us that facts are often irrelevant. Statistics about extinction rates compute, but they don't convert.

We must also ask if there are critical developmental stages in the training of the limbic system for bonding with nature. Just as Harlow's rhesus monkeys must have physical contact with warm, moving bodies if they are ever to breed suc-

---

[2]The distinction between alerting people about a crisis, such as extinction, and motivating them to act constructively, is often forgotten. Though it may sound heretical, our primary objective as conservationists (not as educators) should be to motivate children and citizens, not necessarily to inform them. Research may show that the two objectives are incompatible.

cessfully, so there might be developmental stages for bonding with nature and landscapes (Orians, 1986). Perhaps college-age students are too old to imprint.

Returning to the father-in-law, who is still waiting to be convinced of the importance of biodiversity, we come face to face with the urgency of communications. What is the message that we want to get across? A Buddhist sutra teaches, "Each thing has its own intrinsic value, and is related to everything else in function and position." Ecology affirms it. But what then? How do we convince others? Maybe it begins with the courage to let ourselves describe our private, emotional experience of nature to our father-in-law.

The following chapters reinforce the idea of a multidimensional conceptualization of our place in nature and the importance of subcortical communication. Lovelock's Gaia hypothesis (Lovelock, 1979) is quintessentially scientific *and* holistic. The biosphere (or the entire planet and the Sun) is a dynamic community of relationships in which humans are imbedded. This interpenetration of processes, this network of mutual dependencies endows the biosphere with certain qualities such as homeostasis (Gaia; Lovelock, Chapter 56), unity (God; Cobb, Chapter 55), immediateness and power (meat; McClure, Chapter 53), and spirit (consecration; Littlebird, Chapter 54). Aren't we lucky!

REFERENCES

Lovelock, J. E. 1979. Gaia. A New Look at Life on Earth. Oxford University Press, Oxford. 157 pp.

Naess, N. 1985. Identification as a source of deep ecological attitudes. Pp. 256–270 in M. Tobias, ed. Deep Ecology. Avant Books, San Diego, Calif.

Orians, G. H. 1986. An ecological and evolutionary approach to landscape aesthetics. Pp. 3–22 in E. C. Penning-Rowsell and D. Lowenthal, eds. Landscape Meanings and Values. Allen and Unwin, London.

Wilson, E. O. 1984. Biophilia. Harvard University Press, Cambridge, Mass. 157 pp.

A MAMMAL GALLERY
Five Word Pictures and Three Poems

MICHAEL McCLURE
Poet/Professor, California College of Arts and Crafts, Oakland, California

1.

THE GIANT PANDA, huge mammal, furred in black and white, basks and lolls in the shadiness of the bamboo grove. The panda sometimes sits like a man, on his rump with legs outspread, on an earthy mound covered with moss. Perhaps he looks at his beloved and family. He is surrounded by his nutriment, by the tips of bamboo plants that reach many times his height from the surface of the Earth toward the sun. Perhaps strange, thoughtless philosophies drift across the platens of his sensorium and create and recreate themselves in his limbs and organs. All of his being is an accumulation of his plasm and the activities of his body. He sprang from the matter of the Earth as it was energied by the nearby star that he sees through the sparse places in the glade.

The bamboos about the panda are air creatures. They draw nitrates, some material substance, and water from the Earth through the pores of their searching motile root tips. But much of the substance of the bamboo is drawn from thin air, from the gasses of the atmosphere, which are changed by a chemical cycle and the sun's rays into solid substance. Gasses become the BODY of the panda via the bamboo. The bamboos are threads that reach from the plant toward the star that energies them.

2.

AN INVISIBLE WATCHER is in a room with a man and woman who are arguing—they are a lover and beloved, a man and wife. They are quarrelling about the payment on a car, or about the loss of a laundry ticket. The argument becomes too intensive for so minor an issue. It appears that the man and woman are enacting a rite. If the invisible observer closes his ears to the *meanings* of the words and listens only to the vocalization as *sounds*, a thought occurs to him:

He is listening to two mammals. It might be two snow leopards, two bison, two wolves. It is a *mammal* conversation. The man and woman are growling, hissing, whimpering, cooing, pleading, cajoling, and threatening. The specific rite and biomelodic patterning of meat conversation rises and falls in volume. It makes variations, it repeats itself, it begins again, it grows, diminishes. There is a hiss and counterhiss. There is a reply and new outburst. The game that the man and woman are enacting, and the ritual, is as old as their plasm. It is capable of extremes of nervous modulation because of their neuronic complexity but it is more than ancient—it is an Ur-rite.

If the man and woman are lucky, and if their intelligences are open, then one of them will HEAR that it is a rite—that they are growling and hissing. Then he, or she, will laugh at the comedy and the ridiculousness of the pretext. The other partner will laugh in response, intuiting the same perception. Most likely it is a sexual ritual. They are hungry for contact with each other. Their intellective and emotional processes have been frozen into simulations of indifference by pressures of the surroundings and events. If they are lucky enough, one of them will raise a hand to the other, and touch or stroke, recognizing the other as the universe, the counterpart of a star, a galaxy, a planet, a bacterium, a virus, a leopard.

3.

A MAN IS SITTING CROSS-LEGGED in bright afternoon sunlight. He opens a book of reproductions of Egyptian art. Clear light gleams off the paper. The *alto relievo* statuary is uncanny. The lazy intellectual mind scans the opposite page and finds text describing the statuary in a foreign language. It says, apparently, that this is a Pharaoh and two goddesses. The man's attention returns to the reproduction—a passing perception takes the shape of a fragmentary poem:

THE MESSENGER (RNA)
slides to the ribosome
(to the Constellation).
The beads move.
The Pharaoh, Chacal, & Hathor
are glabrous
perfectly
balanced
arm in arm. The weight
of the Man-God
is on
one foot/or the other.
They create the gleam
of this dimension,
of this single
process,
of perfection.
But who is who? and WHAT?

The words mime the balance of the figures as they stand—Goddess, Pharaoh, Goddess—side by side, touching one another. Their weight is immaculately balanced. The sculptor of the archaic figures had a knowledge difficult to regain, though easy to reperceive thousands of years later.

The sculptor sensed that man-mammal is created from the inside outward. That man begins at the interior of his cells and from their perfect balance the body is created.

((Within the human body the RNA slides through the walls of the cell's nucleus, through infinitesimal tubes in the structure, and finds the pearllike ribosome bodies in the cytoplasm. The bodies MOVE *across* the long threadlike molecules of RNA and create the substances of the cell.))

The three sculpted figures show muscular development that is excellent, generalized, not excessive. The bodies rest naturally in mammal fashion. A wolf can be seen standing in relaxation, peering with interest, involved and yet disinvolved. The carved stone reproduces muscle tone that is healthy and without contradictory strains. The faces of the Pharaoh and goddesses are as interesting, or as uninteresting, as the faces of snow leopards. Their bodies are erect, with the pelvis slightly forward to balance the weight of the head. The Pharaoh stands with one foot a little forward—it is impossible to tell which foot bears his weight, or if both feet do. The goddesses stands in variations of this posture.

4.

I STAND IN FRONT of the cyclone wire cage containing the female snow leopard. My friend has a tape recorder. We have been taping sounds of animals before the zoo opens. I step over the guardrail where the snow leopardess is watching us. She is indifferent to humans when they keep at a distance. Her task is to fight the physical psychosis of encagement. Most of her waking is spent pacing the constricted outlines of her cage. But now it is early morning, and she is resting. When I step over the guard rail she growls in anger without moving—except her head, which swivels to watch me.

No part of her can reach through the mesh of the cyclone wire. I put my face almost to the wire and nearly to her face. There are only a few inches between her mouth and my face. She is enraged, and her face, which seems divine in such proximity, twists into feline lines of rage. The anger and rage are clearer than the conflicting human expressions on the daily streets. She knows the uselessness of pawing or clawing at me.

She puts her face within an inch of the wire and SPEAKS to me. The growl begins instantly and almost without musical attack. It begins gutturally. It grows in volume and it expands till I can feel the interior of her body from whence the energy of the growl extends itself as it gains full volume of fury. It extends itself, vibrating and looping. Then, still with the full capacity of untapped energy, the growl drops in volume and changes in pitch to a hiss. The flecks of her saliva spatter my face. I feel not smirched but cleansed. Her eyes are fixed on me. The growl, without a freshly drawn breath, begins again. It is a language that I understand more clearly than any other. I hear rage, anger, anguish, warning, pain, even humor, fury—all bound into one statement.

I am surrounded by the physicality of her speech. It is a real thing in the air. It absorbs me, and I can hear and feel and see nothing else. Her face and features disappear, becoming one entity with her speech. The speech is the purest, most perfect music I have ever heard, and I am touched on my cheeks, and on my brow, and on the tympanums of my ears, and by the vibrations on my chest.

We play back the several minutes of this growl, and it is more beautiful than any composition of Mozart. Three-quarters of the way into the tape is the clear piercing crow of a bantam rooster making his reply to the *mise-en-scène* about him—to the calls of his ladies, to the sparrows, to the sounds of traffic, to the growling of the leopardess, to the morning sun, to the needs of his own being to vocally establish his territory. The crow of the tiny rooster is smaller but no less perfect or monumental or meaningful than the statement of the leopardess—they make a gestalt. The tape is a work of art as we listen. WE are translated.

5.

TRAVELLING ON A SMALL SHIP to the Farallon Islands near the San Francisco coast, I spoke with a virologist who had just returned from Australia. He was travelling to the Farallons to study the rabbits there. A lighthouse keeper's son had a pair of rabbits that escaped on the island. The rabbits and their progeny devastated the island of every leaf of plant life. The island was left bare rock, without any vestige of higher plant life. The virologist believed that the rabbits—still populous on the island—ate the desiccated corpses of gulls and seabirds. His idea was that only one type of rabbit had the capability of surviving under these conditions.

I wandered on the island—seeing a rabbit and traces of rabbits—but not a blade of grass or a bush. The island is rocky, craggy, like a miniature, eroding crest of the Alps. After climbing the tiny peak, I descended to the beach, which was scattered with boulderlike rocks. I found myself looking down onto a herd of sea lions, the closest no more than 30 feet away. They were drowsing and lolling in the sun. Seeing something comic in the scene, I raised my hand and began speaking as if I were delivering a sermon. The astonished sea lions dived into the ocean. The ones in the ocean swung about to see me. They began a chorus of YOWPS, and huge angered MEAT CRIES, dense in volume and range. I continued my performance, and they carried on their yowping. Perhaps 30 or 40 of the animals were yowling at one time. They were FURIOUS, ENRAGED, ASTONISHED. Like the leopardess, their voices were driven by hundreds of pounds of meat force and energy. I was frightened, worried that they might change about, clamber out, and pursue me. They remained in the water cursing me in a clear ancient language that left little doubt about meaning.

AND THEN I knew that not only were the monster shapes of meat enraged, they were PLEASED. THEY WERE SMILING AS WELL AS ENRAGED! They were overjoyed to be stimulated to anger by a novel—and clearly harmless—intruder. Undoubtedly they enjoyed my astonishment and fear as well as the physical pleasure of their rage. Perhaps they relished my physical reaction to their blitzkrieg of sound. They began to yowp not only at me but to each other.

My ears could not take it any longer and I began walking up the beach. I walked halfway around the island. Five members of the tribe followed in the waves. They watched, taunted, encouraged, scolded, and enjoyed me to the fullest. I have not been in finer company.

GRAY FOX AT SOLSTICE

WAVES AND FLUFF JEWEL SAND
in blackness. Ten feet from his den
the gray fox squats on the cliff edge
enjoying the beat of starlight
on his brow, and ocean
on his eardrums. The yearling
deer watches—trembling.
The fox's garden trails
down the precipice:
ice plant, wild stawberries,
succulents.
Squid eggs
in jelly bags (with moving
embryos) wash up on
the strand.
It is the night of the solstice.
The fox coughs,
"Hahh!"
Kicks his feet—
stretches.
Beautiful claw toes
in purple brodiaea lilies.
He dance-runs through
the Indian paintbrush.
Galaxies in spirals.
Galaxies in balls.
Near stars and white mist swirling.

TO A GOLDEN LION MARMOSET

OH BEAUTIFUL LITTLE FACE,
PEERING THROUGH
THE DAWN
OF TIME,
THE GOLDEN FUR UPON YOUR CHEEKS
is precious as a rhyme.
The April in your gracious snarl
can loose a body to ungnarl

and
stand
upright in the sun.
Come back, I've caught my mind!
Your life is all I find
to prove ours are worthwhile.

The monster caterpillars
and the teeth of fire
that eat your jungle
crunch my house.
ALL
BEASTS
ARE
MEN;
all men are beasts.

BUT
I want you alive
in more than memory!

ACTION PHILOSOPHY

THAT GOVERNMENT IS BEST WHICH GOVERNS LEAST.
Let me be free of ligaments and tendencies
to change myself into a shape
that's less than spirit.
LET ME BE A WOLF,
a caterpillar, a salmon,
or
an
OTTER
sailing in the silver water
beneath the rosy sky.
Were I a moth or condor
you'd see me fly!
I love this meat of which I'm made!
I dive in it to find the simplest vital shape!

AH! HERE'S THE CHILD!!!

WHAT'S LIBERTY WHEN ONE CLASS STARVES ANOTHER?

54

COLD WATER SPIRIT

LARRY LITTLEBIRD

Director, Circle Film Productions Company, Santa Fe, New Mexico

Stories are sacred and real,
and the storyteller is a consecrated person.
from *Breath of Whales*, Circle film.

In the early New Mexico morning, before the yellow light of dawn, the wind sweeps gently through the high mesa flats. I was little, and I remember waking up hearing the battered corrugated tin shade catch in the wind. It was loosely tied with wires to long wooden poles above the doorway of the little stone and mud house that was my grampa's sheep camp.

My bare feet touched the cool dirt floor as I rushed to dress and be out into the new morning before he would call, "Get up! Wash with cold water; it'll make you tough and handsome, like me." In that early Pueblo Indian childhood on the reservation the word was still sacred to me. I grew up in that world of tending sheep and seeing the land, discovering that life for our people continues in the telling of stories, that language is a gift of God, and when a story is told we are given a way to learn and to live.

A child wakes up believing in the world he finds himself in; and for a long time I believed everyone had stories to guide them, like I did. Forty years later I now see much is lost or being forgotten about the telling of stories. We are now living in a time when the storyteller is no longer a human being disciplined by life but can be anyone with a facility for language, anyone who can turn words cleverly for their own ends, never fearing that such willful action is wrong, or that thoughtless words obstruct the life vision. Where I grew up, the storyteller was still a consecrated person.

476

My grampa, in his stern reprimands, the harshness removed by the mischievous twinkle in his clear eye, understood the commands he gave me and applied that knowledge strictly: "Get up! You gonna rot in that bed all day? Get up; run to the river; wash with the cold water; you wanna stay young, like me." I always saw him wash with cold water, and catching me looking he'd say, "The girls can't stay away from me because I'm so young looking!" Grampa knew life was full, never dull, and always new with unexpected surprises. For him life was real. It had dimension and depth and was to be taken seriously to be enjoyed. A man had to be tough; he knew this, and he wanted me to know it also.

His use of language crossed back and forth between our own Keres tongue and the broken English he used for emphasis and added understanding. "Keep your eyes peel it," he'd tell me, softly waving his firm, dark, and weathered hand in front of his face indicating some imagined far horizon or space I was to see. Over and over I heard him tell me, "Keep your eyes peel it." I knew he meant "keep your eyes peeled," stay watchful, be alert. Or so my mind told me, until one day, as I trailed along behind him trying to match his stride by stepping into the prints he made in the soft earth, he stopped abruptly, and in Keres asked me, "Are you tracking me, stepping in my footsteps?" I nodded, yes. We sat down, right there in the sand, and he gestured with his mouth, puckering his lips to point the way we had just come. "See the tracks; it's clear that's the way we got here, to this place, this spot. Stand up. What'dya see ahead, toward the way we're going?" I looked off toward the distance. "No tracks," I said. He responded, "That's right, until we walk that way there are no tracks. And when we go that way we must leave our own trail. You make your tracks, and I make mine." Then in English he added, "Keep your eyes peel it."

He rose from the ground, and his eyes scanned the whole space before us and I saw from the way his eyes moved that his vision took in the entire land and the distance to the far horizon. His words confirmed what I glimpsed in that look, "We follow in the footsteps of the men of old, all those grampas gone on before us. Their paths are here, but we keep our eyes off the ground, to watch ahead and all around us, to seek what they were looking for." In my heart a voice said, "What were they looking for?" The question was caught in the next beat as grampa reminded me, "Keep your eyes peel it," and he began making his tracks across the land once more.

Grampa wasn't the only one to teach me to accept without question the words of elders. During the winter months when he stayed at sheep camp alone, and I was at home in the village going to school, there were nights I will never forget. Looking back on them, I see how important they were in learning to listen carefully.

In the long winter, the people celebrate and come together to hear the songs and see the dances that tell the drama of our lives. There was a special house near the center of the village. It was a large house with a long wide room. All the people would gather at this old house.

When it was dark and after we had eaten our meal, we would put on our coats and caps, and dressed warmly, leave our home, stepping into the brisk wintery night to walk along the pathway through the village to the special house.

My shoes crunching the fine new powder snow made it squeak. The air didn't feel cold, yet tiny flakes swirled about my face. I imagine I could have reached up and pocketed a fist full of stars, they were so close overhead in the night's frozen sky. When we reached the house, people were already there, patiently waiting.

Waiting doesn't come naturally to little boys. In the long clean room, children shuffle restlessly, squirming in their seats. Mothers whisper back and forth visiting with sisters, aunts, and other relations. An older man nods off in seeming sleep, propped against the wall in the back. On the long bench where he droops, two men carry on a steady conversation, first one speaking then the other, each of them punctuating the telling with loud confirmations or acknowledgments. Their voices are the loudest sound in the room. Catching my restlessness, my grandmother motions me to sit beside her, and even better, up on her lap. Quietly, her breath tingling and warming the smooth surface of my ear, she reminds me about patience, with a story. In her voice, time slips away, reaching like a steady hand to pull me effortlessly through the long night.

"Ponci wanted to dance. The grown-ups only laughed, saying he was just a funny little boy. All day long in secret he made up songs, and when no one was around he would dance and sing. Day by day songs grew inside him and stayed in his imagination. Then something wonderful happened that helped Ponci discover the extraordinary.

"His uncle Dan, a man with many songs, and who was always singing, came to visit. Ponci liked Uncle Dan's singing and all the stories he would tell. He could sit on his uncle's lap and bounce to the rhythm of a song or listen quietly for hours as his stories were told."

" 'There's a lizard who lives where it is very dry and dusty. It's so hot where he lives that the sun has scorched two gray dirt lines down his green back. This lizard always has its eyes open hoping to see a cloud so he can run to it and lie cool in its shadow. That's why you see Lizard always hurrying over the parched earth, running here and there. He's seen and found many places where the clouds gather,' Uncle Dan, explained. With brightly shining eyes, Ponci held his breath as Uncle Dan went on. 'When you see him lift his chest up and down and see his throat moving, he's singing, calling clouds.'

" 'But lizards don't make any sound!' Ponci burst forth wide-eyed. 'Oh, sure they do,' Uncle Dan tells him. 'A true-hearted person who wants to hear will be able to hear. The Lizard People make beautiful songs, and they're good dancers. You've got to be careful though; never tease Lizard; he's a powerful person.'

"The next day, Ponci was awake and up without being called. He ate his blue corn atole, never touched his eggs, and before his mother could question his rushing about, he was out of the house running toward the dry, flat lowland.

"In that stretch of long horizons, underneath the spiny, thin branches of a spindly tumbleweed, lay Lizard, watching the running boy approach. His outstretched body had sensed the boy's footfalls, and he'd seen him pass back and forth several times already. Lizard smiled at the boy's seriousness.

"Ponci, disappointed at not finding Lizard as easily as he thought he should have, was turning to go when out of the dust Lizard presented himself. Lizard moved up and down in the pulsing motion Uncle Dan had described, and his throat moved. The boy's whole being concentrated on the lizard before him. He stood still, trying to hear any faint whisperings of what might be a song. He stood like this for a long time. Finally, Lizard sighed; Ponci moved closer. Doubts crept into his thoughts about the truth of Uncle Dan's story. Laughter suddenly burst forth from the lizard on the ground with such force it caused Ponci to jump back. Lizard lay rolling in mighty shakes of laughter, his skinny belly rumbling. At last he quieted down enough to ask in a strangely melodious and resonant voice, 'So you want to dance?' The question set Lizard, twitching and convulsing, into hysterics once more. Gaining control, he wiped tears from his eyes as he spoke again. 'It's funny to me, a small fellow like you, wanting to dance so much that you would come looking for me. No one's been to see me or to ask for songs for so long, I don't know if I can even remember any. You see, this ground is so dry even tumbleweeds are few and small.' He slumped forward sadly, 'I don't know if I can help you.'

"All this time the boy stood frozen, eyes wide open, amazed and afraid of what he saw happening before him. Lizard cocked his head out of the dust, 'You were looking for me, weren't you?' Ponci nodded yes, then stammered, 'But, but I didn't know I'd really find you.'

" 'Well you found me,' Lizard groaned. 'It might be very hard for both of us seeing how small you are and how old I've gotten.' His eyes blinked. 'Let's hear you sing.' Never having sung for anyone before, Ponci became unsure of himself, and as he started, his voice cracked so he stopped. Lizard smiled, gently reminding him, 'Anyone can be born with a beautiful voice. Look at Mockingbird. He makes some of the most beautiful sounds, but not a single song is his own.' Reassured, the boy started again, this time with one of his own songs. The range of his young voice was delicate and varied. His song ended in a high pitch only possible in youthful freedom. Hearing the song inspired Lizard and encouraged him for the task at hand. He didn't say anything about the song or the singing. He didn't have to, for in answer, a tiny white speck of cloud appeared over the top of the distant blue, west mountain. They saw it at the same time.

" 'Quickly!' Lizard commanded. 'Take off your shoes and shirt!' Ponci did as he was told, and at once Lizard ran up the boy's legs, up one side of his back and down the other, leaving two grey dust marks identical to his own. Then he ran back and forth in front of him telling him to do the same and to follow wherever he might run.

"They ran south. After many minutes, Lizard began to sing. 'That's like my song,' thought Ponci, as they continued south while the song lasted. At the end of Lizard's song, Ponci sang one of his own. They turned north. Singing toward the north, they ran and ran. Although Ponci's legs were tiring, and he felt himself gasping for air, he ran and kept on singing, following the dusty little lizard.

"Beneath the desert stillness, out of the parched dirt, a low humming began to rise. Slowly at first, then the unfailing soughing of their voices came together with a sound, becoming one sound emerging from the earth, a singular pulse beating with everything and everybody . . . then, as far as he could see, everything was dancing and everything was singing. The sound went forth, on and on.

"As Ponci sang, he felt a shadow move over his face. Drops of the first cooling rain splashed his skin. He opened his eyes to glance down onto the ground before him. Lizard was gone. Instead, looming out of the west, a huge dark cloud built skyward. Little spurts of dust rose in puffs as the rain began to fall. Ponci looked back toward the village and saw rain pouring in slanting silver sheets onto the thirsty fields in the distance. Their song had been answered. In his weariness, his feet continued to lift as he trotted toward home in the gentle wash of sweet rain."

The breathing in my ear stopped, and slowly the room came back into focus with my nose smelling the moist earth. There's a man with a large watering can sprinkling the dirt floor to settle the dust. He has just finished, and the people in a shuffling of feet settle back again to wait, when a quickened keening brings all of us together at once. The door opens, and from somewhere in the darkness comes the unmistakable sound of metal bells. The deer, eagle, and butterfly dancers are coming. The drums are beating. The singing men's voices can be heard. The people's patience will again be rewarded, and life will be remembered, in the dance, in the song, in the story.

A CHRISTIAN VIEW OF BIODIVERSITY

JOHN B. COBB, JR.

Ingraham Professor of Theology, School of Theology, Claremont Graduate School,
Claremont, California

Most people are distressed by the widespread destruction of species of living things. There is a deep sense that this is a serious loss to the planet. The major problem does not arise from direct approval of the destruction of species and of the simplification of the environment. It arises from the lack of awareness of the consequences of our actions and from the primacy of other concerns. In the pursuit of economic gain, most people do not want to be bothered by questions about biodiversity.

This volume, and the activities of many of its authors, are designed to heighten awareness of what we are doing to our biosphere. The correct assumption is that heightened awareness and intensified attention are the primary needs. People will act more appropriately if they are reminded again and again of the effects of their actions.

The authors of this section have a different role. We were asked to reflect on *why* biodiversity is important. It is not necessary to answer this question for people to recognize its importance. Nevertheless, good answers to our question are urgent, because the intuitive sense of importance is gradually weakened if its justification is not articulated. Also, there are ways of viewing the world that make concern about endangered species appear to be an esoteric or sentimental matter. Indeed, this type of world view is dominant in much of our society. We can all remember many of the disparaging comments that were made about concerns that a species of snail darter interfered with the building of a dam. Staying power in defense of biodiversity probably depends on a world view that grounds it more deeply than sentiment, however natural and healthy that sentiment may be.

The most obvious way to argue for biodiversity is to show how it benefits human beings. In the foregoing sections, much evidence was given for the risk to the human future that would be presented by a drastic simplification of various eco-systems. Hence there is a strong argument that for the sake of the future of our own species, we need to be concerned with biodiversity.

On the other hand, this argument is limited. Human beings have survived the disappearance of thousands of species with relatively little practical loss. If the only reason for preserving a particular species of insect or fish is its value to us, there will be many occasions when other needs will seem far more pressing. Furthermore, our sense of the importance of biodiversity is in fact not adequately reflected in the practical anthropocentric arguments. We *feel* that other species should have their place, even if they do not benefit us. Can we explain or justify this feeling?

One argument, a valid one I believe, is that all living things have intrinsic value. Not only are they of instrumental value to one another and to us, they also have value in and of themselves. They are of value for themselves. Hence, our destruction of other living things, while inevitable, should never be taken lightly. The reasons for destruction may be good ones—our need for food, for example. But we should not underestimate the cost to others. We should tread lightly on the Earth rather than bulldoze away all inconvenient objects.

Whereas this argument is a good one in itself, it does not go very far to explain the specific value of biodiversity. It does explain why we should avoid unnecessary destruction of living things, but it does not explain why a variety of such things is better than a monoculture. If by destroying the biodiversity of a prairie we can bring about the monoculture of a wheat field, and if the total number of insects and animals that are supported is not fewer, then there would seem to be no loss. The value of members of the lost species is made up by the value of more members of the species that is preserved.

Another argument, also valid in my opinion, is based on relations. The human species is not apart from others but is instead intricately and intimately related to the remainder of the web of life. When we experience the whole biosphere in this way, we experience destruction of any of its species as a diminution of ourselves.

The sense of relatedness has two dimensions. One dimension is genetic. We are kin to other living things. We have a common ancestry that has impressed itself in common genetic elements. The same sensibility that gives us a special sense of responsibility toward other human beings who are related to us can operate to give us a sense of responsibility for the other species to which we are also related.

The second dimension is ontological. We are increasingly realizing that indi-vidual entities, including individual human beings, do not exist apart from relations with other beings. We are constituted by our relations. Of course, many of our most important relations are with other human beings. But by no means all. We are related to the whole world of inanimate and animate things. We are part of them, and they are part of us. To feel this relationship with other things is not sentimentality but reality.

Although this is all true, it still does not go far enough to explain our sense of the importance of biodiversity. It does strongly support the sense of the intrinsic value of other living things. It cuts against the widespread Western dualism that places human beings above and outside nature. It works against the dominant

Western ethics that has taught us that only human welfare really matters. It reintegrates us into the web of life and thereby heightens our sense of its importance for us. But it does not tell us specifically why biodiversity has its own inherent value.

The category that comes to mind when we reflect on the value of diversity is aesthetics. At least in traditional art we have thought that the complexity of forms that could be brought into unity and harmony correlated with the greatness of a piece of art. Today, some qualifications would be required, but the general principle still holds. The same applies to experience generally. There is a richness of experience that correlates with the manifold contents that jointly make their contributions.

Much of our negative reaction to the destruction of species seems to stem from this sense that there are possibilities of experience forever lost. We are aware that some of our environments have already been simplified in ways that have impoverished our experience, and we are disturbed at the prospect that such impoverishment continues. Some of the experiences that were possible for us will not be available to our children. We rightly feel this as a loss that we should try to prevent, even at considerable cost in more practical realms of life.

This, too, is a strong and valid argument that goes far to reflect the feelings that are engendered by our awareness of the simplification of the biosphere. Yet it still fails to deal with our total concern. There are myriad species that have lived and died unknown by humans. It is true that their disappearance sets limits on what future generations can experience. But often in ordinary human experience, the ones that are lost do not differ sufficiently from others that remain to affect any but the most perceptive human beings. Judged simply by their potential contribution to the richness of human experience, many species seem to be of limited importance.

There is a deeper sense on our part that even when we are not ourselves able to benefit even aesthetically from the presence of other species, they are still making a contribution to the whole that is irreplaceable. Indeed, in one sense, this is self-evident. Surely the whole is diminished in some way by the loss of any of its parts!

The problem is that it is not so easy to locate this loss. We often try to locate it in human experience of the whole, but we have already seen that this is too limited a locus. It seems to be the whole-as-such that is impoverished. Yet this makes sense only if we can speak of the whole as having its own unity, its own perspective, its own experience.

We theists believe that just such unity, perspective, and experience does characterize the whole. From our point of view, the sense of the importance of biodiversity reflects an often unconscious recognition that the whole is indeed much more than the sum of its parts. Human beings sense that every creature, and especially every species, makes its contribution to the richness of the inclusive or divine experience.

It is this inclusive experience that provides the norm by which all of us are truly evaluated and judged. God knows us better than we know ourselves, and it is this knowledge of us that is the truth about us. For God, I am of no more worth than my neighbor, and hence when I treat my neighbor as a mere means to my own advantage I act wrongly. For God, no one nation is inherently of more worth than others. Hence, we act wrongly when we seek our own national advantage at the expense of other peoples. For God, every species has value. We do wrong when we treat other species as if they existed only for our sake and as if they could be destroyed with impunity when it is convenient for us to do so.

It would be going too far to say that the value of biodiversity is explicitly taught in the Bible. What we mean by this term presupposes much scientific knowledge that is not reflected in the Jewish and Christian scriptures. Nevertheless, the rudiments of the idea are present, and the extension of Biblical teaching into our own time strongly supports the concerns of biodiversity.

Consider the first chapter of Genesis. This account of creation has had profound effects on Western culture. There are features of this story that have been used to justify a mode of human relation to other creatures that has been profoundly destructive. But let us look at the story again.

One point that is striking in this account is that when God created the various plants and animals, God saw that they were good. There is no suggestion here that they were good because they would be useful to human beings. They were good in themselves and thus contributed to the divine satisfaction. Specifically, the story says that God blessed them and told them to be fruitful and to multiply, each according to its kind.

Now it is true that human beings are presented in a special light. We *are* one species among others, but we are also more than that. We are that species that is made in the image of God, and this is closely related to the assertion that God has given us dominion over other living things.

The resulting sense of rightful dominion has been important to the readers of the Bible, and this sense can be reaffirmed today. However, there is no question but that the story has been interpreted to mean that human beings are free to use and destroy other living things at will; and this interpretation needs to be strongly rejected.

Human beings are placed in a position in relation to other creatures much like that of God in relation to the whole of creation. God has dominion over all. We have dominion over the other creatures. God exercises dominion for the sake of those over whom the dominion is exercised. Similarly, the political ruler of Israel is to rule for the sake of those who are governed. A king who uses his power to amass riches for himself at the expense of the suffering of the ruled is a despot, not one who exercises rightful dominion. There is no justification here to suppose that human dominion over other creatures is a sanction of selfish exploitation. The meaning of the dominion given to us is much better expressed in servanthood and stewardship than in exploitation.

This book's content expresses a profoundly biblical view of the relation of human beings to the other species who with us constitute the biodiversity of the world. It recognizes that we human beings do exercise a determinative power over other creatures. Whether hundreds of thousands of species survive depends on the decisions of humans. It would be pointless to deny that we exercise dominion. But unlike so many who have asserted their dominion, we are acknowledging that with power comes responsibility—specifically, responsibility to God. To wipe out unnecessarily whole species of those creatures over whom we exercise stewardship is to betray that stewardship and to impoverish the experience of God. It is a crime against our Creator.

THE EARTH AS A LIVING ORGANISM

JAMES E. LOVELOCK
Launceston, Cornwall, United Kingdom

The idea that the Earth is alive may be as old as humankind. The ancient Greeks gave her the powerful name Gaia and looked on her as a goddess. Before the nineteenth century even scientists were comfortable with the notion of a living Earth. According to the historian D. B. McIntyre (1963), James Hutton, often known as the father of geology, said in a lecture before the Royal Society of Edinburgh in the 1790s that he thought of the Earth as a superorganism and that its proper study would be by physiology. Hutton went on to make the analogy between the circulation of the blood, discovered by Harvey, and the circulation of the nutrient elements of the Earth and of the way that sunlight distills water from the oceans so that it may later fall as rain and so refresh the earth.

This wholesome view of our planet did not persist into the next century. Science was developing rapidly and soon fragmented into a collection of nearly independent professions. It became the province of the expert, and there was little good to be said about interdisciplinary thinking. Such introspection was inescapable. There was so much information to be gathered and sorted. To understand the world was a task as difficult as that of assembling a planet-size jigsaw puzzle. It was all too easy to lose sight of the picture in the searching and sorting of the pieces.

When we saw a few years ago those first pictures of the Earth from space, we had a glimpse of what it was that we were trying to model. That vision of stunning beauty; that dappled white and blue sphere stirred us all, no matter that by now it is just a visual cliché. The sense of reality comes from matching our personal mental image of the world with that we perceive by our senses. That is why the astronaut's view of the Earth was so disturbing. It showed us just how far from reality we had strayed.

486

The Earth was also seen from space by the more discerning eye of instruments, and it was this view that confirmed James Hutton's vision of a living planet. When seen in infrared light, the Earth is a strange and wonderful anomaly among the planets of the solar system. Our atmosphere, the air we breathe, was revealed to be outrageously out of equilibrium in a chemical sense. It is like the mixture of gases that enters the intake manifold of an internal combustion engine, i.e., hydrocarbons and oxygen mixed, whereas our dead partners Mars and Venus have atmospheres like gases exhausted by combustion.

The unorthodox composition of the atmosphere radiates so strong a signal in the infrared range that it could be recognized by a spacecraft far outside the solar system. The information it carries is prima facie evidence for the presence of life. But more than this, if the Earth's unstable atmosphere was seen to persist and was not just a chance event, then it meant that the planet was alive—at least to the extent that it shared with other living organisms that wonderful property, homeostasis, the capacity to control its chemical composition and keep cool when the environment outside is changing.

When on the basis of this evidence, I reanimated the view that we were standing on a superorganism rather than just a ball of rock (Lovelock, 1972; 1979), it was not well received. Most scientists either ignored it or criticized it on the grounds that it was not needed to explain the facts of the Earth. As the geologist H. D. Holland (1984, p. 539) put it, "We live on an Earth that is the best of all possible worlds only for those who are well adapted to its current state." The biologist Ford Doolittle (1981) said that keeping the Earth at a constant state favorable for life would require foresight and planning and that no such state could evolve by natural selection. In brief, scientists said, the idea was teleological and untestable. Two scientists, however, thought otherwise; one was the eminent biologist Lynn Margulis and the other the geochemist Lars Sillen. Lynn Margulis was my first collaborator (Margulis and Lovelock, 1974). Lars Sillen died before there was an opportunity. It was the novelist William Golding (personal communication, 1970), who suggested using the powerful name Gaia for the hypothesis that supposed the Earth to be alive.

In the past 10 years these criticisms have been answered—partly from new evidence and partly from the insight provided by a simple mathematical model called Daisy world. In this model, the competitive growth of light- and dark-colored plants on an imaginary planet are shown to keep the planetary climate constant and comfortable in the face of a large change in heat output of the planet's star. This model is powerfully homeostatic and can resist large perturbations not only of solar output but also of plant population. It behaves like a living organism, but no foresight or planning is needed for its operation.

Scientific theories are judged not so much by whether they are right or wrong as by the value of their predictions. Gaia theory has already proved so fruitful in this way that by now it would hardly matter if it were wrong. One example, taken from many such predictions, was the suggestion (Lovelock et al., 1972) that the compound dimethyl sulfide would be synthesized by marine organisms on a large scale to serve as the natural carrier of sulfur from the ocean to the land. It was known at the time that some elements essential for life, like sulfur, were abundant in the oceans but depleted on the land surfaces. According to Gaia theory, a natural carrier was needed and dimethyl sulfide was predicted. We now know that this compound is indeed the natural carrier of sulfur, but at the time the prediction was made, it would have been contrary to conventional wisdom to seek so unusual a compound in the air and the sea. It is unlikely that its presence would have been sought but for the stimulus of Gaia theory.

Gaia theory sees the biota and the rocks, the air, and the oceans as existing as a tightly coupled entity. Its evolution is a single process and not several separate processes studied in different buildings of universities.

It has a profound significance for biology. It affects even Darwin's great vision, for it may no longer be sufficient to say that organisms that leave the most progeny will succeed. It will be necessary to add the proviso that they can do so only so long as they do not adversely affect the environment.

Gaia theory also enlarges theoretical ecology. By taking the species and the environment together, something no theoretical ecologist has done, the classic mathematical instability of population biology models is cured.

For the first time, we have from these new, these geophysiological models a theoretical justification for diversity, for the Rousseau richness of a humid tropical forest, for Darwin's tangled bank. These new ecological models demonstrate that as diversity increases so does stability and resilience. We can now rationalize the disgust we feel about excesses of agribusiness. We have at last a reason for our anger over the heedless deletion of species and an answer to those who say it is mere sentimentality.

No longer do we have to justify the existence of the humid tropical forests on the feeble grounds that they might carry plants with drugs that could cure human disease. Gaia theory forces us to see that they offer much more than this. Through their capacity to evapotranspire vast volumes of water vapor, they serve to keep the planet cool by wearing a sunshade of white reflecting clouds. Their replacement by cropland could precipitate a disaster that is global in scale.

A geophysiological system always begins with the action of an individual organism. If this action happens to be locally beneficial to the environment, then it can spread until eventually a global altruism results. Gaia always operates like this to achieve her altruism. There is no foresight or planning involved. The reverse is also true, and any species that affects the environment unfavorably is doomed, but life goes on.

Does this apply to humans now? Are we doomed to precipitate a change from the present comfortable state of the Earth to one almost certainly unfavorable for us but comfortable to the new biosphere of our successors? Because we are sentient there are alternatives, both good and bad. In some ways the worse fate in store for us is that of becoming conscripted as the physicians and nurses of a geriatric planet with the unending and unseemly task of forever seeking technologies to keep it fit for our kind of life—something that until recently we were freely given as a part of Gaia.

Gaia philosophy is not humanist. But being a grandfather with eight grandchildren I need to be optimistic. I see the world as a living organism of which we are a part; not the owner, nor the tenant, not even a passenger. To exploit such a world on the scale we do is as foolish as it would be to consider our brains supreme and the cells of other organs expendable. Would we mine our livers for nutrients for some short-term benefit?

Because we are city dwellers, we are obsessed with human problems. Even environmentalists seem more concerned about the loss of a year or so of life expectation through cancer than they are about the degradation of the natural world by deforestation or greenhouse gases—something that could cause the death of our grandchildren. We are so alienated from the world of nature that few of us can name the wild flowers and insects of our locality or notice the rapidity of their extinction.

Gaia works from an act of an individual organism that develops into global altruism. It involves action at a personal level. You well may ask, So what can I do? When seeking to act personally in favor of Gaia through moderation, I find it helpful to think of the three deadly Cs: combustion, cattle, and chain saws. There must be many others.

One thing you could do, and it is no more than an example, is to eat less beef. If you do this, and if the clinicians are right, then it could be for the personal benefit of your health; at the same time, it might reduce the pressures on the forests of the humid tropics.

To be selfish is human and natural. But if we chose to be selfish in the right way, then life can be rich yet still consistent with a world fit for our grandchildren as well as those of our partners in Gaia.

REFERENCES

Doolittle, W. F. 1981. Is nature really motherly? CoEvol. Q. 29:58–63

Holland, H. D. 1984. The Chemical Evolution of the Atmosphere and the Oceans. Princeton University Press, Princeton, N.J. 656 pp.

Lovelock, J. E. 1972. Gaia as seen through the atmosphere. Atmos. Environ. 6:579–580.

Lovelock, J. E. 1979. Gaia. A New Look at Life on Earth. Oxford University Press, Oxford. 157 pp.

McIntyre, D. B. 1963. James Hutton and the philosophy of geology. Pp. 1–11 in Claude C. Albritton, ed. The Fabric of Geology. Addison-Wesley, Reading, Mass.

Margulis, L., and J. E. Lovelock. 1974. Biological modulation of the Earth's atmosphere. Icarus 21:471–489.

EPILOGUE

Young girl with a passion flower, Central Andes above Abancay, on the road to Cuzco, Province of Apurimac, Peru. *Photo courtesy of Hugh H. Iltis*

EPILOGUE

DAVID CHALLINOR
Assistant Secretary for Research, Smithsonian Institution, Washington, D.C.

Ｆor 3 days in late September 1986, a group of about 60 distinguished scholars and scientists addressed capacity audiences at the Forum on BioDiversity at the Smithsonian and the National Academy of Sciences on the importance of biodiversity on our planet. In accordance with their dissimilar professional backgrounds, the speakers approached the topic from many directions. Judging from the active audience participation, the listeners clearly understood the perils to their own well-being presented by a loss of the Earth's biodiversity. This book is based on that forum.

In this chapter I summarize what we heard (what the authors of this book have written), what we learned, and finally, what we might do to slow, or hopefully stop, the rapid, human-induced extinction of the Earth's great variety of plants and animals.

The first keynote speaker (Chapter 1) warned that the diversity of plant and animal species is declining at a much faster rate than generally is realized. This disturbing assessment is exacerbated by our ignorance of just how many different species exist, especially among the terrestrial and marine invertebrates of the tropics. These creatures are both difficult to collect and even harder to identify, since few taxonomists in the world have the expertise to name them. There is a grave danger, therefore, that many as yet uncollected small plants and insects will inadvertently become extirpated without our even knowing that they once existed. Some of these unknown plants and animals could play important roles in our well-being, or even survival, yet they may never be recognized.

All forum participants stressed the need to maintain biodiversity, and a consensus soon emerged that the greatest threat to this goal today is from human activities. Human populations are expanding most rapidly in the tropics—the very area where biodiversity is the greatest. Thus pressure on the land to support even more people results in the transformation of biologically diverse rain forests into pastures and monocultured crops, with only a minimal number of species being grown to sustain the burgeoning population. Altering a natural landscape to satisfy human needs for food and living space has been going on throughout the world for millennia. In most countries of the New World, such activities have accelerated rapidly following European colonization and the importation of modern technology.

Although human activity has directly caused the extinction of many recent species, the geological record indicates that massive destruction of plants and animals occurred episodically at distantly separated intervals long before humans evolved. Despite the enormous volume of debris and noxious fumes spewed by volcanoes into the Earth's atmosphere, pollutants from these relatively few erup-tions dwarf the total amount of devastation caused by mankind. Compared with past natural disasters, the destruction caused by human activity is of quite a different character. Today, species destruction by humans is intense, concentrated in time, and on a global scale not previously experienced in prehistoric times. Sediment loads of rivers, for example, have certainly varied in the past as a result of forest fires and earthquakes, but in time the damaged watersheds recovered and sediments lessened. Today, however, the watersheds, especially in the tropics, are degraded continually by human activity.

Whatever natural extinctions took place before the presence of humans is ir-relevant, because today we are such a dominant species that theoretically we have the understanding and the power to stop or at least to mitigate our destruction of the natural world. The practicability of this approach may be arguable, but several authors give specific examples of how whole new habitats, such as salt marshes, can be restored or how an extirpated subspecies of peregrine falcons can be replaced by a new, artificially created subspecies that seems to fill successfully the niche that was lost.

Although scientists have had isolated successes in restoring habitats or saving endangered species, the key to a more permanent equilibrium and therefore the long-term preservation of biodiversity will depend on effective population control in the tropical Third World. Some governments and many organizations are already working on this goal, but population control is a highly charged political and cultural issue not likely to be resolved until the incentive for large families disappears throughout the world.

One way of mitigating the environmental damage caused by inevitable population growth is to exploit the use of alternative human foods. Upon arrival in the tropics, European entrepreneurs tended to introduce temperate zone pulses and grains into their new colonial territories. Some crops, such as maize, adapted well, but shortly after World War II, in what was then Tanganyika, thousands of hectares were cleared and planted to grow peanuts. With little previous testing, the new crop was a disastrous failure both financially and environmentally. There are, unfor-tunately, many additional examples of such unsuccessful efforts in Africa and tropical Asia. Much innovative plant breeding has been done recently by inter-national centers in Third-World countries. India's wheat harvest, for example, has achieved self-sufficiency for the nation, although this accomplishment is dependent on relatively regular monsoons.

The same scientific advances that have led to increased local food production in the tropics have also inspired imaginative ways to fill the local animal protein needs of the rural people in tropical countries. Projects range from elaborate game ranching in East and South Africa to the development of simple facilities for restocking edible green iguanas in Central America. Not only have scientists successfully increased endangered animal populations through captive and controlled breeding, they have also been able to reintroduce these captive animals into their original habitats with considerable success.

In recent decades, animal breeding techniques have advanced to the point where embryos can be implanted between species within a genus. Thus, the relatively common eland antelope has been implanted with a rare bongo embryo and successfully carried to term. As our knowledge and skills improve, other rare mammals can be born to foster mothers to develop captive populations large enough for eventual reintroduction to their former habitats. Admittedly this is a slow and expensive way to maintain biodiversity and is clearly limited to a few mammal species. Such reintroductions, however, are worthwhile only if a reasonable facsimile of the original habitat exists. There is always the likelihood that all the essential components of the original habitat cannot be reproduced when the landscape is artificially restored.

The reaction of those attending and participating in the forum, however, seemed to support the idea that artificial restoration was well worth undertaking. We do indeed have an obligation to future generations to keep life on the planet as diverse as possible; for this participant, such is the principal message of this book.

What else did the audience learn from all these presentations? The scientists and scholars generally reported on work already published but seldom read by the layman. Judging from the questions from the audience, most were knowledgeable about the subject but undoubtedly learned many new details on all aspects of biodiversity, especially the magnitude of the problem.

Coverage by the major newspapers and journals spread the word beyond the forum attendees. The sessions closed with a national teleconference transmitted via satellite to more than 100 universities and other organizations throughout the nation during which the conferees questioned directly six of the principal participants. (An edited videotape of this conference is available from the National Academy Press.)

Although well publicized by environmental news standards, the critical messages generated at the symposium had to compete for the public's attention with other seemingly more immediate problems, such as the threat of nuclear war, acid rain, and other global nest-fouling threats. World leaders, who must make hard economic choices, may be so overwhelmed by more acute problems that they may not choose to invest in the security of humanity by perpetuating biological diversity.

Despite the bleak picture of the threat of rapid extinctions, the technical advances in biology give hope for alleviation. Perhaps the symposium session most relevant to changing human behavior and to maintaining biodiversity was the last one: Ways of Seeing the Biosphere.

The concluding four speakers addressed the issue of human perceptions of and behavior toward the environment from the viewpoints of a poet, a theologian, a philosopher, and an American Indian folklorist. The Native American gave graphic examples of how his culture views the natural surroundings: it considers human resources to be an integral part of the whole Earth rather than a force to dominate the terrestrial globe. The philosopher reinforced this viewpoint by reminding us that the whole Earth is a living organism and, therefore, we cannot treat humans differently from other living components. The theologian put human beings in the Judeo-Christian perspective as part of God's total creation.

Thus an immediate—as opposed to a geological—solution to the problem of maintaining global biodiversity seems to depend on the collective behaviors and perceptions of people toward their habitat. The Western world in particular has been out of harmony with its environment and through temporary technical superiority has imposed its destructive standards of affluence on the rest of the world. Unfortunately, the Earth's natural resources are finite and inadequate to support a global living standard equivalent to that of the developed countries. To keep the Earth reasonably habitable for humans in the centuries to come, natural forces will have to lower the human population and reduce the indiscriminate exploitation of the natural world. Controlled rational exploitation may be the answer, if the surviving humans have the foresight and sensitivity to carry it out.

INDEX

A

Aardvark (*Oryceteropus afer*), 151
Acacias (*Acacia* spp.), 161, 374
Acadia National Park (Maine), 183
Acid rain, 22, 25, 75
Adaptive radiation, 150, 182
Adzuki bean (*Vigna angularis*), 243
Afforestation, 60
Africa
  biogeographic setting, 249
  conservation approaches, 248–259
  deforestation projections, 64–65
  dry tropical regions, 130
  endemism centers, 249–252
  extinction rates, 64–65
  fragility and vulnerability of ecosystems, 258
  grasslands, 176, 177
  Great Lakes, 252
  heathland shrubs, 251
  introduced species, 251, 252
  monitoring of ecosystems, 257–259
  protected areas, 11, 254–257, 259
  rain forest replacement by savannas, 56
  Sahelian area, 177
  Serengeti, 176
  species diversity, 31, 64–65, 249–252
  see also East Africa; West Africa; *and specific countries*
African eggplants (*Solanum* spp.), 244
African invader (ant), 251
African tilapia (*Tilapia* spp.), 347
African Wildlife Foundation, 147
Afrotropical realm, 249–251, 254
Agency for International Development, U.S.
  conservation strategy, 16, 410, 414, 416
  developmental assistance in conservation, 405, 413-414
  Panama natural resources project, 415
Agouti (*Dasyprocta* spp.), 356, 375
Agriculture
  conversions of grasslands, 176–177, 179
  desert-farming module, 346–351
  developing world, 79–80, 406
  exchange economy growth, 206–210
  intensification of productivity, 394, 406
  peasant, 79–80, 114, 363–366
  slash-and-burn, 141, 151, 327–328, 371, 413
  tropical forest plant contribution, 107
  see also Crops; Cultivation
Agrimony (*Agrimonia eupatoria*), 84
Agroecology, 134, 136, 207, 208, 346–352, 357–358, 361–369
Agroforestry, 373–374, 395–396, 406
Alder (*Alnus* spp.), 171
Algae, red (*Digenea simplex*), 86
Allelic diversity, *see* Genetic diversity
Allelochemicals, 111
Alligators (*Alligator* spp.), 201, 203, 264, 266
Alpacas (*Lama pacos*), 100
Altieri, Miguel A., *361–370*
Amazonia
  Brazilian, 29, 114, 120, 124, 126–127, 138–142, 329, 406
  conservation of tropical forests, 32, 406
  deep ecology, 8
  deforestation projections, 120
  Ecuadorian, 79
  forest fires, 67, 326–327
  history, 124–125
  human disturbances, 327–328
  hunting, 139, 146
  Indians, *see* Amerindians
  natural disturbances, 326–327
  natural pesticides, 111

F